GASDYNAMICS OF DETONATIONS AND EXPLOSIONS

Edited by
J. R. Bowen
University of Wisconsin
Madison, Wisconsin

N. Manson
Université de Poitiers
Poitiers, France

A. K. Oppenheim
University of California
Berkeley, California

R. I. Soloukhin
Institute of Heat and Mass Transfer
BSSR Academy of Sciences
Minsk, U.S.S.R.

Volume 75
PROGRESS IN
ASTRONAUTICS AND AERONAUTICS

Martin Summerfield, Series Editor-in-Chief
Princeton Combustion Research Laboratories, Inc.,
Princeton, New Jersey

Technical papers from the Seventh International Colloquium on Gasdynamics of Explosions and Reactive Systems, Göttingen, Federal Republic of Germany, August 1979.

Published by the American Institute of Aeronautics and Astronautics,
1290 Avenue of the Americas, New York, N.Y. 10104.

American Institute of Aeronautics and Astronautics
New York, New York

Library of Congress Cataloging in Publication Data

International Colloquium on Gasdynamics of Explosions
  and Reactive Systems (7th: 1979: Göttingen, Germany)
  Gasdynamics of detonations and explosions.

  (Progress in astronautics and aeronautics; v. 75)
  Held under the auspices of the Institute of Physical Chemistry of the
University of Göttingen and Max-Planck-Institut
für Strömungsforschung, Aug. 20-24, 1979.
  1. Explosions—Congresses. 2. Gasdynamics—Congresses.
I. Bowen, J. Raymond. II. Universität Göttingen.
Institut für Physikalische Chemie. III. Max-Planck-Institut
für Strömungsforschung. IV. Title. V. Series.
TL507.P75 [QD516]   629.ls[662'.4] 81-1010
ISBN 0-915928-46-9   AACR2
SET 0-915928-45-0

Copyright © 1981 by
American Institute of Aeronautics and Astronautics

All rights reserved. No part of this book may be reproduced in any form or by any means, electronic or mechanical, including photocopying, recording, or by any information storage and retrieval system, without permission in writing from the publisher.

Table of Contents

Preface vii

List of Series Volumes 1-76 xiv

    Plenary Lecture: Particle Sizing in Flames 1
    S. S. Penner and P. H. P. Chang, *University of California, San Diego, La Jolla, California*

Chapter I. Wall and Confinement Effects 31

    Turbulent Flame Propagation and Acceleration in the Presence of Obstacles 33
    I. O. Moen, M. Donato, R. Knystautas, and J. H. Lee, *McGill University, Montreal, Canada* and H. Gg. Wagner, *University of Göttingen, Göttingen, Federal Republic of Germany*

    Detonation of Unconfined Fuel Aerosols 48
    D. C. Bull, M. A. McLeod, and G. A. Mizner, *Shell Research Ltd., Chester, England*

    Initiation of Unconfined Gas Detonations in Hydrocarbon-Air Mixtures by a Sympathetic Mechanism 61
    D. C. Bull, J. E. Elsworth, and M. A. McLeod, *Shell Research Ltd., Chester, England* and D. Hughes, *University College of Wales, Penglais, Aberystwyth, Wales*

    Detection Method for the Deflagration to Detonation Transition in Gaseous Explosive Mixtures 73
    C. Brochet and M. Sayous, *Université de Poitiers, Poitiers, France*

    Influence of the Nature of Confinement on Gaseous Detonation ... 87
    T. V. Bazhenova, V. P. Fokeev, and Yu. Lobastov, *Academy of Sciences, Moscow, U.S.S.R.* and J. Brossard, T. Bonnet, B. Brion, and N. Charpentier, *Université d'Orléans, Bourges, France*

    Mechanical Effects of Gaseous Detonations on a Flexible Confinement 108
    J. Brossard and J. Renard, *Université d'Orléans, Bourges, France*

    Oxyhydrogen Detonations under Surface Catalysis 122
    T. Fujiwara and T. Hasegawa, *Nagoya University, Nagoya, Japan*

**Pressure and Wall Heat Transfer behind a Hydrogen/Azide
Detonation Wave in Narrow Tubes** 134
  C. Paillard, G. Dupre, R. Lisbet, and J. Combourieu, *Université
  d'Orléans, Orléans, France* and V. P. Fokeev, L. G. Gvozdeva, and
  T. V. Bazhenova, *Academy of Sciences, Moscow, U.S.S.R.*

**Pressure Evolution behind Spherical and Hemispherical Detonations
in Gases** 150
  D. Desbordes and N. Manson, *Université de Poitiers, Poitiers, France* and
  J. Brossard, *Université d'Orléans, Bourges, France*

Deflagration Explosion of an Unconfined Fuel Vapor Cloud 166
  S. Taki and Y. Ogawa, *Fukui University, Fukui, Japan*

**Self-Similar Blast Waves Supported by Variable Energy Deposition
in the Flowfield** 178
  R. H. Guirguis and A. K. Oppenheim, *University of California, Berkeley,
  Calif.* and M. M. Kamel, *Cairo University, Cairo, Egypt*

Chemical Kinetics in LNG Detonations 193
  C. K. Westbrook and L. C. Haselman, *Lawrence Livermore National
  Laboratory, University of California, Livermore, Calif.*

Chapter II. Liquid and Solid Phase Phenomena 207

**Molecular Dynamics of Shock and Detonation Phenomena
in Condensed Matter** 209
  J. R. Hardy, *University of Nebraska, Lincoln, Neb.* and A. M. Karo and
  F. E. Walker, *Lawrence Livermore National Laboratory, University of
  California, Livermore, Calif.*

TNT Explosions in a Hard Vacuum 226
  A. L. Kuhl, *R & D Associates, Marina Del Rey, Calif.* and M. R. Seizew,
  TRW Defense and Space Systems, Redondo Beach, Calif.

**Gasdynamic Investigations of Lead Azide/Lead Styphnate Detonation
Processes in Vacuum by Multichannel Mass Spectrometry** 242
  H. Trinks and N. Schilf, *Hochschule der Bundeswehr, Hamburg, Federal
  Republic of Germany*

The Effect of the Shock-Wave Front on the Origin of Reaction ... 253
  A. N. Dremin and V. Yu. Klimenko, *Academy of Sciences,
  Chernogolovka, U.S.S.R.*

Single-Shock Curve Buildup and a Hydrodynamic $p_i^N t_i^f$ Criterion
for Initiation of Detonation.................................. 269
    M. Cowperthwaite, *SRI International, Menlo Park, Calif.*

Induction Delay and Detonation Failure Diameter of Nitromethane
Mixtures... 282
    H. N. Presles and C. Brochet, *Université de Poitiers, Poitiers, France*

Critical Area Concept for the Initiation of a Solid High Explosive
by the Impact of Small Projectiles 296
    H. Moulard, *French-German Institute of Saint-Louis, Saint-Louis, France*

Shock Initiation of Hydrazine Mononitrate 303
    M. Yoshida and T. Yoshida, *University of Tokyo, Tokyo, Japan* and K. Tanaka, S. Fujiwara, and M. Kusakabe, *National Chemical Laboratory for Industry, Ibaragi, Japan*

Action of Charges with Axial Cavities on Rocks 314
    V. V. Mitrofanov and I. T. Bakirov, *Academy of Sciences, Novosibirsk, U.S.S.R.* and G. A. Voroteljak and V. A. Salganik, *Academy of Sciences, Kriyoi Rog, U.S.S.R.*

Characterization of Mass Flow Rates
for Various Percussion Primers............................... 323
    K. K. Kuo, B. B. Moore, and D. Y. Chen, *The Pennsylvania State University, University Park, Pa.*

Chapter III.   Cellular Structure of Detonations 339

Diffraction of a Planar Detonation in Various Fuel-Oxygen Mixtures
at an Area Change.. 341
    D. H. Edwards and G. O. Thomas, *University of Wales, Aberystwyth, Wales* and M. A. Nettleton, *Central Electricity Council, Leatherhead, U. K.*

Reinitiation Process at the End of the Detonation Cell 358
    J-C. Libouton, M. Dormal, and P. J. Van Tiggelen, *Université Catholique de Louvain, Louvain-la-Neuve, Belgium*

Effects of Cellular Structure on the Behavior of Gaseous Detonation
Waves under Transient Conditions 370
    P. A. Urtiew and C. M. Tarver, *Lawrence Livermore National Laboratory, University of California, Livermore, Calif.*

Chapter IV. Detonations at Moderate Pressures............385

Influence of the Heat-Release Function on the Detonation States ..387
H. Guénoche, P. Le Diuzet, and C. Sèdes, *Université de Provence, Marseille, France*

Detonation Characteristics of Gaseous Ethylene, Oxygen, and Nitrogen Mixtures at High Initial Pressures........................408
P. Bauer, S. Krishnan, and C. Brochet, *Université de Poitiers, Poitiers, France*

Detonation Characteristics of Two Ethylene-Oxygen-Nitrogen Mixtures Containing Aluminum Particles in Suspension................423
B. Veyssiere, R. Bouriannes, and N. Manson, *Université de Poitiers, Poitiers, France*

Generation of Detonations by Two-Stage Burning..............439
M. Zalesiński and S. Wójcicki, *Warsaw Technical University, Warsaw, Poland*

PREFACE

This volume and its companion, Volume 76, include the revised and edited papers that were presented at the Seventh International Colloquium on the Gasdynamics of Explosions and Reactive Systems, held in Göttingen, Federal Republic of Germany, in August 1979. These International Colloquia had their origin in 1966 as a consequence of some new advances that had been achieved in the understanding of detonation wave structure. Several of the leading researchers in this field came to the conclusion that a regular forum would be required for discussions of the important new findings in gasdynamics flows associated with exothermic process—the essential feature of detonation waves—but covering a much broader scope of applications.

The subject matter of Gasdynamics of Explosions is concerned principally with the interrelationship between rate processes of energy deposition in a compressible medium and its concurrent nonsteady flow, as it occurs typically in explosion phenomena. Gasdynamics of Reactive Systems is a broader term that includes explosions and deals with the processes of coupling between the dynamics of fluid flow and molecular transformations in reactive media, as they occur in any combustion system. In this connection, besides contributions to the usual topics of explosions, detonations, shock phenomena, and reactive flow, papers were presented at the Seventh Colloquium that dealt especially with gasdynamic aspects of nonsteady flow in combustion systems, fluid mechanic aspects of combustion with particular emphasis on the effects of turbulence, and the diagnostic techniques for the study of combustion phenomena. Of special interest, moreover, were papers that dealt with effects of radiative heat transfer on fluid dynamic features of reactive systems, as in the case of luminous flames, intense fires, and gasdynamic lasers.

The contributions to the Seventh Colloquium have been collected into two volumes; Volume 75 covers Gasdynamics of Detonations and Explosions, and Volume 76 covers Combustion in Reactive Systems. The contributions in Volume 75 have been grouped into chapters on Wall and Confinement Effects (12),* Liquid and Solid Phase Phenomena (10), Cellular Structure of Detonations (3), and

---

*The number in parenthesis after each subtopic indicates the number of papers appearing under that topic.

Detonations at Moderate Pressures (4). The contributions in Volume 76 have been grouped into chapters on Nonequilibrium Processes (5), Ignition (4), Turbulence (10), Asymptotics (7), Detailed Kinetic Modeling (3), and Miscellany (8). In some instances the subject of a particular contribution defies such a neat categorization, and the editors were compelled, more or less arbitrarily, to assign the subtopic, guided by the relevance and the goal of the reported research. A rather straightforward example is the contribution of Moen, Donato, Knystautas, and Lee on the interactions of turbulent flames with obstacles. Since this work is clearly related to the establishment of explosions or detonations in the absence of confinement, it was assigned to the chapter on Wall and Confinement Effects.

"Particle Sizing in Flames," by Penner and Chang, has been given special status since it derives from the plenary lecture given by Penner. This work is concerned with in situ particle sizing, which is central to the observation and understanding of particulate formation in flames. The paper includes a review of optical techniques which employ measurement of scattered laser power spectra and an extended discussion of results which have been obtained principally by the authors and their co-workers from observation of scattered power spectra in a flat flame burner. Their results provide a beautiful example of the utility of laser diagnostics for nonintrusive measurement of mean particle diameter and particle concentration of soot in flames.

Among the sixty-six contributed papers of the Colloquium were many that stimulated exciting discussions. It is not feasible in the short space of this Preface to do justice to all of the stimulating papers, but the more noteworthy among them are highlighted here.

Gasdynamics of Detonations and Explosions (Volume 75)

Wall and Confinement Effects. The papers in this chapter are concerned with the characterization of gas phase detonations or explosions which may occur as a consequence of an accidental release of large quantities of flammable vapor (or a volatile liquid). Such studies are needed to provide assessments of minimal ignition energies, of hazards potential, and of possible preventive measures. One of the central questions is the amount of energy required to initiate explosive burning rates in unconfined media when either substantial levels of turbulence result from the release of the fuel cloud or turbulence-producing obstacles are present. Moen, Donato,

Knystautas, and Lee report their observation and model of the acceleration of methane-air flame in the presence of obstacles. Brochet and Sayous report on a device for observing deflagration-to-detonation transition in large-scale devices, while Bull, McLeod, Mizner, and Elsworth present experimental results on "unconfined" aerosol detonations and sympathetic initiation of unconfined detonations. Westbrook and Haselman combine numerical models for evolution of a blast wave and for the chemical ignition delay with a characteristic time analysis to generate estimates of the minimum amount of explosive to detonate methane/air mixtures.

Liquid and Solid Phase Phenomena. In addition to the phenomenological studies, there is a considerable interest in the potential of condensed phase explosives for damage. Of the ten papers in this area, five are primarily concerned with characterization of the phenomenon, and five deal with effects of explosions. In the former category, Cowperthwaite uses a similarity solution to model the initiation of a detonation in a condensed explosive and to develop a criterion for the onset of detonation. Dremin and Klimenko report a molecular dynamic method for analyzing the effect of a strong shock wave on a solid explosive and suggest that a partial decomposition of the explosive may occur within the shock wave. Trinks and Schilf report an application of multichannel mass spectrometry to observe the products of explosive decomposition.

Of the several papers related to the effects of condensed explosives, the contribution of Hardy, Karo, Walker, and of Kuhl and Seizew are particularly noteworthy. Hardy et al. use computer molecular dynamics to study the response of two-dimensional lattices to shock loading of an exterior wall by an impacting plate. In the several cases studied, the interaction of the lattice with the shock produced highly localized disruption of the opposite wall of the lattice. Kuhl and Seizew report on the modeling of the evolution of spherical and planar TNT explosions in a hard vacuum with a Lagrangian finite difference scheme. Their results show that the asymptotic density and dynamic pressure profiles depend on the mode of energy release, the equation of state for the detonation products, and the flow geometry. These results are useful for estimating blast damage from solid explosives in space.

Cellular Structure of Detonations. The marginal detonation, the one near the pressure or concentration limits of detonability, is sustained by the propagation of a three-dimensional wave complex transverse to the direction of motion of the detonation. The motion of the triple point of this complex produces a cellular pattern on soot records. The study of this cellular structure has been the key to the understanding of marginal detonations. The work of Edwards, Urtiew, and van Tiggelen and their co-workers provides significant new insights. Edwards, Thomas, and Nettleton, and Urtiew and Tarver are concerned with the diffraction of a detonation wave at an abrupt area change, whereas Libouton, Dormal, and van Tiggelen present evidence on the reinitiation process near the end of a detonation cell.

Combustion in Reactive Systems (Volume 76)

Nonequilibrium Processes. The papers in this chapter are concerned with the exploitation of gasdynamic flows to produce population inversions, and possibly laser action, or with the gasdynamic flows which are produced by the discharges of pulsed lasers. Of the former type are the contributions by Fomin, Soloukhin, Golovichev, and Munjee on the calculation of gain coefficients in a gasdynamic mixing layer on the basis of a coupled kinetic relation model with a laminar or turbulent flow model to characterize mixing effects; Meolans, Nicoli, and Burn on the calculation of relaxing flows in nozzles; and Fontaine, Forestier, and Gross on ultraviolet and visible high-power laser action, which is achieved in a supersonic flow subsequent to electron beam excitation. The paper by Ageev, Barchukov, Bunkin, Konov, Korobeinikov, Prokhorov, Putiatin, and Khudiakov is concerned with the gasdynamic effects produced by pulsed laser discharges.

Ignition. The contribution of Guirguis, Karasalo, Creighton, and Oppenheim on methane ignition is concerned with modeling ignition processes associated with the introduction of free radicals into a combustible mixture. This problem is central to the feasibility of proposed jet ignition devices for lean burn engine concepts. Oran and Boris report on a comparison of the predictions of a simplified nonlinear ignition model, based on an induction time concept, with the results of a solution of the full chemical kinetics scheme coupled with the equations of motion and molecular diffusion. Dwyer, Kee, and Sanders report a numerical scheme which employs generalized coordinates and adaptive grids for modeling two-dimensional unsteady combustion, and they present results for ignition and flame propagation with complex internal boundaries. Dabora reports results of an experimental investigation of laser ignition of fuel droplets.

Turbulence. The availability of rapid response and nonintrusive instrumentation and of high-speed computers has led to many advances in the understanding of turbulent flows in reactive systems. The discussions on turbulence at the Colloquium were especially lively. Among the experimental investigations, attention should be drawn to the laser tomographic method which has been developed by Sabathier, Boyer, and Clavin to study weakly turbulent flames. Ramohalli reported his work on measurement of acoustical emissions to deduce the details of turbulent combustion. Boris and Oran review the important considerations and problems that arise in ab initio calculations of turbulent reacting flows. Ashurst presents an interesting vortex simulation of two-dimensional turbulent flows. Studies using phenomenological models of turbulent flows are reported by Parker and Sirignano, by Tamanini and Ahmad, and by Grouset, Esposito, and Candel.

Asymptotics. Perturbation expansion and activation energy asymptotics have been the cornerstone of the approximate analytical solutions of reacting flow models. Various problems have been solved by this method, and the papers given at the Colloquium are typical of the range of problems to which it can be applied. As an example, Clarke gives an exposition on propagation of pressure disturbances in an explosive gas, while Clavin and Williams present an extension of their earlier work on the structure and propagation velocity of premixed turbulent flames.

Detailed Kinetic Modeling. A complete description of the chemical kinetics of a combustion process is required if the concentration of pollutants or particulates is to be predicted or if there is some concern that nonlinear coupling exists between the gasdynamics and kinetic processes. Unfortunately, the variety of chemical constituents and the stiffness of the equations lead to numerical complexity in the solution of models that include a detailed accounting of the chemical kinetics. Kee and Dwyer review the problems of detailed combustion modeling, and Warnatz and Margolis present numerical studies of laminar flames in premixed gases, taking into account a full kinetic scheme.

The Seventh Colloquium also included an informative session on Explosion Hazards, in which a number of recently completed studies were reported. The proceedings of this session have been published as Max Planck Institute Report ISSN 0436-1199. A copy can be obtained by applying to Professor H. Gg. Wagner, Institute of Physical Chemistry, University of Göttingen, Federal Republic of Germany.

The first Colloquium was held in 1967 in Brussels, and Colloquia have been held on a biennial basis since then (1969 in Novosibirsk, 1971 in Marseille, 1973 in La Jolla, 1975 in Bourges, 1977 in Stockholm, and 1979 in Göttingen). They have now achieved the status of a prime international meeting on these topics and attract contributions from scientists and engineers throughout the world. The Proceedings of previous Colloquia have appeared as part of the journal, *Acta Astronautica,* or its predecessor, *Astronautica Acta.* We are pleased now that the Proceedings of the Seventh Colloquium appear as part of the AIAA's *Progress in Astronautics and Aeronautics* series.

Acknowledgments

The Seventh Colloquium, held under the auspices of the Institute of Physical Chemistry of the University of Göttingen and the Max Planck Institut für Strömungsforschung, August 20-24, 1979, was attended by 216 participants and 60 accompanying guests from 20 countries. Arrangements in Göttingen were made by Dr. A. W. Preuss and Prof. Dr. H. Gg. Wagner. Generous financial support for the Colloquium was provided by several German companies and agencies. The publication of the Proceedings has been made possible by a grant from the National Science Foundation (U.S.A.).

Preparations for the Eighth Colloquium are under way. The meeting is scheduled to take place in August 1981 in Minsk, U.S.S.R., under the chairmanship of Prof. R. I. Soloukhin, Institute of Heat and Mass Transfer, Soviet Academy of Sciences.

J. Ray Bowen
Numa Manson
Antoni K. Oppenheim
R. I. Soloukhin
November 1980

Members of the Program Committee (left to right): Numa Manson, Université de Poitiers; Hugh Edwards, University of Wales; Jacques Brossard, Université d'Orléans; Henri Guénoche, Université de Provence; R. I. Soloukhin, Luikov Heat and Mass Transfer Institute; S. S. Penner, University of California, San Diego (Numa Manson Medalist); A. K. Oppenheim, University of California, Berkeley; H. Gg. Wagner, Max Planck Institut für Strömungsforschung; and M. Barrere, ONERA.

Participants in the Seventh Colloquium.

**Progress in
Astronautics and Aeronautics**

Martin Summerfield,
Series Editor-in-Chief
*Princeton Combustion Research
Laboratories, Inc.*

Ruth F. Bryans,
Associate Series Editor
AIAA

Norma J. Brennan,
Director, Editorial Department
AIAA

Brenda J. Hio,
Series Managing Editor
AIAA

VOLUMES

*1. **Solid Propellant Rocket
Research.** 1960

2. **Liquid Rockets and
Propellants.** 1960

3. **Energy Conversion for
Space Power.** 1961

4. **Space Power Systems.**
1961

5. **Electrostatic Propulsion.**
1961

EDITORS

Martin Summerfield
Princeton University

Loren E. Bollinger
The Ohio State University
Martin Goldsmith
The Rand Corporation
Alexis W. Lemmon Jr.
Battelle Memorial Institute

Nathan W. Snyder
Institute for Defense Analyses

Nathan W. Snyder
Institute for Defense Analyses

David B. Langmuir
*Space Technology
Laboratories, Inc.*
Ernst Stuhlinger
*NASA George C. Marshall Space
Flight Center*
J. M. Sellen Jr.
*Space Technology
Laboratories, Inc.*

*Now out of print.

*6. Detonation and Two-Phase Flow. 1962
S. S. Penner
California Institute of Technology
F. A. Williams
Harvard University

7. Hypersonic Flow Research. 1962
Frederick R. Riddell
AVCO Corporation

8. Guidance and Control. 1962
Robert E. Roberson
Consultant
James S. Farrior
Lockheed Missiles and Space Company

*9. Electric Propulsion Development. 1963
Ernst Stuhlinger
NASA George C. Marshall Space Flight Center

*10. Technology of Lunar Exploration. 1963
Clifford I. Cummings and Harold R. Lawrence
Jet Propulsion Laboratory

11. Power Systems for Space Flight. 1963
Morris A. Zipkin and Russell N. Edwards
General Electric Company

12. Ionization in High-Temperature Gases. 1963
Kurt E. Shuler, Editor
National Bureau of Standards
John B. Fenn, Associate Editor
Princeton University

*13. Guidance and Control—II. 1964
Robert C. Langford
General Precision Inc.
Charles J. Mundo
Institute of Naval Studies

14. Celestial Mechanics and Astrodynamics. 1964
Victor G. Szebehely
Yale University Observatory

*15. Heterogeneous Combustion. 1964
Hans G. Wolfhard
Institute for Defense Analyses
Irvin Glassman
Princeton University
Leon Green Jr.
Air Force Systems Command

16. Space Power Systems Engineering. 1966 — George C. Szego, *Institute for Defense Analyses*; J. Edward Taylor, *TRW Inc.*

17. Methods in Astrodynamics and Celestial Mechanics. 1966 — Raynor L. Duncombe, *U. S. Naval Observatory*; Victor G. Szebehely, *Yale University Observatory*

18. Thermophysics and Temperature Control of Spacecraft and Entry Vehicles. 1966 — Gerhard B. Heller, *NASA George C. Marshall Space Flight Center*

19. Communication Satellite Systems Technology. 1966 — Richard B. Marsten, *Radio Corporation of America*

20. Thermophysics of Spacecraft and Planetary Bodies: Radiation Properties of Solids and the Electromagnetic Radiation Environment in Space. 1967 — Gerhard B. Heller, *NASA George C. Marshall Space Flight Center*

21. Thermal Design Principles of Spacecraft and Entry Bodies. 1969 — Jerry T. Bevans, *TRW Systems*

22. Stratospheric Circulation. 1969 — Willis L. Webb, *Atmospheric Sciences Laboratory, White Sands, and University of Texas at El Paso*

23. Thermophysics: Applications to Thermal Design of Spacecraft. 1970 — Jerry T. Bevans, *TRW Systems*

24. Heat Transfer and Spacecraft Thermal Control. 1971 — John W. Lucas, *Jet Propulsion Laboratory*

25. **Communications Satellites for the 70's: Technology.** 1971

Nathaniel E. Feldman
The Rand Corporation
Charles M. Kelly
The Aerospace Corporation

26. **Communications Satellites for the 70's: Systems.** 1971

Nathaniel E. Feldman
The Rand Corporation
Charles M. Kelly
The Aerospace Corporation

27. **Thermospheric Circulation.** 1972

Willis L. Webb
Atmospheric Sciences Laboratory, White Sands, and University of Texas at El Paso

28. **Thermal Characteristics of the Moon.** 1972

John W. Lucas
Jet Propulsion Laboratory

29. **Fundamentals of Spacecraft Thermal Design.** 1972

John W. Lucas
Jet Propulsion Laboratory

30. **Solar Activity Observations and Predictions.** 1972

Patrick S. McIntosh and Murray Dryer
Environmental Research Laboratories, National Oceanic and Atmospheric Administration

31. **Thermal Control and Radiation.** 1973

Chang-Lin Tien
University of California, Berkeley

32. **Communications Satellite Systems.** 1974

P. L. Bargellini
COMSAT Laboratories

33. **Communications Satellite Technology.** 1974

P. L. Bargellini
COMSAT Laboratories

34. **Instrumentation for Airbreathing Propulsion.** 1974

Allen E. Fuhs
Naval Postgraduate School
Marshall Kingery
Arnold Engineering Development Center

35. **Thermophysics and Spacecraft Thermal Control.** 1974

Robert G. Hering
University of Iowa

36. Thermal Pollution Analysis. 1975

Joseph A. Schetz
Virginia Polytechnic Institute

37. Aeroacoustics: Jet and Combustion Noise; Duct Acoustics. 1975

Henry T. Nagamatsu, Editor
General Electric Research and Development Center
Jack V. O'Keefe, Associate Editor
The Boeing Company
Ira R. Schwartz, Associate Editor
NASA Ames Research Center

38. Aeroacoustics: Fan, STOL, and Boundary Layer Noise; Sonic Boom; Aeroacoustics Instrumentation. 1975

Henry T. Nagamatsu, Editor
General Electric Research and Development Center
Jack V. O'Keefe, Associate Editor
The Boeing Company
Ira R. Schwartz, Associate Editor
NASA Ames Research Center

39. Heat Transfer with Thermal Control Applications. 1975

M. Michael Yovanovich
University of Waterloo

40. Aerodynamics of Base Combustion. 1976

S. N. B. Murthy, Editor
Purdue University
J. R. Osborn, Associate Editor
Purdue University
A. W. Barrows and J. R. Ward, Associate Editors
Ballistics Research Laboratories

41. Communication Satellite Developments: Systems. 1976

Gilbert E. LaVean
Defense Communications Engineering Center
William G. Schmidt
CML Satellite Corporation

42. Communication Satellite Developments: Technology. 1976

William G. Schmidt
CML Satellite Corporation
Gilbert E. LaVean
Defense Communications Engineering Center

43. **Aeroacoustics: Jet Noise, Combustion and Core Engine Noise.** 1976

Ira R. Schwartz, Editor
NASA Ames Research Center
Henry T. Nagamatsu, Associate Editor
General Electric Research and Development Center
Warren C. Strahle, Associate Editor
Georgia Institute of Technology

44. **Aeroacoustics: Fan Noise and Control; Duct Acoustics; Rotor Noise.** 1976

Ira R. Schwartz, Editor
NASA Ames Research Center
Henry T. Nagamatsu, Associate Editor
General Electric Research and Development Center
Warren C. Strahle, Associate Editor
Georgia Institute of Technology

45. **Aeroacoustics: STOL Noise; Airframe and Airfoil Noise.** 1976

Ira R. Schwartz, Editor
NASA Ames Research Center
Henry T. Nagamatsu, Associate Editor
General Electric Research and Development Center
Warren C. Strahle, Associate Editor
Georgia Institute of Technology

46. **Aeroacoustics: Acoustic Wave Propagation; Aircraft Noise Prediction; Aeroacoustic Instrumentation.** 1976

Ira R. Schwartz, Editor
NASA Ames Research Center
Henry T. Nagamatsu, Associate Editor
General Electric Research and Development Center
Warren C. Strahle, Associate Editor
Georgia Institute of Technology

47. **Spacecraft Charging by Magnetospheric Plasmas.** 1976

Alan Rosen
TRW Inc.

48. Scientific Investigations on the Skylab Satellite. 1976

Marion I. Kent and
Ernst Stuhlinger
NASA George C. Marshall Space Flight Center
Shi-Tsan Wu
The University of Alabama

49. Radiative Transfer and Thermal Control. 1976

Allie M. Smith
ARO Inc.

50. Exploration of the Outer Solar System. 1977

Eugene W. Greenstadt
TRW Inc.
Murray Dryer
National Oceanic and Atmospheric Administration
Devrie S. Intriligator
University of Southern California

51. Rarefied Gas Dynamics, Parts I and II (two volumes). 1977

J. Leith Potter
ARO Inc.

52. Materials Sciences in Space with Application to Space Processing. 1977

Leo Steg
General Electric Company

53. Experimental Diagnostics in Gas Phase Combustion Systems. 1977

Ben T. Zinn, Editor
Georgia Institute of Technology
Craig T. Bowman, Associate Editor
Stanford University
Daniel L. Hartley, Associate Editor
Sandia Laboratories
Edward W. Price, Associate Editor
Georgia Institute of Technology
James G. Skifstad, Associate Editor
Purdue University

54. Satellite Communications: Future Systems. 1977

David Jarett
TRW Inc.

55. Satellite Communications: Advanced Technologies. 1977
David Jarett
TRW Inc.

56. Thermophysics of Spacecraft and Outer Planet Entry Probes. 1977
Allie M. Smith
ARO Inc.

57. Space-Based Manufacturing from Nonterrestrial Materials. 1977
Gerard K. O'Neill, Editor
Princeton University
Brian O'Leary, Assistant Editor
Princeton University

58. Turbulent Combustion. 1978
Lawrence A. Kennedy
State University of New York at Buffalo

59. Aerodynamic Heating and Thermal Protection Systems. 1978
Leroy S. Fletcher
University of Virginia

60. Heat Transfer and Thermal Control Systems. 1978
Leroy S. Fletcher
University of Virginia

61. Radiation Energy Conversion in Space. 1978
Kenneth W. Billman
NASA Ames Research Center

62. Alternative Hydrocarbon Fuels: Combustion and Chemical Kinetics. 1978
Craig T. Bowman
Stanford University
Jørgen Birkeland
Department of Energy

63. Experimental Diagnostics in Combustion of Solids. 1978
Thomas L. Boggs
Naval Weapons Center
Ben T. Zinn
Georgia Institute of Technology

64. Outer Planet Entry Heating and Thermal Protection. 1979
Raymond Viskanta
Purdue University

65. Thermophysics and Thermal Control. 1979
Raymond Viskanta
Purdue University

66. Interior Ballistics
    of Guns. 1979

Herman Krier
*University of Illinois
at Urbana-Champaign*
Martin Summerfield
New York University

67. Remote Sensing of Earth
    from Space: Role of
    "Smart Sensors." 1979

Roger A. Breckenridge
NASA Langley Research Center

68. Injection and Mixing in
    Turbulent Flow. 1980

Joseph A. Schetz
*Virginia Polytechnic
Institute and State University*

69. Entry Heating and
    Thermal Protection. 1980

Walter B. Olstad
NASA Headquarters

70. Heat Transfer, Thermal
    Control, and Heat Pipes.
    1980

Walter B. Olstad
NASA Headquarters

71. Space Systems and
    Their Interactions
    with Earth's Space
    Environment. 1980

Henry B. Garrett and
Charles P. Pike
Hanscom Air Force Base

72. Viscous Flow Drag
    Reduction. 1980

Gary R. Hough
*Vought Advanced
Technology Center*

73. Combustion Experiments
    in a Zero-Gravity Laboratory.
    1981

Thomas H. Cochran
NASA Lewis Research Center

74. Rarefied Gas Dynamics,
    Parts I and II
    (two volumes). 1981

Sam S. Fisher
*University of Virginia,
Charlottesville*

75. **Gasdynamics of Detonations and Explosions.** 1981

J. R. Bowen
University of Wisconsin
Madison, Wisconsin
N. Manson
Université de Poitiers
Poitiers, France
A. K. Oppenheim
University of California
Berkeley, California
R. I. Soloukhin
Institute of Heat and Mass Transfer
BSSR Academy of Sciences
Minsk, U.S.S.R.

76. **Combustion in Reactive Systems.** 1981

J. R. Bowen
University of Wisconsin
Madison, Wisconsin
N. Manson
Université de Poitiers
Poitiers, France
A. K. Oppenheim
University of California
Berkeley, California
R. I. Soloukhin
Institute of Heat and Mass Transfer
BSSR Academy of Sciences
Minsk, U.S.S.R.

(Other volumes are planned.)

Plenary Lecture

Particle Sizing in Flames

S.S. Penner* and P.H. P. Chang†
University of California, San Diego, La Jolla, Calif.

We describe briefly techniques for in situ particle sizing. All of these have limited utility for application and error bounds tend to be large. Particle-size distributions have probably never been characterized in a quantitative manner without sample withdrawals through probes and possible associated distortions. There is evidence that (mature) particle-size distributions formed from agglomerating solids are self-preserving. These particle-size distributions may be characterized experimentally by a single parameter defining the mean particle radius. The only type of in situ particle-size characterization that we have found to be independent of the complex index of refraction of the (scattering) particles (because it depends only on the particle diffusion coefficient) is the half-width of the scattered laser power spectrum for a monodisperse system. Initially formed particle-size distributions in flames have not been identified successfully. Independent determinations of the two parameters needed to define assumed log-normal distributions for spherical particles involve formidable experimental difficulties. We summarize briefly the problems involved in adapting our procedure, based on determinations of scattered laser power spectra, to measurements of this type. We discuss measurements of 1) mean particle diameters and corresponding number densities and 2) particle-size distributions and corresponding number densities. Experimental results (as functions of equivalence ratio and downstream distance), obtained on a flat-flame burner for fuel-rich mixtures of methane with oxygen, are used for illustrative purposes. We have only limited fundamental understanding of particulate growths in flames. Important research remains to be done on the mechanisms of nucleation, on the shapes and compositions of growing particles that have not been withdrawn through probes, on the rates of agglomeration of smaller particles into larger units before self-preserving distributions obtain, and on particle-size-limiting processes that define the ultimate growths in flames.

Presented at the 7th ICOGER, Göttingen, Federal Republic of Germany, Aug. 20-24, 1979. Copyright © American Institute of Aeronautics and Astronautics, Inc., 1981. All rights reserved.
*Professor of Engineering Physics and Director, Energy Center, Department of Applied Mechanics and Engineering Sciences; 1979 Numa Manson Medalist.
†Research Assistant, Energy Center, Department of Applied Mechanics and Engineering Sciences.

Introduction

IN this paper, we discuss the following topics: methods for in situ particle sizing in flames, self-preserving particle-size distributions, the in situ characterization of initially formed particles, and results on particle-size and number-density measurements using scattered laser power spectra as a diagnostic procedure for log-normal and self-preserving particle-size distributions. The early historical evolution of methods for particle sizing is briefly discussed in the Appendix.

Particle-Size Measurements Using Techniques Other Than Scattered Laser Power Spectra

We comment in this section on some of the more popular experimental procedures that have been used for particle sizing. In all of these (excepting examinations of individual particles), it is tacitly assumed that the particles are spherical.

Particle Sizing Using Light Scattering

Particle sizing using light scattering yields "local" values referring to the scattering volume.

Measurements Using the Angular-Dependence of the Scattered Intensity

For monochromatic incident light, the intensity of the observed scattered radiation depends on scattering angle and particle size. The particle size may be inferred from the measured angular dependence of the scattered intensity by using the Mie theory, provided the index of refraction of the (assumed spherical) particles is known. This method is easy to apply. If monochromatic radiation from a laser is used, the applicable size range generally exceeds 0.1 μm since the readily available short-wavelength lasers have a wavelength around 0.5 μm and Mie scattering theory must be used for $(2\pi r/\lambda) > 1$. If $(2\pi r/\lambda) < 1$, the Rayleigh limit is a good approximation and the angular pattern of the scattered intensity is independent of the particle radius.

Senftleben and Benedikt (1917) measured the size of soot particles formed in a Hefner lamp burning amyl acetate and estimated the mean diameter to be around 1750 Å. They assumed that the soot particles have an index of refraction equal to 1.95-0.66 i, which is the index of refraction of bulk amorphous carbon. Dalzell et al. (1970) used the 4358 Å mercury line to measure the scattered light from soot particles in the smoke above a propane-air flame. The mean soot diameter was estimated to be ~1650-1750 Å for spherical particles with an index of refraction (Dalzell and Sarofim, 1969) of 1.60-0.60 i for $\lambda = 4358$ Å (which was determined for collected soot samples).

Measurements Using the Wavelength Dependence of the Scattered Intensity

The scattered intensity is measured at a fixed scattering angle for various wavelengths. Since the scattering coefficients depend on wavelength differently for different particle sizes, particle-size information may be inferred from the measured scattered intensities. This method is easy to apply for spherical particles but requires information concerning the index of refraction.

D'Alessio et al. (1973) applied this method and two other methods, using a xenon lamp, to measure particle sizes in a premixed CH_4-O_2 flame burning on a flat-flame burner. The wavelength range extended from 2000 to 6000 Å. The mean particle diameters measured were 1000-2000 Å.

Measurements Using the Dependence of Polarization Ratio on Scattering Angle

The ratio of the horizontally polarized component to the vertically polarized component of the scattered light intensity from particles is defined to be the polarization ratio. The angular dependence of the polarization ratio may be used to infer size distributions for spherical particles, provided the index of refraction is known. Numerical calculations are made for selected size distributions. Matijevic et al. (1964) have applied measurements of polarization ratios to the characterization of octanic acid aerosols. Assuming a log-normal size distribution, they obtained values of $\bar{r}$ between 0.1 and 1 µm for $\ln \sigma_g$ between 0.1 and 0.2, depending on the aerosol conditions.

Measurements Using the Dependence of Polarization Ratio on Wavelength

The polarization ratio may be measured at a fixed scattering angle for different wavelengths. Heller and Wallach (1963) used this method to determine the size distribution in a polystyrene latex with mean diameters between 0.7 and 1.3 µm. A mercury vapor lamp and a single-prism monochromator served as light source.

Particle Sizing Using Light Extinction

Particle sizing using light extinction yields average values over the line of sight. Thus, "local" values are obtained only when the extinction volume is appropriately small in size.

Measurements of Light Extinction at Different Wavelengths

The extinction coefficient is proportional to the number of particles in the optical pathlength and to the extinction cross section of the particles. If the index of refraction of (spherical) particles is known, the extinction cross section may be calculated as a function of the particle size using the Mie theory. For a given size distribution, the ratios of the extinction coefficients at two different wavelengths depend only on the particle sizes and not on the number densities. The method yields the mean particle size along the optical pathlength. Measurable diameters are again limited to ranges where the Rayleigh limit is a poor approximation. Carlon et al. (1976) performed experiments on a 3 m diam cloud chamber. Radiation from an He-Ne laser at 0.6328 µm and from an infrared source at 8.5, 10.5, and 12.57 µm was used. Average droplet diameters of 8 and 10 µm were reported for the cloud chamber.

Measurements of Individual Particle Diameters Using Light Extinction

Two intersecting laser beams define a volume uniquely. With this technique, extinction is measured for a particle passing through the interference fringes of the diagnostic volume. In effect, each particle is examined individually. There is no ambiguity in size distribution. However, the allowed number densities and the rates of measurement are limited. Faxvog (1976) used two beams, one of which had a donut-shaped cross-sectional intensity distribution, which served to localize the scattering particle. Tests were performed on the exhaust gases from a diesel engine. It was found that opaque particles with diameters of 0.05-5.0 μm could be sized with a resolution of better than 10% at a sizing rate of about 300 particles/min. The maximum number density was limited to about 10^5 particles/cm^3.

Methods Using Both Scattering and Extinction

For particle sizes smaller than 0.1 μm, neither scattering nor extinction alone will define a particle diameter. The combination of extinction and scattering yields two independent inputs that may be used to determine the average particle size and number density. This technique is especially useful for studies of extensive, homogeneous systems (e.g., the region behind planar shock fronts). Graham et al. (1975) used a 1 W argon-ion laser to study soot formation behind a shock wave at 1600-2500 K at a total carbon concentration of $\sim 2 \times 10^{17}$ atoms/cm^3. Light scattering from Ar was used for calibration.

Particle Sizing Using Fringe-Visibility Changes

Two crossing laser beams form an interference fringe pattern. When a particle passes across these fringes, the scattered light intensity is modulated according to the brightness or darkness of the fringes. The extent of the observed modulation reaches a maximum if the particle is appreciably smaller than the fringe spacing; the modulation in intensity is diminished for larger particles. This method may be used to measure the sizes of individual particles provided the particle concentrations are low. Since the minimum fringe spacing is of the order of the wavelength of light, the minimum discernible particle size is also limited to this range. Farmer (1974) used two perpendicular, polarized light beams from a 5 mW He-Ne laser, the output of which was split into two intersecting beams; glass and aluminum particles with diameters of 10-120 μm were examined. Dos Santos and Stevenson (1977) measured the visibility of laser-induced fluorescent light that is frequency-shifted and isotropic in space; the utility of using fringe-visibility changes for this fluorescence was demonstrated on a solution containing a mixture of 50% benzyl alcohol and 50% ethylene glycol doped with Rhodamine 6G.

Particle Sizing Using Light Diffraction

This procedure is closely related to light scattering for particles larger than the wavelength of light. The diffraction pattern is observed for small scattering angles. The scattering coefficient occurring in the Mie theory is approximated by the simpler results derived from diffraction theory that are independent of the index of refraction. Swithenbank et al. (1976) used the diffraction pattern produced by a 4 g/s pressure jet of kerosene in a nozzle spray illuminated by a 1 mW He-Ne laser; the diffraction pattern was examined in the focal plane and the size distribution of the droplets was determined within 10% for diameters larger than $\sim 1 \mu m$.

Particle Sizing Using Photography

Pictures of particle suspension are taken with a short, pulsed flashlight. Major limitations are the limited depth of field and the poor resolution attainable for small particles. The ultimate resolution achievable with a microscope is of the order of the wavelength of light and is greatly reduced if the depth of the field is increased. Ingebo (1956) used a flash produced from magnesium sparks with an exposure time of about 4 μs; diameter distributions of 0-120 μm were determined, with an accuracy of $\pm 3\mu m$, for isooctane droplets in an air jet.

Particle Sizing Using Holography

The interference of a reference laser beam with the laser light scattered from particles produces a complex wave pattern that may be recorded and later reconstructed. The depth-of-field problems encountered in conventional photography are effectively eliminated. Thompson et al. (1967) used holography to study fogs and aerosols with a TV monitor to display the reconstructed particle images at an amplification factor of about 300; the smallest measurable particle diameter was about $3\mu m$.

Particle-Size Measurements Using Electron Microscopy

Particle samples are collected on plates and are then examined under an electron microscope. The measurable diameters are smaller than 30 Å and the ultimate resolution limit will thus approach molecular dimensions. The effects of the sampling procedure on the particle dimensions are generally not known. Prado et al. (1976) collected soot samples through a water-cooled probe from a turbulent flame and determined number densities, size distributions, and deviations from sphericity for individual particles.

Self-Preserving Particle-Size Distributions

The theoretical description of self-preserving particle-size distributions, formed as the result of coagulation from smaller particles, has been worked out by Lai et al. (1972), Pich et al. (1970), and Friedlander (1961), following earlier work by Smoluchowski (1916) and Tunitskii (1938), as described by Fuks (1964) and Hidy and Brock (1970). The theory of self-preserving particle-size distributions has been developed under the following conditions: 1) the volume fraction of the condensed phase is a constant independent of time; 2) particles are perfectly sticky on collision, an assumption which determines the dynamical equations governing the rates of formation and disappearance of particle populations; 3) the particles are spherical; 4) particulate growths occur either for Knudsen numbers (ratio of the molecular mean free path to the particle diameter) Kn larger than ~ 10 (i.e., in the free molecular flow limit) or for $Kn \ll 1$ (Einstein-Stokes diffusion); the analytical treatment for $Kn \sim 1$ is more difficult to perform and involves many approximations; 5) the collision frequency $\beta(v,\bar{v})$ is a homogeneous function of the volumes v and $\bar{v}$ of colliding particles. Independently of the original particle-size distribution (i.e., for monodisperse particles, polydisperse particle-size distributions, and log-normal particle-size distributions), a self-preserving particle-size distribution is well approximated (Friedlander and Wang, 1966) after elapse of a specifiable induction time τ. The self-preserving distribution $\psi(\eta)$ is independent of the physical properties of the medium (temperature, viscosity, density) and of time. The parameters ψ and η are related to physical variables through the following expressions:

$$\psi(\eta) = n(v,t)\nu/N_\infty^2(t) = n(v,t)\bar{V}/N_\infty(t), \qquad \eta = N_\infty(t)v/\nu = v/\bar{V} \qquad (1)$$

where $n(v,t)$ is the number of particles per unit volume with volume between v and $v + dv$ at time t,

$$\nu = \int_0^\infty v n(v,t) dv = N_\infty(t) \bar{V}(t)$$

is the constant volume fraction of the condensed phase, $N_\infty(t)$ is the total number of particles per unit volume, and $\bar{V}(t)$ the average particle volume.

Mean Particle Volume

In the free molecular flow regime, the rate of change of the total particle concentration with time is governed by the expression

$$\frac{dN_\infty}{dt} = -\tfrac{1}{2} [3/(4\pi)]^{1/6} (6kT/\rho_c)^{1/2} \nu^{1/6} N_\infty^{11/6} \alpha \qquad (2)$$

where

$$\alpha = \int_0^\infty \int_0^\infty (\eta^{1/3} + \tilde{\eta}^{1/3})^2 \left(\frac{1}{\eta} + \frac{1}{\tilde{\eta}}\right)^{1/2} \psi(\eta) \psi(\tilde{\eta}) d\eta d\tilde{\eta}$$

In terms of $\bar{V}$, Eq. (2) becomes

$$\frac{d\bar{V}}{dt} = c_1 \nu^{5/6} \bar{V}^{1/6}$$

with

$$c_1 \nu^{5/6} = \tfrac{1}{2} (3/4\pi)^{1/6} (6kT/\rho_c)^{1/2} \nu \alpha G'$$

where ρ_c is the particle density and the factor G' has been introduced in order to allow for augmentation of the collision rate by dispersion forces (Fuks and Sutugin, 1965). The integral of the preceding expression is

$$[\bar{V}(t)]^{5/6} - [(\bar{V}(t_0)]^{5/6} = (5/6) c_1 \nu^{5/6} (t - t_0) \qquad (3)$$

The functional form of Eq. (3) has been verified by Graham et al. (1975) for soot aerosols formed from aromatic hydrocarbons behind shock fronts. Since the scattering cross section for Ar is known, Ar was used for absolute calibration. Graham et al. (1975) used the following sequence of measurements on uniformly heated, two-phase systems behind incident shock waves: the ratio of transmitted (I_t) to incident (I_0) intensity and the ratios of scattered to transmitted intensities for both soot-forming mixtures and for pure argon. The final relation for the numerical value of $\bar{V}$ involves a function of the complex index of refraction $\tilde{m}$.

Determination of Carbon Conversion to Soot for Monodisperse Systems by Graham et al. (1975)

For spherical particles of radius r in the Rayleigh limit, the ratio of transmitted intensity at time t, I_t, to the initially transmitted intensity I_0 is

$$\frac{I_t}{I_0} = \exp - \left\{ [E(\tilde{m})] \left(\frac{2\pi r}{\lambda}\right)(4\pi r^2 N_\infty \ell) \right\}, \qquad E(\tilde{m}) = \mathrm{Im}\left(\frac{\tilde{m}^2 - 1}{\tilde{m}^2 + 1}\right) \tag{4}$$

where $\tilde{m}$ is again the complex index of refraction and Im denotes the imaginary part of the following quantity, N_∞ is the total number of small $(2\pi r/\lambda < 1)$ absorbing carbon particles per unit volume, and ℓ the absorption pathlength. The total number of particles per unit volume (N_∞) is related to the number of carbon atoms per unit volume in the solid phase (C) through the expression

$$N_\infty = 12(C)/N\rho_c v, \qquad v = 4\pi r^3/3 \tag{5}$$

Using Eqs. (4) and (5), it is found that

$$(C) = N\rho_c \lambda \ln(I_0/I_t)/72\pi \ell E(\tilde{m})$$

But the total number of carbon atoms per unit volume present in any phase, $(C)_{\text{total}}$, is known for a specified fuel-air ratio. Hence the soot conversion ratio $(C)/(C)_{\text{total}}$ may be estimated. Graham et al. (1975) found that $E(\tilde{m})$ had to be changed from 0.224 [i.e., the value of Dalzell and Sarofim (1969)] to 0.254 in order to assure that $(C)/(C)_{\text{total}} < 1$.

Characterization of Mature Two-Phase Systems, Comparisons between Self-Preserving and Log-Normal Distributions

Self-preserving particle-size distributions may be characterized by a single parameter (e.g., the mean volume $\bar{V}$). Thus, after elapse of an appropriate relaxation time to assure the presence of a mature two-phase system, a single measurement will be sufficient to characterize the two-phase distributions. The relaxation or induction times are calculable from the theory of Lai et al. (1972); see the following for details.

It is interesting to compare the functional forms of the log-normal and self-preserving distributions. Let $W(v)\,dv = W(r)\,dr$ or, since $v = 4\pi r^3/3$,

$$W(v) = [W(r)]\frac{dr}{dv} = W(r)/4\pi r^2$$

For the log-normal distribution,

$$W(v)\,dv = \frac{1}{3\sqrt{2\pi}\ln\sigma_g}\frac{1}{v}\left\{\exp\left[-\left(\ln\frac{v}{\bar{v}^*}\right)^2/18(\ln\sigma_g)^2\right]\right\}dv$$

where $\bar{v}^* \equiv (4\pi/3)(\bar{r}^*)^3$ is the particle volume for the most probable particle radius $\bar{r}^*$; it may be shown that

$$\zeta = \bar{V}^*/\bar{v}^* = \exp\left[\frac{9}{2}(\ln\sigma_g)^2\right]$$

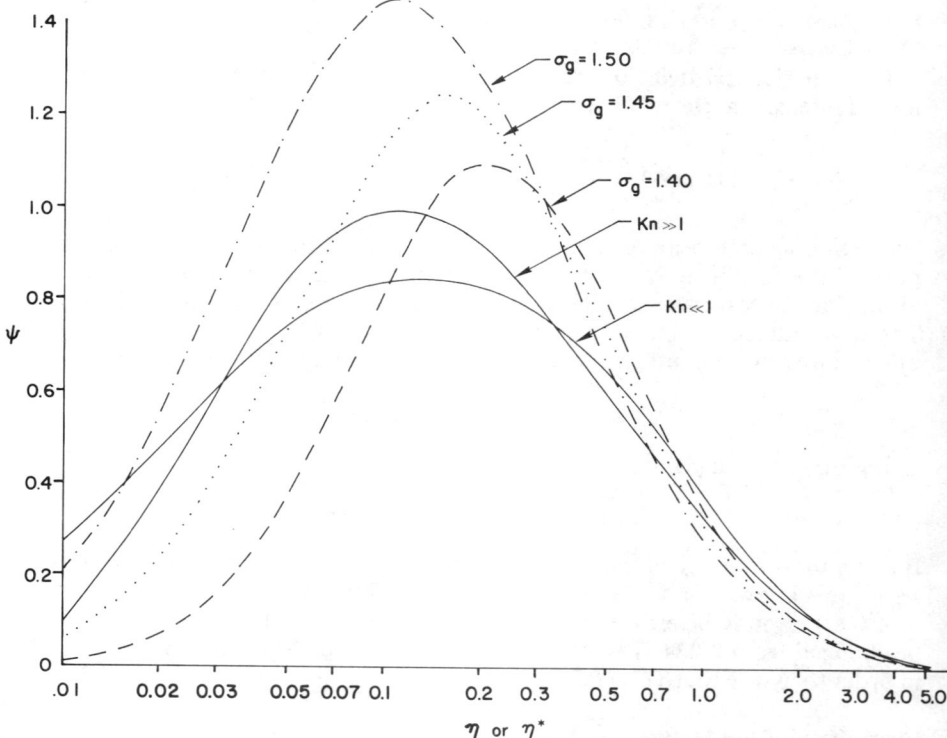

Fig. 1 Plots of the log-normal (broken curves) and self-preserving distributions (solid curves) in the ψ-η and ψ-η^* planes, respectively (the parameter η^* is a function of σ_g).

where $\bar{V}^*$ is the mean particle volume for the log-normal distribution,

$$W(v;\sigma_g) = \frac{1}{3\sqrt{2\pi}\ln\sigma_g} \frac{1}{\bar{V}^*\eta^*} \exp\left\{-[\ln(\eta^*\zeta)]^2/18(\ln\sigma_g)^2\right\}, \qquad \eta^* = v/\bar{V}^* \qquad (6)$$

Equation (6) may be rewritten as

$$\psi(\eta^*;\sigma_g) = \frac{1}{3\sqrt{2\pi}\ln\sigma_g} \frac{1}{\eta^*} \exp\left\{-\left[\ln\eta^* + \frac{9}{2}(\ln\sigma_g)^2\right]^2 \bigg/ 18(\ln\sigma_g)^2\right\} = W(v)\bar{V}^* \qquad (7)$$

In order to compare the log-normal distributions with the self-preserving distributions (for $Kn \ll 1$ and $Kn \gtrsim 10$), as calculated by Lai et al. (1972), we have plotted in Fig. 1 the values of $\psi(\eta^*;\sigma_g)$ for different values of σ_g according to Eq. (7), together with the self-preserving distributions $\psi(\eta)$. We will show (compare the last section) that these distributions, which appear to be quite different, will produce similar light scattering as a function of scattering angle for $\sigma_g \simeq 1.65$ and appropriately chosen particle radii.

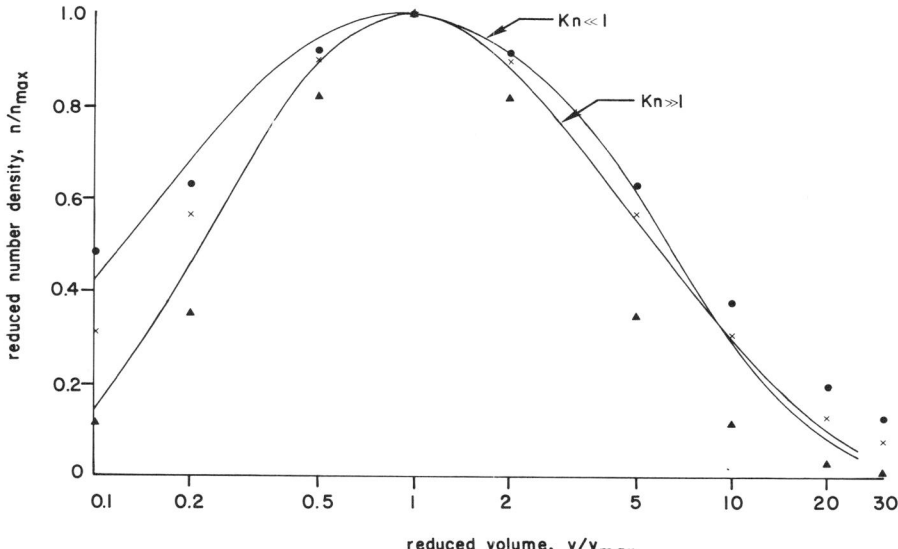

Fig. 2 Comparison of the self-preserving distributions for $Kn \gtrsim 10$ and $Kn \ll 1$ with log-normal distributions for $\sigma_g = 1.45$ (▲), 1.65 (×), and 1.75 (●). The reduced number density (n/n_{max}) and the reduced volume (v/v_{max}) are normalized such that the maximum number density and the corresponding volume are equal to unity for each distribution.

If experimental results are not presented on absolute scales but are instead shown in terms of both number densities and volumes divided by their respective maxima, then we obtain the data shown in Fig. 2. In this particular representation, $\sigma_g \simeq 1.65$ for the log-normal distribution is a fair approximation to the self-preserving distribution. Experimental studies relating to differences between the log-normal and self-preserving distributions are described in the final section of this paper.

Times for Reaching the Self-Preserving Distribution

Lai et al. (1972) have estimated the times required for the moments to reach their asymptotic values. The physical time τ has been shown (Lai et al., 1972) to be

$$\tau \simeq 3/(6kTr_1/\rho_c)^{1/2} N_\infty (t=0) \tag{8}$$

for an initially monodisperse distribution with uniform radius r_1; if the initial distribution is a (discrete) log-normal distribution with $\sigma_g = 1.35$ and the smallest particle radius r_1, then τ was found to be reduced by a factor of 15. The preceding expression for monodisperse systems provides a conservative estimate for τ for any initial particle-size distribution.

In our flat-flame burners, we have assumed a constant volume fraction for the particulate phase, which is consistent with the model of coagulating particles leading to a self-preserving particle-size distribution. Thus,

Table 1 Experimental and derived data for hydrocarbon-oxygen mixtures above a flat-flame burner for selected equivalence ratios φ

	Gas mixture		
	CH_4/O_2, $\varphi=3.15$	C_2H_6/O_2, $\varphi=3.15$	C_2H_4/O_2, $\varphi=3.15$
Height above the burner h, cm	2.5	2.5	2.0
Flame velocity, cm/s	~50	~100	~100
Estimated flame temperature, K	1500	1500	2300
Flow time to the height h, s	0.050	0.025	0.020
Measured particle radius for an assumed monodisperse system, Å	670	1000	1025
Measured number density for an assumed monodisperse system at h, cm^{-3}	3.3×10^8	2.5×10^7	8×10^6
$N_\infty(0)$ for $r_1(0)=20$ Å, cm^{-3}	1.2×10^{13}	3.1×10^{12}	1.0×10^{12}
Relaxation time τ, s [from Eq. (8)]	7.1×10^{-4}	2.7×10^{-3}	6.5×10^{-3}
Ratio of flow time to relaxation time	~70	~9.2	~3.1

measurements of $\bar{r}$ and N_∞ at any height above the flat-flame burner may be used to estimate r_1 and $N_\infty(0)$ on the assumption that

$$[\bar{r}(t)]^3 N_\infty(t) = \text{const}$$

The resulting estimates for $N_\infty(0)$, on the assumption that $r_1(0) = 20$ Å, are listed in Table 1. The corresponding values of τ for various gas mixtures are also shown in Table 1. The values of τ are seen to be generally much smaller than the physical times $t=h/u$; here, h are heights above the burner rim where measurements were made ($h \geq 20$ mm) and u the corresponding flow speeds. Hence, we conclude that the observed particle-size distributions in all gas mixtures should correspond to self-preserving distributions if particle agglomeration occurs according to the assumed model.

Characterization of Incipient (Freshly Formed) Particulate Distributions

Freshly formed particulate distributions require characterizations that do not involve coagulation and subsequent growths to self-preserving distributions. This characterization is readily accomplished only for monodisperse systems, preferably by using our laser-scattering technique that allows sizing without requiring knowledge of the index of refraction for monodisperse systems. For initially polydisperse systems with log-normal distribution functions, our published (Penner and Chang, 1978) methods of measurement yielding high spatial resolution require very demanding precision for proper characterization of the initially formed distribution functions. If the requirement of high spatial resolution is relaxed, characterization of log-normal distribution functions may be readily implemented by using a variety of experimental techniques.

The nature of the initially formed distribution function is closely related to the reaction mechanisms that lead to particulate formations. The elucidation of these processes has been considered by many authors and has been discussed recently, for example, for shocks by Graham et al. (1975) and for flames by Wagner (1978).

Particle-Size and Concentration Measurements Using Scattered Laser Power Spectra

In previous publications, we have described experimental procedures for measurements of particle-size distributions and of number densities using scattered laser power spectra. Following a general description of the methodology involved (Penner et al., 1976a) and of selected experimental findings (Penner et al., 1976b, and Bernard and Penner, 1977), we have considered the possibility of using pulsed lasers and short ($\sim 10^{-4}$ s) observation times (Penner et al., 1977). A general analysis concerning the sensitivities achievable in measurements of particle-size distributions, using a variety of laser-scattering configurations and laser wavelengths, has been given by Penner and Chang (1978).

We shall now show how scattering measurements may be combined with transmission measurements to improve the sensitivity with which particle-size distributions are determined, albeit with the loss of high spatial resolution. Commensurate estimates of number densities for the specified particle-size distributions may then be obtained. Some of our experimental results for fuel-rich mixtures of methane and oxygen, as functions of equivalence ratio and downstream distance, are presented. These are consistent with the view that particulate formations occur very close to the burner rim in flat-flame burners and subsequently grow in size while maintaining an essentially constant total mass of particulate material.

Relations Between Diffusion Coefficients and Particle Sizes

In previously published papers, we related measured or measurable diffusion coefficients to particle diameter through the Stokes-Einstein diffusion relation. This relation is actually applicable only if the molecular mean free path is small compared with the particle diameter (Kennard, 1938), which is generally the case at atmospheric pressure and flame temperatures for particles that are larger than about 1 μm. When the molecular mean free paths are small in comparison with the particle radii, then the observed diffusion coefficients should be related to the particle diameters by using Epstein's equation (Kennard, 1938). Thus, while the entire theoretical apparatus in terms of diffusion coefficients remains unchanged, data interpretation in terms of particle diameters is strongly dependent on the ratio of mean free path to particle diameter. We shall now summarize the applicable expressions for the limiting cases of Stokes-Einstein diffusion (identified by the subscripts SE) and

Epstein diffusion (identified by the subscript E). For Stokes-Einstein diffusion,

$$D = kT/3\pi\mu d_{SE} \quad \text{if} \quad \ell/d_{SE} \ll 1 \tag{9}$$

where ℓ represents the molecular mean free path and the other symbols have their usual meaning. Similarly, for Epstein diffusion,

$$D = 3kT\ell/2\pi\mu d_E^2 \alpha \quad \text{if} \quad \ell/d_E \gg 1 \tag{10}$$

where $1.0 \le \alpha \le 1.44$, depending on the nature of the collision between molecules and particles ($\alpha = 1.0$ for specular reflection, $\alpha = 1.44$ for diffuse reflection). For the same measured value of D, it then follows that

$$d_E = \sqrt{9\ell\, d_{SE}/2\alpha} \tag{11}$$

For $T = 2000$ K and atmospheric pressure, $\ell \simeq 5.9 \times 10^{-5}$ cm in the gas mixture. Hence, for $\alpha = 1$, $d_{SE} = 100$ a should be replaced by $d_E = 1600$ Å, $d_{SE} = 300$ Å by $d_E = 2800$ Å, and $d_{SE} = 500$ Å by $d_E = 3600$ Å. Thus, the values specified for mean particle diameters d_{SE} by Bernard and Penner (1977) should be increased so substantially that they become comparable with values derived from the angular dependence of Mie scattering (D'Alessio et al., 1973, and Penner et al., 1976b). It should be noted, however, that, as d_E increases toward ℓ, it becomes too large and an appropriate correction must be introduced for ℓ neither much larger nor much smaller than d.

Expression for the Determination of the Particle Concentrations in Monodisperse Systems

The number density N_∞ (particles/cm^3) is given by Eq. (26) of Penner et al. (1976a), namely

$$N_\infty = \alpha(0) \Big/ \left(\frac{1}{32\pi^4}\right)\left(\frac{\gamma P_0 \lambda^7}{hc}\right) \left\{ \frac{L}{\left(\sin^2\frac{\theta}{2}\right) a(L\sin\theta + a|\cos\theta|)} \right\} \frac{f}{D} \tag{12}$$

where $\alpha(0)$ is the detectability ratio at zero frequency, γ the quantum efficiency of the photomultiplier (number of electrons produced per photon), P_0 the output laser power (erg/s), λ the laser wavelength (cm), h Planck's constant (erg-s), c the light speed (cm/s), θ the angle between the incident and the scattered beams, a (cm) the width and height of the scattering volume [which is a function of θ such that $a = a_0/(\sin\theta)^{1/3}$ where a_0 is a constant] and L its length (cm), f the Mie scattering coefficient, and D the diffusion coefficient (cm^2/s) for particles in the flame at the location of the scattering volume.

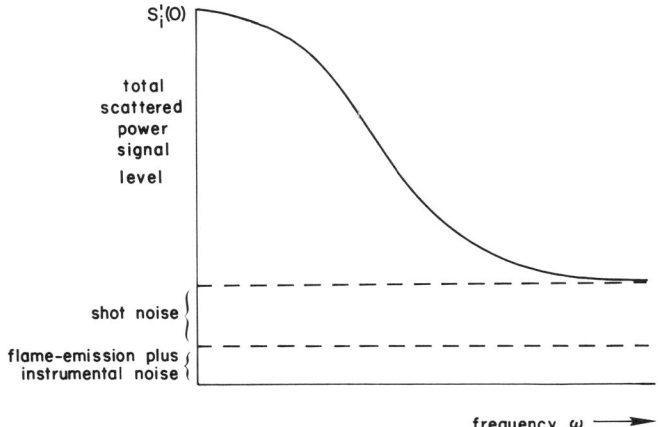

Fig. 3 Schematic diagram showing the scattered power signal level as a function of frequency. The sum of the white background components (instrumental noise and shot noise) is much smaller than the total signal level near the scattered laser frequencies ($\omega \sim 0$).

If we approximate the scattering volume by a cube of width a, Eq. (12) reduces to

$$N_\infty = \alpha(0) \Big/ \left(\frac{1}{32\pi^4}\right)\left(\frac{\gamma P_0 \lambda^7}{hc}\right) \frac{1}{a\left(\sin^2 \frac{\theta}{2}\right)(\sin\theta + |\cos\theta|)} \frac{f}{D} \quad (13)$$

The measurable parameters entering in Eq. (13) are seen to be $\alpha(0)$, γ, P_0, λ, a, θ, and D. The Mie scattering coefficient f is easily calculated after we have deduced the particle diameter d from measured values of D. The number density of particles N_∞ may then be calculated according to Eq. (13).

The experimental procedures have been described by Penner et al. (1976a) and Bernard and Penner (1977). The power spectrum has contributions from three sources, as is shown in Fig. 3. In the most general case, the scattered ac signal $S_i'(\omega)$ is a Voigt profile, the Gaussian part of which is characterized by a half-width v/σ, where v is the flow speed in the scattering volume and $\sigma \approx a$; the Lorentzian part is characterized by a half-width (in Hz) $b_L = 2|k|^2 D/2\pi$ where $|k| = (4\pi/\lambda)\sin(\theta/2)$.

Determination of Parameters and Uncertainty Estimates for the Number Densities in Flames

The wavelength λ is 6328 Å for the He-Ne laser; θ is determined for each experimental measurement and is often chosen to be about 15 deg. The procedure for obtaining b_L and v/σ from the measured ac power spectra $S'(\omega)$ has been described before (Bernard and Penner, 1977). An independent laser Doppler velocimetry measurement is made to determine v. Thus, D is obtained from the relation $D = \pi b_L / |k|^2$ while $\sigma = v/(v/\sigma)$. The total noise level in our power spectrum consists of shot noise, flame-emission noise, and instrumental noise.

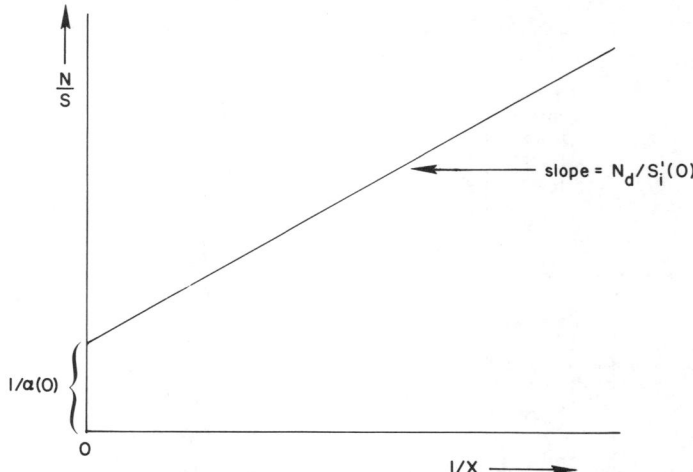

Fig. 4 Schematic diagram showing the total noise-to-signal ratio N/S as a function of the reciprocal of the signal-amplification factor X. The intercept on the N/S axis at $1/X = 0$ is the reciprocal of the detectability ratio $1/\alpha(0)$; the slope $d(N/S)/d(1/X)$ equals $N_d/S_i'(0)$.

The magnitude of the white background signal, which consists of the sum of flame-emission noise, photomultiplier dark noise, and recording and amplifying system noise, may be estimated by recording a power spectrum for flame emission alone, using the same experimental conditions as in the scattering measurements but with the laser turned off. The total detected white background signal is generally small compared with both the observed signal level [at frequencies close to those of the laser source ($\omega = 0$)] and the white shot noise.

The noise contributed by the analyzing system arises primarily in the digitizer. This noise level may be estimated by digitizing the same observed data at different levels of amplification. The resulting power spectra will usually exhibit different signal-to-noise ratios. The observed noise-to-signal ratio may be written in the form

$$\frac{N}{S} = \frac{SN \times X + N_d}{S_i'(0) \times X} = \frac{SN}{S_i'(0)} + \frac{N_d}{S_i'(0) \times X} = \frac{1}{\alpha(0)} + \frac{N_d}{X S_i'(0)} \qquad (14)$$

where N is the total observed noise level, S the total observed signal level, SN the intrinsic shot noise, X the system-amplification factor, N_d all other sources of white noise, $S_i'(0)$ the scattered power level at zero beat frequency, and $\alpha(0) = S_i'(0)/SN =$ detectability. It follows that a plot of the noise-to-signal level against the inverse of the amplification factor X (see Fig. 4) will yield a straight line which intersects the N/S axis for $1/X = 0$ at $1/\alpha(0)$. Thus, the detectability ratio, as well as the inherent noise level of the analyzing system, are easily determined experimentally.

Error Estimates for d_E, d_{SE}, and N_∞ for Monodisperse Particles

We proceed to estimate the errors involved in determining the various parameters entering into Eq. (13).

The errors in λ and θ are very small and may be neglected. The determination of b_L and v/σ from $S_i'(\omega)$ involves a curve-fitting process. Because the observed power spectrum is inherently noisy, we estimate that the errors are about $\pm 15\%$ in deriving b_L from a pure Lorentzian spectrum. Errors of 20% will also be made in estimating the parameter a for a Voigt spectrum [see Penner et al. (1976a) for details]. The line-shape parameter is $a = b_L/(v/\sigma)$; this relation implies an uncertainty of $\pm 35\%$ in v/σ. The measurement of v should be good to 20%. Thus, σ is determined within 55%. The uncertainty in D is expected to be comparable with that of b_L, i.e., $\pm 15\%$.

The principal sources of error in estimating the detectability ratio $\alpha(0)$ arise from uncertainties in $S_i'(0)$ and the shot-noise level. The shot-noise level is obtained by subtracting (see Fig. 3) the flame-emission level from $S_i(\infty)$. Each of the white background contributions to $S_i(\omega)$ can be determined within $\pm 10\%$. Therefore, the shot-noise level should be defined generally within $\pm 20\%$ while $\alpha(0)$ will be determined within $\pm 30\%$.

The attenuation for the laser beam passing to the flame will be known quite accurately. Thus, a conservative error estimate for measurements of γP_0 is $\pm 10\%$. A large uncertainty may arise for the scattering function f because its value is very sensitive to the estimated value of the mean particle diameter [f varies as (diameter)^6 in the Rayleigh limit].

In the Stokes-Einstein limit, the particle diameter is found from the Stokes-Einstein relation $d_{SE} = kT/3\pi\mu D$. Since the viscosity coefficient μ of flame gases is roughly proportional to $T^{2/3}$, the dependence of d_{SE} on T is then about $T^{1/3}$. Therefore, if we assume a $\pm 10\%$ uncertainty in determining the temperature T, we would have a $\pm$ 3-4% uncertainty in d_{SE}. Including the 15% error in D, the total error in estimating d_{SE} is about 20%. With this error in d_{SE}, the calculated scattering coefficient f may be in error by as much as a factor of 1.2 if the index of refraction of the particles is precisely known.

In the Epstein limit, we obtain d_E from Eq. (10). Since $\ell \propto T^{1.3}$ and $\mu \propto T^{2/3}$, $d_E \propto T^{0.8}$ and a $\pm 10\%$ uncertainty in T corresponds to about a $\pm 8\%$ uncertainty in d_E. Including a 15% error in D, d_E is then not determined with an accuracy greater than about 15%.

In estimating the error in f, we should include allowance for uncertainty in the complex index of refraction $\tilde{m}$. The value of $\tilde{m}$ at 6500 Å for soot formed in an acetylene flame is 1.57-0.44 i (Dalzell and Sarofim, 1969), whereas $\tilde{m}$ at 6328 Å for amorphous carbon is 1.95-0.66 i (Senftleben and Benedikt, 1917). For the particle-size ranges of interest to us, the Mie coefficients f calculated with these two values of $\tilde{m}$ differ by a factor of 1.8.

Including all of the specified error estimates, we conclude that the total uncertainty in estimating the number density N_∞ from Eq. (13) is within a factor of 2; for known $\tilde{m}$, N_∞ should be defined within about 65%.

Experimental Results for Methane Burning with Oxygen above a Flat-Flame Burner, Assuming Monodisperse Particles

Preliminary experimental results obtained for methane burning with oxygen above a flat-flame burner are shown in Figs. 5 and 6. All light-scattering

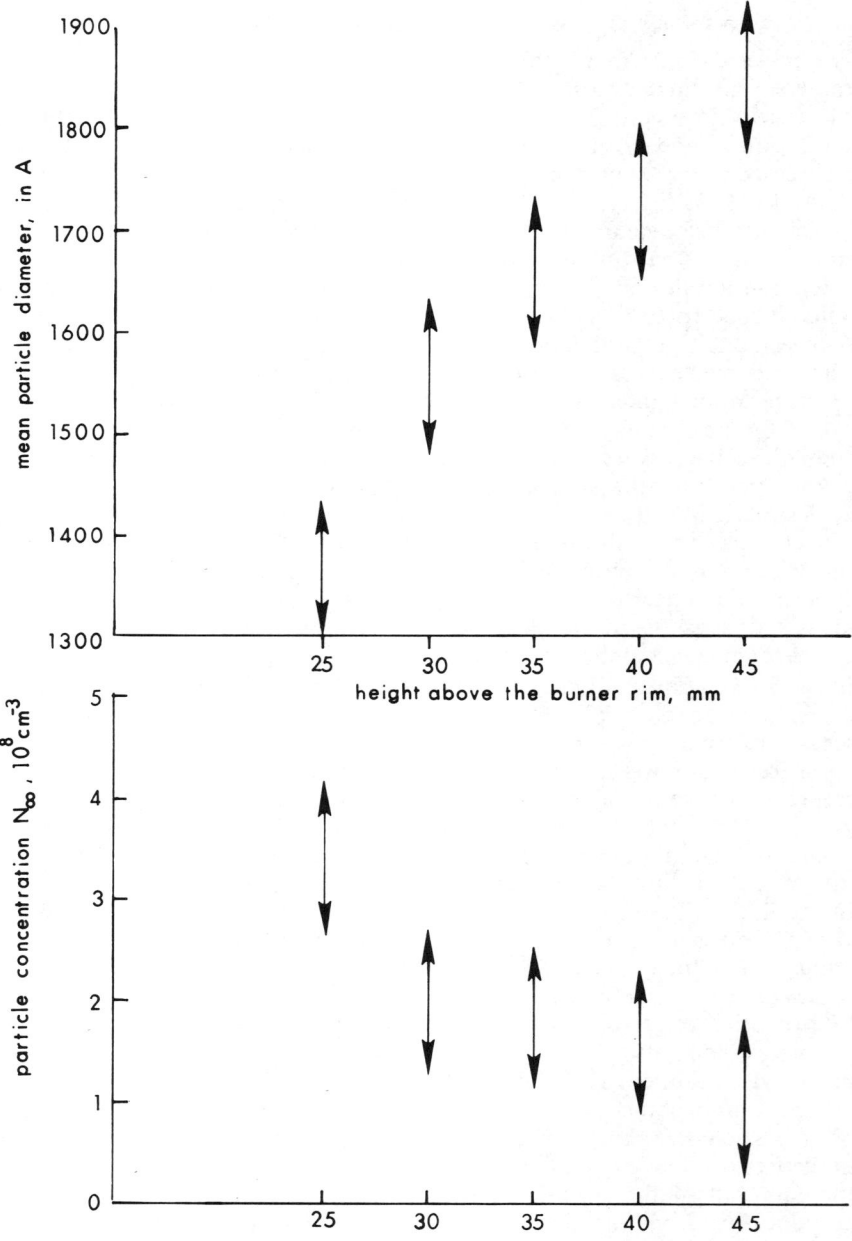

Fig. 5 Mean particle diameters and particle concentrations observed in methane-oxygen mixtures at an equivalence ratio of 3.15 and with an inlet flow speed of 7.7 cm/s, at various heights above the rim of a flat-flame burner. The following numerical values were used in constructing Figs. 5,6,8, and 11: $T = 1500$ K, $\mu = 5.6 \times 10^{-4}$ g-cm/s, $\ell = 3050$ Å, $a_0 = 0.013$ cm, $P_0 = 3.8 \times 10^{-3}$ W, and $\alpha = 1.25$.

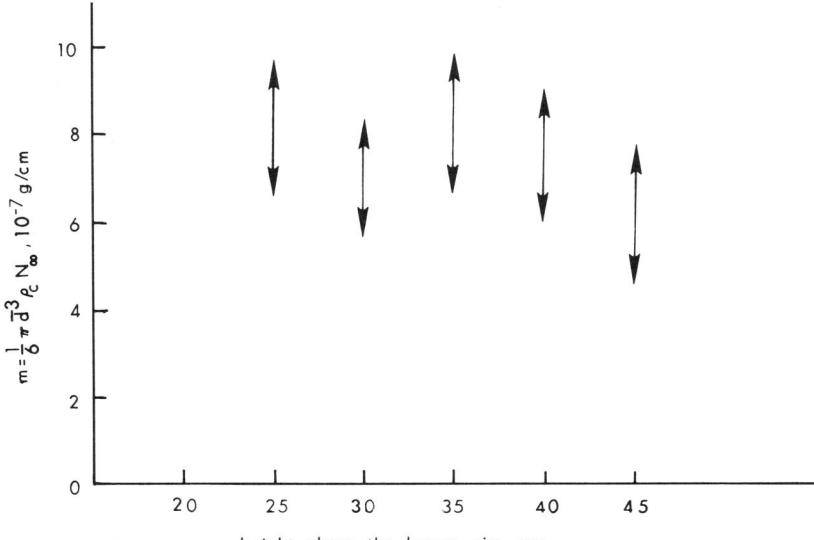

Fig. 6 Total mass m of carbon in particles formed in methane-oxygen mixtures at an equivalence ratio of 3.15 and with an inlet flow speed of 7.7 cm/s, at various heights above the rim of a flat-flame burner.

measurements refer to the axis of the (nearly symmetric) flame formed above the burner. The mean particle diameters are seen to increase at heights of 25-45 mm above the burner rim by about 30% (see the upper part of Fig. 5), while the number density of particles decreases by about a factor of three (see the lower part of Fig. 5) in such a manner that the total particulate mass remains nearly constant (see Fig. 6). This result suggests that the self-preserving distribution (spd) is reasonable to use above the flat-flame burners. A numerical estimate for the concentration is now briefly described. We first rewrite Eq. (13) in the form

$$N_\infty = \{\alpha(0)\left[2(4\pi/\lambda)^2\left(\sin^2\frac{\theta}{2}\right)D\right]a(\sin\theta+\cos\theta)\}/\{(\gamma P_0\lambda^7/32\pi^4 hc)$$
$$\times [2(4\pi/\lambda)^2]f\}$$

where we identify the term $2(4\pi/\lambda)^2[\sin^2(\theta/2)D]$ with $2|k|^2 D = b_L' = 5.91 \times 10^4$ rad/s, i.e., with 2π times the measured Lorentzian half-width of the scattered power spectrum. Also $\lambda = 6328$ Å and

$$(\gamma P_0\lambda^7/32\pi^4 hc) = (\gamma P_0\lambda^6/32\pi^4 h\nu) = (\gamma e P_0/h\nu e)(\lambda^6/32\pi^4)$$

where $(\gamma e/h\nu) = 2.5 \times 10^{-2}$ amp/W for the photomultiplier used by us as detector and $e = 1.6 \times 10^{-19}$ amp-s; also $P_0 = 3.8 \times 10^{-3}$ W for our laser source. The experimentally achieved coherence length a in the scattering region was $a = 0.013$ cm while $\sin\theta + \cos\theta = 1.22$ for forward scattering at $\theta = 15$ deg.

Using the measured value of b_L, we have determined the mean particle diameter $\bar{d}$ and then calculated the Mie function f (which had the value 0.0298 at a height of 40 mm above the burner rim, where the observed value of the detectability ratio was 7.3). In this manner, we find $N_\infty = 2.4 \times 10^8$ cm^{-3}.

Both the number-density and the particle-size estimates, obtained through the use of scattered-laser power spectra and measurements performed with high spatial resolution, are in good accord with experimental results derived from the Mie theory using the known angle dependence of the scattered radiation (D'Alessio et al., 1973).

Number-Density Determinations Using Extinction Measurements in the Rayleigh Limit (Monodisperse Systems)

The fractional transmitted light intensity (I_t/I_0) in the absence of molecular absorption is given by the relation [compare Eq. (4)]

$$I_t/I_0 = \exp(-N_\infty Q_e \ell) \tag{15}$$

where N_∞ is the particle number density per unit volume, the extinction cross section (Q_e) is the sum of the cross sections for absorption (Q_a) and scattering (Q_s), and ℓ is the geometric pathlength through the absorbing medium. In the Rayleigh limit (see van de Hulst, 1957, and Penner and Olfe, 1968),

$$Q_a = (4\pi r^2)(2\pi r/\lambda)\,\mathrm{Im}[(\tilde{m}^2-1)/(\tilde{m}^2+2)] \tag{16}$$

and

$$Q_s = (8/3)(2\pi r/\lambda)^4 (\pi r^2)\,|(\tilde{m}^2-1)/(\tilde{m}^2+2)|^2 \tag{17}$$

It is apparent from Eqs. (16) and (17) that $Q_a \gg Q_s$ for $r/\lambda \ll 1$. Equation (15) may be used to estimate N_∞ according to the relation

$$N_\infty = -(1/Q_e\ell)\ln(I_t/I_0) \tag{18}$$

Simultaneous Measurements of Extinction and Scattering to Determine Log-Normal Size Distributions and Number Densities for Polydisperse Systems; Comparisons with Self-Preserving Distributions

For the log-normal distribution, we have previously used (Penner and Chang, 1978) the relation

$$\frac{1}{2} = S_i'(b_L')/S_i'(0) = \frac{\sum_{q=1}^{30}\sum_{p=1}^{30}\varphi_q\varphi_p f_q f_p \dfrac{|k|^2(D_q+D_p)}{[|k|^2(D_q+D_p)]^2 + (b_L')^2}}{\sum_{q=1}^{30}\sum_{p=1}^{30}\varphi_q\varphi_p f_q f_p / |k|^2(D_q+D_p)} \tag{19}$$

to construct universal plots (for scattering angles of 15, 20, and 30 deg) showing the observed Lorentzian half-widths

$$b_L' = 2\pi b_L = 2|k|^2 D_L, \qquad k = (4\pi/\lambda)\sin(\theta/2)$$

as functions of $\bar{r}^*$ for $1.0 \leq \sigma_g \leq 4.0$. Here we employ again the notation introduced in previous publications for the power spectrum (S_i'), the Mie scattering functions (f_q, f_p), the fractional numbers of particles (φ_q, φ_p), and the wave vector $\mathbf{k}$. In addition to measuring b_L', we also determine the detectability (which is the ratio of the scattered laser power at zero frequency to the time-averaged value of the shot noise)

$$\alpha(0)_{\lambda,\theta} = \frac{\gamma_\lambda \lambda^7 (I_{0,\lambda} N_{\text{to}}) \sum_q \sum_p [\varphi_q \varphi_p f_q f_p / (D_q + D_p)]_{\lambda,\theta}}{16\pi^4 hca [\sin^2(\theta/2)](L\sin\theta + a|\cos\theta|) \sum_q (\varphi_q f_q)_{\lambda,\theta}} \quad (20)$$

where $I_{0,\lambda}$ is the incident spectral laser intensity, N_{to} the total number of particles in the scattering volume, γ_λ the photomultiplier (detector) quantum efficiency at the wavelength λ, a the diameter and L the length of the illuminated cylinder, and the other symbols have their usual meanings. Finally, we measure the total extinction for the geometric path length ℓ, which determines the total number of particles per unit volume according to the relation

$$N_\infty = \left[-1 / \left(\sum_p \varphi_p Q_{e,p}\right) \ell\right] \ell n (I_t / I_0) \quad (21)$$

where $Q_{e,p}$ = extinction cross section for particles of type p.

The problem under discussion deals with three unknowns, namely, $\bar{r}^*$, σ_g, and N_∞. In principle, these three unknowns are determinable from three appropriate measurements using b_L' and $\alpha(0)$ or b_L' and I_t/I_0 (by using two or more wavelengths or two or more scattering angles). Penner and Chang (1978) found that a variety of scattering measurements, using b_L' and $\alpha(0)$ for various wavelengths and scattering angles, did not yield accurate values for the unknowns because the required measurement sensitivity could not be achieved since the same functional group, namely

$$\sum_q \sum_p \varphi_q \varphi_p f_q f_p / |\mathbf{k}|^2 (D_q + D_p)$$

dominated the magnitudes of the observables. However, a different functional group, namely

$$\sum_p \varphi_p Q_{e,p}$$

relates N_∞ to the observed fractional transmission. For this reason, we find an increase in the sensitivity of the observable parameters to $\bar{r}^*$, σ_g, and N_∞ when half-width measurements are combined with extinction measurements.

It is convenient to eliminate N_∞ from Eqs. (20) and (21) by forming the ratio

$$\xi = \frac{\alpha(0)}{-(1/\ell)\ln(I_t/I_0)}$$

$$= \frac{a^2\gamma_\lambda\lambda^7(I_{0,\lambda}) \sum_q \sum_p \varphi_q\varphi_p f_q f_p/(D_q+D_p)}{16\pi^4 hc[\sin^2(\theta/2)](L\sin\theta+a|\cos\theta|)\left(\sum_q \varphi_q f_q\right)\left(\sum_p \varphi_p Q_{e,p}\right)} \quad (22)$$

for fixed values of λ and θ. The ratios ξ (in cm) in the Epstein limit are shown in Figs. 7 and 8 for selected scattering angles for the log-normal and self-preserving distributions, as functions of $\bar{r}^* = (3\bar{v}^*/4\pi)^{1/3}$ (for $1.2 \leq \sigma_g \leq 2.0$) or $\bar{r}$, respectively. The half-widths b_L (in Hz) in the Epstein limit are shown as functions of $\bar{r}^*$ (in Å) for $1.0 \leq \sigma_g \leq 2.0$ in Fig. 9 for the log-normal

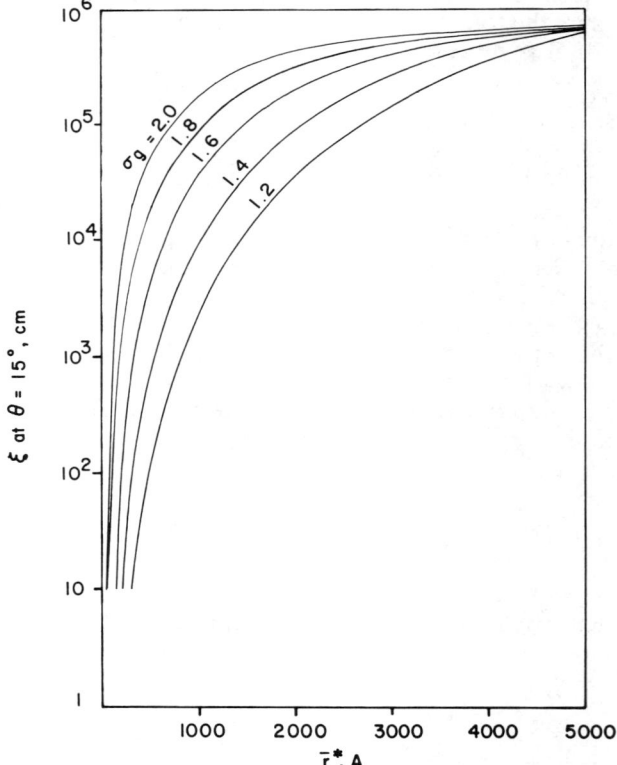

Fig. 7 Ratio ξ as a function of $\bar{r}^*$ in the Epstein limit for the log-normal distribution with $1.2 \leq \sigma_g \leq 2.0$, at a scattering angle of 15 deg. The following numerical values were used in Figs. 7, 9 and 10: $T = 1500$ K, $\mu = 5.6 \times 10^{-4}$ g-cm/s, $\ell = 3050$ Å, $a_0 = 0.013$ cm, $\alpha = 1.25$, $\tilde{m} = 1.57 - 0.44\,i$ at 6328 Å.

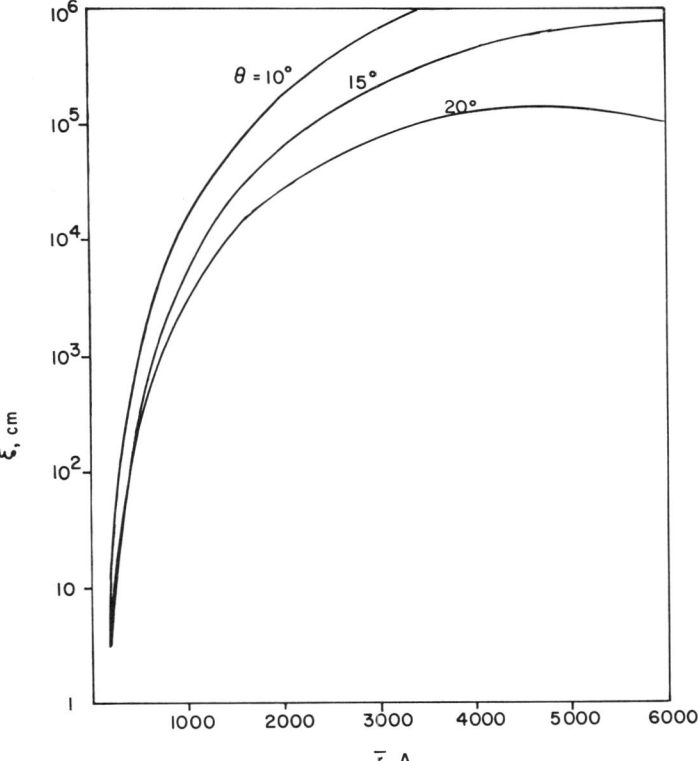

Fig. 8 Ratio ξ (in cm) as a function of $\bar{r}$ in the Epstein limit for the self-preserving distribution at $\theta = 10, 15,$ and 20 deg.

distribution for $\theta = 15$ deg; in Fig. 10, they are shown as functions of $\bar{r}$ for $\theta = 10, 15,$ and 20 deg for the self-preserving distribution (spd).

Experimentally observed data at $\theta = 15$ deg for CH_4/O_2 mixtures at $\varphi = 3.1$ have been used to construct Fig. 11 for $25 \leq h \leq 45$ mm on the assumption that a log-normal distribution obtains. The intersections of the curves derived from measurements of b_L (solid curves) with those derived from measurements of ξ (dotted curves) show the applicable values of σ_g and $\bar{r}^*$ at the indicated heights above the burner rim (see Fig. 11) for the log-normal distribution. These results are summarized in Table 2. Also shown in Table 2 are the values of $\bar{r}$ for the self-preserving distribution, obtained from measured values of b_L (Hz) and ξ (cm) by using Figs. 8 and 10, respectively.

It is apparent that the values of $\bar{r}$ determined from b_L and from ξ for the spd represent, respectively, applicable values for point measurements and for average values over the line of sight, which will be different for nonhomogeneous systems. On the other hand, the estimates of σ_g, $\bar{r}^*$ for the log-normal distribution correspond to a special type of weighting over local and mean values. Consistent predictions of scattered intensities from the

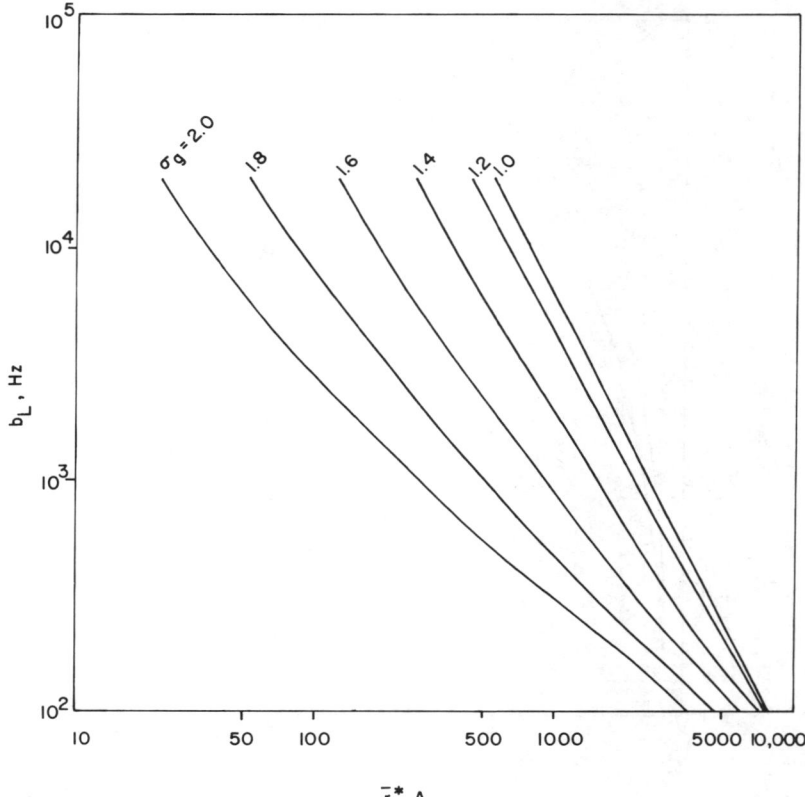

Fig. 9 Half-width b_L (Hz) in the Epstein limit as a function of $\bar{r}^*$ (in Å) with $1.0 \leq \sigma_g \leq 2.0$ for the log-normal distribution at $\theta = 15$ deg.

values of σ_g and $\bar{r}^*$ and the two values of $\bar{r}$ can be expected only if observations are restricted to completely homogeneous flames.

Reference to the data listed in Table 2 shows that $\bar{r}^*$ increases from about 53 to 92 Å while σ_g remains in the range 1.85 ± 0.05 as h increases from 25 to 45 mm. Thus, the experimental measurements at a scattering angle of 15 deg suggest that the log-normal distribution remains similar as the gases flow downstream. For the spd, we find (see Table 2) that the measured values of b_L and ξ yield generally similar estimates for $\bar{r}$ when $h \gtrsim 30$ mm. The estimates for the mean radius $\bar{r}$ of the spd are seen to be roughly three times as large as the values of the radius $\bar{r}^*$ for the maximum in the distribution curve of the log-normal distribution.

We have thus found two different distributions (namely, the spd and the log-normal distribution) that are consistent with the observed data at $\theta = 15$ deg.

Using (compare Table 2) $\sigma_g = 1.85$, $\bar{r}^* = 53$ Å (corresponding to the observed data for the log-normal distribution at $h = 25$ mm) and $\bar{r} = 425$ Å

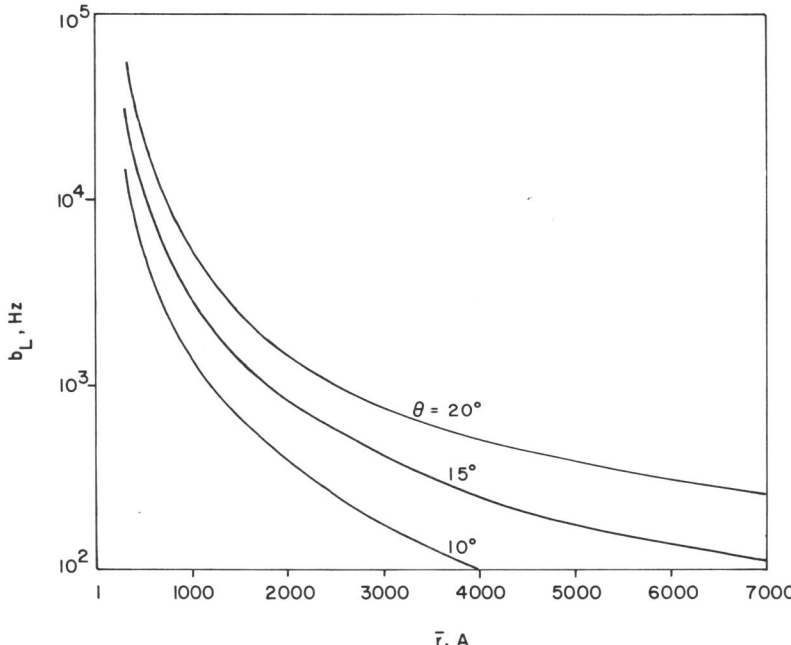

Fig. 10 Half-width b_L (in Hz) for the self-preserving distribution as a function of $\bar{r}$ (in Å) in the Epstein limit for $\theta = 10$, 15, and 20 deg.

(corresponding to the spd at $h = 25$ mm as defined by measurements of b_L), we may construct the predicted curves of $I(\theta)\sin\theta/I(\theta = 90 \text{ deg})$ as functions of θ (see Fig. 12). We find that these predicted scattered intensity ratios are experimentally indistinguishable. Similar results are obtained at $h = 45$ mm for $\sigma_g = 1.84$, $\bar{r}^* = 92$ Å, and $\bar{r} = 610$ Å for the two distributions, respectively. We conclude that, for the specified particle-size ranges, it is not possible to distinguish between a log-normal distribution and the spd by using light-scattering measurements as a function of angle.

We have estimated the values of N_∞ from Eqs. (21) and (20) for the spd and the log-normal distribution, using the values of $\bar{r}$ determined from b_L. These are compared in Table 3 with estimates of N_∞ derived on the assumption that the particulate systems are monodisperse (see Fig. 5).

Reference to Table 3 shows that diffusion-scattering spectroscopy leads to values for N_∞ that are much larger in polydisperse systems than in an assumed monodisperse system. This result follows because, in a polydisperse system, there are very many small particles which contribute little to the total scattered intensity. These small particles also account for only a small fraction of the total volume of the solid-phase carbon. Values for the total carbon volume, determined by using the half-width and the detectability (see columns 2 and 4 on the bottom of Table 3) are comparable for monodisperse systems and for the spd, in spite of somewhat larger differences in the total number densities

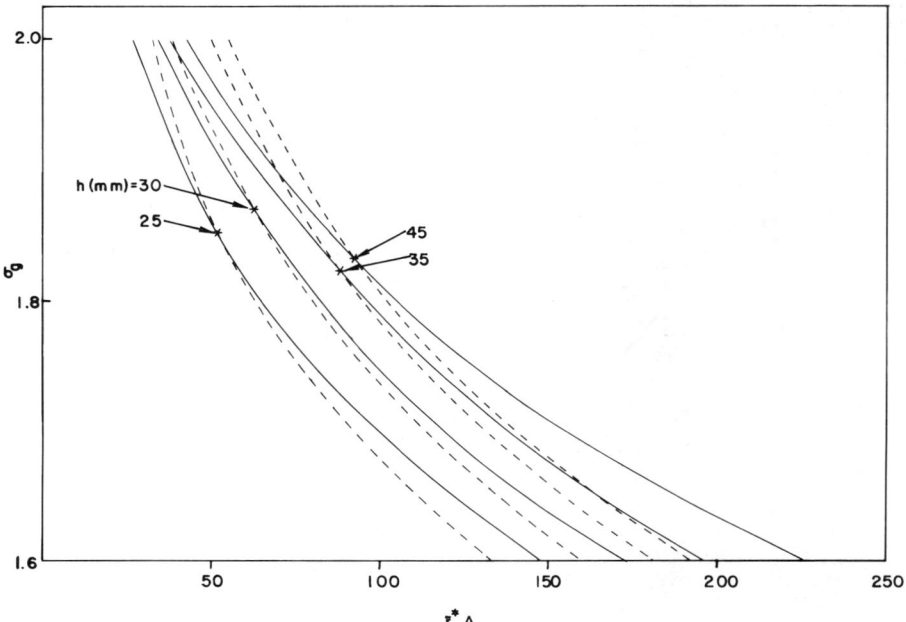

Fig. 11 Experimentally observed values at $\theta = 15$ deg of half-widths (corresponding to solid lines) and ξ (corresponding to dotted lines), for CH_4/O_2 mixtures at $\varphi = 3.15$, have been used, in conjunction with Figs. 7 and 9, to show allowed values of α_b and $\bar{r}^*$; $25 \leq h$, mm ≤ 45; these curves apply to the log-normal distribution).

(see columns 2 and 4 in the upper part of Table 3). On the other hand, the combined use of extinction and half-width measurements leads to substantially larger values of N_∞ and of ν for the spd (see the third columns in Table 3); this difference may reflect nonuniformities in the flame structure.

The log-normal distribution requires very much larger number densities in order to account for the measured half-width and ratio of detectability to extinction (see the fifth column in the upper part of Table 3) and also

Table 2 Values of $\bar{r}^*$, σ_g as functions of height (h) for the log-normal distribution and of $\bar{r}$ for the self-preserving distribution obtained from scattering measurements at 15 deg for a CH_4/O_2 mixture at $\varphi = 3.1$ using measured values of b_L and ξ in the Epstein limit

h, mm	Log-normal distribution		Self-preserving distribution	
	$\bar{r}^*$, Å	σ_g	$\bar{r}$, Å (from b_L)	$\bar{r}$, Å (from ξ)
25	53 ± 6	1.85 ± 0.05	425 ± 50	330 ± 30
30	67 ± 6	1.87 ± 0.05	500 ± 60	370 ± 40
35	88 ± 7	1.82 ± 0.05	535 ± 60	400 ± 40
40	90 ± 8	1.84 ± 0.05	560 ± 70	410 ± 40
45	92 ± 8	1.84 ± 0.05	610 ± 70	420 ± 40

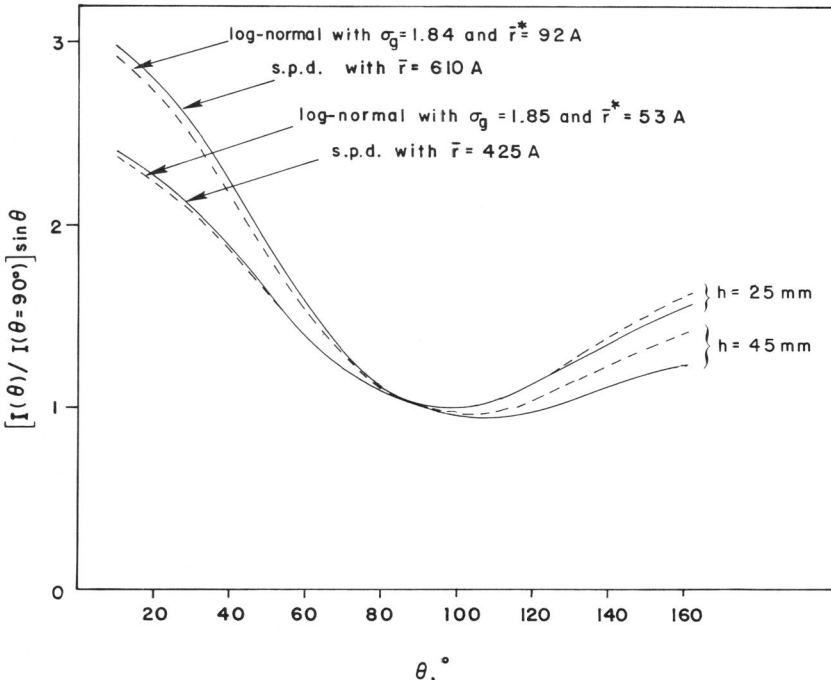

Fig. 12 Calculated curves of $[I(\theta)/I(\theta = 90 \text{ deg})] \sin \theta$ vs θ at $h = 25$ and 45 mm for the log-normal distribution (dotted curves) and the spd, using the data of Table 2, with $\bar{r}$ determined from b_L (solid curves).

Table 3 Estimates of number density N_∞ and total particle volume fraction v for the monodisperse system, for the spd, and for the log-normal distribution; the observed variabilities in the results are also indicated

h, mm	Monodisperse system	spd using Eq. (21) and $\bar{r}$ obtained from b_L	spd using Eq. (20) and $\bar{r}$ obtained from b_L	Log-normal distribution using Eqs. (21) or (20)
		$N_\infty \times 10^{-8}$, cm^{-3}		
25	3.3 ± 0.7	32 ± 6	9.5 ± 0.9	3900 ± 500
30	1.8 ± 0.4	20 ± 4	4.7 ± 0.6	1900 ± 300
35	1.8 ± 0.4	18 ± 4	4.8 ± 0.7	1500 ± 300
40	1.6 ± 0.3	16 ± 3	3.8 ± 0.5	1100 ± 200
45	0.9 ± 0.1	12 ± 2	2.3 ± 0.3	1000 ± 200
		$v \times 10^7$, cm^3 of particles per cm^3 of volume		
25	4.2 ± 0.8			13 ± 3
30	3.6 ± 0.6			14 ± 3
35	4.2 ± 0.8	11 ± 2	2.7 ± 0.4	22 ± 3
40	4.1 ± 0.9			18 ± 4
45	3.1 ± 0.7			18 ± 4

somewhat larger volume fractions of carbon particles (see the fifth column in the lower part of Table 3). The preceding statement holds because, especially for a log-normal distribution with σ_g as large as 1.85, the influence of the small particles on the observables is negligibly small.

The particulate volume fractions for the monodisperse system and the spd derived from measurements of b_L are seen to agree within about a factor of 2, as are the particulate volume fractions for the monodisperse system and the log-normal distribution; corresponding values derived from extinction measurements are again seen to be relatively large.

Conclusion on In Situ Particle Sizing

The evidence presented in this paper and the results obtained by other investigators suggest that particulate sizes rapidly approach the spd above regions of active combustion in flames. On the assumption that this is the case, the local values of the particle distributions are conveniently characterized by measurements of b_L (e.g., at $\theta = 15$ deg) and determination of $\bar{r}$ from Fig. 9. The corresponding number densities may then be obtained by using either Eq. (20) or (21).

Proper experimental measurements of the particle-size distributions that are initially formed in combustion systems will require applications of laser-scattering techniques and spectroscopic measurements to low-pressure burners. Soot formation may occur through the condensation of aromatic rings or through agglomeration of C_yH_x fragments (Graham et al., 1975; Wagner, 1978; Blazowski, 1979). Calcote (1979) has suggested that charged ions may play an important role in the initial nucleation processes. The rates of particulate growths depend on all of the process variables, including fuel type, equivalence ratio φ, combustor design, etc. In view of the complexity and variability of the initial nucleation processes, it is apparent that very detailed measurements are required to gain significant insight into the starting mechanisms for soot formation.

Summary Statements Concerning Mechanisms of Particulate Formations and Growths in Flames

We summarize in Table 4 observational data on particle formations and growths in flames and our interpretation of their physical importance or origin. The experimental bases for the data given in Table 4 involve results obtained by us and not described in this paper, as well as experimental findings of other investigators.

Appendix: Historical Evolution of Particle-Size Measurement Techniques

Observations of small objects date from the discovery of the first magnifying glass. A landmark was the development of the microscope. Andreas Vesalius published his superb anatomical studies in 1543, under the title *De humani corporis fabrica,* in which he showed the fine structure of the tissue. With this instrument, observations were made of small structures of tissues, fibers, small particles, cells, etc.

Table 4 Qualitative interpretation of experimental data on particles in flames using monodisperse-particle results

Observational Fact	Interpretation
1) $\bar{d}$ increases rapidly with height h at first and subsequently only very slowly.	1) Nuclei for particle growth form very rapidly at first; these particles grow by agglomeration with dehydrogenation and carbonization. The particle size is ultimately limited by particle burn-up. But particle burn-up removes the smallest particles first, thus accounting for seeming growth in $\bar{d}$ after the early nucleation and growth processes have been terminated. Furthermore, individual particles may actually grow in size if internal oxidation occurs.
2) The values of m_t (total particle mass per unit volume) increase rapidly at first and then level off at pseudo-steady values.	2) The initial nucleation is accompanied by rapid particle growth while subsequent particle agglomeration occurs at nearly constant total soot concentration.
3) $\bar{d}$ increases with φ for $\varphi \geq 1$ while m_t may actually decrease under these same conditions.	3) The nucleation rate increases rapidly with T and particles grow on the nuclei. As φ increases, T decreases and hence fewer particles are formed. But these particles tend to be of larger size $\bar{d}$ because the particle-size limiting processes (e.g., burn-up, which is reduced by both smaller T and less O_2) are less effective.
4) $\bar{d}_{CH_4} < \bar{d}_{C_2H_6} < \bar{d}_{C_2H_4}$ for fixed values of φ and T.	4) The soot particle-size ordering is determined by chemical processes involved in the combustion reactions.
5) Particles grow very rapidly to about 10^3 Å in size in most of the fuel-air mixtures studied.	5) Processes which limit the ultimate growths of particle sizes (e.g., particle burn-up) become important at late times.
6) The initial nucleation processes and particle-growth rates occur very rapidly (i.e., very close to the burner rim).	6) The incipient nucleation rates are closely coupled to chemical rate processes and occur at commensurate rates.
7) Self-preserving particle-size distributions are observed in mature flames.	7) Particle growths by agglomeration (collisions) between particles are important for well-defined minimum particle number densities.
8) Particles are nearly spherical.	8) Spherical particles are formed from a) molten spheres or b) very rapidly growing nuclei.

The light-scattering properties of particles were first considered in connection with studies of the rainbow. Aristotle developed a theory according to which the rainbow is observed when sunlight is reflected from a dark cloud. Descartes observed that rainbows were produced in fountain sprays and that sprays are composed of small droplets of water. These studies led to the conclusion that a rainbow is seen only when a bright red point appears in each of the droplets. Furthermore, lines drawn from the droplet to the observer make an angle of about 42 deg with respect to the incident light (Descartes, 1637).

The application of light scattering to estimate the size of extremely small particles started with Rayleigh's work to explain the blue color of the sky. Newton (1706) had previously suggested that the blue of the sky was produced in the same manner as colors of thin films and resulted from the interference of rays reflected from the front and rear surfaces of suspended water droplets. However, such specular reflection should have resulted in complete polarization at the Brewster angle for water (74 deg) rather than at the observed angle of reflection (45 deg). Additional evidence against Newton's theory was provided by Clausius (1848) who demonstrated that a cloud of droplets large enough to exhibit interference effects should cause a star to appear enormously magnified. Tyndall (1869) showed in his famous experiment on light scattering from vapors that "whenever the precipitated particles are sufficiently fine, no matter what the substance forming the particles may be, the direction of maximum polarization is at right angles to the illuminating beam." Tyndall disproved an earlier suggestion of Brewster that perfect polarization would occur at right angles with respect to the incident beam if there is reflection in air of light passing through air. Rayleigh (1871a) gave a full theoretical description of scattering by small particles in the sky. A rigorous solution to the problem of light scattering by small particles was deduced by Rayleigh in 1871 (1871b). Following an earlier suggestion by Maxwell (1873), Rayleigh succeeded (1881, 1899) in estimating the scattering of sunlight by molecules in the atmosphere and concluded that this scattering alone, even in the absence of water or ice or salt particles, was sufficient to cause the blue sky.

The modern theory of light scattering, as it is applied for measurements of particle sizes and number densities, is closely linked with the work of Clebsch (1863), Debye (1909), Lorenz (1898), and Mie (1908). The evolution of these ideas is discussed in detail by M. Kerker (1969).

Acknowledgments

This research was supported by the National Science Foundation under Grant ENG-7411494. This paper has been prepared as the invited inaugural lecture for presentation at the Seventh International Colloquium on Gasdynamics of Explosions and Reactive Systems.

References

Bernard J.M. and Penner S.S. (1977) Determination of particle sizes in flames from scattered laser power spectra. *Experimental Diagnostics in Gas Phase Combustion Systems, Progress in Astronautics and Aeronautics,* Vol. 53, pp. 411-420. AIAA, New York.

Blazowski W.S. (1979) Dependence of soot production on fuel blend characteristics and combustion conditions. ASME Paper 79-GT-155.

Calcote H.F. (1979) Soot formation in flames, a review. Paper 79-0296, presented at 17th AIAA Aerospace Sciences Meeting, Jan. 1979.

Carlon H.R., Milham M.E., and Frickel R.H. (1976) Determination of aerosol droplet size and concentration from simple transmittance measurements. *Appl. Optics* **15**, 2454-2456.

Clausius R. (1848) Über die Lichtzerstreuung in der Atmosphäre und die Intensität des durch die Atmosphäre reflektierten Sonnenlichts. *Crelle's J.* **36**, 185-215.

D' Alessio A., Di Lorenzo A., Sarofim A.F., Baretta F., Masi S., and Venitozzi C. (1973) Optical and chemical investigation on fuel-rich methane-oxygen premixed flames at atmospheric pressure. *14th Symposium (Intl.) on Combustion*, pp. 941-953. The Combustion Institute, Pittsburgh, Pa.

Dalzell W.H. and Sarofim A.F. (1969) Optical constants of soot and their application to heat-flux calculations. *J. Heat Trans.* **91**, 100-104.

Dalzell W.H., Williams G.C., and Hottel H.C. (1970) A light scattering method for soot concentration measurements. *Combustion and Flame* **14** 161-170.

Descartes R. (1637) Discours de la méthode pour bien conduire sa raison et chercher la verité dans les sciences, plus la dioptique, les météores et la géométrie. *Météors*, Leyden, the Netherlands.

dos Santos R. and Stevenson W.H. (1977) Aerosol sizing by means of laser-induced fluorescence. *Appl. Phys. Letters* **30**, 236-237.

Farmer W.M. (1974) Observations of large particles with a laser interferometer. *Appl. Optics* **13**, 610-622.

Faxvog F.R. (1976) Instrument for sizing submicron airborne particles. Paper THHH4, p. 98. Conference on Laser and Electro-Optical Systems, San Diego.

Friedlander S.K. (1961) Theoretical considerations for the particle size spectrum of the stratospheric aerosol. *J. Atmo. Sci.* **18**, 753-759.

Friedlander S.K. and Wang C.S. (1966) The self-preserving particle size distribution for coagulation by Brownian motion. *J. Colloid and Interface Sci.* **22**, 126-132.

Fuks N.A. (1964) *The Mechanics of Aerosols*. (Trans.) E. Lachowicz. Pergamon Press, New York.

Fuks N.A. and Sutugin A. (1965) Coagulation rate of highly dispersed aerosols. *J.Colloid Sci.* **20**, 492-500.

Graham S.C., Homer J.B., and Rosenfeld J.L.J. (1975) The formation and coagulation of soot aerosols generated by the pyrolysis of aromatic hydrocarbons. *Proc. Roy. Soc.* (London) **A344**, 259-285.

Heller W. and Wallach M.L. (1963) Experimental investigation on the light scattering of colloid spheres, V: Determination of size distribution curves by means of spectra of the scattered ratio. *J. Phys. Chem.* **67**, 2577-2583.

Hidy G.M. and Brock J.R. (1970) *The Dynamics of Aerocolloidal Systems*. Pergamon Press, New York.

Ingebo R.D. (1956) Atomization, acceleration, and vaporization of liquid fuels. *6th Symposium (Intl.) on Combustion*, pp. 684-687. The Combustion Institute, Pittsburgh, Pa.

Kennard E.H. (1938) *Kinetic Theory of Gases*, Chaps. VII and VIII. McGraw-Hill Book Co., New York.

Kerker M. (1969) *The Scattering of Light and Other Electromagnetic Radiation*. Academic Press, New York and London.

Lai F.S., Friedlander S.K., Pich J., and Hidy G.M. (1972) The self-preserving particle size distribution for Brownian coagulation in the free-molecule regime. *J. Colloid and Interface Sci.* **39**, 395-405.

Matijevic E., Kitani S., and Kerker M. (1964) Aerosol studies by light scattering, II. *J. Aerosol Sci.* **19**, 223-237.

Newton I. (1706) *Opticks* (reprint). Dover, New York, 1952.

Penner S.S., Bernard J.M., and Jerskey T. (1976a) Power spectra observed in laser scattering from moving, polydisperse particle systems in flames—I: Theory. *Acta Astronautica* **3**, 69-91.

Penner S.S., Bernard J.M., and Jerskey T. (1976b) Laser scattering from moving, polydisperse particles in flames—II: Preliminary experiments. *Acta Astronautica* **3**, 93-105.

Penner S.S., Bernard J.M., and Chang P. (1977) Determination of particle sizes (or temperatures) using the power spectra of scattered laser radiation from pulsed lasers. *Physichochemical Hydrodynamics (V.G.Levich Festschrift)*, pp. 229-246. Advance Publications, Ltd., London.

Penner S.S. and Chang P. (1978) On the determination of log-normal particle-size distributions using half widths and detectabilities of scattered laser power spectra. *J. Quant. Spectros and Rad. Trans.* **20**, 447-460.

Penner S.S. and Olfe D.B. (1968) *Radiation and Reentry*, Chap. 4. Academic Press, New York.

Pich J., Friedlander S.K., and Lai F.S. (1970) The self-preserving particle size distribution for coagulation by Brownian motion—III. *Aerosol Sci.* **1**, 115-126.

Prado G.P., Lee M.L., Hites R.A., Hault D.P., and Howard J.B. (1976) Soot and hydrocarbon formation in a turbulent diffusion flame. *16th Symposium (Intl.) on Combustion,* pp. 649-661. The Combustion Institute, Pittsburgh, Pa.

Rayleigh Lord (1871a) On the light from the sky, its polarization and colour. *Phil. Mag.* **41**, (Ser. 4) 107-120 and 274-279.

Rayleigh Lord (1871b) On the scattering of light by small particles. *Phil. Mag.* (Ser. 4), **41**, 447-454.

Rayleigh Lord (1881) On the electromagnetic theory of light. *Phil. Mag.* **12**, (Ser. 5), 81-101.

Rayleigh Lord (1899) On the transmission of light through an atmosphere containing small particles in suspension, and on the origin of the blue of the sky. *Phil. Mag.* **47** (Ser. 5), 375-384.

Senftleben H. and Benedikt E. (1917) Über die optischen Konstanten und die Strahlungsgesetze der Kohle. *Annalender der Physik* **54**, 65-78.

Smoluchowski M. V. (1916) Three discourses on diffusion, Brownian movements and the coagulation of colloid particles. *Physikalische Zeitschrift* **17**, 557-571 and 585-599.

Swithenbank J., Beer J.M., Taylor D.S., Abbot D., and McCreath G.C. (1976) A laser diagnostic technique for the measurement of droplet and particle size distribution. Paper 76-69 presented at 14th AIAA Aerospace Science Meeting, Washington, D.C.

Thomspon B.J., Ward J.H., and Zinky W.R. (1967) Application of hologram techniques for particle size analysis. *Appl. Optics* **6**, 519-526.

Tunitskii N. (1938) Coagulation of polydisperse systems. *J. Exp. and Theo. Phys.* (U.S.S.R.) **8**, 417-424.

Tyndall J. (1869) On the blue colour of the sky, the polarization of skylight, and on the polarization of light by cloudy matter generally. *Phil. Mag.* **36** (Ser. 4), 384-394.

van de Hulst H.C. (1957) *Light Scattering by Small Particles*. John Wiley and Sons, New York.

Vesalius A. (1543) *De humani corporis fabrica*. Basel, Switzerland.

Wagner H. Gg. (1978) *Soot Formation in Combustion.* Plenary lecture at the 17th (Intl.) Combustion Symposium, Max-Planck-Institut für Ströhmungsforschung, Göttingen, Federal Republic of Germany.

I. Wall and Confinement Effects

Turbulent Flame Propagation and Acceleration in the Presence of Obstacles

I. O. Moen,[*] M. Donato,[†] R. Knystautas,[‡] and J. H. Lee[§]
McGill University, Montreal, Canada
and
H. Gg. Wagner[¶]
University of Göttingen, Göttingen, Federal Republic of Germany

> This paper reports on an investigation of the influence of obstacles on the propagation of freely expanding stoichiometric methane-air flames. The investigation confirms the dramatic influence of obstacles on the flame speed reported previously and shows that with repeated obstacles flame speeds in excess of 400 m/s can be attained over a distance of only 120 cm. The shock waves generated by these fast flames have overpressures up to 0.64 bar. A simple model of the flame acceleration process based on a feedback coupling between the propagating flame and the flow ahead of the flame is proposed. The model predicts an exponential increase in flame speed with distance propagated which is in good agreement with the experimental results.

Introduction

VIOLENT flame acceleration and enhanced transition to detonation when turbulence producing obstacles are placed in the path of a flame have been observed by various authors (Shchelkin, 1940; Chapman and Wheeler, 1926; and Dörge et al., 1976). However, at the present time, the questions of maximum flame speed and flame acceleration in unconfined or partially confined situations with turbulence-producing obstacles and/or initial turbulence are still far from being resolved. From the practical point of view it is important to determine the flame acceleration and maximum flame speed for a wide range of turbulence conditions since the turbulence in a cloud of combustible mixture, which may result from an accidental spill of explosive gases and liquids, can vary widely depending on the nature and location of the spill.

Presented at the 7th ICOGER, Göttingen, Federal Republic of Germany, Aug. 20-24, 1979.
Copyright © American Institute of Aeronautics and Astronautics, Inc., 1981. All rights reserved.
[*]Research Associate, Mechanical Engineering.
[†]Graduate Student, Mechanical Engineering.
[‡]Associate Professor, Mechanical Engineering.
[§]Professor, Mechanical Engineering.
[¶]Professor, Institut für Physikalische Chemie.

Even if the turbulence level is low prior to ignition, the presence of obstacles in the form of buildings, vehicles, pipes, etc., can lead to large flame speeds and potentially dangerous blast waves.

At the present time, flame laws relating the flame speed of a freely propagating flame to the turbulence field or obstacle configuration are not available. Such flame laws are required as input to existing numerical gasdynamic calculations to evaluate the resulting pressure waves, so that the blast effects due to a fuel-air explosion in a turbulent environment can be predicted (Kurylo et al., 1979; Strehlow et al., 1979; and Guirao et al., 1976). We have undertaken a program the aim of which is to develop such flame laws.

In a previous paper we reported on the dramatic acceleration of cylindrically expanding methane-air flames when repeated obstacles were placed in the flame path (Moen et al., 1980). With repeated obstacles of the appropriate sizes and separations, flame speeds up to 130 m/s were obtained. This was approximately 24 times the speed observed with no obstacles. The mechanism proposed for the rapid acceleration of the flame from an initial flame speed of about 3.3 m/s to speeds in excess of 100 m/s over a distance of 30.5 cm was based on the positive feedback coupling between the flame and the upstream flow produced by the specific volume increase across the flame. With repeated obstacles in the flame path, gradients in the upstream flowfield produce "flame folding" which increases flame area and hence the rate of burning, leading to a larger flow velocity and stronger flowfield gradients, which then further increases the rate of burning, etc. This positive feedback mechanism can thus lead to a flame which, without some damping or limiting factors, will continue to accelerate as long as the obstacle configuration remains the same.

Most of the results reported by Moen et al. (1980) were based on experiments in a chamber which provided only 30.5 cm of flame travel. A few experiments were performed in a larger chamber providing 61 cm of flame travel. These latter experiments confirmed the dramatic influence of obstacles on the flame propagation, but the results were inconclusive on the questions of continued flame acceleration and maximum flame speed. To investigate these questions further we constructed a larger chamber so that cylindrical flame propagation could be observed over a distance of 120 cm. This paper reports on the results of this investigation. Based on these results and spark Schlieren photographs of flame propagation over obstacles in a rectangular channel, a simple model of the flame acceleration process is proposed. The model predicts an exponential increase in flame speed with distance propagated over obstacles which is in good agreement with the experimental results.

Experimental Details

The investigation of flame propagation over obstacles was performed in a 2.5×2.5 m chamber consisting of two wooden parallel plates separated by spacers at the corners. The plate separation could be adjusted by changing the spacers. The chamber was sealed at the periphery by plastic bands while it was filled with the appropriate gas mixture. A spring mechanism was used to remove these plastic bands prior to ignition so that the flame could expand freely in the radial direction. The gas mixture of stoichiometric methane-air

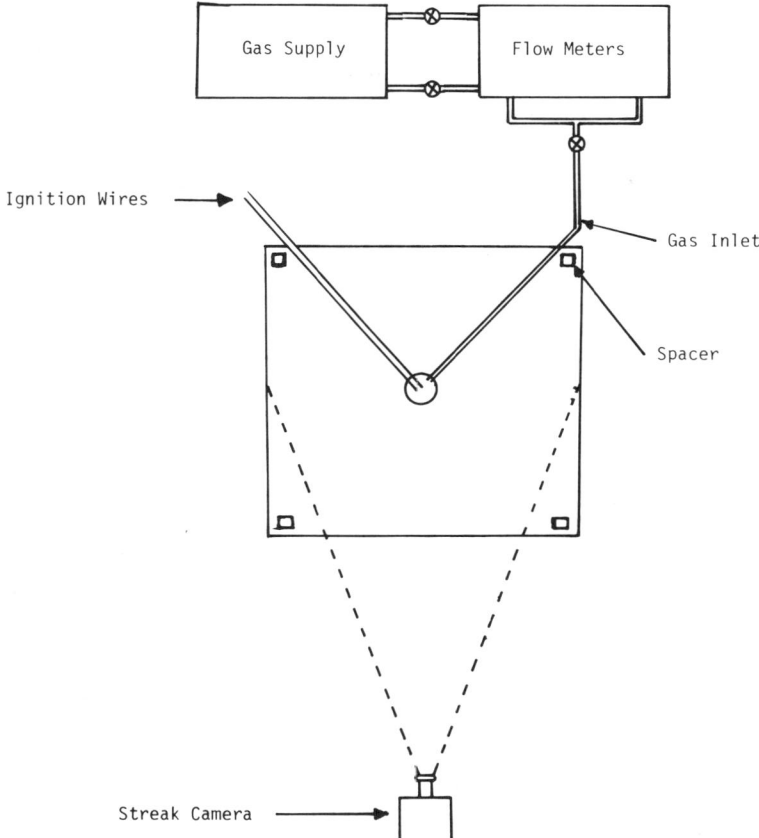

Fig. 1 Schematic diagram of experimental setup for flame propagation experiments.

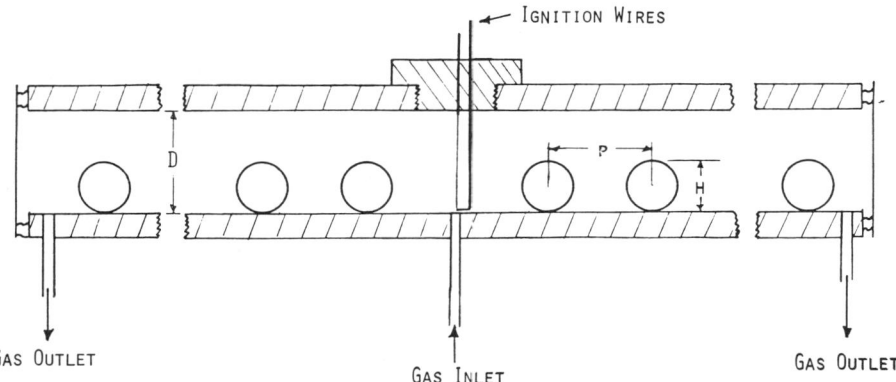

Fig. 2 Schematic cross-sectional view of experimental chamber showing the obstacle configuration.

was prepared by continuous flow, and the chamber was purged at least 10 times in order to insure a good mixture in the chamber. Ignition was achieved by a fast-burning wire at the center of the chamber, and the subsequent radial propagation of the flame was recorded by a rotating-drum streak camera. A schematic diagram of the experimental setup is shown in Fig. 1.

The obstacle configurations were identical to those used in the laboratory experiments reported by Moen et al. (1980), with obstacles consisting of spirals of copper or plastic tubing. The spiral obstacles were placed on the top or bottom plate so that a radially expanding flame would encounter a series of obstacles, corresponding to the windings of the spirals, which extended all the way to the periphery of the chamber. A cross-sectional slice through the center of the chamber illustrating the obstacle configurations is shown in Fig. 2. The obstacle height H, pitch p, and plate separation D are also defined in this figure. For these particular experiments two types of obstacle spirals were used: a spiral of 1.25 cm diameter copper tubing with a pitch of 3.8 cm and a spiral of 4 cm plastic tubing with a pitch of 10 cm. The plate separation was varied between 3.5 and 20 cm.

To investigate in more detail the mechanism of flame propagation over obstacles Schlieren photographs were also taken of a methane-air flame propagating over obstacles in a rectangular channel of length 30 cm. This channel was constructed by placing wooden bars in the 30.5 cm radius chamber described by Moen et al. (1980), and the Schlieren photographs were taken through windows in the two plates of the chamber. A schematic diagram of the channel with obstacles is shown in Fig. 3, where the relevant parameters, pitch p, obstacle diameter H, and channel width W are also defined. The height of the channel was 2.54 cm. For some of the experiments in the channel one or more of the obstacles were filled with dry ice. Three small holes in these obstacles allowed CO_2 to escape. The escaping CO_2 provides an excellent Schlieren tracer for the flowfield ahead of the flame.

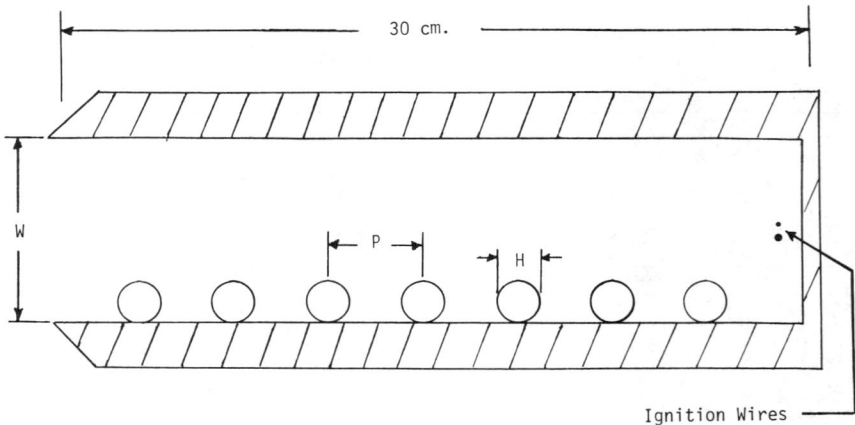

Fig. 3 Schematic diagram of the flame propagation channel with obstacles.

General Considerations

As discussed by Moen et al. (1980), the dramatic influence of repeated obstacles on the flame propagation can be understood in terms of the positive feedback coupling between the flame itself and the flow ahead of the flame. When obstacles are placed in the flame path, the upstream flowfield will be that characteristic of the obstacle configuration. For repeated obstacles placed on the bottom and/or top plates of the chamber, the main features of the flow ahead of the flame are a standing eddy in the wake region behind the obstacles

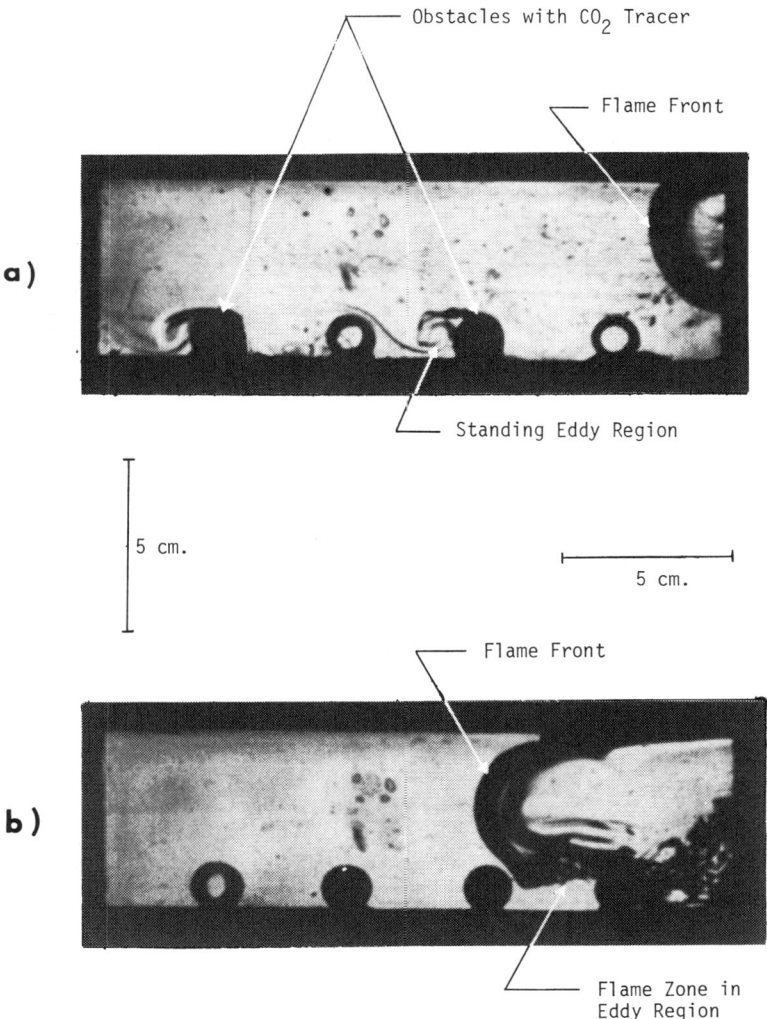

Fig. 4 Spark Schlieren photographs of stoichiometric methane-air flame propagation in a channel with obstacles (obstacle height $H = 1.3$ cm, obstacle separation $p = 3.8$ cm): a) channel width $W = 5.3$ cm (dry ice in second and fourth obstacles); b) channel width $W = 5.1$ cm (no dry ice used).

which is separated from the outer flow region by a shear layer. The flow pattern ahead of the flame and the entrainment into the standing eddy for low flame velocities is illustrated by the spark Schlieren photograph shown in Fig. 4a. In this case the second and the fourth obstacles have been filled with dry ice and the resulting CO_2 tracer clearly shows a standing eddy region between the second and third obstacle. As the flame encounters this flowfield it becomes "folded" due to the gradient in the flowfield, leading to a flame consisting of a curved leading flame front in the outer flow region with a trailing flame in the standing eddy between the obstacles. This is illustrated in the spark

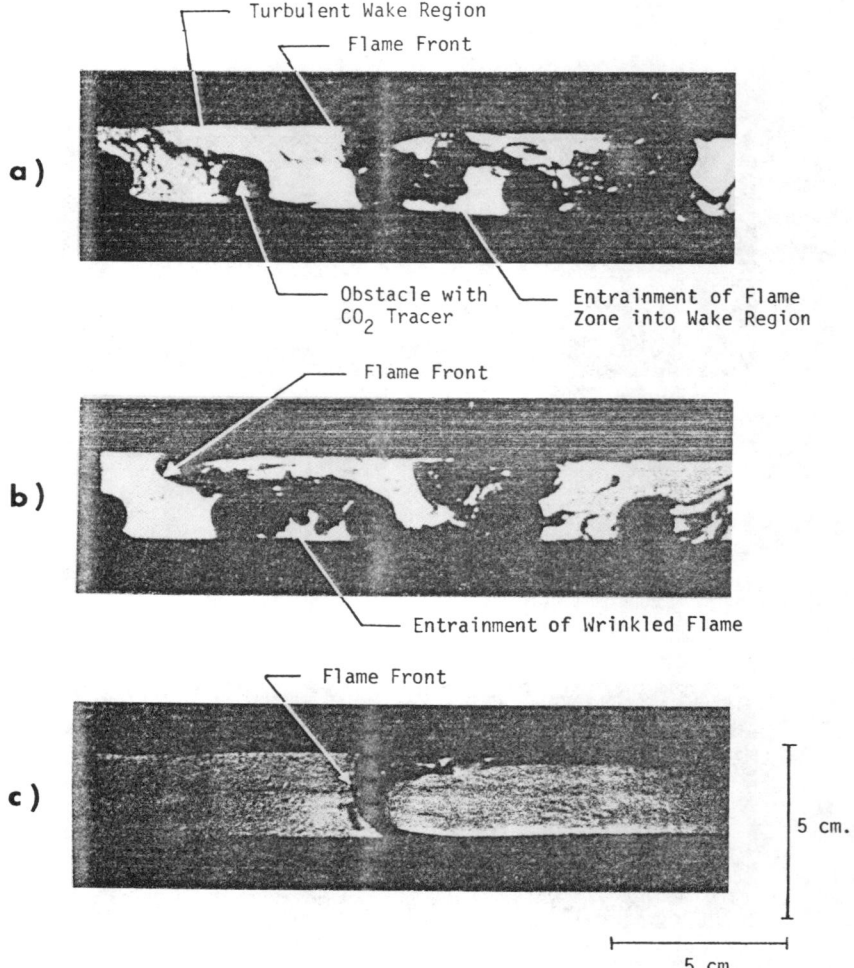

Fig. 5 Spark Schlieren photographs of stoichiometric methane-air flame propagation in a channel of width 2.5 cm: a) obstacle height $H = 1.3$ cm, obstacle separation $p = 3.8$ cm (dry ice in the next to last obstacle); b) obstacle height $H = 1.3$ cm, obstacle separation $p = 3.8$ c, (no dry ice used); c) no obstacle.s

Schlieren photograph shown in Fig. 4b. This photograph also shows the entrainment of the flame zone into the eddy. For larger flame speeds and thus larger flow velocities ahead of the flame, the shear layers and the wakes of the obstacles will become turbulent and, as the velocity increases, the turbulent flow region will encompass more and more of the flowfield. This is illustrated in Fig. 5a where dry ice in the fourth obstacle provides the CO_2 tracer. Turbulent wrinkling and entrainment of the flame zone into the wake region can also be seen in this figure. The smaller scale turbulent wrinkling of the flame is further illustrated by the spark Schlieren photograph shown in Fig. 5b. The flame structure with no obstacles is shown in Fig. 5c.

These spark Schlieren photographs confirm the picture of flame propagation over obstacles proposed by Moen et al. (1980). A schematic diagram illustrating this physical picture is shown in Fig. 6. As discussed by Moen et al. (1980), the turbulent burning velocity S_T based on the average heat release rate per unit projected or frontal flame area is given by

$$S_T = \frac{1}{2\pi r D} \int S \, da \qquad (1)$$

where the integral is over the area of the flame fold shown in Fig. 6, S the local burning velocity at the flame-fold surface, D the plate separation, and r the radial position of the flame front. Assuming zero flow velocity in the unburned gases behind the flame, the displacement velocity of the unburned gases just ahead of the flame averaged over the plate separation is given by

$$\langle V_u \rangle = \left(\frac{\rho_u}{\rho_b} - 1\right) S_T \qquad (2)$$

where ρ_u and ρ_b are the average densities of the unburned gas ahead of the flame and the burned gas behind the flame zone, respectively. Thus, assuming that the advance of the flame is uniform across the plate separation and that transient effects can be neglected, the corresponding flame speed is given by

$$\dot{R}_f = \frac{\rho_u}{\rho_b} S_T \qquad (3)$$

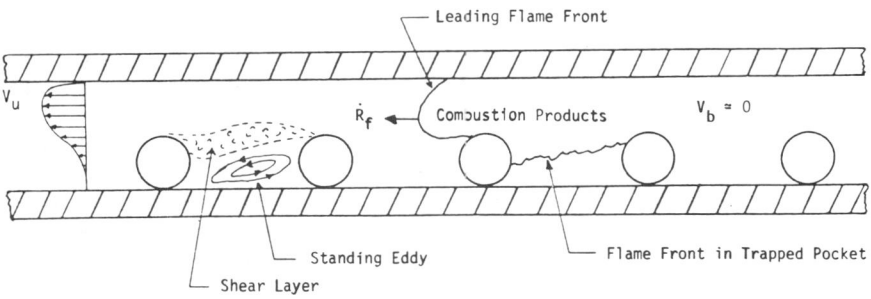

Fig. 6 Schematic diagram illustrating flame propagation over obstacles.

For low flame speeds in the initial stages of the flame acceleration, the flame acceleration mechanism is clearly flame folding due to the gradient in the mean flowfield produced by the obstacles. The turbulent burning velocity S_T is enhanced by an increase in relative flame surface area (i.e., relative to the projected flame area $2\pi rD$) as the flame is stretched in the mean flow gradient field. As the burning velocity increases the flow velocity ahead also increases, leading to more intense flame folding and thus to a larger burning velocity, etc. Superimposed on the large-scale flame fold is the smaller scale turbulent flowfield that leads to flame wrinkling and increased rate of turbulent transport of heat and mass across the flame-fold surface. The resulting increase in burning velocity at the flame-fold surface or increase in flame-fold burning rate leads to a further increase in turbulent burning velocity. During these initial stages the flame acceleration will be determined mainly by the rate of flame folding. However, as the flame accelerates, the smaller scale turbulent flowfield in the wakes and shear layer regions will begin to play a more important role in the acceleration process. As observed in Fig. 5, the burning in the wake regions will become less well-defined due to the smaller scale and larger intensity turbulent entrainment and mixing, and the main flame front will become wrinkled as the turbulence in the wake regions spreads into the main flow region.

Transition to turbulence in the wake of a cylinder occurs at Reynolds numbers above 200 and the location of the transition point moves toward the cylinder as the Reynolds number increases (Roshko, 1954; Papailiou and Lykoudis, 1974). Even for very low flow velocities of ~10 m/s with 1.25 cm obstacles, the Reynolds number is more than 8000 so that the wake of the obstacles will always be turbulent. The acceleration of the flame thus depends on both the rate of growth of the flame fold relative to the projected area $2\pi rD$ and on the increase in the flame-fold burning rate. In the initial stages of the flame acceleration, where the mean flow velocity is low, the dominant mechanism is the rate of growth of the flame fold. But as the flame speed increases, the folded area for a given obstacle configuration will approach some asymptotic maximum since the flame cannot be stretched out indefinitely. Beyond this, the flame acceleration will depend more on the small-scale turbulence intensity which governs the burning rate of the flame folds. The small-scale turbulence derives its energy from the shear flow gradient of the mean flowfield, thus the flame speed will be governed by the rate at which the fine-scale turbulence can derive its energy from the mean shear flow. For the initial stages of the flame acceleration where the development of the flame fold dominates, a computer simulation of the transient inviscid flow structure should be adequate for determining the flame acceleration due to flame-fold area increase. For the later stages of the flame acceleration, when the folded structure approaches some steady value and small-scale turbulence dominates subsequent flame acceleration, it should be possible to model the acceleration analytically by postulating a dependence of the burning velocity on small-scale turbulence. In this paper we propose a simple feedback mechanism for the latter stages of the flame acceleration based on the assumption that the increase in the rate of burning of the flame fold controls the acceleration process. The details of the initial flame acceleration leading to the establishment of the flame fold will not be considered.

Results and Discussion

According to Townsend (1976) the turbulence intensity u'/u_0 in a plane wake is of the order of 0.4, where u' is the rms turbulent velocity and u_0 is the mean velocity variation across the wake. If we take u_0 as one-half the maximum flow velocity ahead of the flame, we obtain a turbulent velocity $u' \simeq 0.2\dot{R}_f$ in the wake of the obstacle ahead of the flame. Townsend's conclusions are based on fully developed turbulent flow in the far-wake region of an obstacle in an infinite stream, but the estimate of $u' \simeq 0.2 U_{max}(\sim 0.2\dot{R}_f)$ in the near-wake regions of bluff obstacles in a boundary layer is supported by measurements by Chang (1966); Plate (1971); Counihan, Hunt, and Jackson (1974); and by recent measurements in our laboratory. Townsend also quotes a value of $u_e \simeq 0.21\, u_0$ (i.e., $\simeq 0.1 R_f$) for the entrainment velocity, i.e., rate at which nonturbulent fluid flows through the bounding surface of the wake and becomes turbulent. Thus as the speed of the flame increases, both the rms turbulent velocity and the rate at which the unburned gas ahead of the flame is made turbulent increases.

To model the feedback mechanism based on increase in flame-fold burning rate due to this increase in turbulence with flame speed, let us focus on the region around the nth obstacle and denote the average speeds of the flame front through the regions just before and just after this obstacle by $\dot{R}_f(n-1)$ and $\dot{R}_f(n)$, respectively. These flame speeds will be determined by the respective turbulent burning velocities, which according to our previous discussion depend on both the relative size of the flame fold and on the flame-fold burning rate. If the relative size of the flame fold (i.e., relative to the projected area $2\pi rD$) remains the same, the increase in flame speed is due solely to an increase in flame-fold burning rate, i.e., increase in effective burning velocity at the flame-fold surface. Making the assumption that this burning rate is proportional to $(u')^\beta$ (for the correlation proposed by Abdel-Gayed and Bradley (1977) for high-intensity isotropic turbulence, for example, $\beta = 0.238$) we find that

$$\dot{R}_f(n) = \dot{R}_f(n-1)\left[1 + \beta \frac{\delta u'}{u'}\right] \qquad (4)$$

where $\delta u'/u'$ is the relative increase in turbulence due to the presence of the obstacle, and we have assumed that $\delta u'/u' \ll 1$.

The increase in turbulence in the developing wake region will not be uniform throughout the region between the obstacles, but will develop from an initial transition point, which depends on the Reynolds number of the flow and the initial turbulence, and spread as more of the outer stream fluid is engulfed into the wake (Roshko, 1954; Papailiou and Lykoudis, 1974). For the flame speeds we are concerned with, the initial turbulence and Reynolds numbers are sufficiently large so that transition can be considered to occur right at the obstacle itself. Hence, $\delta u'/u'$, which represents the relative increase in turbulence averaged over the entire region between the obstacles, will depend on the separation of the obstacles and on the blockage ratio H/D. If we assume a linear growth of the region with increased turbulence in the near wake and a power dependence on H/D we obtain a relation of the form

$$\dot{R}_f(n) = \dot{R}_f(n-1)\left[1 + (\beta p/L)(H/D)^\alpha\right] \qquad (5)$$

between the average flame speeds before and after the nth obstacle, where α is a constant and L is a length which characterizes the growth of the turbulent region. The assumption of linear growth of the turbulent region in the near wake region is rather arbitrary, but as we shall see this assumption does give the correct scaling from obstacle configurations with $p = 3.8$ cm to obstacle configurations with $p = 10$ cm. Although Eq. (5) has been obtained by focusing

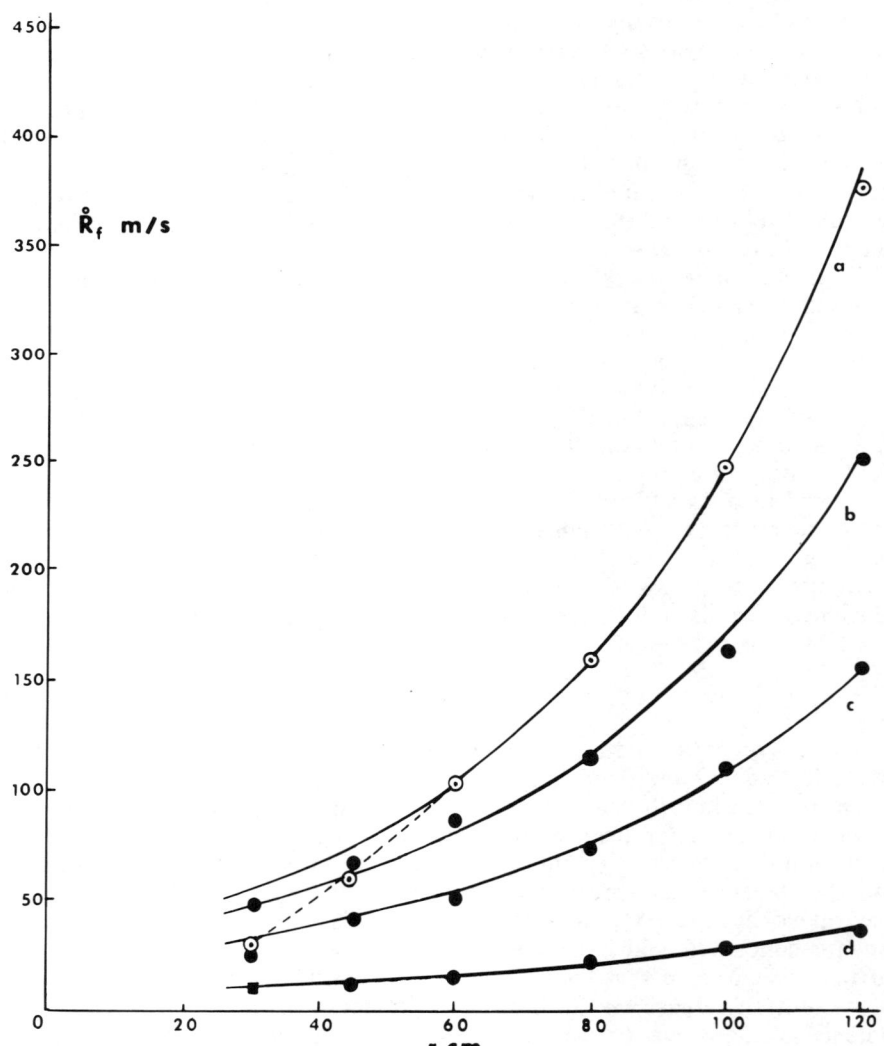

Fig. 7 Flame velocity $\mathring{R}_f$ vs distance of propagation r for stoichiometric methane-air flames (solid curves correspond to Eq. (7) with $\alpha = 0.31$ and $p_0 = 37.5$ cm): a) copper tube spiral, $H = 1.25$ cm, $p = 3.8$ cm on bottom plate, plastic tube spiral; $H = 4$ cm, $p = 10$ cm on top plate; $H/D = 0.52$; b) copper tube spiral on bottom plate, $H/D = 0.34$, c) copper tube spiral on bottom plate, $H/D = 0.25$; d) copper tube spiral on bottom plate, $H/D = 0.13$ [the experimental point at 30 cm is from the laboratory results of Moen et al. (1980)].

on the regions just before and just after the nth obstacle, the result does not depend on the relative size of the flame fold. As long as the relative size of the flame fold remains the same and the relative increase in turbulence depends on the obstacle configuration only, exactly the same arguments apply for any size flame fold.

Since the above mechanism does not include the contribution to flame acceleration due to increase in the relative size of the flame fold, it cannot describe the initial acceleration. However, if we assume that the increased flame-fold burning rate dominates the flame acceleration process after the flame has propagated over n_0 obstacles, then for $n > n_0$ we can use Eq. (5) to obtain the following relation

$$\dot{R}_f(n) = \dot{R}_f(n_0)\left[1 + \frac{p}{p_0}\left(\frac{H}{D}\right)^\alpha\right]^{(n-n_0)} \tag{6}$$

for the flame velocity after the nth obstacle, where we have defined $p_0 = L/\beta$. For small p the above relation becomes

$$\dot{R}_f(r) = \dot{R}_f(r_0)\exp\left[\frac{(r-r_0)}{p_0}\left(\frac{H}{D}\right)^\alpha\right] \tag{7}$$

In other words, it is predicted that the flame velocity increases exponentially with r for $r > r_0$. As can be seen in Fig. 7, this prediction is supported by the experimental results. If r_0 is chosen at 60 cm with $\alpha = 0.31$ and $p_0 = 37.5$ cm, Eq. (7) provides a good correlation of the experimental results from all the experiments with obstacle configurations which include the 1.25 cm copper tube spiral. In fact, for those experiments with copper tube spiral obstacles only, the correlation is good down to about 30 cm of flame travel. The prediction for the larger 4 cm obstacles with the same parameters α and p_0 is shown in Fig. 8, using Eq. (6) rather than Eq. (7) since the exponential approximation is not valid in this case. Again, agreement with the experimental results is quite good for $r \gtrsim 60$ cm (i.e, $n \gtrsim 5$), but significant deviations are clearly observed for $r < 60$ cm.

The transition from flame acceleration dominated by the rate of flame-fold growth to acceleration dominated by the increase in flame-fold burning rate is clearly not a sharp one. There will be a transition region where both mechanisms are equally important, and in this region the flame acceleration is expected to be greatest. That the initial acceleration of the flame is greater than that predicted by Eqs. (6) and (7) is clearly evident in both Figs. 7 and 8. Note that all the curves should converge to an initial flame velocity of about 3.3 m/s independent of plate separation and obstacle configuration as observed by Moen et al. (1980) in their laboratory experiments. Once the increase in flame-fold burning rate begins to dominate the acceleration process, a simple model of the feedback mechanism based on a positive feedback coupling between the turbulence produced by the obstacles and the flame speed predicts an exponential increase in flame speed which agrees well with the experimental results. In fact, Eq. (4) which relates the change in flame speed to the change in turbulence can be applied more generally. It would predict, for example, that the flame would slow down if there were no obstacles beyond a certain

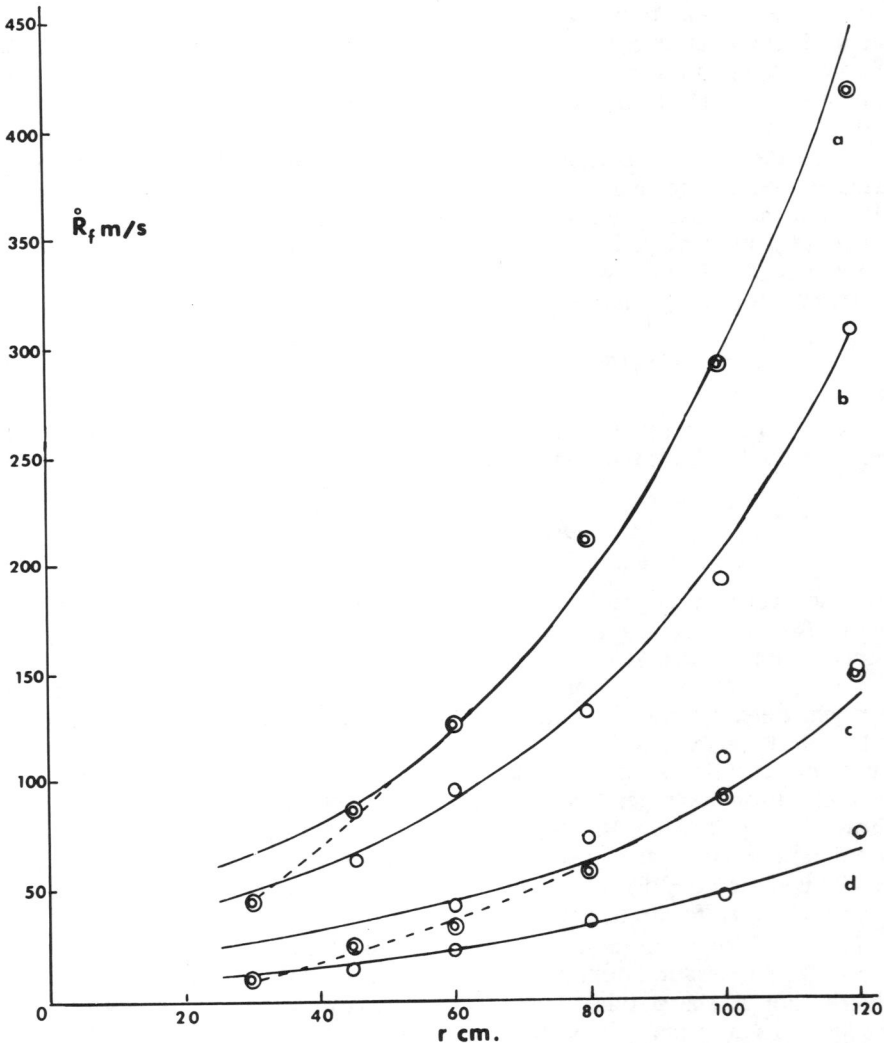

Fig. 8 Flame velocity $\dot{R}_f$ vs distance of propagation r for stoichiometric methane-air flames (solid curves correspond to Eq. (6) with $\alpha = 0.31$ and $p_0 = 37.5$ cm, dashed curves are drawn through data points for $x < 60$ cm): a) Plastic tube spirals, $H = 4$ cm, $p = 10$ cm on both top and bottom plates, $H/D = 0.67$; b) plastic tube spiral on bottom plate, $H/D = 0.57$; c) plastic tube spirals, ○ top plate only, ⊙ top and bottom plate, $H/D = 0.4$; d) plastic tube spiral on bottom plate, $H/D = 0.33$.

distance; the change in turbulence intensity $\delta u'$ would then eventually become negative and the flame would decelerate. Furthermore, the parameters p_0 and α, in particular p_0, would be expected to depend on geometry (i.e., planar, cylindrical, or spherical) and to be constant only in a certain range of p/H.

The predicted exponential increase in flame speed cannot continue indefinitely. As the flame speed increases the strength of the shock waves produced will increase and shock-wave heating combined with turbulent mixing may lead to autoignition near the leading shock wave, in which case a coupled shock-wave reaction zone complex may continue to propagate at a constant velocity characteristic of the obstacle configuration. Such coupled complexes with velocities depending on the obstacle configuration but much smaller than the Chapman-Jouguet detonation velocity have been observed by many authors (Shchelkin, 1940; Guenoche and Manson, 1949; and Brochet, 1966). Another possibility is that the rate of chemical reaction becomes the limiting factor. As the flame speed increases, the intensity of the turbulence in the regions between obstacles increases, which in turn leads to the observed exponential flame acceleration. However, there is a maximum mass rate at which a given volume of a combustible gas at a given pressure can burn, which corresponds to the well-stirred reactor rate. Thus at a certain turbulence intensity chemistry begins to play a role. If the rate of turbulent mixing exceeds the rate at which the mixture can burn, the flame may be temporarily blown out or locally quenched. Such quenching has been observed by Wagner and collaborators (Dörge et al., 1979), for example. Our experimental results (which include flame speeds larger than 400 m/s for stoichiometric methane-air mixtures) show no indication that the flame speed is approaching a maximum value. Thus we would expect the flame to continue to accelerate to even larger flame speeds if the obstacle configuration was continued.

The maximum flame speed observed with no obstacles in the chamber was 9 m/s, and no acceleration was observed in the last 20 cm of propagation. This is almost twice the value observed in the 30.5 cm radius chamber with steel upper and lower plates (Moen et al., 1980). The difference can be attributed to additional surface roughness of the wooden plates and slight variation in plate separation due to sagging of the top plate. The plate separation varied by as much as 1 cm over the area of the chamber. This variation of plate separation also accounts for the fact that the flame speeds at 30 cm were typically larger than those observed in the smaller laboratory chamber for the same obstacle configuration. For $H/D = 0.33$ with 1.25 cm copper tubing obstacles, for example, the speed at 30 cm is 45 ± 5 m/s to be compared with about 30 m/s observed in the laboratory experiments with the same obstacles at this H/D. The variation in D due to sagging leads to an uncertainty in H/D of 0.06. If this uncertainty is taken into account, the two results are in agreement.

The strength of the shock waves produced by these fast turbulent flames were not measured in a systematic manner, but two pressure transducers at the periphery of the chamber were used to record the shock-wave overpressures in some of the experiments. For the fastest flame with a maximum flame speed of 415 m/s a shock-wave overpressure of 0.64 bar was obtained. According to the steady-state analysis of Guirao et al. (1976), this would correspond to a constant burning velocity of about 35 m/s to be compared with a maximum turbulent burning velocity of 55 m/s obtained using Eq. (3), with a density ratio of 7.5. Similarly, for the flame with maximum velocity 305 m/s the measured shock overpressure was 0.27 bar, which according to Guirao et al. (1976) corresponds to a constant burning velocity of about 28 m/s to be compared with a maximum turbulent burning velocity of 41 m/s obtained from Eq. (3). In view of the dramatic acceleration of the flames observed in both of these cases the disagreement with steady-state calculations based on maximum burning velocity is not surprising. Strehlow et al. (1979) have recently shown that calculations using constant velocity flames based on the maximum effective burning velocity of accelerating flames can lead to an overestimate of the maximum overpressure. Furthermore, Eq. (3) which we have used to calculate the turbulent burning velocity may have to be modified for violently accelerating flames. The agreement with steady-state analysis is much better for the slower flames. For the flame with maximum speed of 150 m/s, for example, the measured shock overpressure is 0.06 bar, corresponding to a steady-state burning velocity of about 17 m/s to be compared with 20 m/s obtained from Eq. (3).

Conclusion

The influence of repeated obstacles on freely expanding stoichiometric methane-air flames in cylindrical geometry has been investigated in a chamber where the flame propagation could be observed over a distance of 120 cm. This investigation confirms the dramatic influence of obstacles on the speed of the flame reported previously and shows that with repeated obstacles the flame speed continues to increase up to speeds in excess of 400 m/s. The shock waves generated by these fast turbulent flames have overpressures up to 0.64 bar, which could lead to extensive blast wave damage.

The mechanism responsible for the rapid acceleration of the flame in an obstacle environment can be understood in terms of the positive feedback coupling between the flame and the upstream flow produced by the specific volume increase across the flame. The physical picture of this positive feedback mechanism proposed in a previous paper by Moen et al. (1980) is confirmed by spark Schlieren photographs of the flame propagation over obstacles in a rectangular channel.

The acceleration of the flame depends on both the rate of growth of the flame fold relative to projected flame area and on the increase in the flame fold burning rate. In the initial stages of the flame acceleration the dominant mechanism is the rate of growth of the flame fold, but as the flame speed increases the rate of increase in the flame-fold burning rate begins to dominate the flame acceleration process. A simple feedback mechanism is proposed for the latter stages of the flame acceleration, based on the assumption that the increase in the rate of burning of the flame fold due to the turbulence produced by the obstacles controls the acceleration process. This model assumes that a flame fold (which does not continue to grow) has already been established, and that the rate of burning of this flame fold is determined by the turbulence field which the flame propagates into. The details of the turbulent combustion or the nature of the combustion zone are not considered. The model predicts an exponential increase in flame speed with distance propagated over obstacles, which is in good agreement with the experimental results. Although this exponential increase cannot continue indefinitely, no evidence that the flame seped is approaching a limiting value was found in the present experiments.

The details of the initial acceleration of the flame (which involves both flame folding and increase in flame-fold burning rate due to turbulence or the critical size of the flame fold for which the rate of growth of the flame fold can be neglected) cannot be determined from the present analysis. A more detailed model of the flow-field produced by the accelerating flame propagation over obstacles and of the flame turbulence interaction is required to describe the initial flame acceleration and to estimate the size of the flame fold. Investigations with the aim of developing such a model are now in progress.

Acknowledgments

This research was supported by the Natural Sciences and Engineering Research Council of Canada, Atomic Energy of Canada Ltd., U. S. Air Force Office of Scientific Research, and the Quebec Department of Education. The help of K. J. Dörge and Dr. P. Thibault with the experiments is greatly appreciated.

References

Abdel-Gayed R. G. and Bradley D. (1977) Dependence of the burning velocity on turbulent Reynolds number and ratio of laminar burning velocity to RMS turbulent velocity. *16th Symposium (Intl.) on Combustion,* p. 1725. The Combustion Institute, Pittsburgh, Pa.

Brochet C. (1966) Contribution a l'étude des détonations instables dans les mélange gazeux, Theses présentées a la Faculté des Sciences de L'Université de Poitiers, Chap. VI, pp. 107-121.

Chapman W. R. and Wheeler R. V. (1926) The propagation of flame in mixtures of methane-air, Part IV. *J. Chem. Soc.* **1926,** 2139.

Chung S. C. (1966) Velocity distributions in separated flow behind a wedge-shaped model hill. Colorado State Univ., Denver (unpublished).

Counihan J., Hunt J. C. R., and Jackson P. S. (1974) Wakes behind two-dimensional surface obstacles in turbulent boundary layers. *J. of Fluid Mech.* **64,** 529.

Dörge K. J.,Pangritz D., and Wagner H. Gg. (1976) Experiments on accelerations of flames by grids. *Acta Astronautica* **3,** 1069.

Dörge K.J., Pangritz D., and Wagner H. Gg., (1981) Influence of obstacles on the propagation of flames. Paper presented 7th International Colloquium on Gasdynamics of Explosions and Reactive Systems, Göttingen, Fed. Rep. of Germany.

Guénoche H. and Manson N. (1949) Influence des conditions aux limites transversales sur la propagation des ondes de choc et de combustion.*Revue de L'Institut Francais du Petrole* **2,** 53.

Guirao C. M., Bach G. G., and Lee J. H. (1976) Pressure waves generated by spherical flames. *Combustion and Flame* **27,** 341.

Kurylo J., Dwyer H. A., and Oppenheim A. K. (1979) Numerical analysis of flow fields generated by accelerating flames. Paper 79-0290 presented at AIAA 17th Aerospace Sciences Meeting, New Orleans, La.

Moen I. O., Donato M., Knystautas R., and Lee J. H. (1980) Flame acceleration due to turbulence produced by obstacles. *Combustion and Flame* **39,** 21

Papailiou D. D. and Lykoudis P. S. (1974) Turbulent vortex streets and entrainment mechanism of the turbulent wake. *J. Fluid Mech.* **62,** 11.

Plate E. J. (1971) The aerodynamics of shelter belts. *Agric. Meteorol.* **8,** 203.

Roshko A. (1954) On the development of turbulent wakes from vortex streets. NACA Rept. 1191.

Shchelkin K. I. (1940) Effect of roughness of the surface in a tube on origination and propagation of detonation in gases.*J. of Exp. and Theo. Phys.* (U.S.S.R.) **10,** 823.

Strehlow R. A., Luckritz R. T., Adamczyk A. A., and Shimpi S.A. (1979) The blast wave generated by spherical flames. *Combustion and Flame* **35,** 297.

Townsend A. A. (1976) *The Structure of Turbulent Shear Flow* (2nd Ed). Cambridge University Press, Cambridge, England.

Detonation of Unconfined Fuel Aerosols

D. C. Bull,* M. A. McLeod,† and G. A. Mizner*
Shell Research Ltd., Chester, England

Experiments have been performed to determine the conditions marginal to detonation in unconfined fuel/air aerosols. Aerosols were generated in a 5 m^3 volume by sonic air-blast atomizers producing measured particle sizes of 5-30 μm. Detonation of the fuel/air mixture was initiated by the method of overdriving with the blast wave from plastic high-explosive charges. Particular emphasis was placed upon the influence of fuel volatility on the detonability of aerosols; n-hexane and dodecane were used to represent extremes of very high and low vapor pressure fuels. Fuels having intermediate volatilities were synthesized by mixing the two fuels in appropriate proportions. Hexane aerosols have been shown to detonate within concentration limits similar to those expected for gaseous detonations. As with planar two-phase detonations, a velocity deficit compared with Chapman-Jouguet values has been observed in the spherical mode, even with droplets <50 μm diam. Dodecane and decane have not been detonated, even by relatively large (0.5 kg) initiator charges. This has led to the conclusion that a minimum effective equivalence ratio necessary for detonation propagation is not being attained behind the shock wave within the limiting induction period. It is further concluded that propagation of self-sustained detonations through n-paraffin hydrocarbon aerosols in the spherical mode, with droplet sizes ≳ 10 μm, may require the presence of a certain quantity of fuel vapor prior to ignition.

Introduction

THIS paper presents the first results of a series of experiments designed to investigate the detonation characteristics of unconfined fuel/air aerosols. Using the method of direct initiation of the detonation by high-explosive charges, fuels with a high and low volatility are investigated as aerosols with droplet sizes of less than 100 μm diam.

While understanding of the combustion mechanisms for unconfined explosions in gaseous systems has increased considerably in the past few years,

---

Presented at the 7th ICOGER, Göttingen, Federal Republic of Germany, Aug. 20-24, 1979. Copyright © American Institute of Aeronautics and Astronautics, Inc., 1980. All rights reserved.
*Senior Scientist, Thornton Research Center.
†Senior Technician, Thornton Research Center.

the hazards posed by two-phase systems are less well defined. The study of detonations of fuel mists and sprays has been restricted to relatively few workers and much of this work has been directed toward controlling the instabilities that may occur in liquid rocket motors. All of the reported work has been performed either as planar waves in shock tubes using pure oxygen or oxygen-enriched atmospheres (Dabora et al., 1966; Bowen et al., 1971; and Kauffman and Nicholls, 1971), or as cylindrical waves in a pie-shaped sector (Nicholls et al., 1974). With the exception of Bowen et al. (1971), who used 2 μm decane mists, the investigations have all concentrated on drops of the size encountered in sprays rather than in mists, i.e., with diameters between 200 and 2000 μm. This work has demonstrated that detonations can propagate through two-phase systems and has determined the mechanisms whereby this may occur—through the rapid shattering and evaporation of the droplets as a result of the convective flow behind the leading shock wave. The effect of drop size on the detonation characteristics has been shown quite clearly and it is now known that detonations may propagate through two-phase systems at velocities much below the Chapman-Jouguet (C-J) values, and with considerably longer induction periods than are experienced in gas detonations.

Here the detonability of a cloud of fuel droplets in air is investigated in which the droplet sizes are <100 μm diam. To conveniently distinguish this work from studies with fuel sprays we use the term aerosol to describe the cloud, while noting that the term is normally used for systems where the droplets are small enough to show a considerable degree of stability in suspension, i.e., <10 μm diam.

For this investigation, fuels giving rise to two distinctly different types of aerosol were chosen: one a highly volatile fuel that results in a hybrid mixture of liquid fuel droplets/fuel vapor/air, and the other a fuel with a very low volatility at ambient temperatures that represents a strictly two-phase liquid fuel droplet/air mixture. Experiments have been performed permitting comparison betweeen these types. To minimize behavior differences arising from differing chemical reactivity, only the n-paraffin hydrocarbons have been used.

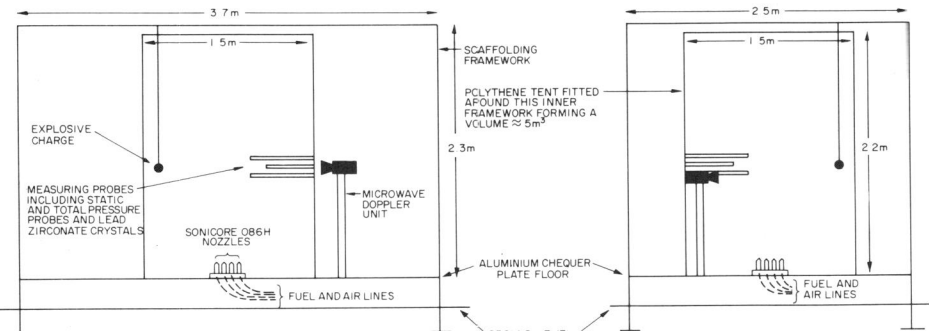

Fig. 1 Schematic elevations of the experimental facility.

Experimental Method

The outline of the experimental facility is schematically illustrated in Fig. 1. Essentially it consists of a framework structure situated 50 m from a bunker that is used as the control room. A 125 μm thick polythene tent is fitted over the inner framework forming a volume of approximately 5 m^3 in which the aerosol is formed. Although it would be preferable to conduct the experiments without any form of bag, preliminary experiments showed the necessity of protecting the aerosol cloud from unsteady ambient conditions.

In order to quantify the role of particle size on aerosol detonation characteristics more clearly, one would ideally wish to generate droplets of known size and with a narrow size distribution. Of the available devices for droplet generation, the ultrasonic quartz crystal atomizer and the vibrating capillary needle technique as used by Dabora (1967) both produce very precise droplet sizes, but they are not readily adaptable to the high liquid flow rates ($\sim 23 \times 10^{-6}$ m^3/s^{-1}) required in these experiments. A pressure jet atomizer, on the other hand, while allowing for high flow rates, typically produces a broad size distribution. For these experiments, Sonicore nozzles were chosen which atomize by the process of an air blast through a sonic convergent-divergent nozzle. These atomizers represent a good compromise between the conflicting requirements of high liquid flow rates and narrow droplet size distribution. They permit considerable flexibility in that the global fuel/air concentration in the tent can be easily varied by spraying for different time intervals, while droplet sizes can be altered within fairly narrow limits by adjusting fuel and air flow rates, or within wider limits by selecting different nozzle sizes.

As Fig. 1 shows, four 086H Sonicore nozzles are positioned in the center of the tent, on the floor pointing upward. They are operated simultaneously to give the required flow rate with the fuel being supplied using an RS/40 Danfoss domestic heating pump. The atomizing air is supplied from a compressed air bottle at approximately 4×10^{-3} m^3/s^{-1} through each nozzle. Aerosols from these nozzles have been characterized using the Malvern particle size analyzer, which employs a laser light diffraction technique to obtain the droplet size distribution. For purposes of the characterization of the nozzle, commercial grade hexane (purity $\sim 90\%$) and kerosene were used with fuel flow rates between 1×10^{-6} and 7×10^{-6} m^3/s^{-1} and airflow rates between 0.3×10^{-3} and 4×10^{-3} m^3/s^{-1}. A few check distributions were also obtained for the actual hydrocarbons used in the detonation experiments. These were hexane (purity 99% from BDH), decane (purity 97% from Koch Light), and dodecane (purity 99% from Koch Light), all of which were used without further purification.

Ignition of the aerosol is by the use of plastic explosive type PE4 fired by a Nobel short-delay 8 star detonator. PE4 contains 88% RDX (Hexogen), 1% pentaerythritol dioleate, and 11% plasticizer. It has an energy of approximately 4900 kJ/kg^{-1}. The explosive charges were located near one corner of the volume as indicated in Fig. 1, in order to maximize the path length available over which the detonation wave could develop and, hence, be monitored. The technique of initiating a spherical detonation away from the center of the bag has been established by Bull et al. (1976), and by a few check experiments in this study, as giving detonations the properties and critical initiation regimes which were indistinguishable from those in which initiation was central.

Piezoelectric pressure transducers were used to measure the total and static pressures and lead zirconate crystals were used to obtain further time-of-flight information of the detonation. The microwave Doppler equipment for continuously monitoring the advancing reaction front, as described in Bull et al. (1976), was also deployed. All transducer outputs were recorded on Datalab DL922 twin-channel transient event recorders and later transferred to a digital tape cassette or chart recorder. For certain experiments a Hycam K1001 high-speed camera was used to obtain color films of the detonations at approximately 10^4 frames/s.

Results

The results of Lorenzetto and Lefebvre (1977) for a plain jet air-blast atomizer indicated that the mean droplet diameter would not vary by very much over the relatively small range of fuel physical properties used here. Several particle sizing experiments on the pure paraffin hydrocarbons used in the combustion experiments confirmed this to be the case and indicated that the much cheaper fuels, commercial grade hexane and kerosene, could be used for the nozzle characterization studies.

Two typical particle size distributions for commercial grade hexane and kerosene are shown in Figs. 2a and 2b for flow rates very similar to those used in the detonation experiments. Figure 2a shows a mass distribution for hexane. Clearly the majority of the mass in the aerosol occurs in the sizes between approximately 6–17 μm. Mass distribution represents a more relevant parameter than number distribution in this investigation, because the breakup and evaporation of the droplets necessary to achieve the correct fuel concentration for ignition involves mainly those droplets containing the largest proportion of the fuel mass. Figure 2b shows a distribution for kerosene. Here the mass is concentrated mainly between droplet sizes of 9–30 μm.

In order that these experiments may be related to those in which other initiators are used, the characteristics of the blast wave in free air produced by the plastic explosive were measured using the piezoelectric pressure transducers. The results are presented in Figs. 3 and 4 as peak overpressures and shock arrival times. The broken curve on both graphs represents the characteristics of Tetryl as obtained by Bull et al. (1976), and, as can be seen, they are very close to those for PE4. The solid curve is the TNT data of Peters (1968).

In the combustion experiments both static and stagnation pressures were measured, although owing to "ringing" in the transducers only the pressure across the leading shock wave was reliably measured in the case of the static pressure. Where detonation was initiated, the characteristic sharp pressure rise across the leading shock wave was observed and, from the stagnation pressure records, appeared to be followed by an expansion and another smaller pressure rise toward what was probably the reaction front.

Although the microwave Doppler equipment was deployed, no meaningful signals were obtained from it. Detonation was, however, readily characterized

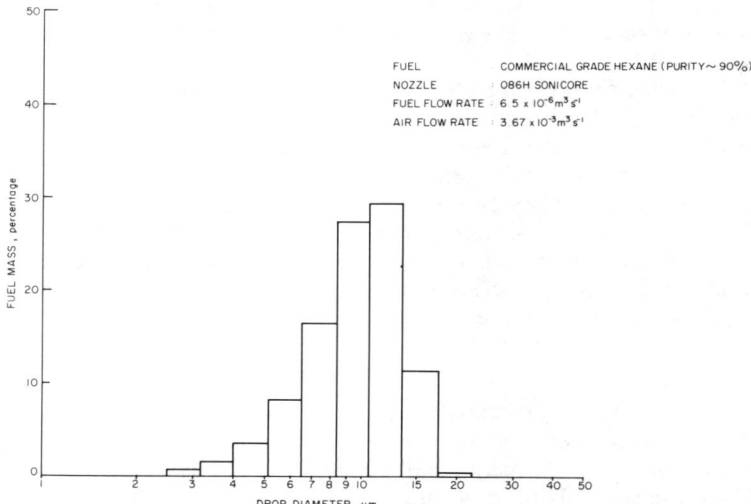

Fig. 2a Droplet size mass distribution for commercial grade hexane using 2a 086H Sonicore nozzle.

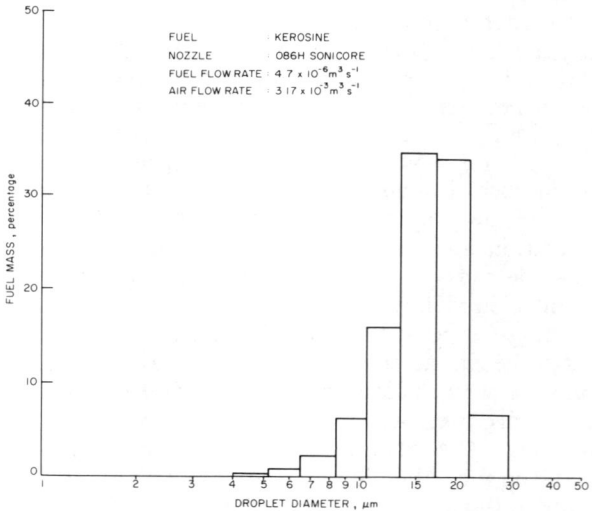

Fig. 2b Droplet size mass distribution for kerosene using 2a 086H Sonicore nozzle.

and distinguished from experiments in which detonation failed, on the basis of the pressure records, velocity information from the time-of-flight data and remnants from the polythene tent were always much smaller when detonation had occurred than when it had not.

The results of the experiments with hexane are presented in Fig. 5 in terms of the initiator charge weight for detonation as a function of the concentration λ, where λ is the equivalence ratio defined as the ratio (fuel/oxygen)

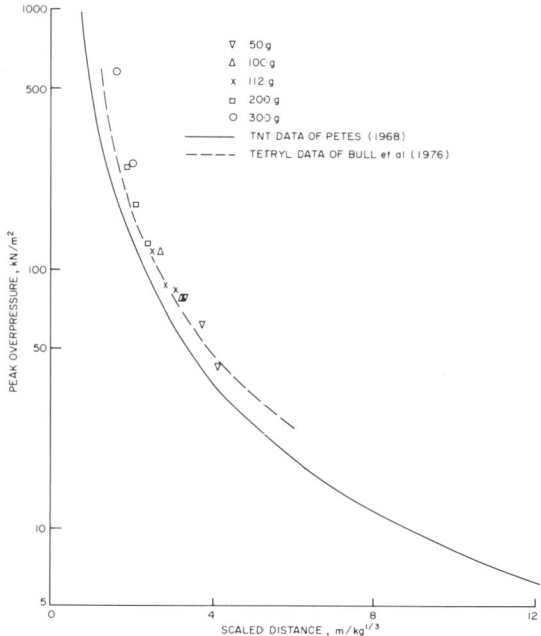

Fig. 3 Peak overpressure produced by plastic explosive charges in air.

÷(fuel/oxygen for a stoichiometric mixture), and may be represented as $\lambda C_6H_{14} + 9.5O_2 + 35.7\ N_2$. A limit curve has been drawn between the detonation and no-detonation points within the range of charge weights used. The most readily detonable concentration is seen to lie between $\lambda = 1.1$ and 1.2, and occurs at a charge weight of approximately 0.025 kg. These values compare with those of $\lambda = 1.18$ at a charge weight of 0.0182 kg of Tetryl obtained for ethane by Bull et al. (1979).

Also shown in Fig.5 is a curve derived using the phenomenological model suggested by Bull et al. (1979) to predict initiation energies relative to the stoichiometric value. The model is based on a modification of Zeldovich's criterion $E \propto \tau^3$, where E is the initiation energy and τ an appropriate high-temperature induction period. It requires a sufficient knowledge of the chemical kinetics of the fuel/oxidant system to compute τ for the temperature and density conditions appropriate to the initiation process. High-temperature oxidation of hexane/air has not yet been studied sufficiently rigorously to describe the behavior by even the global pseudo-Arrhenius type of term that has been derived for the lower paraffin hydrocarbons (Bull et al., 1979). Some data of Barnard (1978), however, do give an indication of activation energy, and by analogy with other hydrocarbon oxidations at high temperatures a first-order dependence upon oxygen concentration is taken as an approximation. Thus,

$$\tau = A\ [O_2]^{-1} \exp\ (121.4/RT)$$

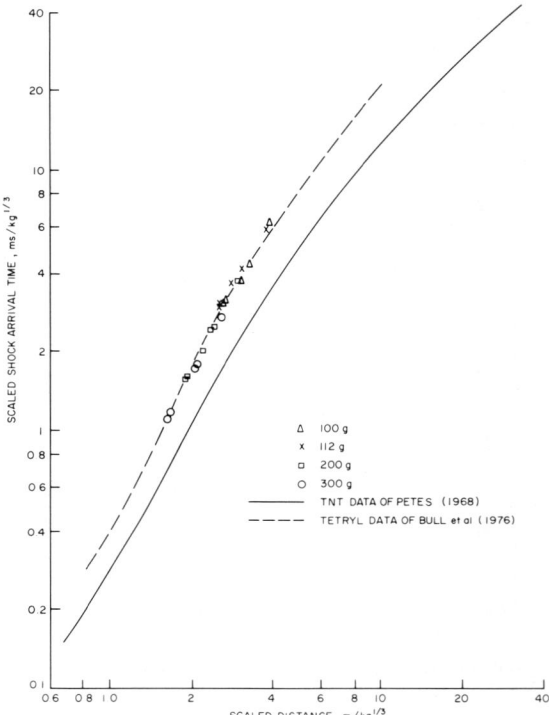

Fig. 4 Free air shock arrival times for plastic charges.

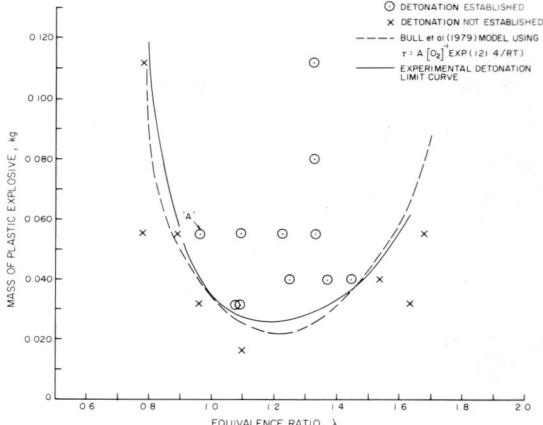

Fig. 5 Initiator energy to detonate hexane/air aerosols as function of concentration (drop size range 6-17 μm).

where the activation energy is in kJ/mole^{-1}, T is the temperature in Kelvin and R the gas constant. Sensitivity of the detonability model has been discussed in detail elsewhere (Bull et al., 1979) but, essentially, as it is required only to compute the ratio τ/τ_s (where s denotes the stoichiometric value), some uncertainty in the kinetic expression can be tolerated.

Detonation velocities obtained from time-of-flight measurements for a few experiments are presented in Fig. 6, with error bars of ±5% which arise from the relatively short path length of approximately 0.15 m between the transducers. Even with such small droplet sizes and with a highly volatile fuel, the measured velocities lie between 6-13% below the gas-phase C-J values. One additional experiment was carried out with hexane in which the droplet sizes were increased by reducing the air flow rate through each nozzle to approximately 1.2×10^{-3} m^{3}s^{-1}. The majority of the mass in this case was concentrated between 50-90 μm compared with 6-17 μm in the experiment reported in Fig. 5. A charge mass of 0.055 kg was used, which was in excess of the minimum required, to detonate 6-17 μm aerosols (~0.03 kg) at the concentration in question, $\lambda = 1.1$. Detonation was achieved in the 50-90 μm aerosol but the velocity of propagation was about 1300 m/s, compared with the value of ~1580 m/s observed for the finer aerosol.

With dodecane ($C_{12}H_{26}$) and decane ($C_{10}H_{22}$), both of which have very low saturation vapor pressures at ambient temperatures, detonations could not be initiated over a wide range of conditions, as shown in Fig. 7. These conditions

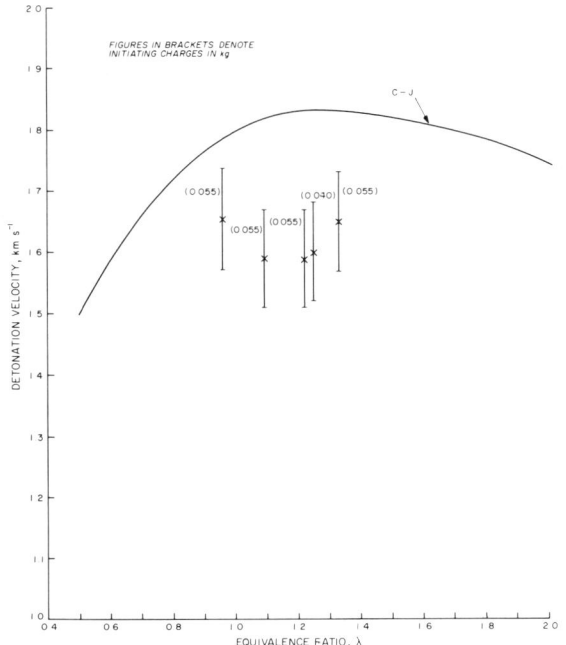

Fig. 6 Detonation velocities as a function of concentration for hexane/air aerosols.

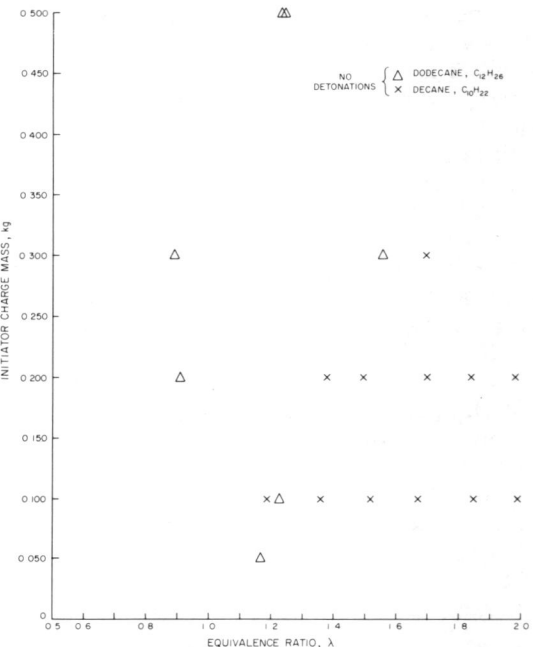

Fig. 7 Results of attempts to detonate dodecane and decane aerosols (drop size range 9-30 μm).

included initiator charge masses of up to 0.5 kg, although strictly speaking the path length of the experiment, 1.83 m, would have been too short to be certain that any detonation wave initiated was truly self-sustained. Nevertheless, it clearly sufficed to show, from the blast wave time of arrival data, that the wave was not only decaying but had decayed to such an extent that no detonation had occurred or could have occurred even if more aerosol path had been available. The droplet sizes for the experiments with dodecane and decane are very similar to those shown in Fig. 2b for kerosene, with the majority of the mass lying between 9-30 μm. For these fuels, however, no evaporation takes place prior to the arrival of the blast wave and so the droplet sizes remain intact and are not reduced by vaporization as is the case with hexane.

In an attempt to control the volatility of the fuel, hexane and dodecane were mixed in varying proportions. The results of these experiments, which were carried out mainly at a charge weight of 0.12 kg but with a few at 0.10 kg, are presented in Fig. 8. It can been seen that as the proportion of dodecane in the mixture is increased, the minimum equivalence ratio for detonation increases, and reaches a value of approximately $\lambda = 2$ for a 50% mixture. The equivalence ratio here, as elsewhere, is based on a global value of the concentration computed from the volume of the enclosure and the quantity of fuel dispensed. The cases shown in Fig. 8 as delayed detonations were so characterized because in these experiments the blast wave from the initiator

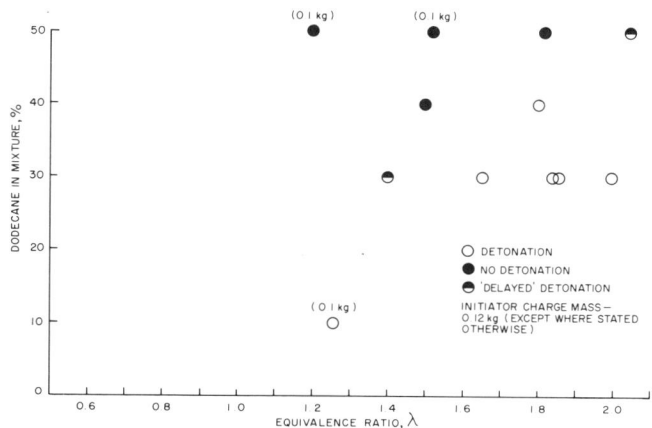

Fig. 8 Detonation of mixtures of dodecane and hexane in varying proportions.

arrived at the measuring probes before the detonation wave. High-speed film evidence showed that in such cases the detonation had commenced at some point remote from the initiator charge.

Discussion

When the results of the experiments using mixtures of hexane and dodecane (Fig. 8) are compared with those using pure hexane (Fig. 5) it appears that detonations of these mixtures are effectively just hexane detonations, since the quantity of hexane in each of the mixtures in Fig. 8 is at least that required for the lean detonation limit, at 0.10-0.11 kg charge mass, in Fig. 5. This means that the dodecane plays little or no part in the combustion, with the hexane presumably being distilled out of the droplets. This accounts for the increasing quantity of fuel that has to be dispensed as the percentage of dodecane is increased.

In some of the above experiments the arrival of the blast wave from the initiating charge occurred ahead of the detonation, a phenomenon labeled delayed detonation in Fig. 8. High-speed color films confirm that in these cases the initiation occurs first at a point remote from the initiating charge. This seems to imply some degree of inhomogeneity in the initiation process in these cases. Although no measurements were made to test the concentration homogeneity of the aerosols dispersed, the repeatability of experiments in this study has been such that, overall, the mixing of the aerosol seemed acceptably good. However, close to the concentration limits the homogeneity of the mixture will play a more crucial role. The delayed detonation phenomenon seems to be an indication both that slight local concentration variations do occur and that conditions are close to those limiting the initiation of detonation.

The mechanism whereby a detonation may propagate through a two-phase medium has been fairly well established by previous workers studying planar and cylindrical waves. It involves the rapid breakup and evaporation of the droplets behind the leading shock wave followed by the usual chemical induction period. This results in a total induction period considerably longer than that experienced in gaseous detonations. Failure to receive sufficient reflected signals when using the microwave Doppler equipment may be attributable to the thick reaction front associated with two-phase burning which in turn poses a diffuse reflector for microwaves.

It is expected that there will be a critical maximum induction period τ_{max} for any given system beyond which the energy liberated at the reaction front cannot sustain the shock wave ahead of it, and the detonation will then fail or die. Consequently, for self-sustained propagation a minimum effective equivalence ratio must be attained within τ_{max}. This poses a severe limitation on initiation by the rapidly decaying blast wave from an initiating charge in the spherical mode, in sharp contrast to shock-tube experiments where the decay of the planar waves is one-dimensional and, hence, much less rapid. Conditions for initiation of spherically divergent detonations in aerosols will thus be much more stringent than those for planar wave initiations. Many cases can probably be described where planar detonations can be obtained, because the shock tube permits establishment of a long induction zone, while spherical detonations cannot be achieved under realistically attainable initiation conditions.

The relative ease with which hexane detonated indicates that sufficient fuel had been vaporized within τ_{max}. Hexane is, of course, a highly volatile fuel with a very high saturation vapor pressure at ambient temperatures, so that, under equilibrium conditions, there will be a flammable concentration in the vapor phase. In these experiments the liquid experiences cooling as it is atomized by the supersonic airstream from the convergent–divergent nozzle, and considerable further cooling as part of the aerosol evaporates during the spraying period. This cooling, together with the fact that the time required for droplet evaporation to reach equilibrium is in excess of the delay between aerosol generation and ignition, results in there being a hybrid mixture of fuel vapor/liquid fuel/air at the time of initiation, in proportions that depend on the initial ambient temperature and the spraying time.

When dodecane was used, the only really significant difference from the hexane case was the fact that negligible evaporation occurred before the arrival of the blast wave, so that there was no fuel vapor but a true two-phase mixture. Further, the droplets would have been slightly larger than the hexane droplets, simply because there was no evaporation. Studies on the break up of droplets in shock tubes, (Kauffman and Nicholls, 1971; Ranger and Nicholls, 1969), have suggested that for droplets ≲ 100 μm diam. subjected to shock waves of up to Mach number ~4 the break up time is only a few microseconds and is thus comparable to the chemical induction period. Here the indications are that τ_{max} is not long enough for dodecane with droplet sizes considerably below 100 μm (Fig. 2b).

Although the pressure records were disturbed by gage ringing, the shock and reaction fronts could easily be discerned as previously described, and it is possible to obtain from their separation some idea of a total induction period. For example, a detonation of hexane was obtained with a charge weight of

0.055 kg at $\lambda = 0.96$, at point A which, as can be seen in Fig. 5, lies fairly close to the limit curve. The measured detonation velocity in this experiment was 1655 m/s^{-1}. For this case a value of the induction period of $\tau = 100$ μs is estimated. Other records for hexane detonations that are believed to be near the detonation limit have produced values of τ up to 180 μs.

One of the interesting features of two-phase detonations is their ability to propagate at velocities considerably below the C – J value. The deficit from the C – J value has been shown to increase with droplet size, and the evidence for this has come from experiments in which the droplets were large compared with those used in this study, for example, Dabora et al. (1969) and Nicholls et al. (1974). In this study evidence has been produced that for spherical detonations the velocity of propagation will be below the gas-phase *C-J* values even with the small droplets used here (Fig. 6). The effect of droplet size on the detonation velocity was also observed in the single experiment where the droplet size range was increased from 6-17 to 50-90 μm, and the measured velocity dropped from 1580 to about 1300 m/s. Such a large effect is surprising in this droplet size range, but can only be attributed to an increasing drop breakup time as a result of the larger sizes, leading to a longer induction period.

The results presented here indicate that the conclusions reached elsewhere regarding two-phase detonations obtained under confined conditions prevail in the spherical mode but are extended so as to make the criteria for initiation and propagation of the detonation more stringent. This is believed to be a result of the divergence and decay of the initiating blast wave and the energy released at the reaction front in the spherical mode, which supports the preceding shock wave. Thus, even with droplets of less than 50 μm diam. velocities well below the gas-phase C-J values are obtained for detonations well within the limit curve, while with the larger droplets, 50-90 μm, an even greater velocity deficit is observed.

The failure to detonate dodecane and decane suggests that for propagation of unconfined detonations certain quantities of fuel vapor may be required prior to ignition.

Further investigation will no longer be aided by using mixtures of fuels because the lighter molecules tend to be distilled and react by themselves. Instead, paraffin hydrocarbon fuels of volatilities intermediate between those of hexane and dodecane, such as heptane and octane, will be used. The results of these investigations will be published in a subsequent paper.

Conclusions

1) It has been shown that a fuel with a high volatility but in aerosol form, such as hexane, will detonate within limits very similar to those expected for gaseous fuels.

2) As with planar two-phase detonations, a velocity deficit compared with C-J values is observed for unconfined two-phase detonations, even with droplets < 50 μm of diam.

3) The inability to detonate dodecane and decane even with small droplet sizes has led to the conclusion that with initiator charges of up to 0.5 kg, a

minimum effective equivalence ratio necessary for propagation, is not being attained behind the initiating blast wave within τ_{max}.

4) Self-sustained detonations of n-paraffin hydrocarbon aerosols in the spherical mode may require a certain quantity of fuel vapor to be present prior to ignition.

References

Barnard J.A. (1978) Private communication.

Bowen J. R., Ragland K. W., Steffes F. J. and, Loflin, T. G. (1971). Heterogeneous detonations supported by fuel fogs or films. *13th Symposium (Intl) on Combustion,* pp. 1131-1139. The Combustion Institute, Pittsburgh, Pa.

Bull D. C., Elsworth J. E., Hooper G., and Quinn C. P. (1976) A study of spherical detonation in mixtures of methane and oxygen diluted by nitrogen, *J. Phys. D: Appl. Phys.* **9,** 1991-2000.

Bull D. C., Elsworth J. E., and Hooper G. (1979) Concentration limits to unconfined detonation of ethane/air. *Combustion and Flame* **35,** 27-40.

Dabora E. K. (1967) Production of monodisperse sprays. *Rev. of Scientific Instru.* **38,** 502-506.

Dabora E. K., Ragland K. W., and Nicholls J. A. (1966) A study of heterogeneous detonations. *Astronautica Acta* **12,** 9-16.

Dabora E. K., Ragland K. W., and Nicholls J. A. (1969) Drop-size effects in spray detonations. *12th Symposium (Intl.) on Combustion,* pp. 19-25. The Combustion Institute, Pittsburgh, Pa.

Kauffman C. W. and Nicholls J. A. (1971) Shock-wave ignition of liquid fuel drops. *AIAA J.* **9,** 880-885.

Lorenzetto G. E. and Lefebvre A. H. (1977) Measurements of drop size on a plain-jet airblast atomizer. *AIAA J.* **15,** 1006-1010.

Nicholls J. A., Sichel M., Fry R., and Glass D. R. (1974) Theoretical and experimental study of cylindrical shock and heterogeneous detonation waves. *Acta Astronautica* **1,** 385-404.

Petes J. (1968) Blast and fragmentation charasteristics. *Ann. Acad. Sci.* **152,** 283-316.

Ranger A. A. and Nicholls J. A. (1969) Aerodynamic shattering of liquid drops. *AIAA J.* **7,** 285-290.

Initiation of Unconfined Gas Detonations in Hydrocarbon-Air Mixtures by a Sympathetic Mechanism

D. C. Bull,* J. E. Elsworth,† and M. A. McLeod‡
Shell Research Ltd., Chester, England
and
D. Hughes§
University College of Wales, Penglais, Aberystwyth, Wales

Decaying blast waves from established spherical "donor" gas detonations have been shown only to traverse narrow air gaps and re-establish as detonations in "acceptor" volumes of gas mixtures. The reactive gases examined were well-mixed stoichiometric mixtures of either ethylene-air or propane-air and were prepared in cylindrical polythene bag/steel hoop structures on a common longitudinal axis. Gaseous detonation was initiated in the donor volumes by overdriving with high explosive charges and, where induced detonation in the acceptor section occurred, X-band microwave interferometry showed it to be self-sustaining. Measured detonation velocities were in agreement with calculated C-J values. The geometry of the fuel-air/air/fuel-air interfaces was defined by the metal hoops. Immediately prior to initiation of the high explosive charges, diaphragms comprising the two facing ends of the cylindrical volumes were removed rapidly by pneumatic piston action. Successive adjustments of the hoop separation enabled the critical air gap, just preventing detonation in the acceptor, to be established. Interpretation of blast wave decay measurements across the air gap indicates that, in the critical case, the shock entering the acceptor charge is of a strength consistent with that observed in other critical gas detonation initiation processes.

Presented at the 7th ICOGER, Göttingen, Federal Republic of Germany, Aug. 20-24, 1979.
Copyright © American Institute of Aeronautics and Astronautics, Inc., 1980. All rights reserved.
*Senior Scientist, Thornton Research Center.
†Scientist, Thornton Research Center.
‡Senior Technician, Thornton Research Center.
§Postgraduate Student, Department of Physics.

I. Introduction

THIS paper describes preliminary experiments to study factors influencing detonation propagation in a gas of heterogeneous composition. Decaying blast waves from established "donor" gas detonations were only able to traverse very small air gaps and re-establish gas detonations in "acceptor" volumes of identical gas. By successive adjustments, the critical air gaps, which just prevented reinitiation of gas detonations in the acceptor volumes, were established for stoichiometric mixtures for both ethylene-air and propane-air systems.

It has now been firmly established (e.g., Collins, 1974; Sichel, 1977; and Bull et al., 1979) that concentration limits to spherical detonation are dependent upon the strength of the initiation process. Thus, if a spherical gas detonation is once established in a mixture of composition a) but passes into a mixture of composition b) showing a lower detonation susceptibility then, dependent upon the overall wave energy, it may either continue to propagate or it may begin to uncouple into shock and reaction wave components and detonation will cease. In the extreme cases of a detonation wave encountering either pure fuel vapor or air, the transmitted shock will eventually decay to sonic velocity.

In the general case we may consider a heterogeneous (but single phase) fuel-air mixture to comprise a random series of gas pockets of any composition ranging from pure air to pure fuel vapor. We need to assess the ability or failure of a spherically divergent detonation wave to traverse successfully the various randomly distributed concentration gradients and interfaces. Because this general case is too difficult for meaningful practical study we have selected a simpler system which can be considered as a *sector* of an imaginary sphere consisting of concentric spherical shells of different gas concentration. With this arrangement 1) the inherent concentration randomness of the general case is reduced and 2) detonation in the donor can be end-initiated, a technique that has been shown by us previously (1976) as giving identical detonations to the spherical case. Finally, though our study relates to the behavior of a divergent wave in such a sector, for ease of construction and manipulation, the experimental containment geometry we have adopted is actually cylindrical.

The experiment we describe here assesses the ability of a blast wave, emerging from a donor gas detonation and crossing an air gap, to initiate detonation in a second, similar, acceptor gas mixture. The processes parallel a study of solid explosives behavior, known as "sympathetic" detonation. Unlike the latter case, however, we know of no previous study of sympathetic detonation involving gas mixtures.

We expected, from our previous studies of detonation initiation (1978), that chemical kinetic reactivity plays a major role in the propensity of gas mixtures to detonate via a sympathetic mechanism. Accordingly, we chose ethylene and propane as hydrocarbon gases having differing high temperature oxidation rates.

II. Experimental

Gas mixtures were prepared with air from an oil-free compressor supply mixed with either ethylene (>99.8%) or propane (>99%) gas taken from Air

INITIATION OF SYMPATHETIC GAS DETONATIONS

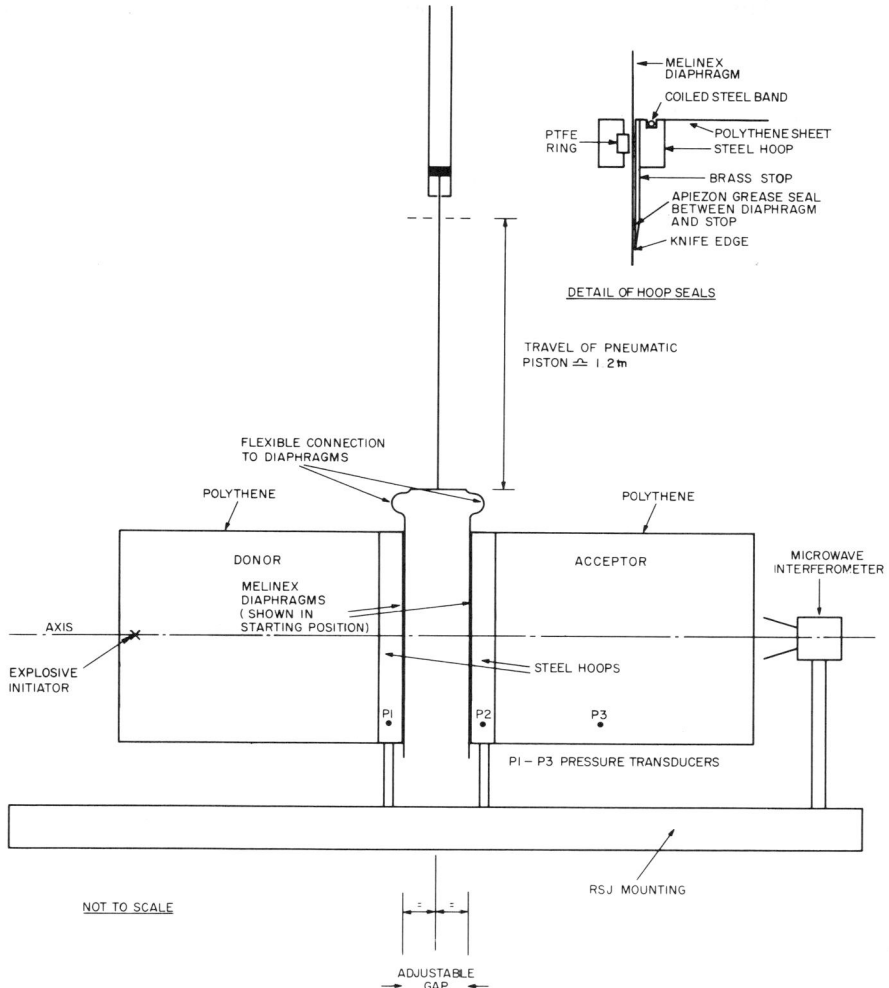

Fig. 1 Apparatus for "sympathetic" detonation experiments.

Products cylinders without further purification. The gas mixture was injected continuously into previously evacuated polythene/Melinex/steel hoop structures as shown in Fig. 1. When completely filled the bag structures assumed cylindrical forms of donor length either 1.35 m or 3 m, and acceptor length either 1.5 m or 2.1 m. The lengths of the donor sections were based on previous experience (Bull et al, 1978), but our original intention that donor and acceptor charge lengths should be equal was found to be impossible for the propane experiments because of space constraints. In all experiments the cylindrical diameter was 0.75 m, the polythene sheet was 55 μm thick and of specific mass 0.034 kg m^{-2}, and the Melinex was 70 μm thick. Mixture proportions and homogeneity were ensured by the filling system consisting of needle control valves, integrating (wet) gas meters, float (rate) indicators, and a mixer/flame trap assembly. As in previous experiments, fuel concentrations were accurate to ±0.4% of the hydrocarbon component. All gas mixtures were made to stoichiometric concentration.

The donor and acceptor volumes were on the same cylindrical axis, the "gap" geometry being defined by the steel hoops and Melinex diaphragms. Leakage of gas mixture was minimized by the use of coiled steel bands and PTFE seals used in conjunction with low vapor pressure Apiezon "M" grease. The hoop assembly incorporated circular "stops" made from a 7 mm brass sheet (tapered to a knife edge finish on the internal surfaces) to minimize spurious shock reflectance that might otherwise have occurred off the steel hoops.

Consistent initiation of gaseous detonation in the donor sections was achieved by using 0.054 kg and 0.083 kg plastic explosive (PE4) for ethylene and propane, respectively. The plastic charges were of roughly spherical form and were initiated by Nobel's No. 8 star instantaneous electric detonators. The performance of this explosive is discussed in detail elsewhere (Bull, McLeod, and Mizner, 1980), but is not unlike Tetryl in its effect. The explosive charge was inserted at the end of the donor volume remote from the air gap on the central axis.

Immediately prior to initiation of the explosive charge the Melinex diaphragms were removed by a pneumatic piston. The electric detonators were fired by the closure of a microswitch activated when the pneumatic piston had removed the Melinex diaphragms. The stroke of the pneumatic piston was 1.2 m, which enabled the piston to accelerate over 0.45 m *before* the diaphragms started moving. Typically, diaphragm removal took ≤ 0.2 s. Successive adjustments of the hoop separation distance allowed the critical air gap just causing detonation in the acceptor to be determined.

The velocity of the strongly ionized reaction front was monitored continuously with a microwave interferometer (10.687 GHz) equipped with a combined transmitting/receiving horn antenna having a nominal gain of 21 dB. In all experiments, time of arrival and static pressure measurements were made at representative locations with piezoelectric gages (Kistler 601A). All velocity and pressure signals were recorded on transient event recorders (Datalab 922).

Primary evidence of acceptor gas detonation was made by reference to the microwave interferogram. Secondary evidence was provided by the piezo gages, which indicated a larger signal and earlier arrival time for a gas detonation than with a deflagration or nonignition. The polythene fragments were very small when detonation had occurred, while with a deflagration the acceptor bag fragments were typically larger and fused together. Distinction between donor and acceptor fragments was achieved by prior spray painting of the donor and acceptor bags with different colors.

III. Results

Typical experimental records are shown in Fig. 2. Trace a) is for an ethylene-air mixture in which gaseous detonation was re-established in the acceptor

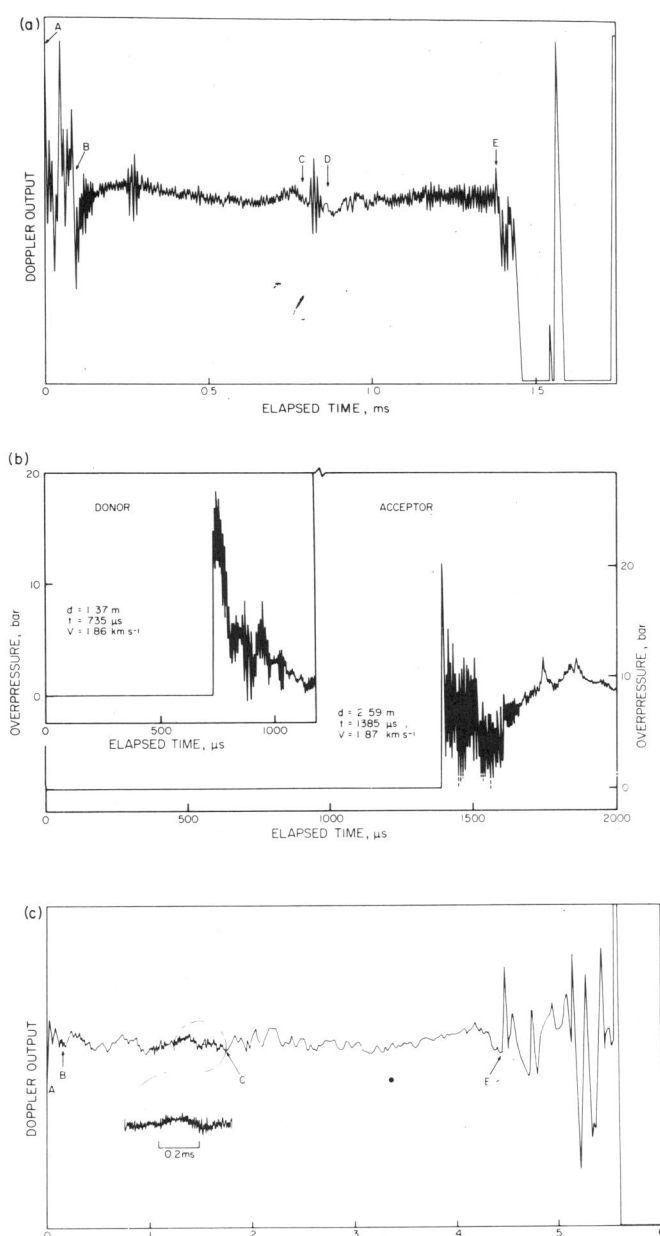

Fig. 2 Typical experimental records: a) microwave interferogram from an ethylene-air experiment showing sympathetic gas detonation established in acceptor, b) pressure recordings from locations in donor and acceptor corresponding to a), c) microwave interferogram from a propane-air experiment in which sympathetic detonation failed to re-establish across a 0.175-m air gap.

section. In the initial period AB there is a decaying disturbance associated with the explosive charge; period BC shows a frequency corresponding to gaseous detonation in the donor section. A discontinuity can be seen at C-D attributable to the air gap. Between D and E the steady Doppler frequency *indicates* re-establishment of gaseous detonation. At E the wave ruptures the polythene bag and thereafter enters the atmosphere.

Trace b) shows pressure records corresponding to Fig. 2a. The two traces show sudden rises in overpressure of about 18-20 bars, *confirming* that detonation occurred in both donor and acceptor sections. The average speeds deduced from times of arrival of the wave at the pressure transducers are 1.86 km/s^{-1} and 1.87 km/s^{-1} for the donor and donor-gap-acceptor, respectively. These values are only 2-3% higher than Chapman-Jouguet values calculated for water-saturated gases at the experimental temperature (283 K) and pressure (0.96 bar), from JANAF Thermochemical Data in the usual way (e.g., Eisen et al., 1960). The oscillations shown *after* arrival of the detonation wave are attributable to pressure transducer resonances.

Similarly, trace c) indicates gaseous detonation in the donor section which ceases at C. There is no evidence of any steady Doppler signal beyond C, indicating that gaseous detonation did not re-establish in the acceptor section. Large discontinuities are observed beyond E, indicative of bag rupture. The time of transit between C and E is 2.7 ms for the *decaying* wave to travel a nominal 2.35 m; the mean speed is therefore about 0.87 km/s^{-1}.

The effects of varying the air gap length between the donor and acceptor volumes for stoichiometric mixtures of air with ethylene and propane are shown in Figs. 3a and 3b, respectively. With ethylene mixtures, detonations are only re-established if the air gap is ≤ 0.150 m using a nominal donor radius of 1.35 m; detonation wave velocities were deduced from between 29 and 91 Doppler cycles, indicative of the sustained nature of the re-established detonations. Similarly for propane mixtures, detonations are re-established only with air gaps ≤ 0.125 m from a donor radius of 3.0 m; detonation wave velocities were based on between 47 and 98 Doppler cycles. For both ethylene and propane systems, when detonation failed to re-establish, the average velocity of the *decaying* wave in the acceptor volume is shown. These values were deduced from wave arrival times at the microwave horn after making allowances for time of flight across the donor plus gap.

Further experiments will be required to determine the variation of these gap lengths as a function of donor radius.

IV. Discussion

Although the gaseous detonation was induced in the donor sections by overdriving, the contribution of the initiator charge energy to the total energy per unit solid angle released in the wave had reached less than 0.5% (0.1%) for ethylene-air (propane-air) by the time the wave entered the air gap. In both donor and acceptor sections the velocities derived from the microwave interferometer were both mutually in good accord and also with calculated C-J values. In the acceptor sections where detonation was re-established, velocity was monitored over path lengths of up to 1.25 m (1.4 m), thus confirming that

INITIATION OF SYMPATHETIC GAS DETONATIONS

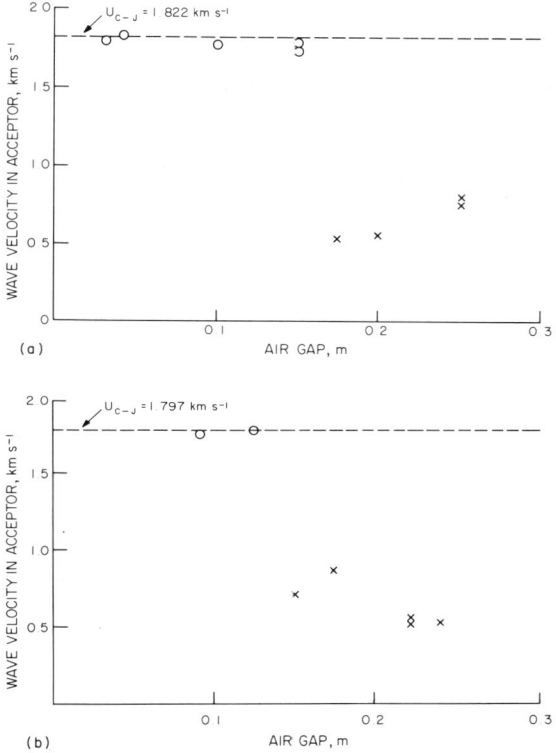

Fig. 3 Experimental wave velocity values in acceptor sections as a function of air gap length: a) stoichiometric ethylene-air, b) stoichiometric propane-air. In both cases ○ represents sustained velocity deduced from interferograms for re-established detonations and × represents the average values of decaying waves in the experiments in which sympathetic detonation failed to occur.

two of the basic detonation criteria were met, viz., 1) the self-sustaining nature and 2) the constant velocity of the reaction wave.

The removal of the adjacent end walls between donor and acceptor was necessary because prior experiments with the gas charges contained by 25 μm polythene sheet had shown that subsequent gas detonation did not occur in the acceptor sections even with the minimum practicable gap of 2-3 mm. This observation agrees with the findings of Nettleton (1976) and Bull (1978) that shock reflection losses are appreciable even at such very thin plastic interfaces Though the necessity of using removable diaphragms introduces doubts about the geometry of the actual gas/air/gas interfaces, this problem has been recently studied by Edwards (1979) using identical gas mixtures to ours. His results indicate, for the density gradients in question, that the turbulent mixing in the wake of the very thin Melinex diaphragms, and the buoyancy-driven mixing and diffusive mixing processes, are negligibly small during the period (≤ 0.2 s) the gas sections are open prior to initiation.

Our initial step in analyzing the interactions that occur when a *diverging* detonation wave encounters a concentration interface neglects the transverse wave structure and uses the ZND model approximation.

First, it is clear that when the detonation wave enters an air gap then chemical reaction must cease, leaving the uncoupled shock as a decaying blast wave. As with any shock encountering an acoustic interface, however, there must be normal shock splitting into transmitted and reflected components. In the present work, there follows a second splitting of the decaying shock at the acceptor (air/fuel-air) interface. When detonation is re-established in the acceptor this second transmitted component must be sufficiently strong to reinitiate a chemical reaction front which ultimately recouples with the now increasingly stronger shock to produce stable, sympathetic detonation. Alternatively, if the gap exceeds a critical length then the strength of the decaying shock is insufficient to stimulate sufficient chemical reaction in the time and distance available; thus detonation fails to be re-established.

Figure 4a is a distance-time history of an ethylene experiment in which gaseous detonation was re-established across an air gap of 0.15 m. Superimposed on the linear plots derived from microwave Doppler information are the experimental points P1, P2, P3, and H appropriate to wave arrival times. The experimental information available does not enable us to deduce the precise behavior of the wave in the region B-D, which is made up of the air gap plus the distance in which no microwave reflection was observed. An *average* wave speed of ~ 1.12 km/s^{-1} was measured from P1 to P2 (straddling the air gap) and an *average* speed of ~ 1.89 km/s^{-1} can be deduced for the region C-D. This latter speed indicates a rapid re-establishment of detonation soon after the wave re-enters reactive mixture at C.

We have applied Brinkley-Kirkwood (1947) theory to the decay of the blast wave in the air gap. This theory describes (on a one-dimensional basis) the shock wave decay caused by energy transfer to the fluid (in this case, air) when crossed by the shock front. To apply the theory it is necessary to make an initial assumption concerning the strength of the blast wave which arises immediately the detonation wave encounters the air boundary. The assumption we adopted was to choose a blast wave strength such that the conditions (T, ρ) which obtain behind it match those obtaining for the burned gas at the C-J plane in a detonation. This assumption accounts for the steep jump B→B′ in Fig. 4b which is the corresponding velocity-time wave history of the same ethylene experiment. At B (the first acoustic interface), with no further chemical heat release, the detonation wave uncouples and a *transmitted* shock wave of velocity ~ 1.26 km/s^{-1} emerges. This blast wave decays in the region B′ to C, when the second acoustic interface is reached. The average velocity of the decaying blast predicted is ~ 1.20 km s^{-1} from B to D. We have not as yet made any experimental observation to elucidate the processes occurring between C and D, but within this distance of 0.26 m and a time span of 137 μs, the wave has re-established stable detonation.

Similarly rapid jumps in velocity have been observed previously in other marginal detonation re-establishment contexts by Edwards and Morgan (1977) and Bull et al. (1978). In the former experiment, a detonation tube study of galloping planar waves, a velocity change from 1.16 km/s^{-1} to 2.25 km/s^{-1}

INITIATION OF SYMPATHETIC GAS DETONATIONS

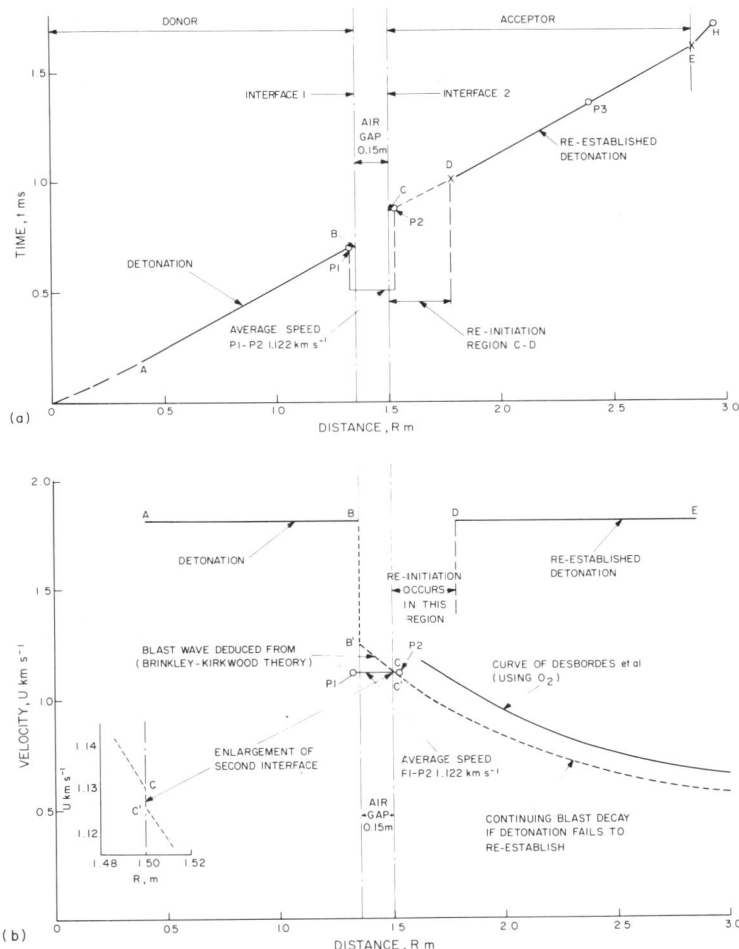

Fig. 4 Sympathetic ethylene-air detonation across 0.15 m air gap. a) distance-time history, time of arrival data are indicated by ○ and limits of Doppler velocity measurement by ×; b) velocity-time plot showing changes of blast wave velocity at the two interfaces.

over 0.13 m was measured. In the latter, a study of the marginal initiation of spherical detonation in ethylene-air, a wave recovery was observed from 0.78 km/s^{-1} to 1.816 km/s^{-1} over 0.33 m. Simple application of homogeneous auto ignition theory, using, e.g., shock tube derived global kinetic expressions, cannot satisfactorily account for such a rapid reignition and suggests that we are not justified in completely neglecting the transverse waves (although they are decaying) present in the original detonation front.

No previous studies are reported on the structural changes that must occur at a concentration interface, but it is clear that the frontal structure of a detonation wave on encountering a boundary between fuel-air and air must widen and eventually disappear as the transverse waves degenerate. Some record of this was observed in a supplementary experiment in which the decaying detonation wave, emerging to air from the donor volume of ethylene-air, was allowed to traverse a sharpened (polished stainless steel) soot plate. The characteristic cellular pattern of the detonation structure was recorded right from the sharpened front edge. At the edge itself the cell dimensions (length ~0.045 m, width ~0.03 m) were close to those recently observed within ethylene-air detonation (Bull, Elsworth, Metcalfe, and Shuff, 1980). The pattern thereafter widened and degenerated over a distance of some 0.13 m.

It is clear that, in the course of traversing the 0.15 m critical air gap we measured for the ethylene-air detonation, transverse shocks continue to play a role. Although we are not yet able fully to model this behavior, transverse shock effects clearly may account for the occurrence of reinitiation under conditions where it is not predicted by considering the plane shock alone.

Generalized Implications

In order to compare the results for stoichiometric mixtures of propane-air with ethylene-air and to consider the general case for other systems, we now discuss the specific interactions at and beyond each interface. At the first interface (fuel-air/air), the strength of the transmitted shock wave will depend on the relative acoustic impedance at the boundary. Thus, referring to Fig. 4b, the velocity drop B-B′ would be smaller for mixtures of higher sound speed—e.g., hydrogen-air, methane-air—but larger for those of lower sound speed—e.g. propane-air, butane-air. Additionally, for the denser gases the velocity drop would increase considerably for fuel-rich mixtures. The subsequent rate of blast wave decay in the gap (B′-C) will also depend on the degree of divergence of the wave (wave radius) at B.

At the second acoustic interface (air/fuel-air), again the instantaneous transmitted shock strength will depend on whether this is a slow/fast or a fast/slow interface. In the former case a weaker wave will be transmitted, in the latter a stronger wave. The ability of the transmitted shock wave to kindle detonation in the acceptor gas charge will then depend on its strength and the kinetic reactivity of the mixture. Thus, if a certain transmitted blast wave in the region C′-D just succeeds in kindling detonation in ethylene-air, we would expect that, all other things being equal, a stronger shock would be required to reinitiate detonation in the less reactive propane-air. The implications of this are 1) that an ethylene/air detonation of given wave radius could traverse the same gap as a propane-air detonation of larger radius, and 2) for detonation waves of the same radius, ethylene-air detonation will re-establish over a larger air gap than will propane-air. The experimental results in this study are wholly consistent with this reasoning.

The blast lines shown in Fig. 4b were calculated by the Brinkley-Kirkwood method in which, for the purposes of computation, the shock radii were nondimensionalized by reference to the radius of the divergent detonation wave at a point B, the first interface with air. It follows from the nature of the solution that, if we are concerned initially only with the wave velocity at C′, the fuel-air side of the second interface, we can generalize from our experimental results to predict the air-gap size at which the wave attains the same velocity as at C' for detonations of any radius. Figure 4b also shows the blast wave decay curve of Desbordes et al. (1978) obtained by measurement of blast arrival times in the air surrounding detonations induced in soap bubbles containing either stoichiometric ethylene-oxygen or acetylene-oxygen mixtures. The shape of the experimental blast decay curve is very similar to our predicted curve, but the higher level of the Desbordes curve is attributable to the use of reactive mixtures prepared with oxygen.

While we have limited the discussion to the experimental configuration we used, similar reasoning can be applied to the general case of:

detonable mixture [concentration 1]	nonreacting mixture (within the transit-time)	detonable mixture [concentration 2]

In this way the conditions preventing spherical wave propagation in a medium heterogeneous in one dimension may be delineated.

The results imply that unless stringent precautions are taken to ensure thorough mixing between fuel and oxidant, inhomogeneity will tend to arrest the propagation of a detonation wave. This suggests simple methods for deliberately arresting unconfined detonation waves.

The study is also germane to the appraisal of the hazard posed by the interaction between a detonation wave and the inhomogeneous "cloud" of fuel-air vapor formed upon leakage or spillage of volatile combustibles.

VI. Conclusions

We conclude the following from our results:

1) Using the experimental procedures described, stoichiometric mixtures of both ethylene-air and propane-air exhibit "sympathetic" gas detonation only across small air gaps.

2) Conditions critical to sympathetic gas detonation accord with predictions of a simple theory taking account of the net shock decay occurring across two acoustic interfaces bounded by an air gap. Only if the strength of the shock upon entering the acceptor exceeds a threshold value for the particular gas mixture does sympathetic detonation occur.

3) Reinitiation of detonation is not satisfactorily explained by planar blast wave decay and autoignition considerations. There are indications that transverse components of the decaying shock contribute to the reignition process.

4) The passage of unconfined detonation will be arrested if the wave encounters a region comprising air, fuel vapor, or a nondetonable mixture, provided that it extends beyond a critical dimension.

Acknowledgments

The authors thank J. Swithenbank and staff of the Sheffield University Outstation, Harpur Hill, Buxton for the site, workshop, and other facilities without which this work would have been impossible. The authors also thank L. R. Cairnie of the Mathematics Division, Thornton Research Center for assistance with blast wave computations.

References

Brinkley S. R. and Kirkwood J. G. (1947) Theory of the propagation of shock waves. *Phys. Rev.* **71**, 606-611.

Bull D. C. (1978) Contribution to discussion. *17th Intl. Combustion Symposium*, Leeds, England.

Bull D. C., Elsworth J. E., and Hooper G. (1978) Initiation of spherical detonation in hydrocarbon-air mixtures. *Acta Astronautica* **5**, 997-1008.

Bull, D. C., Elsworth J. E., and Hooper G. (1979) Concentration limits to unconfined detonation of ethane-air. *Combustion and Flame* **35**, 27-40.

Bull D. C., Elsworth J. E., Hooper G. and Quinn C. P. (1976) A study of spherical detonation in mixtures of methane and oxygen diluted by nitrogen. *J. Phys. D: Appl. Phys.* **9**, 1991-2000.

Bull D. C., Elsworth, J. E., Metcalfe E., and Shuff P. J. (1980) Detonation cell structures in fuel-air mixtures. (To be submitted to *Combustion and Flame*.)

Bull D. C., McLeod M. A., and Mizner G. A. (1979) Detonation of unconfined fuel aerosols. Paper presented at *Seventh Intl. Colloquium on Gas Dynamics of Explosions and Reactive Systems*.

Collins P. M. (1974) Detonation initiation in unconfined fuel-air mixtures. *Acta Astronautica* **1**, 259-266.

Desbordes D., Manson N., and Brossard J. (1978) Explosion dans l'air de charges sphériques non confinées de mélanges réactifs gazeux. *Acta Astronautica* **5**, 1009-1026.

Edwards D. H. (1979) Private communication.

Edwards D. H. and Morgan J. M. (1977) Instabilities in detonation waves near the limit of propagation. *J. Phys. D: Appl. Phys.* **10**, 2377-2387.

Eisen C. L., Gross R. A., and Rivlin T. J. (1960) Theoretical calculations in gaseous detonation. *Combustion and Flame* **4**, 137-147.

Nettleton M. A. (1976) *J. Occup. Accidents* **I**, 3.

Sichel M. (1977) A simple analysis of the blast initiation of detonations. *Acta Astronautica* **4**, 409-424.

Detection Method for the Deflagation to Detonation Transition in Gaseous Explosive Mixtures

C. Brochet* and M. Sayous†
Université de Poitiers, Poitiers, France

A technique to study deflagration to detonation transition for a range of experimental conditions is presented. Because of the large difference between the electrical properties of deflagrating and detonating gases, the continuous measurement of the electrical conductivity during the propagation of the combustion wave permits one to detect the onset of the detonation. The transition time (the interval between ignition and detonation buildup) and the transition distance are measured and deduced in various mixtures of H_2 or hydrocarbons with oxygen and nitrogen contained in a long tube. Under constant and accurate experimental conditions (source of ignition, mixture, and tube wall roughness), transition time has been found to be fairly reproducible and dependent on mixture parameters (e.g., pressure and composition), which may give a measure of the relative ease in which the explosive gaseous mixture can be detonated.

Introduction

VAPOR cloud explosion following the release of flammable gases or liquids in the air as well as the propagation of flames in large, confined explosive mixtures (i.e., flammable gases in galleries, dust in grain elevators) have received considerable attention in recent years.

From a weak ignition, combustion takes place first as a deflagration wave, which may accelerate to generate a detonation. This problem has stimulated a renewal of interest in the so-called initiation problem and in the deflagration to detonation transition (DDT) studies. The need for new experimental data in this field has led to large-scale experiments.

To take the best advantage of these new investigations, further improvement in techniques or diagnostics are needed (especially to characterize the transition to detonation phenomenon). After the early streak photograph and ionization probe methods, more refined experiments such as stroboscopic laser Schlieren cinematography (Urtiew and Oppenheim, 1966) or Doppler shift of microwaves reflected by the detonation front (Edwards et al., 1970, Brossard et

---

Presented at the 7th ICOGER, Göttingen, Federal Republic of Germany, Aug. 20-24, 1979.
Copyright © American Institute of Aeronautics and Astronautics, Inc., 1980. All rights reserved.
*Maitre de Recherche, Laboratoire d'Energétique et de Détonique.
†Ingenieur, Laboratoire d'Energétique et de Détonique.

al., 1974) have been successfully applied to study the DDT or the wave velocity over an appreciable propagation distance in gaseous mixtures for various geometries. However, those techniques were very often not suitable or too sophisticated for use in large-scale experiments.

The purpose of this paper is to present preliminary results of a DDT detection method which may be carried out for wide, experimental conditions. That detection method is based on the electrical properties of product gases, because of the large difference in electrical conductivity between the products of deflagration and those of detonation, deflagration or detonation can be discriminated.

The method has been applied to determine the transition time for DDT in a long tube containing H_2 or hydrocarbon-O_2-N_2 mixtures ignited by a weak spark. The influence of the following parameters has been investigated—initial pressure and the N_2 concentration in the stoichiometric mixtures.

Ionization in Flame and Detonation

The electrical properties of a flame have been studied for a long time. A useful review of this subject is contained in the monograph of Lawton and Weinberg (1969).

Studying the ionization of premixed flames has been the purpose of many studies. One of the important results deals with the presence of chemi-ionization occurring within the flame thickness and giving ion densities that are far in excess of thermal equilibrium (Calcote, 1957). Chemi-ionization processes in gaseous detonations were observed by Kelly et al. (1965) in hydrocarbon-oxygen-nitrogen mixtures. The electron density in the reaction zone at low initial pressure ($\simeq$ 50 Torr) was found to be 10^{18} e/m^3, which is the same order of magnitude for hydrocarbon-air or hydrocarbon-oxygen flames (10^{15} or 10^{18} e/m^3) (Lawton and Weinberg, 1969).

Green and Sugden (1962) have shown that in the reaction zone of a hydrogen flame, where the gases have been highly purified, there is little ionization present. They conclude that ionization in such systems arises entirely from hydrocarbon impurities. This effect of the small amount of

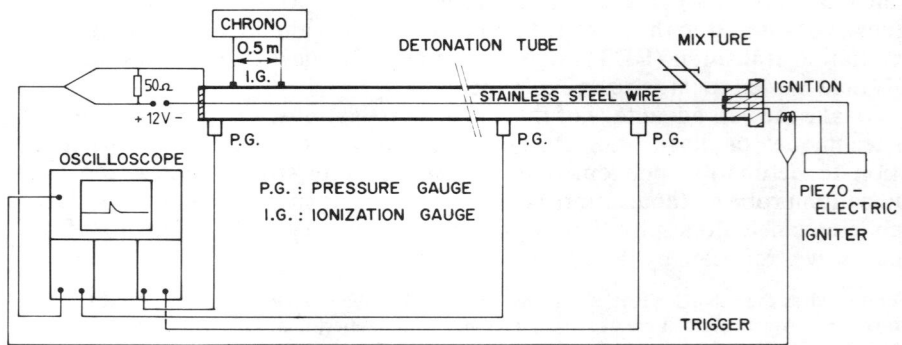

Fig. 1 Experimental setup.

hydrocarbon on the chemi-ionization of the reaction zone of hydrogen-oxygen has been confirmed experimentally in the case of a detonation (Brochet, 1975).

For pure hydrogen-oxygen-nitrogen mixtures, it seems that chemi-ionization processes are insignificant and the ionization would be essentially thermal. The degree of ionization in flame and in detonation gases may then be calculated from the Saha equation. The only component whose ion should appear in appreciable concentrations is NO because of its low ionization potential.

From the degree of ionization α, the electrical conductivity σ for weakly ionized gases at temperature T may be calculated (Chapman and Cowling, 1963).

$$\sigma = K\alpha Q^{-1} T^{-1/2}$$

where Q is the electron-atom collisional cross section and K a constant. Such calculations (Veyssière, 1968; Roux, 1965) show that α (and thus σ) are two orders of magnitude higher in detonation than for a flame propagating in the same mixture. Experimental results with propane-oxygen and propane-air mixtures are in qualitative agreement with this prediction (Veyssiere and Manson, 1967; Roux, 1965).

Due to the small thickness of the reaction zone, chemi-ionization phenomena in hydrocarbon-oxygen gases may be masked by thermal ionization, which is expected to make the most important contribution to the electrical conductivity measured behind the deflagration or detonation fronts.

The principle of the method proposed to study DDT is based on the fact that for given initial conditions of a mixture, the temperature of the detonation gases is significantly higher than that of the burned product behind a deflagration wave. As thermally ionized products with a low ionization potential such as NO (or an alkali metal acting as a "seed") are highly sensitive to the temperature (Chinitz et al., 1959), the electrical conductivity of the detonation products should be substantially higher than that of the deflagration products.

Experimental Apparatus

The explosive mixture (H_2 or hydrocarbon-O_2-N_2, higher than 99.9% purity) was contained at room temperature (280-295 K) and at an initial pressure of 0.1-1.5 bar in a conducting cylindrical tube (stainless steel, 52 mm in diameter, up to 7.3 m long) provided with an axial electrode (1 mm stainless steel wire), see Fig. 1. Some experiments were conducted in the tube with a carefully constant cross section and a smooth wall surface especially close to the ignition.

The tube and the electrode are connected to a power supply (dc $E = 12$ V) in such a manner that when the flame (ignited at the bottom) propagates in the tube, a discharge goes through the ionized reaction zone and the following burned gases. The total current I through a load resistance R_0 is recorded by an oscilloscope.

The properties of the total impedance Z of the conducting combustion products behind a detonation wave were studied previously (Tanaka et al., 1971; Brochet and Veyssiere, 1974). The total current I is given by $I =$

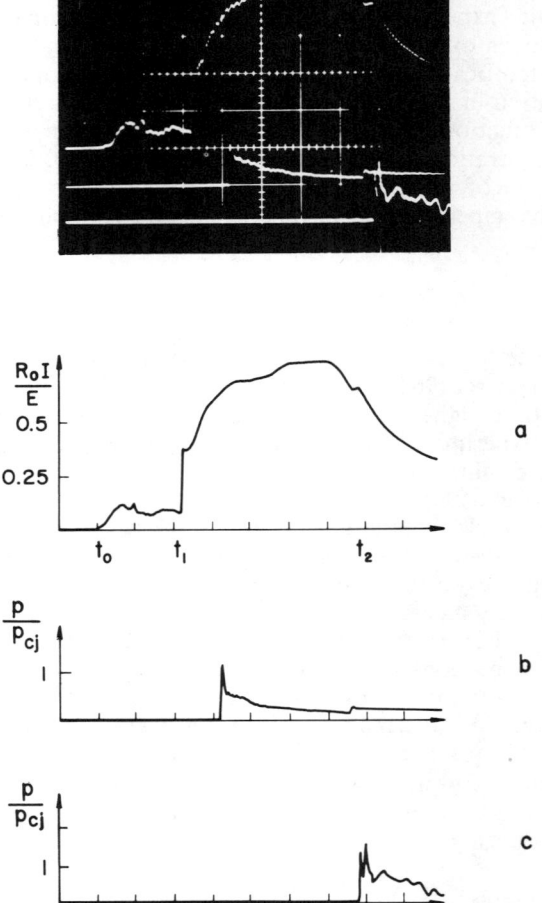

Fig. 2 Typical oscilloscope traces: a) electrical signal; b) pressure gage signal at $X = 2.2$ m; c) pressure gage signal at $X = 7.3$ m ($2H_2 + O_2 + 0.04\ N_2$), initial pressure $= 0.8$ bar, Chapman-Jouguet pressure $p_{C-J} = 14.8$ bar, time scale 0.5 ms/division).

$E\ (R_0 + Z)^{-1}$, so that Z behaves as a resistance which depends on σ and the geometric parameters. It has been observed that per unit length of the flame, Z is always greater behind a deflagration (Z_{def}) than behind a detonation (Z_{det}). R_0 is chosen such that $Z_{det} < R_0 \ll Z_{def}$.

Pressure gages and ionization probes are placed along the tube to observe the pressure profiles in the deflagration or detonation regime and also to measure the detonation velocity.

The ignition of the mixture was produced by a weak spark generated by a piezoelectric device which released an electrical energy of several tens of

DETECTION METHOD FOR DDT IN GASEOUS MIXTURES 77

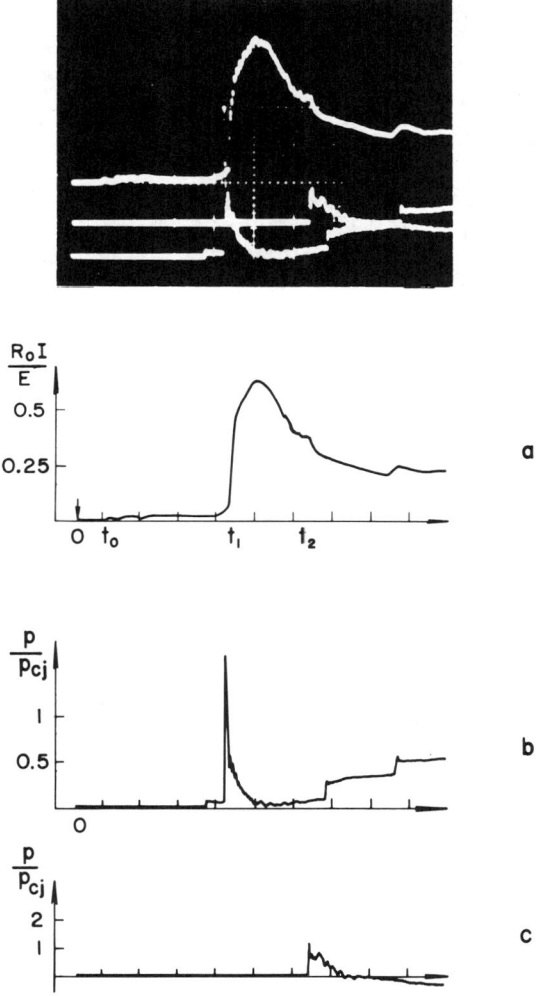

Fig. 3 Typical oscilloscope traces: a) electrical signal; b) pressure gage signal at $X = 2.2$ m; c) pressure gage signal at $X = 7.3$ m ($2H_2 + O_2 + 0.04N_2$, initial pressure $= 0.4$ bar, Chapman-Jouguet pressure $p_{C-J} = 7.3$ bar, time scale 1 ms/division).

millijoules during approximately 1 µs. Simultaneously, the four-channel oscilloscope was triggered to record, from the ignition time, the total current I as well as two or three pressure gage signals.

Results

The oscilloscope records displayed in Figs. 2-4 are typical of the results observed in stoichiometric H_2-O_2 mixtures.

The probe voltage ($R_0 I/E$) vs time (trace a) exhibits a steep rise after a delay

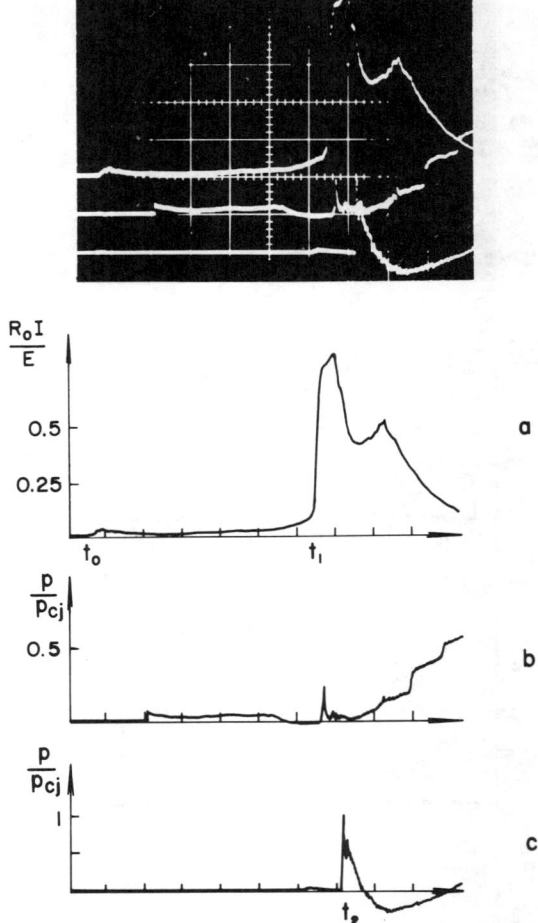

Fig. 4 Typical oscilloscope traces: a) electrical signal; b) pressure gage signal at $X=2.2$ m; c) pressure gage signal at $X=7.3$ m ($2H_2 + O_2 + 0.04N_2$, initial pressure = 0.3 bar, Chapman-Jouguet pressure $p_{C-J} = 5.4$ bar, time scale 2 ms/division).

t_1. At time t_1 the actual transition to detonation occurs and the amplitude of this signal increases toward values of the order of one.

During this delay t_1, the signal exhibits generally a zero or low-amplitude signal, up to t_0, followed by a fluctuating amplitude signal during $t_1 - t_0$. The onset of the detonation wave associated with the steep rise of the electrical signal, was proved by pressure gages which recorded various profiles according to the location of the DDT in the tube. The traces b of the pressure profile show, respectively:

1) A fully developed detonation starting between the bottom of the tube and the pressure gage (Fig. 2b).

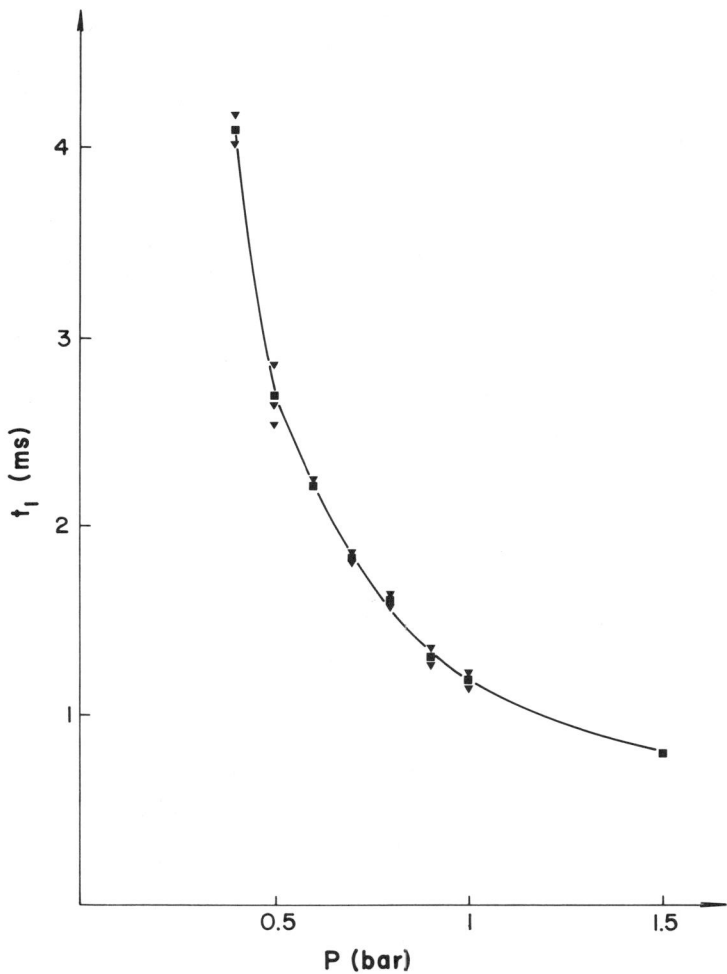

Fig. 5 Transition time in terms of initial pressure for constant experimental condition ($2H_2 + O_2 + 0.04N_2$, $T = 290$ K; ▼ = experimental results).

2) The onset of the detonation close to the pressure gage (Fig. 3b);
3) The onset of the detonation beyond the position of the gage (Fig. 4b).

The reflection of the detonation is indicated at time t_2 by a second pressure gage located at the closed end of the tube (trace c). From the records, it is simple to deduce the transition time t_1 and the propagation time of the detonation $(t_2 - t_1)$.

Figure 5 shows the variation of t_1 for the stoichiometric mixture $2 H_2 + O_2 + 0.4 N_2$ in terms of the initial pressure. These results, which were obtained with constant experimental conditions (samples taken from a mixture prepared and stored in a large tank, and constant tube length and wall

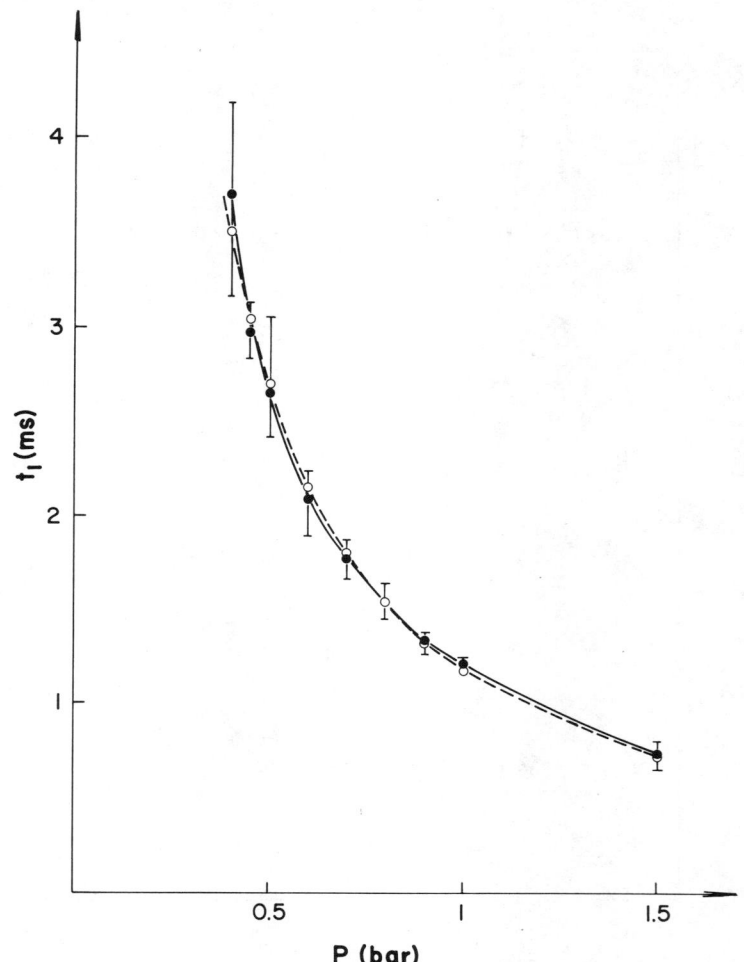

Fig. 6 Transition time in terms of initial pressure for several mixtures ($2H_2$-O_2-$0.04N_2$, $T = 280$-293 K; Φ = mean value and deviation; $\circ = t_I = 1.18(P)^{-1.19}$).

roughness), show good reproducibility. For several series of experiments carried out with more than one mixture and slightly changing initial conditions (temperature and position of the gages), the deviation of the results increases, especially for the low initial pressure of the mixture (Fig. 6).

The results shown in Fig. 6 illustrate the sensitivity of the parameters t_I to the experimental conditions that must be controlled accurately in order to get reproducible data.

The DDT distance X_I may be deduced from our results with either: 1) the average velocity measured for the distance of propagation of the detonation (using the data given by pressure gages b and c), i.e., usually 5.1 m, or 2) the

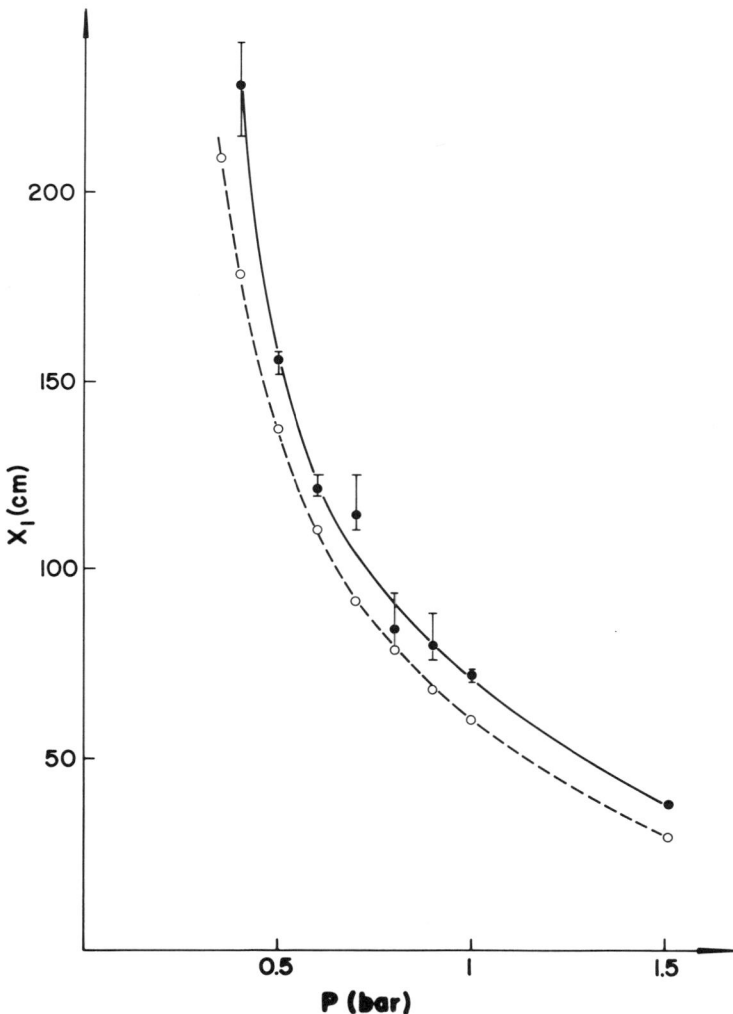

Fig. 7 Transition distance in $2H_2$-O_2-$0.04N_2$ ($\circ = X_I$ calculated using average detonation velocity on 5.1 m; $\bullet = X_I$ calculated using stable detonation velocity).

stable detonation velocity measured on the last part of the tube by the ionization probes.

The discrepancy between the two sets of results (Fig. 7) is due to the fact that the average detonation velocity is slightly greater than the stable velocity, the detonation regime always being overdriven just after the onset of detonation.

Nitrogen dilution of the H_2-O_2 mixture has a strong influence on the DDT parameters. This is evidenced in Fig. 8 which shows the effect of N_2 content on X_I. For lower initial pressure (<0.5 bar) and higher dilutions, X_I is expected to increase up to values close to the length of the tube (maximum, 7.3 m). With

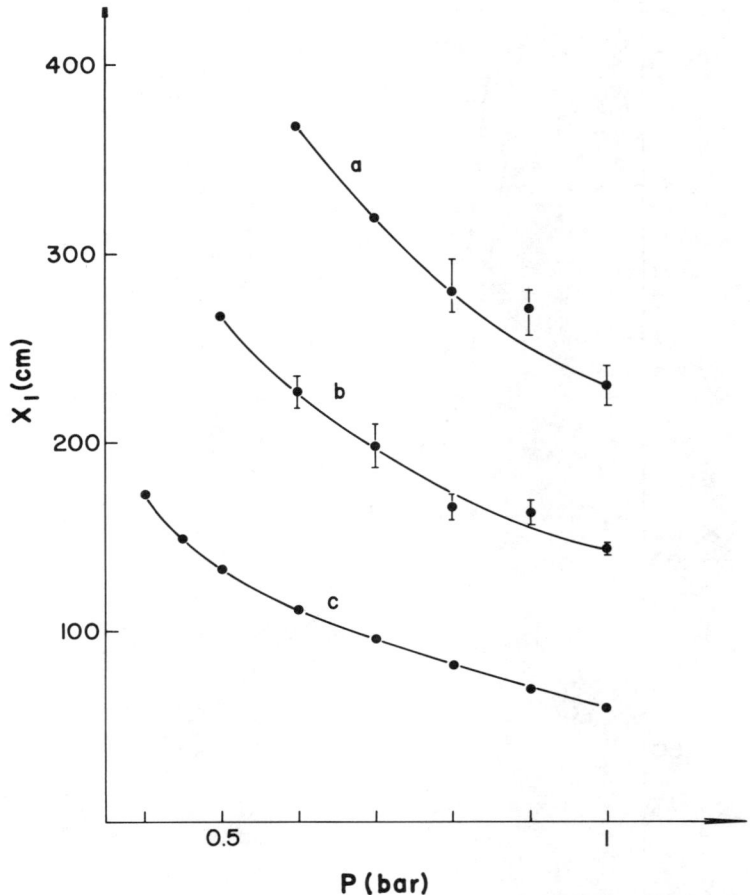

Fig. 8 Influence of N_2 content on transition distance ($2H_2 + O_2 + ZN_2$ mixture, curve a, $Z=2$; b, $Z=1$; c, $Z=0.04$).

these conditions, detonation or "explosion" is observed close to the tube end, but then the significance of X_t is doubtful and it is no longer correlated with the mixture parameters. Thus the transition distance is confirmed to be an irreproducible parameter as soon as it becomes comparable to the length of the tube. As shown on Fig. 4 (trace c), the pressure gage located at the closed end of the tube indicates a significant increase of the pressure before the arrival of the detonation front. For long transition distances, the mixture far ahead of the reaction zone is always disturbed by the propagation of shock waves generated during the long process of flame acceleration.

When the initial pressure of the H_2-O_2 mixture was reduced to the limit of propagation ($\simeq 0.1$ bar), it was observed that the transition time t_t increased drastically, as is shown in Fig. 9.

For low initial pressure (<0.3 bar), the probe signal no longer exhibits the

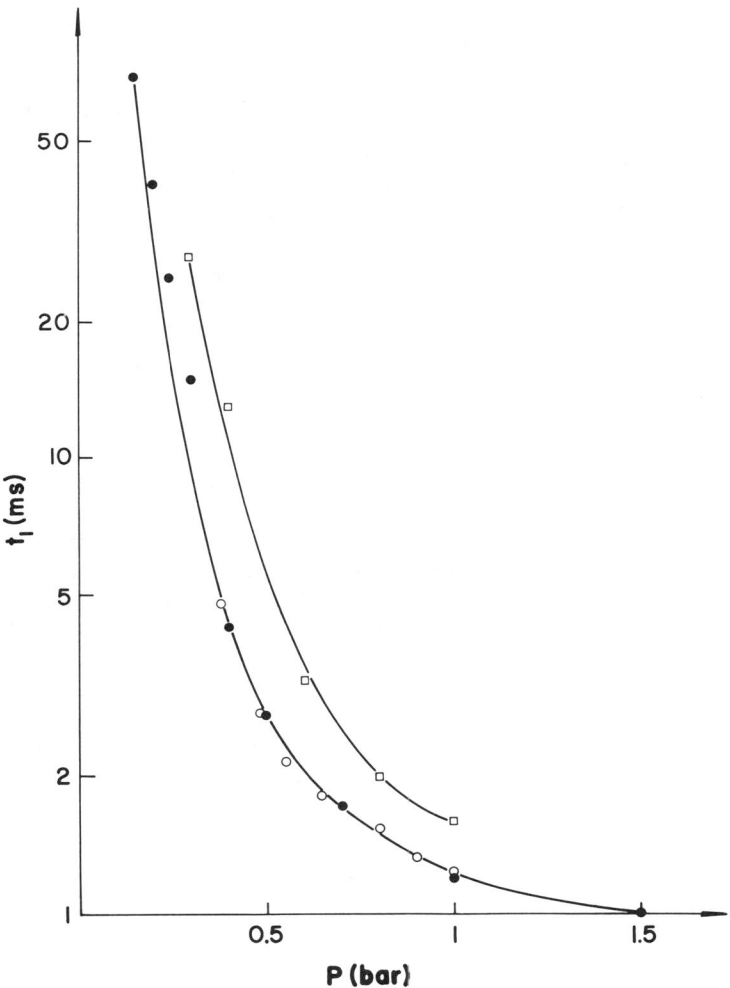

Fig. 9 Influence of initial pressure for several experimental conditions on transition time ($2H_2 + O_2 + 0.04 N_2$; ● = tube length 7.3 m, ○ = tube length 3.4 m, □ = smooth wall surface).

same profile. During t_1, only a low-voltage amplitude is observed. From these observations we may propose the following interpretation of the shape of the signal:

1) During the delay t_0, the low-amplitude signal is associated with the "laminar" flame propagation.

2) During $t_1 - t_0$, the wrinkling of the front surface, the onset of turbulent front structure, and the following turbulent flame propagation are responsible for the fluctuating amplitude.

These conclusions may be drawn from results presented by Oppenheim and Soloukhin (1973) for similar experimental conditions. The duration of the first stage has been found to be very sensitive to the boundary conditions (e.g., wall roughness or change in the cross-section area of the tube). To estimate the effect of these conditions, some experiments were performed in the same tube with a smooth wall and accurate constant cross-section area conditions.

With the smooth wall, the time t_0 was multiplied by a factor of 2 or 3 compared with previous results. Consequently, the transition time t_1 increased significantly. These data are given for the smooth wall conditions and for several mixture compositions in Table 1. The transition time t_1 given for two series of experiments are definitely apparatus-dependent as far as boundary conditions on the wall are concerned (Fig. 9). However, t_1 appears to be insensitive to the tube length (Fig. 9) and more reproducible in given experimental conditions than the transition distance. The data listed on Table 1 give a measure of the relative detonation hazards in several fuel-oxygen-nitrogen mixtures.

Discussion and Conclusion

The experimental results given here to illustrate the possibility of the technique are in good agreement with those obtained for H_2-O_2 mixtures with similar experimental conditions. Bollinger and coworkers (1961) found that transition distances varied 50-100 cm at 1 bar initial pressure in tubes of variable diameters. Moreover, the addition of N_2 to H_2-O_2 mixtures increased the transition distance in the same range of values as observed in the present study.

Oppenheim's velocity V_{opp}, the ratio of transition distance to transition time, was found to be (in the range of 10% accuracy) independent of the pressure (except for the lowest initial pressure where V_{opp} decreases drastically). A similar result was observed by Pawell et al. (1969) who found

Table 1 Transition parameters for several mixtures composition

Mixture composition	p, bar	t_1, ms	X_1, m	V_{opp}, m/s
$C_2H_2 + 2.5O_2$	1	0.2	0.10	500
	0.6	0.3	0.16	533
	0.4	0.5	0.24	480
	0.15	1	0.30	300
$C_2H_4 + 3O_2$	1	0.6	0.19	316
	0.6	1	0.30	300
	0.3	1.9	0.50	263
$2H_2 + O_2 + 0.04 N_2$	1	1.6	0.85	531
	0.8	2	1.10	550
	0.6	3.2	1.40	437
	0.4	13	~3	230
	0.3	28	~4	143
$C_2H_4 + 3O_2 + 3N_2$	1	4	1.10	275
	0.6	5	1.25	250
$C_2H_4 + 3O_2 + 6N_2$	1	11	2.60	236
	0.6	14	3.80	271

for $2H_2$-O_2 mixtures, $V_{opp} \simeq 800$ m/s, which is higher than our results of 500 m/s. This velocity difference may be considered an indication of the effect of the boundary conditions and/or the source of ignition (a pilot flame was used by Pawell et al.) in the acceleration process during the initial stage of the development of the flame.

The transition parameters measured carefully in several mixtures depend strongly on the mixtures as evidenced by some preliminary results gathered in Table 1 (for comparison with the H_2-O_2 mixture). The sensitivity of stoichiometric acetylene-oxygen mixture is particularly evident, and the relative ease in which fuel-oxygen-nitrogen mixtures can be detonated may be compared. These first results are qualitatively in accord with the classification given by Matsui and Lee (1979).

In addition to the data obtained by the techniques used on transition parameters, one may emphasize that electrical probe observation coupled with pressure gage records can supply useful information on the transition phenomenon from the ignition to the DDT. The shape of the signal before the onset of the detonation is interpreted as due to the turbulent flame propagation, the effect of which is playing an important part in the deflagration to detonation transition in gaseous mixtures ignited by a weak source.

Acknowledgment

The authors are very grateful to Y. Sarrazin for his assistance in carrying out the experiments.

References

Bollinger L.E., Fong M.C., and Edse R. (1961) Experimental measurements and theoretical analysis of detonation induction distances. *ARS J.* 31, 588-595.

Brochet C. (1975) Sur la conductivité d'origine chimique et thermique des détonations dans les mélanges gazeux. *2nd European Symposium on Combustion,* Orléans-France, 460-466.

Brochet C. and Veyssière M. (1974) Mesures électriques des propriétés gazodynamiques des détonations. *Acta Astronautica* 1, 267-285.

Brossard J., Desbordes D., and Manson N. (1974) Célérité de la détonation sphérique divergente en fonction du rayon. *Acta Astonautica* 1, 873-884.

Calcote H.F. (1957) Mechanisms for the formation of ions in flames. *Combustion and Flame* 1, 385-403.

Chapman S. and Cowling T.G. (1963) *Mathematical Theory of Nonuniform Gases.* MIT Press, Cambridge, Mass.

Chinitz W.E., Eisen C.L., and Gross R.A. (1959) Aerothermodynamic and electrical properties of some gas mixtures to Mach 20. *ARS J.* 29, 573-579.

Edwards D.H., Hooper G., and Meddins R.J. (1970) Microwave velocity measurements of marginal detonation waves. *J. Phys. D.: Appl. Phys.* 3, 1130-1133.

Green J.A. and Sugden T.M. (1962) Some observations on the mechanism of ionization in flames containing hydrocarbons. *9th Symposium (Intl) on Combustion*, pp. 607-621. Academic Press, New York.

Kelly J.R., Toong T.Y., and Tung C.C. (1965) Chemi-ionization in gaseous detonations. *26th AGARD Propulsion and Energetics Panel Meeting on Fundamental Studies of Ions and Plasma,* Pisa, Italy, pp. 139-159.

Lawton J. and Weinberg F.J. (1969) *Electrical Aspects of Combustion.* Clarendon Press, Oxford, England.

Matsui M. and Lee J.H. (1978) On the measure of the relative detonation hazards of gaseous fuel-oxygen and air mixtures. *17th Symposium (Intl.) on Combustion,* pp. 1269-1280. The Combustion Institute, Pittsburg, Pa.

Oppenheim A.K. and Soloukhin R.I. (1973) Experiments in gasdynamics of explosions. *Annual Review of Fluid Mechanics* **5**, 31-58.

Pawell D., Vasatko H., and Wagner H.G. (1969) Observations on the initiation of detonation. *Acta Astronautica* **14**, 509-510.

Roux J. (1965) Mesure de la conductivité électrique des produits de combustion d'un mélange propane-air. Thèse 3e cycle, Poitiers, France.

Tanaka K., Brochet C., and Veyssière M. (1971) Conductivité électrique et vibration dans les gaz en écoulement produit par une détonation. *J. de Physique* **32** (C56), 22-23.

Urtiew P.A. and Oppenheim A.K. (1966) Experimental observation of the transition to detonation in an explosive gas. *Proc. Roy. Soc.* (London) **A295**, 13-28.

Veyssière M. and Manson N. (1967) Conductivité électrique des produits de détonation du mélange propane-oxygène-azote. *C. R. Acad. Sci.* (Paris) **264B**, 199-202.

Veyssière M. (1968) Calcul des caractéristiques thermoélectriques des produits de détonation des mélanges propane, oxygène, azote. *Revue Générale de Thermique* **81**, 863-869.

Influence of the Nature of Confinement on Gaseous Detonation

T.V. Bazhenova,* V.P. Fokeev,† and Yu. Lobastov‡
Academy of Sciences, Moscow, U.S.S.R.
and
J. Brossard,§ T. Bonnet,¶ B. Brion,¶ and N. Charpentier**
Université d'Orléans, Bourges, France

The effect of the nature of confinement material on the propagation parameters of the detonation wave in propane-oxygen mixtures at various initial pressures and dilutions was investigated. The experiments were conducted on a constant rectangular cross-sectional tube provided with a test section (1.2 m in length) in which the internal removable walls were aluminum or commercial polymers such as polyethylene, plexiglass, polypropylene, polyvinylchloride, and polyamide. The measured parameters were the detonation front velocity and pressure, the pressure profiles, the temperature and the electrical conductivity of the burned gases near the wall, the heat transfer to the wall, and the boundary-layer thickness. The measured velocity was slightly modified by the nature of the confinement. The pressure just behind the detonation front and the pressure profiles were disturbed when polymers were substituted for aluminum; the higher the thermal product ($k\rho c$) of the material, the higher the gradients in the rarefaction zone and the smaller the pressure of the detonation front. The thin-film temperature variation and the heat flux were very sensitive to the nature of the wall. They increased when polymers were substituted for aluminum and when initial pressure increased. When the thermal product of each material is considered instead of the Pyrex one, however, the relative position of the heat-transfer curves becomes coherent with the pressure profile curves. From the measurements of the electrical conductivity by a technique with two electromagnetic gages, the boundary-layer thickness was deduced. This boundary layer was not perceptible with aluminum but was clearly developed for polymers and the amplitude was confirmed by strioscopic visualization. The chemical reactivity of the material with the lean detonating mixtures was perceptible, especially for confinement by polyethylene or plexiglass.

Presented at the 7th ICOGER, Göttingen, Federal Republic of Germany, Aug. 20-24, 1979. Copyright © American Institute of Aeronautics and Astronautics, 1981. All rights reserved.
*Professor, Department of Mechanics, Institute of High Temperatures.
†Candidate in Physical Sciences, Institute of High Temperatures.
‡Candidate in Technological Sciences, Institute of High Temperatures.
§Professor, Department of Mechanical Engineering.
¶Research Scientist, Laboratoire de Recherche Universitaire.
**Assistant, Department of Mechanical Engineering.

Nomenclature

C	= heat capacity
C_f	= friction coefficient
D	= detonation front velocity
K	= thermal conductivity
p	= pressure
q	= rate of heat transfer
T	= temperature
t	= time
X	= mole number
x	= abscissa
$\Delta p = p - p_f$	= overpressure
δ	= boundary-layer thickness
ρ	= specific mass
σ	= electrical conductivity

Subscripts

b	= values of parameters on detonation front
f	= values of parameters in front of wave
S	= values on the surface of confinement material

Material Symbols

Al	= aluminum
PA	= polyamide
PE	= polyethylene
PMMA	= polymethylmetacrylate
PP	= polypropylene
PVC	= polyvinylchloride

Introduction

IN numerous gaseous detonable systems, the measured parameters of the detonation front are different from the values calculated by the simple ideal detonation-wave theory (Brochet et al., 1970). In the case of plane detonation confined in rigid tubes, numerous experimental investigations show the influences of the charge diameter (Manson et al., 1963) of the cross-sectional geometry (Boisleve, 1970), and of the nature and the roughness of the metallic wall (Renault, 1972; Brossard and Charpentier, 1976) on the front-wave characteristics. The difference between experiment and theory is frequently attributed to the effects of the complex multidimensional structure of the detonation wave and the wall boundary layer. From this point of view, the pressure profile and the heat transfer to the confining walls at the rear of the detonation front have been recently investigated (Edwards et al., 1970; Veyssière, 1971; Paillard et al., 1979) with the hope of characterizing the boundary layer. Models have been proposed by Manson (1957), Edwards et al. (1970), Tsugé (1971), and Dove et al. (1974) to characterize the interaction between the boundary layer and the extended reaction zone. It seems clear that the coupling between the reaction zone and the rear flow must be dependent on the boundary conditions, and particularly on the energy losses to the walls.

NATURE OF CONFINEMENT ON GASEOUS DETONATION   89

The purpose of this paper is to provide new experimental results that give more precise information regarding the existence and the height of the boundary layer, the relative effects of the initial conditions (pressure, dilution of the detonating gas, and nature of the confining material) on the energy transfer rate to the walls, and consequently on the detonation front parameters. The investigated parameters of the detonation wave are front velocity, pressure profile, temperature, and electrical conductivity of the burned gases near the wall; the heat losses; and the structure of the flowfield by the Schlieren technique. The height of the boundary layer was measured by two different techniques, one based on the electrical conductivity of burned gases and the other on the optical properties of the burned gases.

Experimental

A detailed description of the experimental apparatus is given by Bonnet (1979) and Brion (1979); only the main characteristics are summarized in this paper. The detonable mixture was prepared with pure propane and oxygen, premixed in a large cylinder. The dilution was defined by the mole number X of oxygen in the range of 2-18. The mixture was introduced into the previously evacuated detonation tube. The initial conditions (subscript f) were the room temperature $T_f = 293$ K and the pressure p_f in the range of 0.2-0.5 bar. The detonation was ignited by an exploding wire, which was located at the closed end of the tube and which released a nominal stored energy of 250 J. The detonation tube, made of aluminum, had a cross section of 46×21 mm^2 and was 8 m long. The internal wall of the tube was smooth but unpolished. The test section, 1.25 m long, was located 5 m from the ignitor. The walls of this section were removable and different confinement materials could be used. In these experiments the materials were aluminum (Al), high-density polyethylene (PE), polymethylmetacrylate (PMMA), polypropylene (PP), polyvinyl chloride (PVC), and polyamide (PA). All the polymers were commercial products and not chemically pure substances. A second test section with identical geometric dimensions was provided with two opposite glass windows for optical investigation. The detonation front velocity D ($\pm 0.3\%$) was deduced from transit time between seven ionization probes. Velocity measurements were made at four positions: the first (D_1) on the driver section of the detonation tube just before the test section, the second and third (D_2 and D_3) on the test section, and the last (D_4) after the test section.

The pressure profiles p/t ($\pm 4\%$) of detonation waves were monitored by three kinds of piezoelectric pressure gages. The first was the 113 A24 model (PCB Piezotronics), the second was the IVTAN model developed at the Institute of High Temperatures at Moscow (Bazhenova et al., 1968), and the third was the ENSMA model developed at Poitiers (Veyssière, 1971). The rise time of each sensor was approximately 1-2 μs and the sensitive surface diameter was about 1-3 mm. The three pressure gages were located at each observation station on the tube. The observation stations were at distances $x = 5.4$ and 5.9 m from the ignitor.

The temperature profiles $Ts(t)$ ($\pm 2\%$) of the burned gases near the wall and the heat transfer $q(t)$ were obtained by a thin-film platinum resistance thermometer developed by Zhilin (1976) who adapted the design of previous investigators (Laderman et al., 1962; Edwards et al., 1970; Paillard et al., 1979).

The electrical conductivity $\sigma(t)$ of the burned gases was deduced from the voltage difference measured on the driver coil of a radio frequency oscillator (10-30 MHz), a technique used in previous investigations (Brossard et al., 1965; Bazhenova et al., 1968). The spiral plane coil, which served as the conductivity gage was made flush with the inside wall of the tube. The electromagnetic field of the oscillator coil penetrated the plasma to a depth of approximately one radius of the coil (2-5 mm). With two different probe diameters and an appropriate calibration, the value of the conductivity and the thickness of the cold boundary layer was deduced. The rise time of the electrical circuit, including the video amplifier, was less than 2 μs. Further details are given by Brion (1979). The detonation wave structure was investigated optically by a Schlieren system; the Schlieren image was recorded with an electronic image converter and an appropriate delay line.

Results and Discussion

The parameters of the experiments were the dilution X, the initial pressure p_f, and the nature of the confinement. Extensive data were obtained with Al, PE, and PMMA materials; a few experiments were conducted with PP, PVC, and PA to characterize the relative influence of these materials in comparison with Al, PE, and PMMA.

With aluminum, the measured velocities D_2 and D_3 on the test section are slightly different from the calculated ones. The velocities increase linearly with the logarithm of the initial gas pressure, but the slopes ($\partial D/\partial \log p_f$) of the experimental curves are higher than the predicted ones. The difference decreases with the increasing pressure, and is more sensitive (-1.2%) for lean ($X=15$) and rich ($X=2.5$) mixtures than for the stoichiometric one. In the last case ($X=5$) and $p_f=0.5$ bar, the experimental velocity is higher ($+0.5\%$) than the calculated one. These results agree well with previous works (Boislève, 1970; Renault, 1972) concerning the influence of initial conditions. When polymers are substituted for aluminum, the average data from a large number of runs lead to the following conclusions: the velocities are slightly increased; and the lower the thermal product $(k\rho c)^{\frac{1}{2}}$, the higher the velocity (see Table 1). In connection with this thermophysical property of the material, the velocity increases successively with Al, PE, and PMMA, and the maximum deviation is about 0.5%.

The pressure profiles for each type of gage were found to be the same at the two observation stations. For the same value of x, the records obtained simultaneously by the three shock-tube calibrated pressure sensors have been compared. The three values of the detonation front pressure p_b coincide well, but a nonnegligible deviation appears in the compared pressure profiles. The difference, shown in Fig. 1 with IVTAN and PCB gages, is clearly a consequence of the specific transfer function of the gage. However, we observe the same relative position of the pressure curves with the influence of the material.

When the experimental scatter and the average data of the numerous records are taken into account, it appears that the measured values of p_b with Al

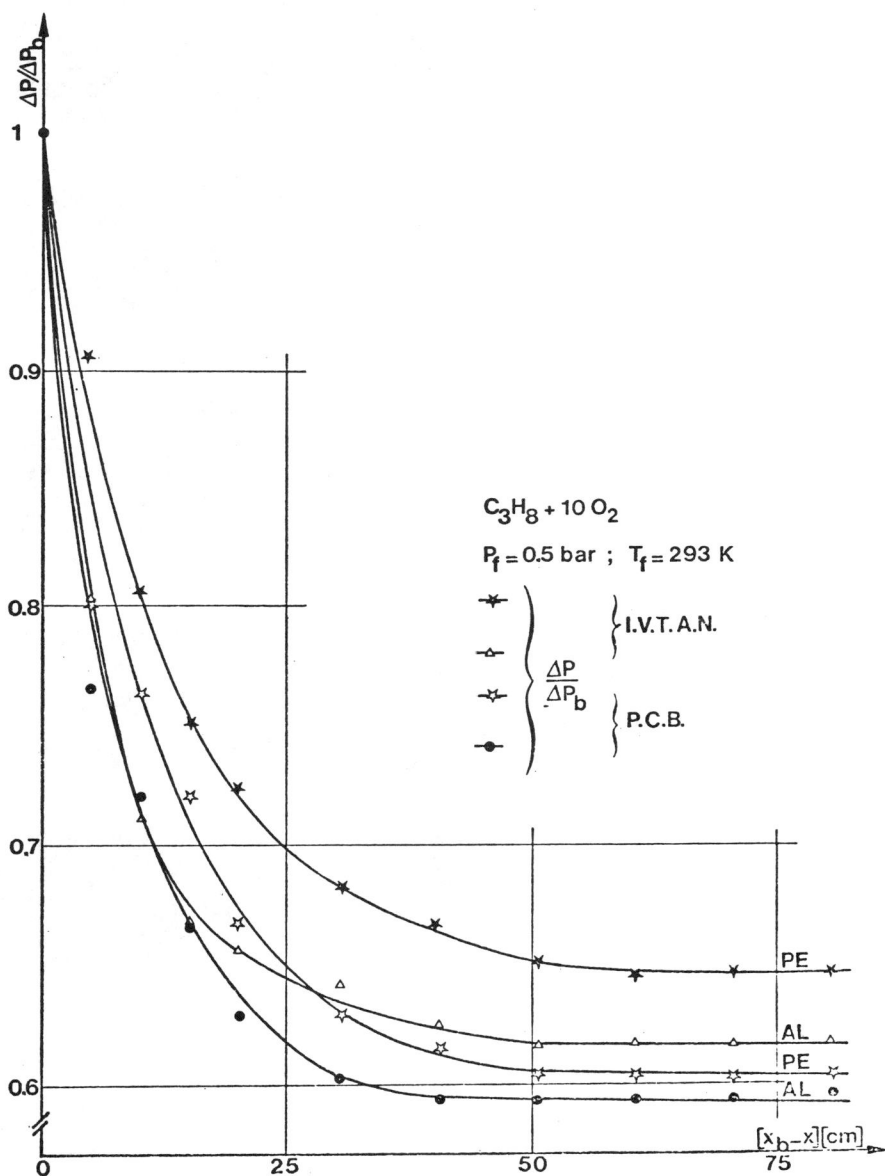

Fig. 1 Relative overpressure profiles vs distance behind detonation front for different types of pressure gages and materials of confinement.

p_f, bar		Knife position
0.2	$(C_3H_8 + 10O_2)$	Horizontal
0.2	$(C_3H_8 + 10O_2)$	Vertical
0.5	$(C_3H_8 + 18O_2)$	Vertical
0.2	$(C_3H_8 + 18O_2)$	Vertical
0.5	$(C_3H_8 + 18O_2)$	Vertical

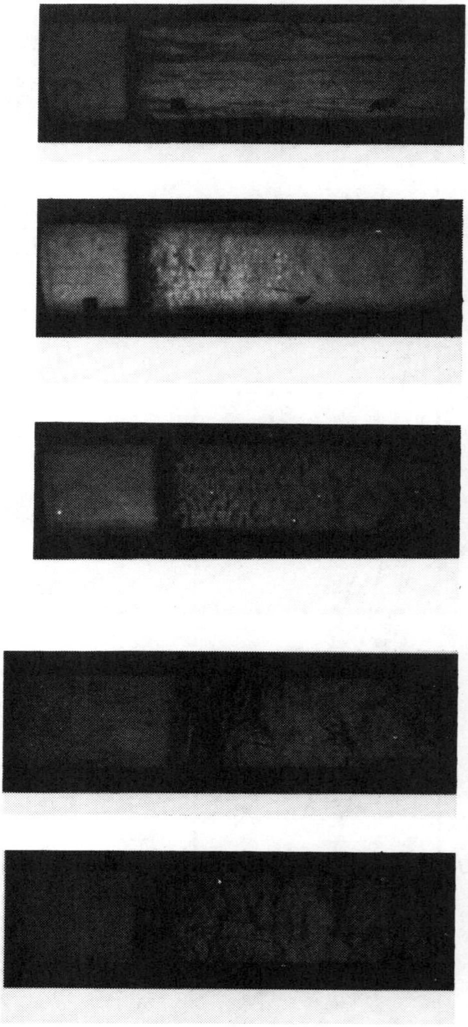

Fig. 2 Strioscopic records of detonation wave structure in $C_3H_8 + XO_2$ (PMMA surface).

Table 1 Thermal product of various materials

Material	Specific mass ρ, kg/dm^3	Heat capacity C, kj/kg·°C	Heat conductivity k, W/m·°C	Thermal product $(k\rho c)^{1/2}$, W·s$^{1/2}$/m^2·°C
Al	2.8	0.92	129	18,200
PE	0.95	2.30	0.43	970
PMMA	1.19	1.47	0.18	560
PVC	1.25	1.47	0.15	525
Pyrex	2.5	0.84	0.90	1375

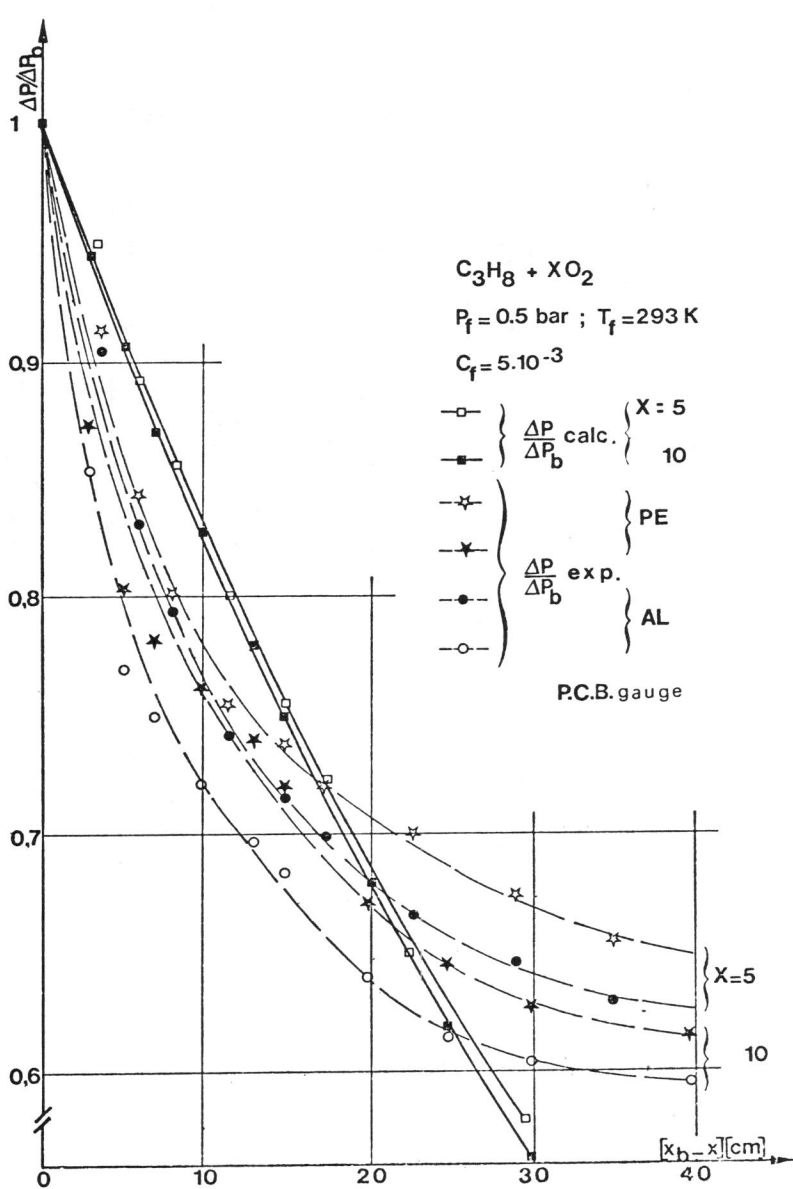

Fig. 3 Relative overpressure profiles vs distance behind detonation front with dilution and confinement materials as parameters (solid lines represent the nonsteady-state solution proposed by Edwards (1970).

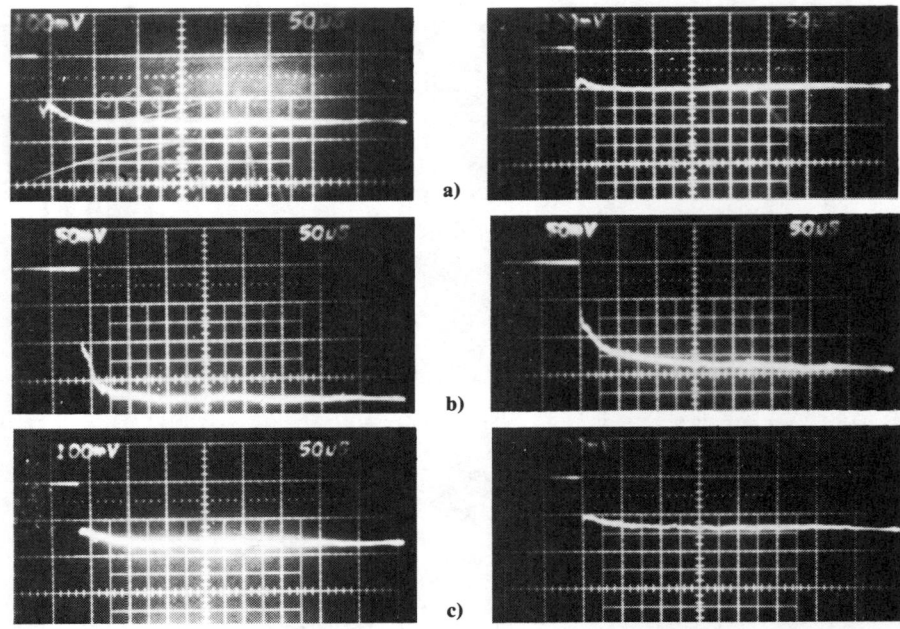

Fig. 4 Typical oscilloscope traces (50 μs/div) of thin-film temperature of $C_3H_8 + XO_2$ mixture (left side $X = 5$, right side $X = 10$, $p_f = 0.5$ bar; $T_f = 293$ K): a) new PE, 83 K/div; b) old PE, 34 K/div; and c) PMMA, 92 K/div.

confinement are lower than the predicted values (-6%), and with PE and PMMA confinement are located between the Al and theoretical curves. The pressure p_b increases successively with Al, PE, and PMMA and the difference is more apparent between Al and PMMA than Al and PE. The curves relative to PMMA are very close to the theoretical curves (-1%). These remarks are valid for both the parameters, the dilution X and the initial pressure p_f, in the domains investigated. Near the limits with dilution ($X \sim 2$ or 18) and initial pressure ($p_f \sim 0.2$ bar), the pressure records are perturbed strongly by the multidimensional structure of the detonation front, and the p_b values are not easily defined. Several photographic records of the detonation structure are presented in Fig. 2. The aspect of the picture is modified with the knife position in the strioscopic system. In the case of PMMA confinement, we observe the influences of X and p_f. With the increase of X and the decrease of p_f, the detonation front thickness is enlarged and the complex structure becomes more evident, but the confinement apparently does not influence this structure. To eliminate the gage transfer function effect on the pressure profile investigation, the results hereafter presented are deduced from the PCB gage measurements only. However, excluding the exact amplitude of the pressure in the flowfield, we have verified that the results were similar to those obtained from the other gages. This is confirmed in Fig. 1, but in this case, the deviation of the amplitudes of the quasiconstant levels for the distances larger than 50

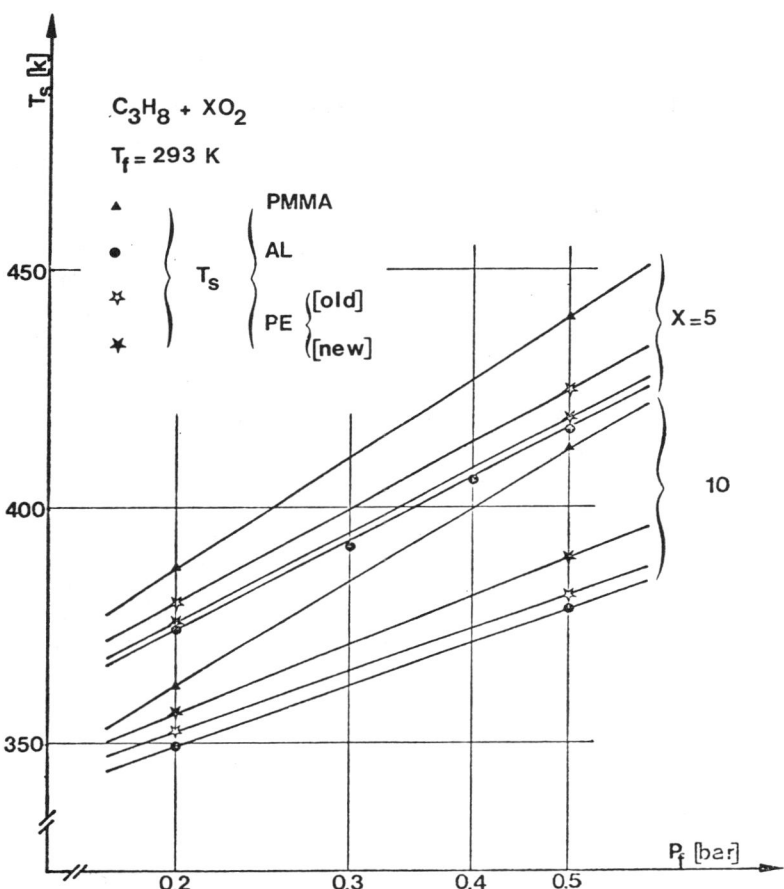

Fig. 5 Thin-film temperature curves as functions of initial pressure, dilution, and nature of confinement material.

cm is approximately 6%. Figures 1 and 3 present a few examples of the measured pressure profiles as functions of the dilution X and the confinement materials. It seems clear that: 1) the higher the thermal product $(k\rho c)^{1/2}$ of the material, the higher the gradient of rarefaction behind the detonation front; and 2) the higher the temperature of the burned gases flowfield, the smaller the gradient of the rarefaction. A very sensitive effect of the material is noted, especially on the quasiconstant level in the range 50-100 cm of the distance. With PMMA, the curves are located 1% above the PE curves. The nonsteady-state solution (Edwards, 1970) for the friction coefficient $C_f = 5.10^{-3}$ is also shown in Fig. 3.

Several typical records of the measured temperature $T_s(t)$ by the thin-film gage are presented in Fig. 4. Only in the case of PE confinement has it been necessary to distinguish between material used for the first time and that used for the second time or more. The difference in the evolution of the temperature T_s is probably a consequence of the thermal degradation of the surface of the commercial material during the first run. Except in the case of old PE material, a quasiconstant level T_s was attained with a short delay after the impact of the detonation front on the thin-film gage. In Fig. 5, the measured variations of T_s with the initial pressure p_f, the dilution X, and the nature of the confinement are plotted. For a given dilution, the temperature increased as the initial pressure was increased and as materials of lower thermal product were used. The effect of the material became more significant as the initial pressure increased. For a given initial pressure p_f, the T_s variation with the dilution X is parallel with the predicted Chapman-Jouguet temperature T_b. In the particular case of PE confinement, a singularity is observed under the oxidizing effect of the flowfield. The experimental curves (Fig. 5) for the old and the new PE are inverted with stoichiometric and lean mixtures.

The instantaneous rate of heat transfer $q(t)$ in the Pyrex support of the gage has been deduced from the observed output voltage of the temperature gage circuit. The technique has been described by Zhilin (1976). The computed results are shown in Figs. 6-8 as a function of the distance $(x_b - x)$ behind the detonation front. The respective influence of the initial pressure, dilution, and confinement are very clear. In Fig. 7, the heat transfer appears greater for the stoichiometric mixture than for the lean and rich mixtures. In fact, comparing the heat transfer with the chemical energy release by the detonation process shows that the relative heat losses are more important for the lean and rich mixtures than for the stoichiometric one. In Fig. 8, the heat transfer appears more important for polymers than for Al material. This result disagrees with the relative situations of the pressure profiles. In fact, the calculated heat transfer $q(t)$ is the exclusive heat flow into the Pyrex glass support of the gage. Therefore, if we consider that the film temperature T_s gives an actual representation of the common temperature of the walls and the burned gases, it is easy to obtain an order of magnitude of the heat transfer to the walls, taking into account the thermal product $(k\rho c)^{1/2}$ (Table 1) of each material instead of the Pyrex one. The curves are presented in Fig. 9, and the results agree well with the previous results concerning velocities and pressure profiles. The heat losses are 10 times greater for metallic confinement than for polymers.

Typical oscillograms of the electrical conductivity $\sigma(t)$ (see Fig. 10) present two signal records with opposite signs simultaneously obtained on the two opposite sides of the test section. When the signals are parallel with two constant mean levels (Fig. 10) the measured conductivity of the burned gases near the walls, given by the two gages, is the same. The boundary-layer is so thin that it is not detectable. When the signals are divergent (Fig. 10), one with a constant mean level (large-field gage) and the other with a decreasing level (small-field gage), the difference indicates the cold boundary-layer thickness. With the appropriate calibration runs in the shock tube, the sensitivity of the sensor is approximately 0.1 mm for the boundary-layer thickness. Typical results are shown in Figs. 11 and 12 for the lean mixture $X=10$. Just behind the detonation front, the electrical conductivity near the wall is about 1 S/m.

NATURE OF CONFINEMENT ON GASEOUS DETONATION

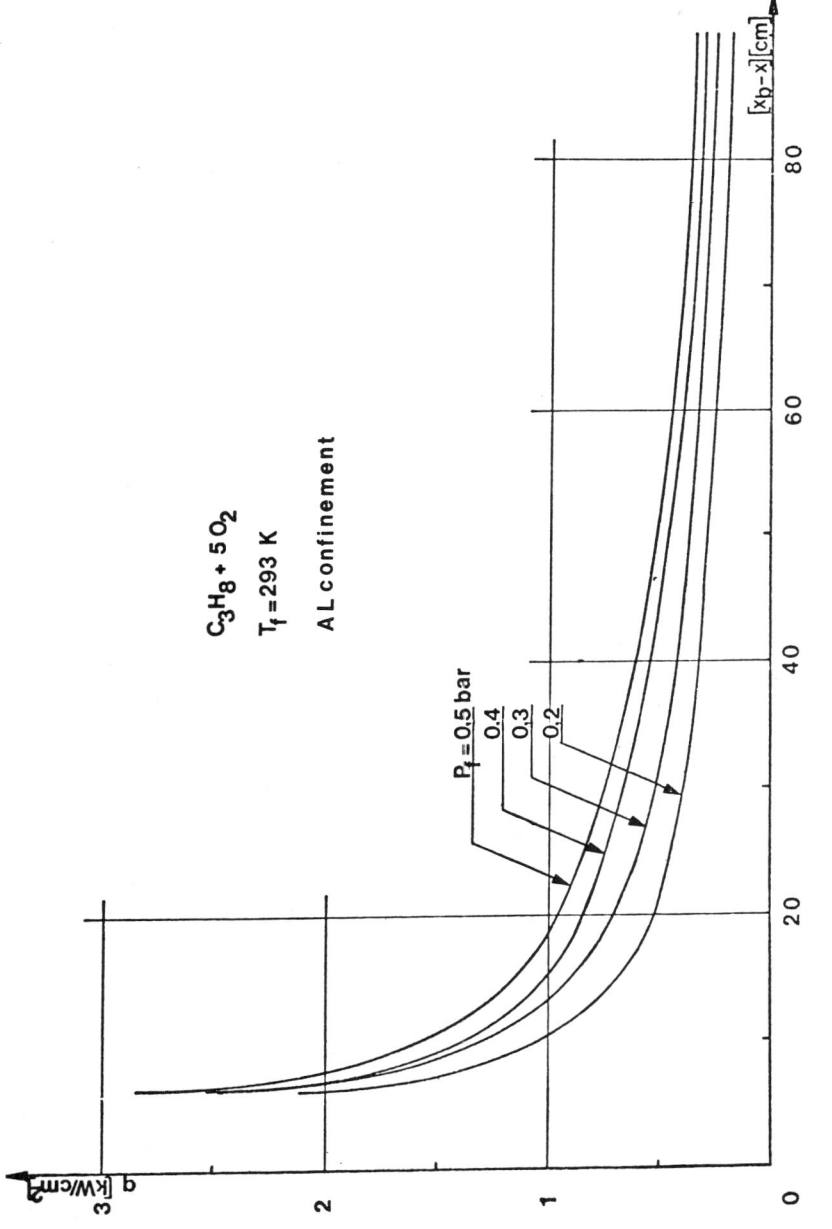

Fig. 6 Rate of heat transfer behind detonation front: initial pressure effect.

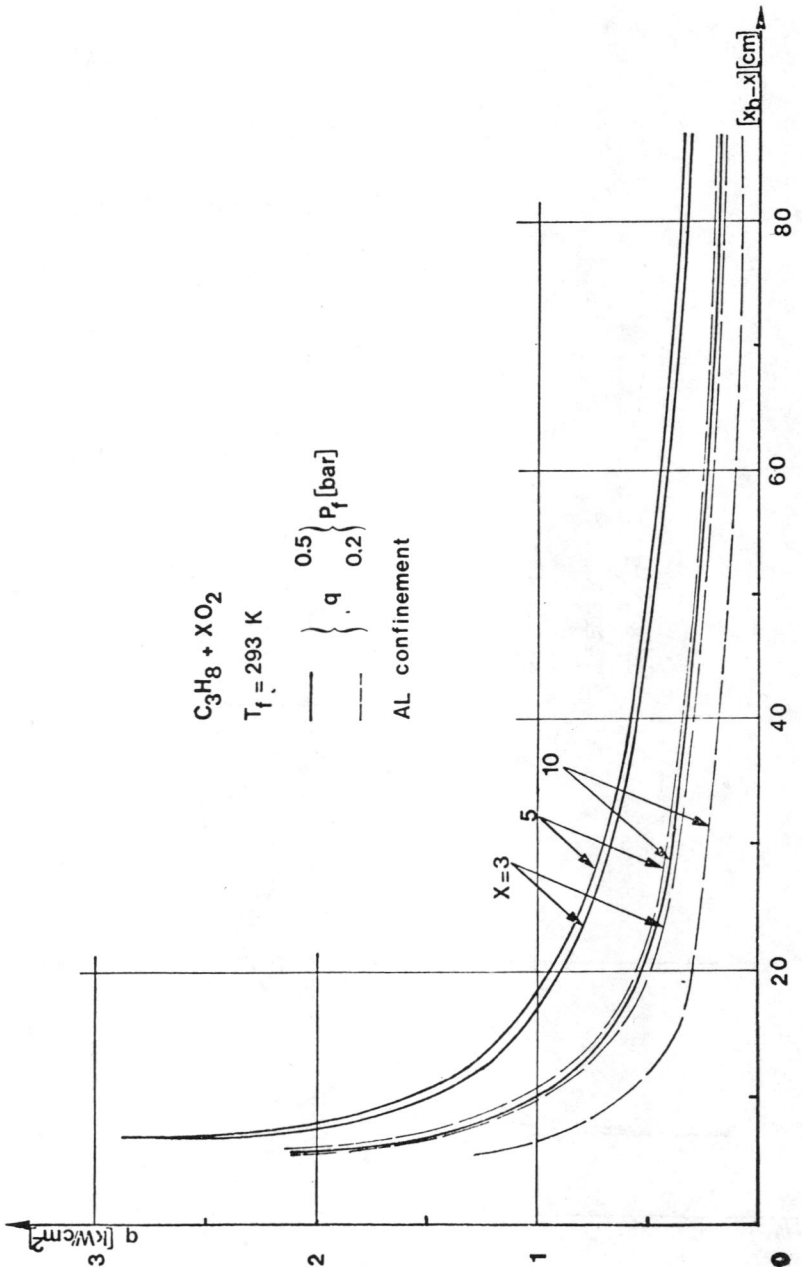

Fig. 7 Rate of heat transfer behind detonation front: dilution effect.

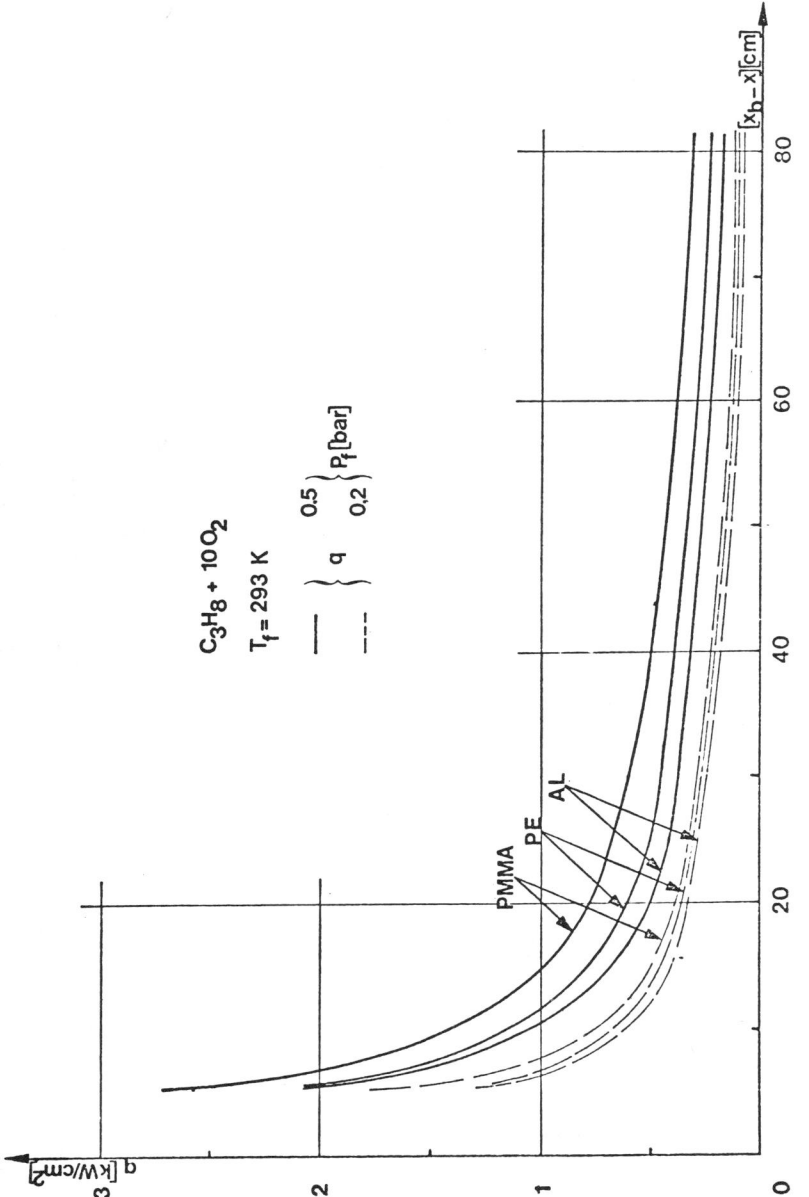

Fig. 8 Rate of heat transfer behind detonation front: confinement material effect.

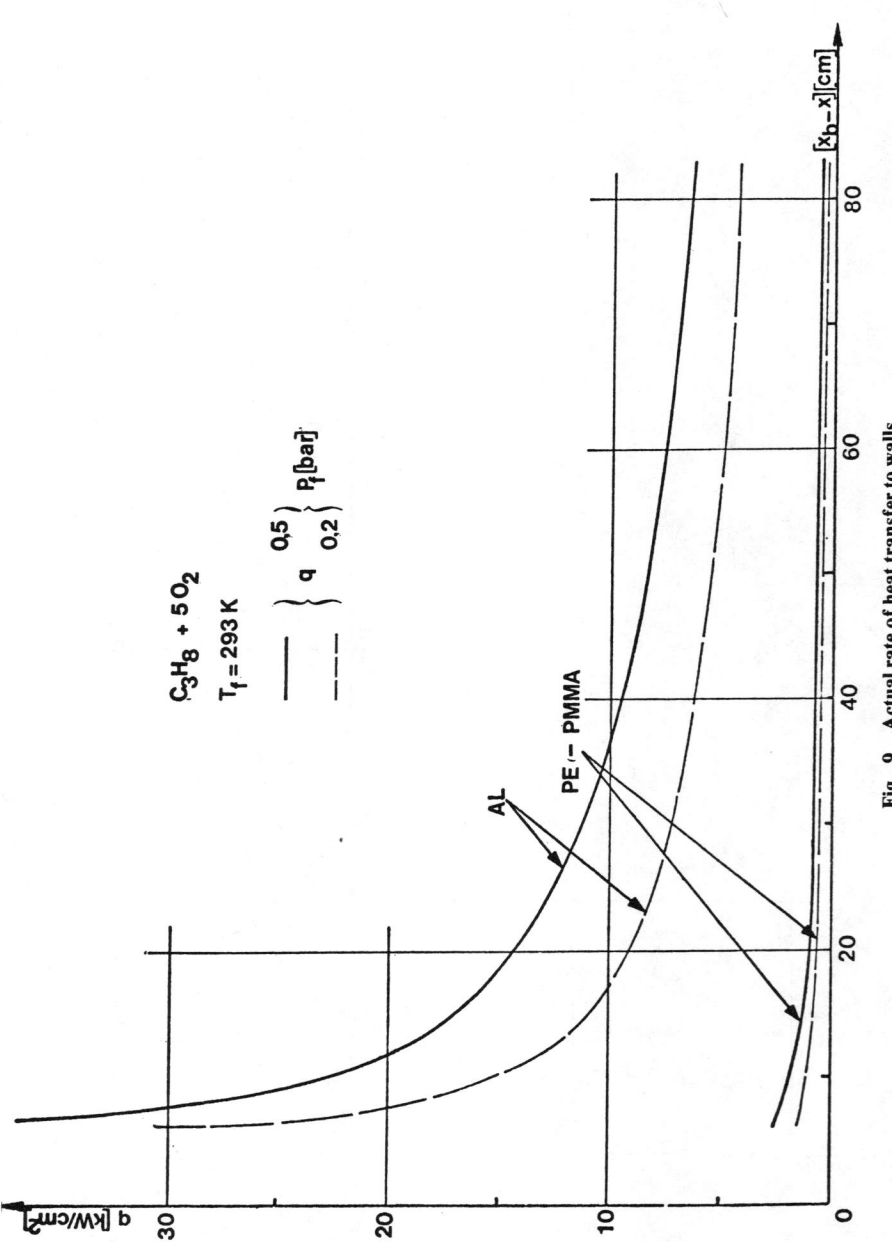

Fig. 9 Actual rate of heat transfer to walls.

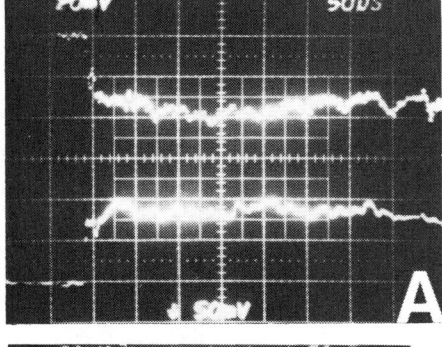

a) $C_3H_8 + 5\ O_2$, $p_f = 0.2$ bar, Al material.

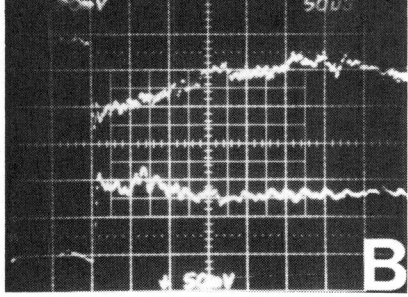

b) $C_3H_8 + 10\ O_2$ $p_f = 0.5$ bar, PE material.

Fig. 10 Typical oscilloscope traces (50 μs/div) of electrical conductivity (upper trace, 0.110 (S/m)/ div; lower trace, 0.092 (S/m)/div.

whatever the initial conditions may be. In the rear flow, but outside the boundary layer, the conductivity is constant for Al and new PE confinements. A slight evolution is detectable for old PE and PMMA confinements. In this case, the order of magnitude of σ near the wall is approximately 100 times the values reported by Veyssière (1971). The difference is probably due to the impurities contained in the commercial confinement materials.

The thickness $\delta(t)$ of the boundary layer is more sensitive to the initial parameters. With Al confinement the boundary layer was not detectable, whatever the initial conditions were. With polymers the thickness δ was generally detectable in the first 25 μs but was very sensitive to the initial conditions. It is clear that the lower the initial pressure, the lower the thickness. For a stoichiometric mixture at $p_f = 0.2$ bar, the boundary layer was less than 0.1 mm for all confinement materials. For a lean mixture $X = 10$ at $p_f = 0.2$ bar the boundary layer was detectable only with old PE confinement. The boundary layer was fully developed approximately 150-250 μs after the passing time of the detonation front, and the maximum thickness was about 1.5 mm.

When the structure of the detonation front was investigated by the strioscopic visualization (Fig. 2), the boundary layer did not appear on the records. With a appropriate location of the horizontal knife with respect to the light path, however, the boundary layer became detectable. Some pictures

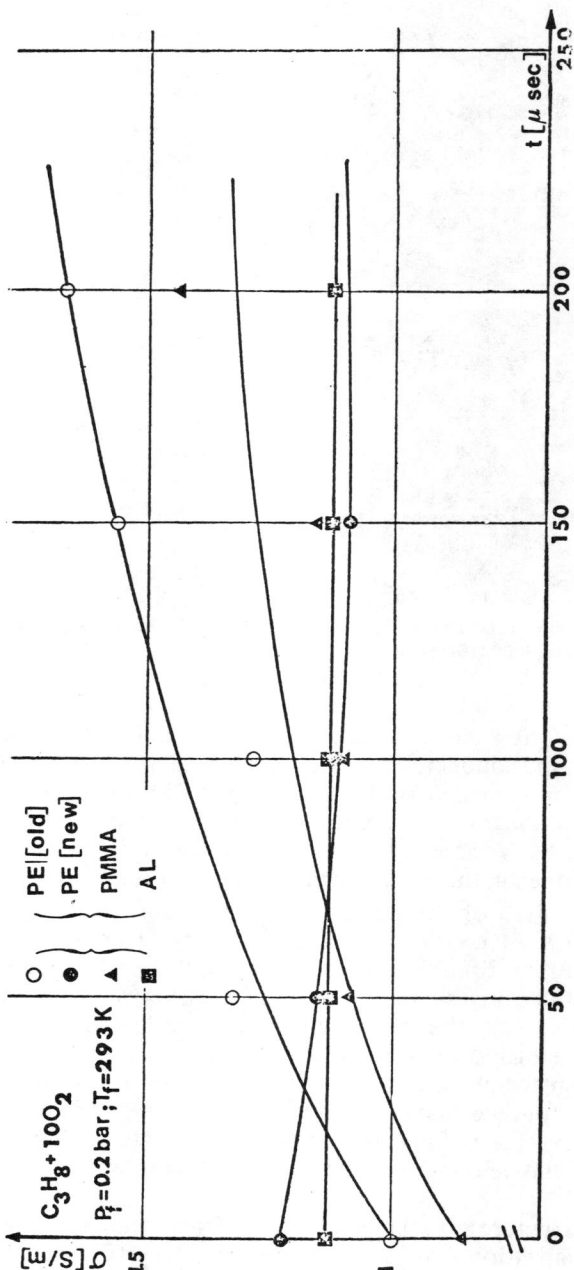

Fig. 11 Electrical conductivity of burned gases near walls, outside boundary layer.

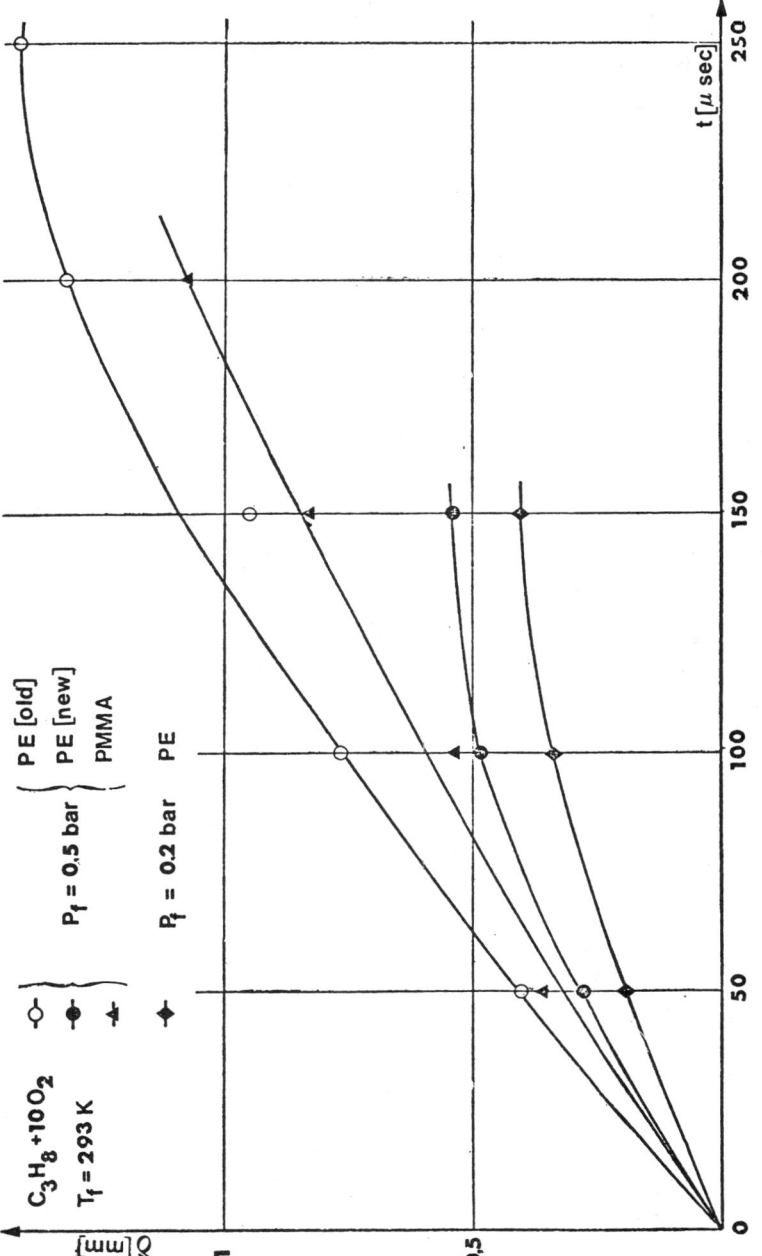

Fig. 12 Boundary-layer thickness behind the detonation front: confinement material effect.

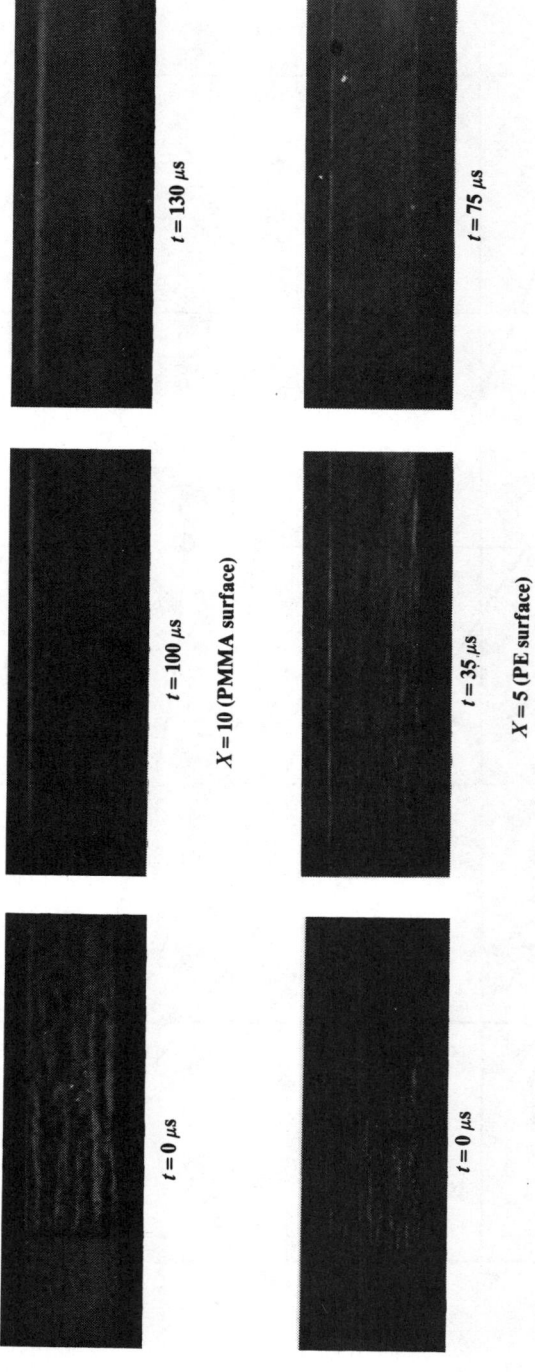

Fig. 13 Typical pictures of visualized boundary layer by the strioscopic system: $C_2H_8 + XO_2$ mixture, $p_f = 0.5$ bar, $T_f = 293$ K, horizontal knife, full scale (21 mm) of test section.

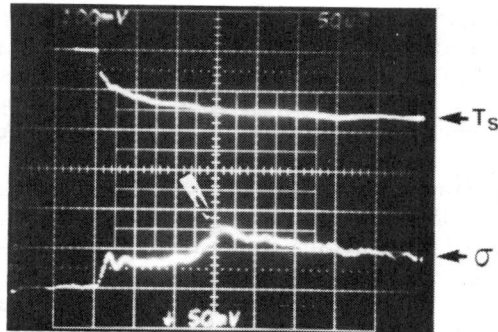

Fig. 14 Thin-film temperature (upper trace) and electrical conductivity (lower trace) for PMMA confinement of lean mixture, $X=10$, 50 μs/div); transition point from laminar to turbulent structure of the boundary layer is indicated.

are given in Fig. 13, and all the previous results were confirmed by optical investigation. By a detailed examination of the original records, in particular the case $X=10$, $p_f=0.5$ bar, and PMMA surface, we discovered the multidimensional structure of the flowfield inside the boundary layer at times larger than 130 μs. This observation was clearly confirmed by the perturbation, which appeared simultaneously in the conductivity signal (Fig. 14) and the pressure record, and is supported by the remark made by Edwards et al. (1970).

The experimental results concerning the influence of PP, PVC, or PA confinement were similar to those obtained with the PE confinement but without the run-order effect.

Conclusions

It has been shown that however great the influence of the confinement material on the heat flux to the walls, the pressure and velocity of the detonation front are slightly modified by the boundary conditions.

This first important result is a consequence of the weak coupling that takes place between, on one hand, the very short space-time scale which characterizes the chemical energy release in the reaction zone of the premixed gas detonation and, on the other hand, the large space-time scale that governs the heat losses to the walls.

The second important result is the very clear correlation that appears between the pressure profiles and heat transfer to the walls for a given confinement: the higher the Chapman-Jouguet pressure and temperature of the burned gases through the initial pressure or dilution, the greater the heat transfer and, consequently, the pressure gradients behind the detonation front. The determinating characteristic of the material on the interaction between the flowfield and the confinement seems to be the thermal product $(k\rho c)^{1/2}$, which governs the heat losses and, consequently, the gas temperature. Respectively, they increase and decrease with the increasing values of $(k\rho c)$.

The third significant and original result is the detection of the boundary layer by two independent techniques. One is based on the electrical conductivity, and the other on the optical index of the burned gases near the wall. Within the pressure domain studied, the boundary layer is not detectable for aluminum, but becomes perceptible for the polymer confinement; moreover, the thickness seems dependent on the chemical reactivity of the material in the presence of the oxidizing gaseous flow. This thickness increases slightly with the increasing initial pressure. This effect is probably a consequence of the thermal degradation of the material and must be confirmed.

Unfortunately, all these results are relative to only one geometry of the test section. It would be interesting to investigate the effect of the equivalent radius on the heat losses and the boundary-layer thickness for different confinement materials.

Acknowledgment

A part of this work has been carried out in the framework of the cooperation between France and U.S.S.R. on high-temperature physics.

References

Bazhenova T.V., Gvozdeva L., Lobastov Y., Naboko I., and Nemkov R. (1968) *Shock Waves in Real Gases.* Nauka, Moscow.

Boisleve J. (1970) Propagation des détonations dans les mélanges gazeux contenus dans des tubes rectangulaires. *Thèse 3e cycle,* Poitiers, France.

Bonnet T. (1979) Influence de la nature du confinement rigide sur les caractéristiques de la détonation des mélanges gazeux. *Thèse 3e cycle,* Orléans—Bourges, France.

Brion B. (1979) Etude de l'écoulement des produits de détonation des mélanges gazeux: Influence de la nature du confinement sur le développement de la couche limite. *Thèse 3e cycle* (en cours de rédaction), Orléans—Bourges, France.

Brochet C., Brossard J., Manson N., Cheret R., and Verdes G. (1970) A comparison of spherical, cylindrical and plane detonation velocities in some condensed and gaseous explosives. *Fifth (Intl.) Symposium on Detonation.* ONR ACR **184,** 41-46.

Brossard J. Fauchais P., and Moreau M. (1965) Mesure de la conductivité électrique d'un jet de plasma. *Congrès IUPAC,* Moscow.

Brossard J. and Charpentier N. (1976) Effets d'un confinement souple sur la détonation des mélanges gazeux. *Acta Astronautica* **3,** 971-981.

Dove J.E., Scroggie B.J., and Semerjian H. (1974) Velocity deficits and detonability limits of hydrogen-oxygen detonations. *Acta Astronautica* **1,** 345-359.

Edwards D.H., Brown D.R., Hooper G., and Jones A.T. (1970) The influence of wall heat transfer on the expansion following a C.J. detonation wave. *J. Phys. D: Appl. Phys.* **3,** 365-376.

Laderman A.J., Hecht G.J., and Oppenheim A.K. (1962) Thin film thermometry in detonation research. *Temperature—Its Measurement and Control in Science and Industry* **3,** 943-947.

Manson N., Brochet C., Brossard J., and Pujol Y. (1963) Vibratory phenomena and and instability of self-sustained detonations in gases. *9th Symposium (Intl.) on Combustion,* pp. 461-469. Academic Press, New York.

Paillard C., Dupré G., Lisbet, R., Combourieu J., Fokeev V.P., and Gvozdeva L.G. (1979) A study of hydrogen azide detonation with heat transfer at the wall. *Acta Astronautica* **6,** 227-242.

Renault G. (1972) Propagation des détonations dans les mélanges gazeux contenus dans des tubes de section circulaire et rectangulaire; influence de l'état de surface interne des tubes. *Thèse 3e cycle,* Poitiers, France.

Tsugé S. (1971) Effect of boundaries on the velocity deficit and the limit of gaseous detonations. *Combustion Sci. Technol.* **3** (4), 195-205.

Veyssiere, M. (1971) Contribution à l'étude des caractéristiques physiques des produits de détonation dans les mélanges gazeux. *Thèse doctorat d'*état, Univ. de Poitiers, France.

Zhilin Yu.V. (1976a) Influence des propriétés thermophysiques du support sur la réponse des thermomètres à résistance. IVTAN, N.T. 2-004, Moscow,

Zhilin Yu.V. (1976b) Méthode de mesures des transferts de chaleur transitoires à l'aide des sondes à résistance. IVTAN, N.T. 2-005, Moscow.

Mechanical Effects of Gaseous Detonations on a Flexible Confinement

J. Brossard* and J. Renard*
Université d'Orléans, Bourges, France

To analyze the interaction between a gaseous detonation wave and its confinement in a straight circular elastic tube, the dynamic response pattern of the tube is determined. Numerical solutions are compared with experimental results that were obtained for two confinement materials, polyvinylchloride (PVC) and stainless steel. In the first case the detonation velocity is higher than the dilatational wave velocity in a thin shell of the material, in the second case it is lower. Experimental results provide information on the dependence of the transverse and longitudinal strains in the external skin of the tube on the detonation velocity for propane-oxygen-nitrogen stoichiometric mixtures (initial conditions at standard temperature and pressure) and on the geometrical features of the tube (mean diameter in the range of 16-33 mm; wall thickness varying 1.2-2 mm for PVC, 0.2-0.3 mm for stainless steel). The precursor effects, the oscillations and their frequencies, the strain ratio, the dynamic amplification factors, and the end effects agree well with theory. Measured propagation velocities of the detonation, lower than that of an ideal wave, are in agreement with values already reported in a previous paper. However, the observed increase of the transverse strain behind the detonation front and its amplitude do not agree with the predictions of the model with the ideal pressure step of the detonation for a rigid confinement. In spite of some difficulties with pressure measurements at high temperatures of the detonation products, their results compare well with those deduced from observed deflections. However, the observed pressures differ significantly from that of the ideal pressure step for a PVC test section, while they are closer for a stainless-steel test section. The discrepancy may be due to the fact that the pressure distribution and viscosity effects of the material are not properly recognized in the model. The effects of mechanical stiffness of the confinement are exhibited.

Presented at the 7th ICOGER, Göttingen, Federal Republic of Germany, Aug. 20-24, 1979. Copyright © American Institute of Aeronautics and Astronautics, Inc., 1981. All rights reserved.
*Laboratoire de Recherche Universitaire.

Nomenclature

$[A]$	=	mass matrix of the equilibrium system
a	=	mean radius of the tube
Ax	=	amplifying dynamic factor for longitudinal strain
$A\theta$	=	amplifying dynamic factor for transverse strain
$[C]$	=	rigidity matrix of the equilibrium system
$[D]$	=	forcing pressure matrix
E	=	Young's modulus of the tube
h	=	wall thickness of the tube
$H(x)$	=	Heaviside's distribution
K	=	constant of integration
$P(x,t)$	=	pressure distribution
P_f	=	pressure in fresh gaseous mixture
r	=	radial position
t,τ	=	time coordinates
T_f	=	temperature in fresh gaseous mixture
$u(x,t)$	=	longitudinal displacement
V	=	velocity of the pressure front
V_b	=	dilatational wave velocity in a bar
V_p	=	dilatational wave velocity in a thin shell
$w(x,t)$	=	radial displacement
X,x	=	coordinates along the axis of the tube
$[\epsilon]$	=	strains vector in the equilibrium system
ϵ_x	=	longitudinal strain
ϵ_o	=	transverse strain
θ	=	angular position of a point of the tube
ν	=	Poisson's ratio
ρ	=	mass of the tube per unit volume
σ_x	=	longitudinal stress
σ_θ	=	transverse stress
ω	=	vibrational frequency of the tube

Introduction

AMONG the physical conditions which can modify the propagation characteristics of the detonation of a gaseous mixture, the confinement behavior is fundamental. In the case of straight tubes, besides the geometry, the state of the internal surface, the nature, the chemical reactivity with the flowing gases, and the mechanical stiffness obviously affect the velocities and limits of stable detonation. The effects of expansion of confinement were considered by Zeldovich and Kompaneets (1955) and Manson (1957). The case of a confinement made of a surrounding inert gaseous medium was investigated by Sommers (1961), Dabora et al. (1965), Shchelkin (1968), and more recently by Tsuge et al. (1972, 1974). On the other hand, until now, few papers have dealt with the effects of the detonation on the confinement. Oppenheim et al. (1966) and Brossard and Charpentier (1976) pointed out that the detonation could produce deformations and vibrations of a confining shell. To the pressure disturbance produced by the detonation correspond longitudinal and transverse strains, the amplitude of which is determined by the contribution of mechanical energy from the gaseous medium to the confining wall. Sometimes, this transferred energy leads to the brittle rupture of the material (Brossard and Charpentier, 1976) and, consequently, to the destruction of the detonation. It is also shown that the stable detonation of the mixture with a great dilution in nitrogen have more destructive mechanical consequences for the material than a nondiluted mixture, that is to say than a more energetic mixture.

A study with two different soft confinements such as steel and viscoelastic (polymers) tubes, offers a possibility of analyzing the interaction of the confinement material with the detonation. As a matter of fact, an analysis of the theory of the wave propagation in the materials shows that in the first case the detonation front is slower than the propagation of the waves in the material and in the second case, faster. The response of the confinement to pressure oscillations induced by the detonation will be appreciably different.

To characterize the interaction between the confinement and the detonation, a theoretical analysis of the deformation of the tube was developed, the experimental results were examined to assess the accuracy of the model, and estimates were obtained for the amplitudes of the deformations and for their consequences on the detonation characteristics.

Governing Equations

The confinement is a circular tube whose wall thickness h is small compared to the mean radius a. Because of axial symmetry, the dependent variables in the shell deformation are expressed in terms of the independent variables x, θ, r, t, which are the axial position, the angular position, the radial position, and the time, respectively. The shell deformation is then defined by the radial displacement $w(x,t)$ (an expansion of the tube is taken as a positive value) and the longitudinal displacement $u(x,t)$. As $h \ll a$, the stress distribution in the material of the shell of the tube is assumed to be uniform. Besides, the tube is long enough to suppose that the end effects are negligible. The material is homogeneous, isotropic, with a specific mass ρ. Its rheological behavior is described by Hooke's law. The detonation wave is plane and stable. It propagates with a constant velocity V and produces a pressure perturbation $p(x,t)$ which acts on the internal surface of the confinement.

From these hypotheses have flowed numerous studies of the mechanical equilibrium equations. As an example, Sing-Chin-Tang (1965) accounted for the bending moment effect, the shear forces, and the rotary inertia. As the influences of such effects, from an a posteriori analysis, appear to be quite negligible, mechanical equilibrium can be expressed by the following simplified equations which relate the longitudinal and transverse stresses σ_x and σ_θ to the displacements u and w.

$$\frac{\partial \sigma_x}{\partial x} = \rho \frac{\partial^2 u}{\partial t^2} \quad \text{and} \quad \frac{h}{a} \sigma_\theta + \rho h \frac{\partial^2 w}{\partial t^2} = p(x,t)$$

The strains, with their linear expressions $\epsilon_x = \partial u/\partial x$ and $\epsilon_\theta = w/a$, are related to the stresses for Hookean material with plane stresses by:

GASEOUS DETONATIONS ON A FLEXIBLE CONFINEMENT

$$\sigma_\theta = \frac{E}{1-\nu^2}\left(\frac{w}{a} + \nu\frac{\partial u}{\partial x}\right)$$

$$\sigma_x = \frac{E}{1-\nu^2}\left(\frac{\partial u}{\partial x} + \nu\frac{w}{a}\right)$$

E is Young's modulus and ν Poisson's ratio.

The following set of partial differential equations describes the coupling between the deformation of the tube and the pressure perturbation $p(x,t)$ resulting from substitution of Hooke's law into the mechanical equilibrium conditions:

$$\frac{E}{1-\nu^2}\frac{a}{h}\left(\frac{w}{a} + \nu\frac{\partial u}{\partial x}\right) + \rho a\frac{\partial^2 u}{\partial t^2} = \frac{a}{h}p(x,t) \tag{1}$$

$$\frac{E}{1-\nu^2}\left(\frac{\partial^2 u}{\partial x^2} + \frac{\nu}{a}\frac{\partial w}{\partial x}\right) = \rho\frac{\partial^2 u}{\partial t^2} \tag{2}$$

Far from the origin of the tube, it is possible that the tube deformation may be considered as a traveling wave which moves with the velocity V. Consequently, in the coordinates (x, r, θ) which are fixed on the wavefront, the displacements u and w and the pressure perturbation p are functions of the single variable $X = x - tV$. For this case Eqs. (1) and (2) become:

$$\frac{E}{1-\nu^2}\left(\frac{w}{a} + \nu\frac{du}{dx}\right) + \rho h V^2\frac{d^2 w}{dx^2} = p(x) \tag{3}$$

$$\frac{E}{1-\nu^2}\left(\frac{d^2 u}{dx^2} + \frac{\nu}{a}\frac{dw}{dx}\right) = \rho V^2\frac{d^2 u}{dx^2} \tag{4}$$

Integration of Eq. (4), and substitution of the relations $\epsilon_x = du/dx$ and $\epsilon_\theta = w/a$ yields:

$$\frac{\nu E}{1-\nu^2}\epsilon_x + \frac{E}{1-\nu^2}\epsilon_\theta + \rho a^2 V^2\frac{d^2\epsilon_\theta}{dx^2} = \frac{a}{h}p(x) \tag{5}$$

$$\left(\frac{E}{1-\nu^2} - \rho V^2\right)\epsilon_x + \frac{\nu E}{1-\nu^2}\epsilon_\theta = \left(\frac{E}{1-\nu^2} - \rho v^2\right)\epsilon_{x0} + \frac{\nu E}{1-\nu^2}\epsilon_{\theta 0} \tag{6}$$

where $(du/dx)_0$ and w_0 are evaluated at the detonation front.

The independent variable is transformed to $\tau = X/V$, and the system is written in matrix notation to yield:

$$\begin{bmatrix} \rho a^2 & 0 \\ 0 & 0 \end{bmatrix}\begin{bmatrix} \epsilon_\theta'' \\ \epsilon_x'' \end{bmatrix} + \begin{bmatrix} \dfrac{E}{1-\nu^2} & \dfrac{\nu E}{1-\nu^2} \\ \dfrac{\nu E}{1-\nu^2} & \dfrac{E}{1-\nu^2} - \rho V^2 \end{bmatrix}\begin{bmatrix} \epsilon_\theta \\ \epsilon_x \end{bmatrix} = \begin{bmatrix} \dfrac{a}{h}p(\tau) \\ K \end{bmatrix}$$

or
$$[A][\epsilon''] + [C][\epsilon] = [D] \tag{7}$$

K represents the right-hand side of Eq. (6). The general solution $[\epsilon] = (\epsilon_0) + [\epsilon_I]$ of this system is the sum of $[\epsilon_0]$, the particular solution of the complete system, and of $[\epsilon_I]$, the general solution of the homogeneous associated system. The solution of the characteristic equation of Eq. (7) is the following:

$$\omega^2 = -\frac{E}{\rho a^2} \frac{E - \rho V^2}{E - \rho V^2 (1 - \nu^2)}$$

or in another form

$$\omega^2 = -\frac{V_p^2}{a^2} \frac{V^2 - V_b^2}{V^2 - V_p^2}$$

where

$$V_b = (E/\rho)^{1/2} \text{ and } V_p = (E/\rho(1-\nu^2))^{1/2} \tag{8}$$

are the dilatational wave velocities in a bar and in a thin shell, respectively, of the material. The form of the solution $[\epsilon_I]$ will depend on the relation of V to V_b and V_p

$$[\epsilon_I] = [\epsilon_{I1}]e^{-\omega\tau} + [\epsilon_{I2}]e^{-\omega\tau} \text{ for } V_b < V < V_p$$

and

$$[\epsilon_I] = [\epsilon'_{I1}]\cos\omega\tau + [\epsilon'_{I2}]\sin\omega\tau \text{ for } V > V_p \text{ or } V < V_b$$

The case $V > V_p$ will be discussed in greater detail because it corresponds to our experimental conditions.

When $V > V_p$, no mechanical perturbation can occur ahead of the detonation front. For every point ahead of the front

$$u(x,t) = w(x,t) = \frac{\partial u}{\partial t} = \frac{\partial w}{\partial t} = 0$$

At the front:

$$\epsilon_{x0} = 0 \text{ and } \epsilon_{\theta 0} = 0 \tag{9}$$

The solution of Eq. (7) requires that the function $p(\tau)$ be specified. As a first approximation $p(\tau)$ is idealized as a step function:

$$p(\tau) = P \cdot H\left(t - \frac{x}{V}\right) \text{ with } H = 1 \text{ when } x < V \cdot t$$

$$\text{and } H = 0 \text{ when } x > V \cdot t$$

The right-hand side of Eq. (7) takes the form:

$$D = \begin{bmatrix} \frac{a}{h} P \cdot H\left(t - \frac{x}{V}\right) \\ 0 \end{bmatrix}$$

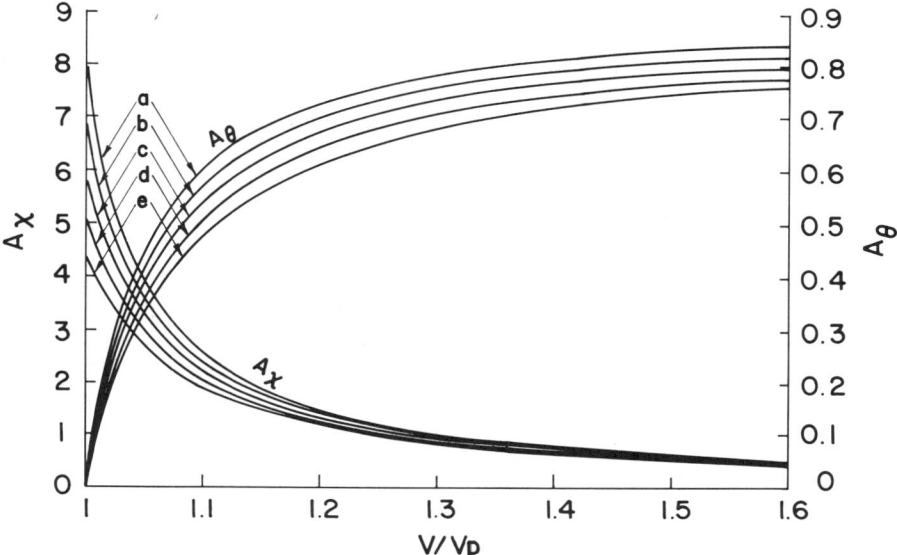

Fig. 1 Dynamic amplifying factors of deformation in function of V/V_p and ν: for curves a, b, c, d, and e, $\nu = 0.32, 0.34, 0.36, 0.38$, and 0.4, respectively.

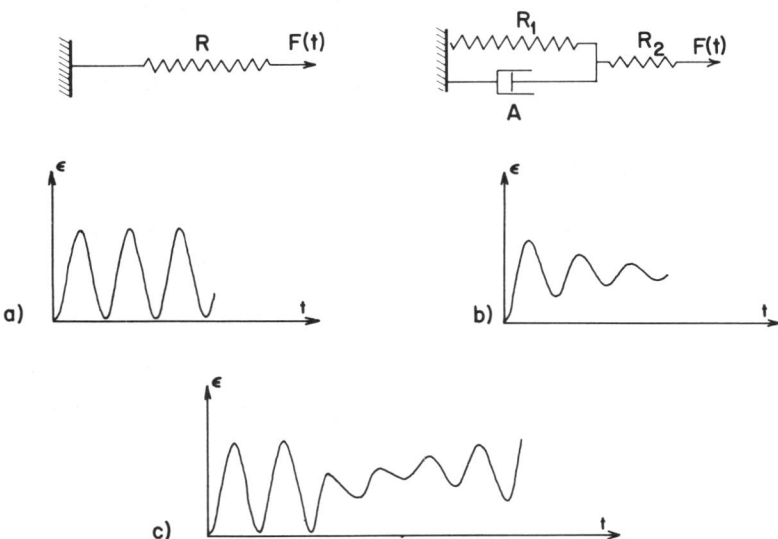

Fig. 2 Pattern of deformation ϵ according to different models: a) elastic, linear behavior of the material; b) viscoelastic behavior similar to Kelvin-Voigt's model; c) linear behavior including the extremity effect.

and the solution becomes

$$\epsilon_\theta = \frac{P}{E} \frac{a}{h} (1 - \cos\omega\tau) \frac{V_b^2}{V_p^2} \frac{V^2 - V_p^2}{V^2 - V_b^2} \qquad (10)$$

$$\epsilon_x = \frac{\nu P}{E} \frac{a}{h} (1 - \cos\omega\tau) \frac{V_b^2}{V_p^2} \frac{V_b^2}{V^2 - V_b^2} \qquad (11)$$

with

$$\omega = \frac{V_p}{a} \left(\frac{V^2 - V_b^2}{V^2 - V_p^2} \right)^{1/2}$$

The coefficients Pa/Eh and $\nu Pa/Eh$ correspond to the static deformation of the cylinder under the pressure P. The dynamic deformations due to the traveling wave oscillate about the static deformations which are increased by dynamic amplifying factors in the longitudinal and transverse directions

$$A_x = -\frac{V_b^2}{V_p^2} \frac{V_b^2}{V^2 - V_b^2}$$

$$A_\theta = \frac{V_b^2}{V_p^2} \frac{V^2 - V_p^2}{V^2 - V_b^2}$$

The variations of the coefficients A_x and A_θ with the ratio V/V_p and the constant ν are shown in Fig. 1.

The following observations concerning the effects of V and P of the ideal detonation are significant:

1) The transverse strain increases with increased V due to A_θ and P.
2) The longitudinal strain increases when V decreases due to the evolution of A_x and P.
3) The ratio $\epsilon_x/\epsilon_\theta = \nu V_b^2/(V^2 - V_p^2)$ is positive does not depend on P, and decreases when V increases.
4) If the elastic behavior of the material is modeled by a spring, the deformation looks like Fig. 2a.

To judge the influence of the material behavior laws on the one hand and the end effect on the other hand, the solutions of Eq. (7) have been computed with a viscoelastic model similar to Kelvin-Voigt's. In a typical deformation

Table 1 Characteristics of the confinement materials

Material	Polyvinylchloride				Steel	
ρ, kg/m^3	1,300				7,900	
E, hbar	340				20,300	
ν	0.38				0.29	
V_b, m/s	1,617				5,069	
V_p, m/s	1,748				5,297	
Internal diameter ϕ, mm	16	20	25	33	15.6	19.4
Wall thickness h, mm	1.2	1.4	1.7	2	0.2	0.3
Mean radius	8.6	10.7	13.35	17.5	7.9	9.85
a/h	7.17	7.64	7.85	8.75	39.5	32.8

pattern (Fig. 2b) the damping effect is clearly evident. Near the end of the tube the phenomenon can no longer be considered as a quasisteady state, and the deformation will be a function of both x or τ. The solution of the partial differential equations will also reflect the boundary conditions at the end of the tube, when $x=0$. The numerical solution of Eqs. (1) and (2) has been computed by the finite-difference method for several different values of the abscissa x, and a diagram of the deformation is shown in Fig. 2c.

When $V < V_b$ a deformation occurs before the detonation front. With the steady-state hypothesis the strains are no longer zero, and the ratio $\epsilon_x / \epsilon_\theta$ is negative. The numerical solution [Eq. (7)] becomes more critical.

Experimental Procedure

The experiments were made with detonations in gaseous mixtures of $C_3H_8 + 5O_2 + ZN_2$ in straight thin-walled circular tubes of either PVC or stainless steel. The initial conditions were ambient ($T_f = 290$ K and $P_f = 1$ bar). The ignition and the stabilization of the detonation were achieved in a stainless-steel tube (4 m long) connected to the 3 m long test section. The mechanical characteristics and mean sizes of the test sections which were used are shown in Table 1. The velocities of the detonations were measured at several places on the tube and appeared to agree with those already mentioned by Brossard and Charpentier (1976). The transverse and longitudinal deformations ϵ_θ and ϵ_x were measured with strain gages on the external skin of the tube. The sizes of longitudinal gages (6.35×3.18 or 0.79×0.81 mm) and the width of transverse gages (3.18 or 0.71 mm) were chosen to be considered weak in respect to the wave length (about 50 or 80 mm) of the oscillations of the tube. The transverse gage was placed at the same abscissa x as the longitudinal one, and when possible its length was nearly equal to the circumference of the tube in order to minimize the effects of a nonideal circular tube. The two recorders have a bandwidth of about 1 MHz. Typical records obtained with PVC and stainless-steel test sections are shown in Fig. 3. The pressure was measured with piezoelectric gages of 3 mm of diameter and with resonance frequency of about 1 MHz.

Results and Discussion

The important dynamic parameter is the velocity V of the detonation in relation to V_b and V_p (see Table 1). This parameter appears directly in the deformation through the dynamic amplifying factors A_θ and A_x, and indirectly through the pressure p which is produced by the detonation. When the dilution Z with inert increases from 0 to 9, the velocity V of the stable detonation decreases from 2415 m/s to about 1940 m/s. In the case of PVC the range of A_θ and A_x (Fig. 1) will be those given for V/V_p between 1.4 and 1.1.

The important geometrical parameters are, on the one hand, the mean radius a acting directly on the vibrational frequency ω and, on the other hand, the ratio a/h showing the effect of the tube wall thickness on the amplitude of the deformation. Because of the variation in the sizes of commercial tubes, only the very specific values a, h, and V of each experiment have been considered.

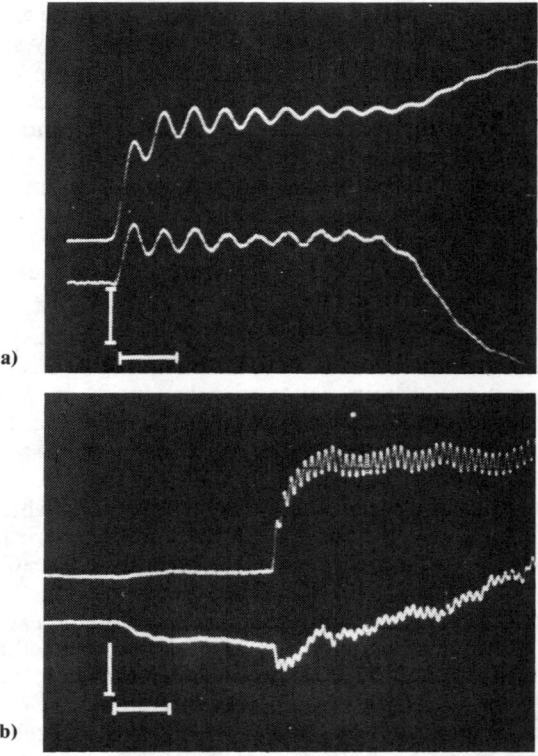

Fig. 3 Oscillograms of transverse (upper trace) and longitudinal (lower trace) deformations (detonating mixture $C_3H_8 + 5O_2$, $p_f = 1$ bar and $T_f = 290$ K): a) PVC tube, $2a = 17.35$ mm, $h = 1.25$ mm, $V = 2390$ m/s, $\epsilon_\theta = 1180$ μm/m, and $\epsilon_x = 1200$ μm/m per division, sweep rate = 50 μs/division; and b) stainless-steel tube, $2a = 19.7$ mm, $h = 0.3$ mm, $V = 2358$ m/s, ϵ_θ and $\epsilon_x = 240$ μm/m per division, sweep rate = 100 μs/division.

PVC Tube $(V > V_p)$

Typical results for various dilutions and test section diameters and wall thicknesses are shown on Fig. 4. The analysis of the numerous experimental results leads to the following conclusions:

1) As predicted, no deformation can occur before the detonation front (see Fig. 3) and the space-time correlation allows us to confirm that the beginning of the deformation coincides with the arrival of the detonation front at the point of observation.

2) The two deformations ϵ_x and ϵ_θ have the same sign and phase, their ratio agrees well with the theoretical estimates which depend only on the detonation velocity (see Fig. 5).

3) The set of curves in Fig. 4 shows that the amplitude of ϵ_θ decreases and that of ϵ_x increases when the detonation velocity decreases. It appears that ϵ_θ decreases more quickly than ϵ_x increases.

GASEOUS DETONATIONS ON A FLEXIBLE CONFINEMENT

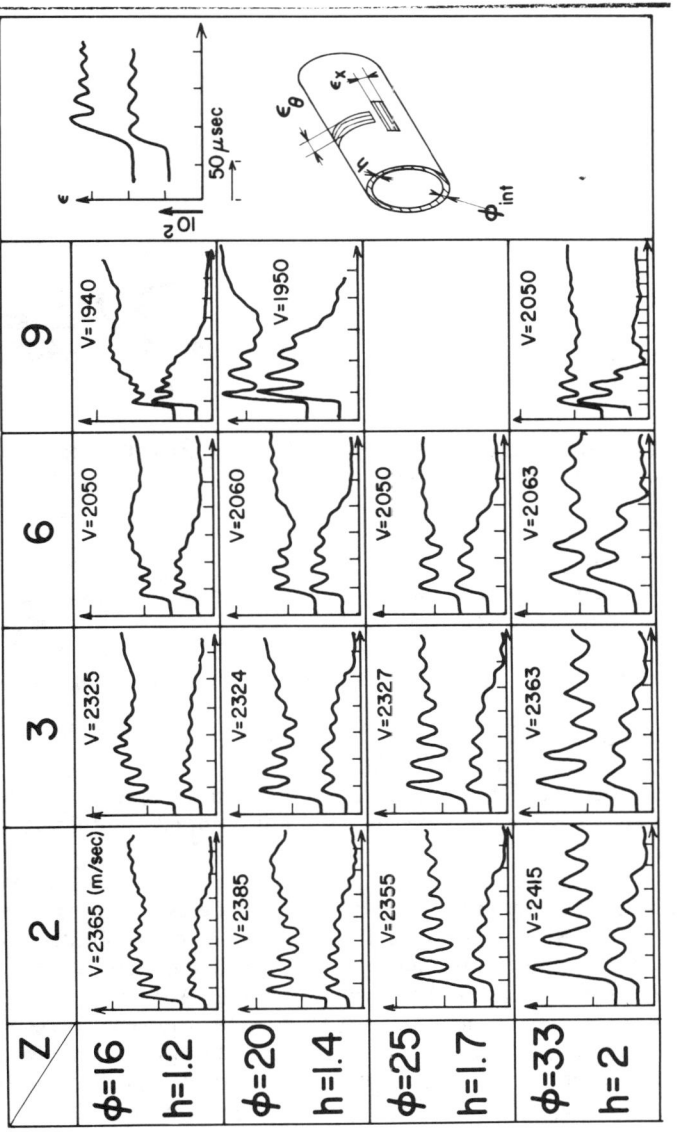

Fig. 4 Longitudinal and transverse deformation in PVC tubes (detonating mixture $C_3H_8 + 5O_2 + ZN_2$, $p_f = 1$ bar, and $T_f = 290$ K).

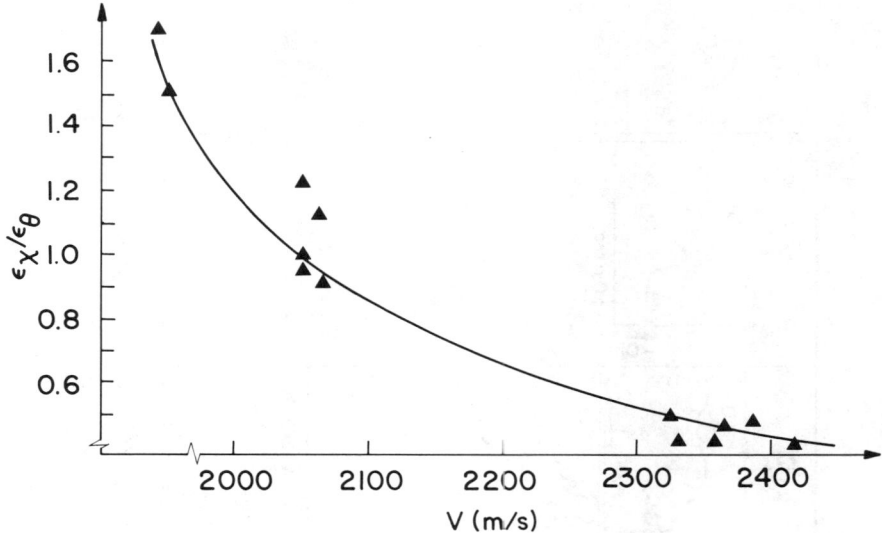

Fig. 5 Ratio of longitudinal and transverse deformations as function of the detonation velocity V for PVC tubes (—— prediction, ▲ observation).

4) The amplitude of ϵ_θ which can be considered in fact as the expansion of the test section remains lower than 0.4%. The disturbance of the gas flow because of deformation should remain lower than 0.8%.

5) The decrease of ϵ_x behind the wave front is faster for the weaker detonations.

6) The measured period of the oscillations around the mean deformation agrees well with the calculated value (Fig. 6).

7) The amplitude of the oscillations is the more important, the ratio a/h is the greater; it decreases like the wave velocity; it is damped out quickly.

Nevertheless a doubt remains about the amplitude of the mean deformation. The oscillograms are different from the calculated curves for an ideal Chapman-Jouguet pressure step with elastic or viscoelastic material behavior (Figs. 2a and 2b). Two important differences have to be pointed out: the gradual increasing to a level, and the amplitude of this level. If a deformation record, ϵ_θ for instance, is digitized and the numerical results may be used with the differential equations to calculate, at each time, the real corresponding pressure. An example of the results of such a calculation is shown in Fig. 7. The calculated pressure is very different from the ideal pressure step but the level value seems to agree well with the measured value for $t > 50$ μs. The measured ϵ_θ and ϵ_x were used with Eqs. (10) and (11) to calculate the actual pressure p. These results are compared with those of the ideal wave in a rigid confinement in Table 2. The ratio included between 0.45 and 0.80 decreases with the detonation velocity, and increases with the mean diameter of the tube.

Stainless-Steel Tube ($V < V_b$)

The observed phenomena are appreciably different from those of the previous case. Noteworthy observations are as follows:

1) A deformation with a nonnegligible amplitude propagated upstream of the detonation front (see Fig. 3).
2) The deformations ϵ_θ and ϵ_x have different signs, as predicted.
3) The transverse deformation ϵ_θ changes gradually and remains steady, but ϵ_x is always increasing.
4) The oscillations damp out very slowly and their frequencies about 80 KHz agree well with the theory.
5) Contrary to the PVC case, the amplitude of the level of transverse deformations is about 90% of the calculated value corresponding to the ideal pressure step in a rigid confinement.
6) For every tube we have used, the amplitude of ϵ_θ remains lower than 0.2%.

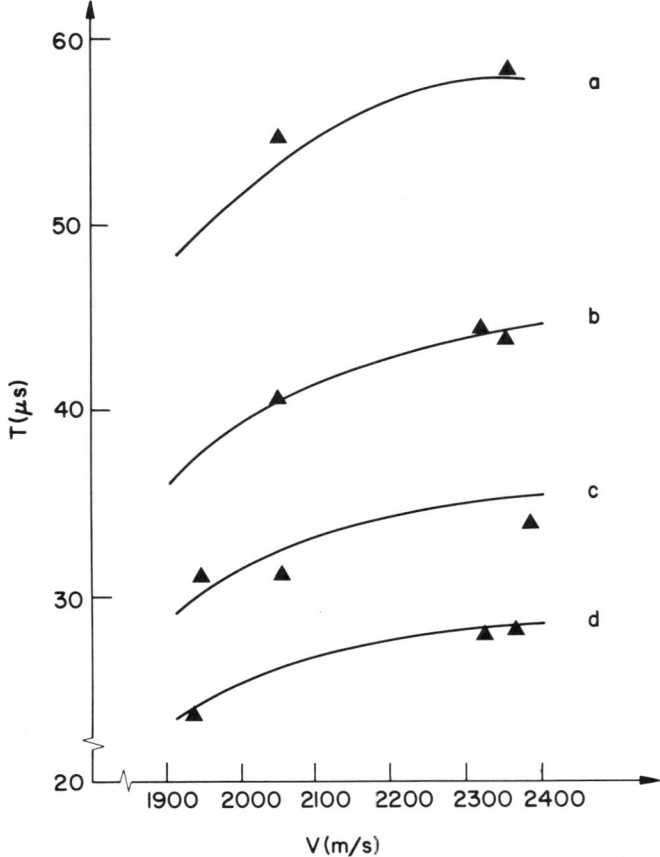

Fig. 6 Variations of period T of oscillation of PVC tubes as function of detonation velocity V for different values of the diameter of the tube (—— prediction; ▲ observation; for curves a, b, c, and d, $\phi = 33, 25, 20$, and 16 mm, respectively).

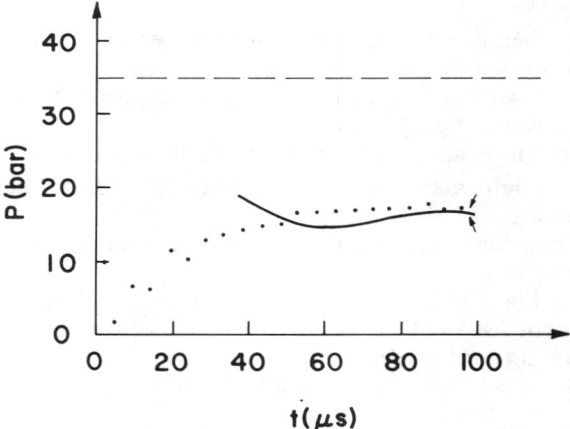

Fig. 7 Evolution of calculated and measured pressure at rear of detonation front for PVC confinement: $C_3H_8 + 5O_2$, $p_f = 1$ bar, $T_f = 290$ K, $V = 2380$ m/s, $2a = 17.35$ mm, and $h = 1.25$ mm (———— pressure step for a rigid tube, ——— pressure observed by transducer, ····· pressure calculated from deformations).

Table 2 Ratio of the pressure deduced from the mean deformation over the ideal pressure step (in PVC tubes)

Internal diameter	Pressure ratio			
	ON_2	$3N_2$	$6N_2$	$9N_2$
16	0.50	0.48	0.46	0.43
20	0.61	0.59	0.56	0.52
25	0.60	0.55	0.59	–
33	0.71	0.55	0.66	0.80

In the experiments the detonation waves were presumed to be stable and the analysis of the confinement expansion was made in respect to the ideal elastic behavior of the material. Nevertheless, this behavior appeared very different with the material confinement. In the case of PVC the brittle rupture of the tube could be obtained either by decreasing the detonation velocity (large Z) or by increasing the ratio a/h of the tube. This fact agrees well with the foregoing theoretical analysis. Although the expansion of the section of the tube remains weak under the effects of the detonation pressure, the flow of the burned gases can be really disturbed by the presence of expansion waves in the flowfield. So it is perhaps possible to explain the troubles in the pressure measurements and the appearance of "galloping" detonations in PVC tube with a diameter greater than 33 mm. With the expansion of the section it is also possible to associate a mechanical deformation energy. In an extreme case it does not exceed 3% of the thermochemical energy of the detonation.

Conclusions

With the objective of displaying the effect of a detonating gaseous mixture on its elastic circular confinement, a mathematical model was developed and the solutions compared with experimental results. Confinement materials with two different deformation patterns were investigated: polyvinylchloride and stainless steel. In the first case the detonation velocity is greater than the dilatational wave velocity in a thin shell of the material and in the second case, smaller.

The measurements of transverse and longitudinal deformations on the external skin of the test sections for several detonation velocities and for different geometrical and mechanical characteristics of the test sections allow us to characterize some features of these deformations: the precursor effect, the oscillation and their frequencies, the deformation ratio, and the dynamic amplifying factors. Nevertheless, two important observations remain without explanation: the gradual increasing of the transverse deformation to a certain level and, in the case of PVC, the amplitude of this level. Though the difficulties occurring in the pressure measurement were overcome, the correlation with the deformation amplitude is not quite satisfying. As a conclusion the simplified pressure distribution of the model on the one hand and the neglect of the effects of the viscosity of the material on the other are probably at the origin of the disagreement between model results and observation.

In the case of a polymer, a slow detonation may produce more serious mechanical damage than a faster one.

References

Brossard J. and Charpentier de Coysevox N. (1976) Effets d'un confinement souple sur la détonation des mélanges gazeux. *Acta Astronautica* **3**, 971-981.

Dabora E.K., Nicholls J.A., and Morrison R.B. (1965) The influence of a compressible boundary on the propagation of gaseous detonations. *Proc. 10th Symposium on Combustion,* pp. 817-832. The Combustion Institute, Pittsburgh, Pa.

Dawson T.H. (1972) Weak pressure wave in a gas filled elastic tube. *J. Sound and Vibration* **24**, 241-246.

Manson N. (1957) La théorie hydrodynamique et le diamètre limite de propagation des ondes explosive. *Zeit. für Elektrochemie* **61**, 586-592.

Niordson F. (1952) Transmission of shock waves in thin walled cylindrical tubes. *Acta Polytechnica* **2**, 3-23.

Oppenheim A.K., de Malherbe M.C., Wing R.D., and Laderman A.J. (1966) Response of a cylindrical shell to internal blast loading. *J. Mech. Eng. Sci.* **8**, 91-98.

Shchelkin K.I. (1968) Effect of a compressible boundary on the propagation of a gas detonation. *Phys. Comb. Explos.* **4**, 39-45.

Tang S.-C. (1965) Dynamic response of a tube under moving pressure. *J. Eng. Mech. Div.* **EM5**, 97-122.

Sommers W.P. (1961) Gaseous detonation waves interactions with non rigid boundaries. *ARS J.* **31**, 1780-1782.

Tsuge S. and Fujiwara T. (1972) Quasi-one dimensional analysis of gaseous free detonations. *J. Phys. Soc. Jap.* **33**, 237-241.

Tsuge S. and Fujiwara T. (1974) On the propagation velocity of a detonation shock combined wave. *Zeit. Ang. Math. und Mech.* **54**, 157-164.

Zeldovich Ya. and Kompaneets A.S. (1955) Théorie de la détonation. G.I.T.T.L. Moscow.

Oxyhydrogen Detonations under Surface Catalysis

T. Fujiwara* and T. Hasegawa†
Nagoya University, Nagoya, Japan

Oxyhydrogen detonations in a small-diameter tube have been observed experimentally and compared with theoretical calculations, taking into account two different phenomena independently—heterogeneous reactions and condensation of water vapor on the surface of the tube. To confirm the influence of surface catalytic reactions or water condensation, which was pointed out in a previous study, the experimental conditions were varied more extensively. Along with stoichiometric oxyhydrogens, three additional mixtures, H_2+O_2, H_2+2O_2, and $3H_2+O_2$, were investigated for pressure range $p_1 = 0.5 \sim 1.0$ atm and T_1 = room temperature, using an alumina tubing of 4 mm i.d. In contrast with about $4 \sim 5\%$ slowdown (about 100 m/s) in stoichiometric oxyhydrogen detonations, the deviations reached $9 \sim 11\%$ ($200 \sim 250$ m/s) in H_2+O_2 detonations, $17 \sim 20\%$ ($330 \sim 370$ m/s) in H_2+2O_2 detonations, and $12 \sim 15\%$ ($350 \sim 450$ m/s) in $3H_2+O_2$ detonations, while approximately steady modes of propagation were still retained. The results of theoretical calculations can be summarized as follows: 1) in H_2+O_2 mixtures, simultaneous surface reactions $O+OH \rightarrow HO_2$ and $H+H \rightarrow H_2$ provided best correlation with experimental data; 2) in H_2+2O_2 mixtures, $H+OH \rightarrow H_2O$ and $O+OH \rightarrow HO_2$; 3) in $3H_2+O_2$ mixtures, $H+H \rightarrow H_2$ and $H+OH \rightarrow H_2O$, which was similar to stoichiometric oxyhydrogens; and 4) the observed velocity deficit was explained also by assuming H_2O vapor adsorption with the adsorption coefficient 0.3 for all but $3H_2+O_2$ observed mixtures. Thus, the velocity deficit is likely to be attributed to surface catalysis rather than adsorption.

Introduction

THE influence of tube wall on the propagation velocity of gaseous detonations is strengthened by the presence of shock waves and flows in transverse directions. Slowdown of oxyhydrogen detonations was observed by Munday et al. (1968) even for a tube of 28 mm i.d. More conspicuous slowdown was observed by Fujiwara (1977) who used very thin tubes of about 4 mm i.d. To explain the considerable observed velocity deficits, several

---

Presented at the 7th ICOGER, Göttingen, Federal Republic of Germany, Aug. 20-24, 1979. Copyrifht © American Institute of Aeronautics and Astronautics, Inc., 1981. All rights reserved.
*Associate Professor, Department of Aeronautical Engineering.
†Graduate Student, Department of Aeronautical Engineering.

possible mechanisms were considered: 1) increased viscous and heat conduction effects caused by the presence of microscopic roughness on the surface of alumina tubing, 2) heat loss to the tube wall due to the catalytic destruction of radical species, and 3) heat loss to the tube wall by the adsorption of water vapor. Among these three possible mechanisms, the loss of chemical energy represented by 2) exceeds the phase change energy 3), followed by the thermal energy 1), when measured per unit mole of associated molecules.

In view of such a consideration, more observations were performed in the present study. Steady and unsteady propagation of detonations were measured for hydrogen-lean ($H_2 + O_2$, $H_2 + 2O_2$) and hydrogen-rich ($3H_2 + O_2$) mixtures under various initial pressures ($p_I = 0.5 \sim 1.0$ atm) using an alumina and a Pyrex tube. Also carried out are theoretical calculations of the propagation velocity in small tubes; these calculations take into account the surface catalytic reactions and water vapor adsorption independently.

Experiments

As shown in Fig. 1, the detonation tube used in the present experiment consists of a 28 mm i.d. stainless-steel driver section with an ignition plug mounted at one end and two long and thin tubes, one 4 ± 0.1 mm i.d. smooth Pyrex glass tube and the other 3.9 ± 0.1 mm i.d. aluminum, the inner surface of which was converted into gamma-alumina using a conventional method. In other words, an aluminum tube was treated with a solution of sodium-hydroxide and later by dry ammonia gas for 10 min, followed by baking at 300°C for 8 h. The microscopic observation of the inner surface of the tube revealed that the surface was rough and silver shiny with a roughness about 50 μm in average and believed to be gamma-alumina.

Tubes with an i.d. of 4 mm were almost ideal for the present particular purposes. In tubes of smaller diameters, even stoichiometric oxyhydrogen mixtures allowed no steady propagation at initial pressures less than 1 atm; whereas for larger diameters, no substantial velocity decrease was obtained. If the distance between the leading shock and the Chapman-Jouguet surface in a detonation is of the order of 1 mm, for example, and the transverse and longitudinal flow velocities are of the same order, 75% of the gas in the region comes under the influence of the wall surface. As is frequently observed, in real detonations the flowfield behind the leading shock wave is far from one dimensional, causing the reaction length to be much longer than that of a quasione-dimensional process. As a result, surface effects show up much more readily than one can expect from one-dimensional calculations.

The detonation tube is divided into two sections by a 0.2 mm thick aluminum diaphragm. An equimolar oxyhydrogen mixture is introduced into the driver section at 0.6 atm, while a test gas is admitted to the driven section at $p_I = 0.5 \sim 1.0$ atm and $T_I = 280 \sim 310$ K, after the entire system has been evacuated to 10^{-2} Torr. The gas is ignited by the discharge of 0.1 μF at 6 kV. The detonation velocities in the two parallel tubes are simultaneously detected through the photodiodes output. Photodiode 2 is used as a trigger, while 3 and 4 (1091 mm apart) are used to measure the propagation velocities. In the opaque alumina tube, a short Pyrex glass tube 60 mm in length is inserted at each photodiode location to allow radiation detection.

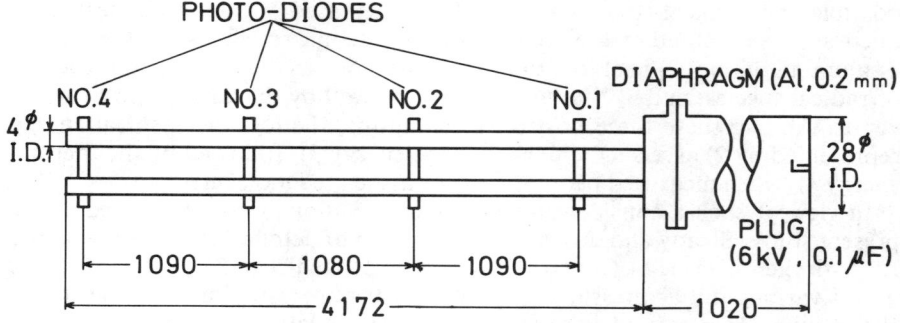

Fig. 1 Experimental apparatus.

Typical examples of oscillograms of photodiode outputs are illustrated in Fig. 2, where the outputs from diodes 3 and 4 are introduced into the same channel of the oscillograph. In Fig. 2a, for example, the arrival of a detonation front at diodes 3 and 4 is marked by A and B for the Pyrex tube, and C and D for the alumina tube, respectively, representing a fairly large difference in the passage times, i.e., 478 and 544 μs. Such a difference becomes most marked in Fig. 2b, where the velocity deficit reaches 27%. The different times of arrival at diodes 3 are utilized to identify steady propagation. If the detonation has been steady since entering the thin tube, the propagation velocity in the alumina tube prior to the arrival at location 3 can be obtained as

$$(D')_{Al} = \frac{(D)_{Pyrex}}{1 + \frac{\Delta \tau}{L}(D)_{Pyrex}} \tag{1}$$

where $\Delta \tau$ denotes the difference in arrival times, L the tube length up to location 3, and $(D)_{Pyrex}$ the detonation velocity in the Pyrex tube. The measured detonation velocities for three different oxyhydrogen mixtures are shown in Figs. 3-5 along with calculated detonation velocities, assuming different mechanisms for heat loss to the tube surface.

Before going into the detailed theoretical analysis and formulation of the problem, it may be interesting to consider possible mechanisms of observed retardation of detonations. The retardation can arise from either/or a combination of the following causes:

1) The drag or heat-transfer effects due to a thick boundary layer caused by surface roughness (Zeldovich and Kompaneets, 1960).
2) The diameter difference between the Pyrex and alumina tubes.
3) The heat loss due to vaporization of surface alumina.
4) The heat loss originated from catalytic destruction of radical molecules.
5) The heat loss caused by the adsorption of water vapor on the tube surface.

Although the heat conductivity of alumina is 30 times as high as that of Pyrex glass, the dominant factor in heat transfer is obviously the gaseous counterpart, which is irrelevant to surface material. Even the characteristic heat conduction time, $\rho C_p d^2/4\lambda$, is 10^{-2} s for the alumina tube, much longer

OXYHYDROGEN DETONATIONS UNDER SURFACE CATALYSIS

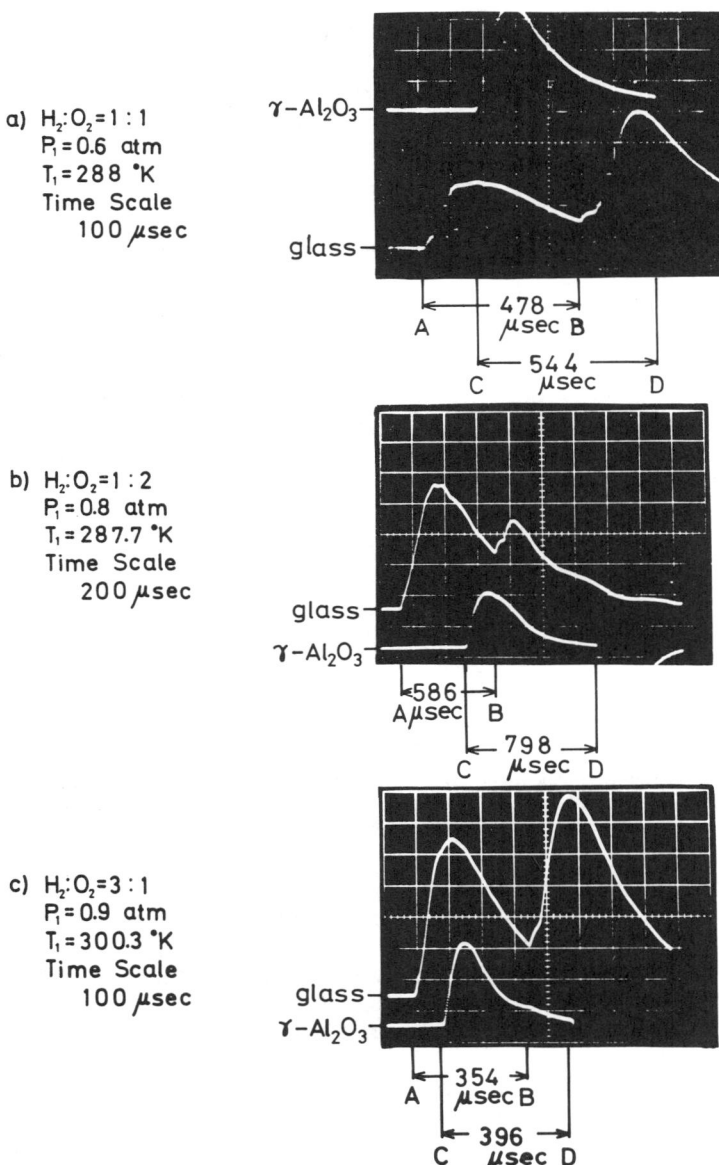

a) $H_2 : O_2 = 1 : 1$
  $P_1 = 0.6$ atm
  $T_1 = 288$ °K
  Time Scale
  100 µsec

b) $H_2 : O_2 = 1 : 2$
  $P_1 = 0.8$ atm
  $T_1 = 287.7$ °K
  Time Scale
  200 µsec

c) $H_2 : O_2 = 3 : 1$
  $P_1 = 0.9$ atm
  $T_1 = 300.3$ °K
  Time Scale
  100 µsec

Fig. 2 Typical oscillograms of photodiode outputs: signals from two different locations along the same tube are introduced into one channel.

than the time for chemical reaction (10^{-6} s). According to our previous work (Fujiwara, 1977), about 1% velocity deficit was observed for an artificially roughened Pyrex tube with about 20 µm roughness (the roughness/radius ratio

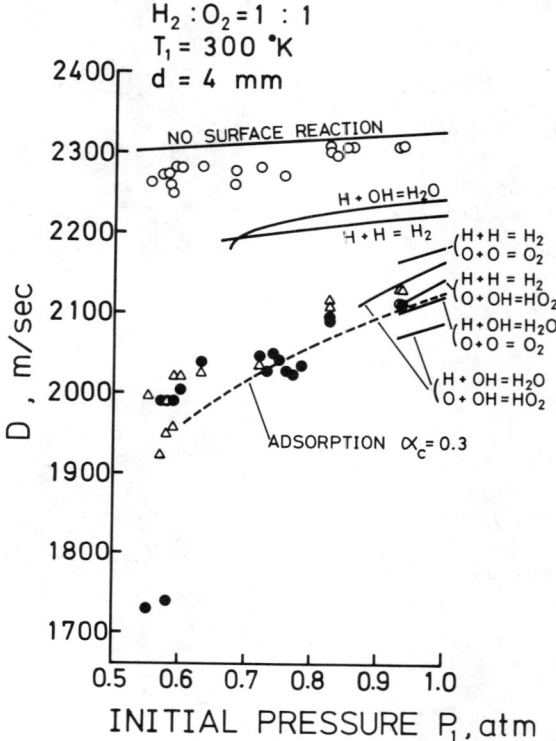

Fig. 3 Observed and calculated detonation velocities for $H_2 + O_2$ mixtures at T_I = room temperature in tubes of 4 mm i.d. (initial pressures are varied between 0.5 and 1.0 atm).

was 1.5% at most). In connection with roughness, Shchelkin (Zeldovich, 1960, p. 187) reported 2.4% deficits for $2.07H_2 + O_2$ in a tube of 4.5 mm i.d. where a helical wire was located to produce a roughness/radius ratio as high as 0.13. From the foregoing observations, it is reasonable to eliminate the effects of thickened boundary layer on increased viscosity and heat transfer from possible mechanisms. Note, however, that the influence of increased surface area may be important.

The diameter difference and the alumina vaporization were discussed and found trivial in our previous work.

In contrast with the foregoing several mechanisms, both the surface catalysis and the water adsorption can explain large velocity deficits up to more than 20%, which are often observed in Figs. 3-5. Adsorption can be, for example, of importance from the following experimental results:

1) The alumina tube has stronger affinity with water and a few orders of magnitude larger surface area than the Pyrex one, as confirmed by much longer evacuation time.

2) In the case of insufficient initial evacuation, the firing of detonation makes subsequent evacuation more difficult, which suggests that physisorp-

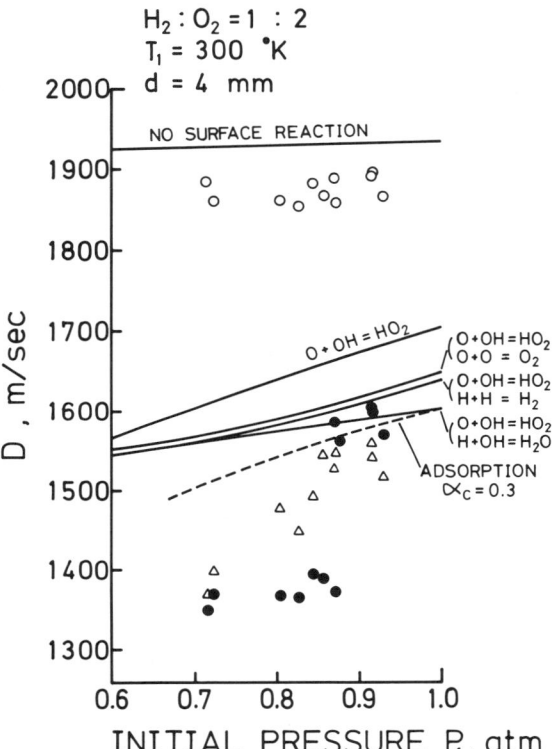

Fig. 4 Observed and calculated detonation velocities for $H_2 + 2O_2$ mixtures at T_1 = room temperature in tubes of 4 mm i.d. (initial pressures are varied between 0.7 and 1.0 atm).

tion has been converted into chemisorption by the action of detonation.

3) The observed velocity deficits are nearly irrelevant to the extent of initial vacuum.

In other words, the existence of water molecules up to 1 Torr in the initial gas phase essentially does not change the molecular structure of adsorbed species. According to infrared spectroscopy of alumina surface (Hair, 1967), physisorbed H_2O molecules are removed after evacuation at room temperature, while chemisorped ones still remain.

To make the situation more complicated, the structural differences between physically and chemically adsorbed H_2O and radicals and their behavior as catalysts are not known. Therefore, it is highly difficult to decide which out of the two possible mechanisms is occurring on alumina surface. Subsequent theoretical analysis will be performed to answer this question more clearly.

Theoretical Prediction of Slowdown

Surface catalytic destruction of radical species or water vapor adsorption will be included in the following theoretical formulation as a mechanism of

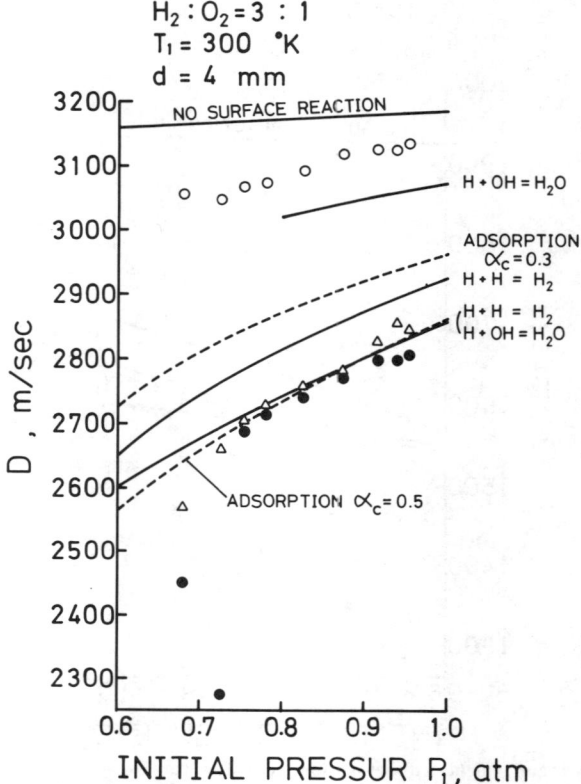

Fig. 5 Observed and calculated detonation velocities for $3H_2 + O_2$ mixtures at $T_I =$ room temperature in tubes of 4 mm i.d. (initial pressures are varied between 0.6 and 1.0 atm).

heat loss to the tube wall. This effect leads to the slowdown of the detonations in a thin alumina tube.

Eleven gas-phase reactions listed on Table 1 are taken into account throughout the analysis. In the formulation of a surface catalysis model, on the other hand, at most two out of the seven surface reactions listed in Table 2 are simultaneously included in one run of the calculations, while no surface reactions other than water vapor adsorption are considered in the adsorption model. For simplicity, the following assumptions are made in connection with the phenomena occurring near the surface:

1) Convection of reaction products to and from the surface is fast enough to maintain a perfectly reaction-controlled process. Then, a quasione-dimensional formulation becomes possible; for example, the rate of reaction 1 in Table 2 can be written as

$$W_I^* = -[OH]k_I^* \frac{4}{d} \qquad \text{mole} \cdot \text{cm}^{-3} \cdot \text{s}^{-1} \qquad (2)$$

Table 1 Gas-phase reactions and associated rate constants utilized in the calculation [a]

Elementary reactions in gas phase	Forward reaction rates, $cm^3 \cdot mole^{-1} \cdot s^{-1}$ or $cm^6 \cdot mole^{-2} \cdot s^{-1}$	Backward reaction rates
$H_2 + O_2 \rightleftharpoons OH + OH$	$k_1 = 2 \times 10^{11} \exp(-20/RT)$	$k_2 = k_1/K_1$
$H + O_2 \rightleftharpoons OH + O$	$k_3 = 8.08 \times 10^{13} \exp(-16.6/RT)$	$k_4 = k_3/K_2$
$O + H_2 \rightleftharpoons OH + H$	$k_5 = 2 \times 10^{13} \exp(-9.2/RT)$	$k_6 = k_5/K_3$
$OH + H_2 \rightleftharpoons H_2O + H$	$k_7 = 3.5 \times 10^{13} \exp(-5.9/RT)$	$k_8 = k_7/K_4$
$OH + OH \rightleftharpoons H_2O + O$	$k_9 = 3.8 \times 10^{12} \exp(-1.0/RT)$	$k_{10} = k_9/K_5$
$H + H + X \rightleftharpoons H_2 + X$	$k_{11} = 9 \times 10^{17}/T$	$k_{12} = k_{11}/(K_6 \cdot RT)$
$H + OH + X \rightleftharpoons H_2O + X$	$k_{13} = 9 \times 10^{18}/T$	$k_{14} = k_{13}/(K_7 \cdot RT)$
$H + O + X \rightleftharpoons OH + X$	$k_{15} = 1 \times 10^{15}$	$k_{16} = k_{15}/(K_8 \cdot RT)$
$O + O + X \rightleftharpoons O_2 + X$	$k_{17} = 1.05 \times 10^{15}/\sqrt{T}$	$k_{18} = k_{17}/(K_9 \cdot RT)$
$H + O_2 + X \rightleftharpoons HO_2 + X$	$k_{19} = 8.25 \times 10^{15}$	$k_{20} = k_{19}/K_{10} \cdot RT)$
$HO_2 + H_2 \rightleftharpoons H_2O + OH$	$k_{21} = 1.3 \times 10^{14} \exp(-22/RT)$	$k_{22} = k_{21}/K_{11}$

[a] $R = 1.988 \times 10^{-3}$ kcal/mole/K; $K_j = (K_A \cdot K_B \cdot K_C ...)/(K_a \cdot K_b \cdot K_c ...)$ for jth reaction $a+b+c+... \rightleftharpoons A+B+C+...$, where K_A, K_B, etc., are equilibrium constants of species A, B, etc., respectively, found in JANAF tables.

Table 2 Possible surface reactions occurring on gamma-alumina and associated rate constants [a]

Reaction No.	Elementary reactions on surface	Reaction rates k_j^*, cm/s [b]
1	$OH + surface \rightarrow \frac{1}{2}H_2O + \frac{1}{4}O_2 + surface$	$C_{OH}/4\alpha_1$
2	$O + surface \rightarrow \frac{1}{2}O_2 + surface$	$C_O/4\alpha_2$
3	$H + surface \rightarrow \frac{1}{2}H_2 + surface$	$C_H/4\alpha_3$
4	$OH + surface \rightarrow \frac{1}{2}O_2 + \frac{1}{2}H_2 + surface$	$C_{OH}/4\alpha_4$
5	$OH + surface \rightarrow \frac{1}{2}HO_2 + \frac{1}{4}H_2 + surface$	$C_{OH}/4\alpha_5$
6	$OH + H + surface \rightarrow H_2O + surface$	$C_{OH}/4\alpha_6$ or $C_H/4\alpha_6$
7	$OH + O + surface \rightarrow HO_2 + surface$	$C_{OH}/4\alpha_7$ or $C_O/4\alpha_7$

[a] Two reactions at most are included simultaneously in one calculation.
[b] $C_i = \sqrt{8kT/\pi m_i}$ = thermal velocity of ith species molecule; α_j = catalytic efficiency of jth surface reaction.

where $k_j^* = C_{OH}\alpha_j/4$, C_i is the thermal velocity of ith species, α_j the catalytic efficiency of jth reaction, and d the tube diameter. The rate of water adsorption can be expressed similarly as

$$W_{ad} = \alpha_c \frac{C_{H_2O}}{4} [H_2O] \frac{4}{d} \quad mole \cdot cm^{-3} \cdot s^{-1} \qquad (3)$$

where α_c is the condensation coefficient.

2) Energy differences between reactants and products in surface reactions, or the heat of condensation, are perfectly transferred to the surface although the surface temperature remains at ambient. Then, the heat loss term in the energy equation is given for surface catalysis as

$$\frac{dQ}{dx} = \sum_{j=1}^{7} \frac{1}{\rho V} W_j^* [\overline{H}_j(T_1) - H_j(T)] \quad cal \cdot cm^{-1} \cdot g^{-1} \qquad (4)$$

and for water adsorption as

$$\frac{dQ}{dx} = \frac{1}{\rho V} W_{ad} [H_{H_2O}(T_l) - H_{H_2O}(T)] \quad \text{cal·cm}^{-1}\cdot\text{g}^{-1} \tag{5}$$

where j denotes the seven reactions listed in Table 2, and W_j^* the rate of reaction typically represented by Eq. (2). The enthalpies $\overline{H}_j(T_l)$ and $H_j(T)$ are defined as

$$\overline{H}_1(T_l) = \tfrac{1}{2} H_{H_2O}(T_l) + \tfrac{1}{4} H_{O_2}(T_l)$$
$$H_1(T) = H_{OH}(T) \tag{6}$$

for reaction 1 in Table 2, for example, where H_i is the molar enthalpy of ith species. Although a complete loss of released heat to the wall is questionable because a significant amount of "surface reactions" can occur in the cold gas-phase boundary layer, the order of magnitude presumably remains unchanged.

Thus, the fundamental equations governing the flow behind a leading shock wave are written as:

Mass conservation

$$\rho \frac{dV}{dx} + V \frac{d\rho}{dx} = \begin{cases} 0 & \text{(surface catalysis)} \quad (7) \\ -W_{ad} M_{H_2O} & \text{(water adsorption)} \quad (8) \end{cases}$$

Momentum conservation

$$\frac{dp}{dx} + \rho V \frac{dV}{dx} = 0 \tag{9}$$

Energy conservation

$$\frac{d}{dx}\left(h + \frac{V^2}{2}\right) = \frac{dQ}{dx} \tag{10}$$

$$h = \sum_i \frac{\rho_i}{\rho} h_i(T) \tag{11}$$

$$\rho_i = [i] M_i \tag{12}$$

$$\rho = \sum_i \rho_i \tag{13}$$

where ρ is the density, V the gas velocity in the wave coordinate, T the temperature, h the enthalpy per unit mass, $[i]$ the molar concentration, and M_i the molar weight.

Equation of state

$$\frac{dp}{dx} = R_0 \left(X \frac{dT}{dx} + T \frac{dX}{dX} \right) \tag{14}$$

$$X = \sum [i] \tag{15}$$

Table 3 Species concentrations and temperatures at Chapman-Jouguet surface for purely one-dimensional detonations without surface effects[a]

Mixing ratio	$H_2 + 2O_2$	$H_2 + O_2$	$2H_2 + O_2$	$3H_2 + O_2$
O, %	2.29	5.47	3.86	1.30
H, %	0.32	2.57	8.12	9.65
OH, %	6.63	14.2	13.6	7.94
HO_2, %	0.02	0.03	0.01	0.02
H_2O, %	34.6	47.5	53.1	48.6
H_2, %	0.58	3.94	16.5	31.7
O_2, %	55.6	26.2	4.92	0.79
T_{C-J}, K	3020	3460	3672	3593

[a] Initial pressure $p_I = 1$ atm, initial temperature $T_I = 300$ K.

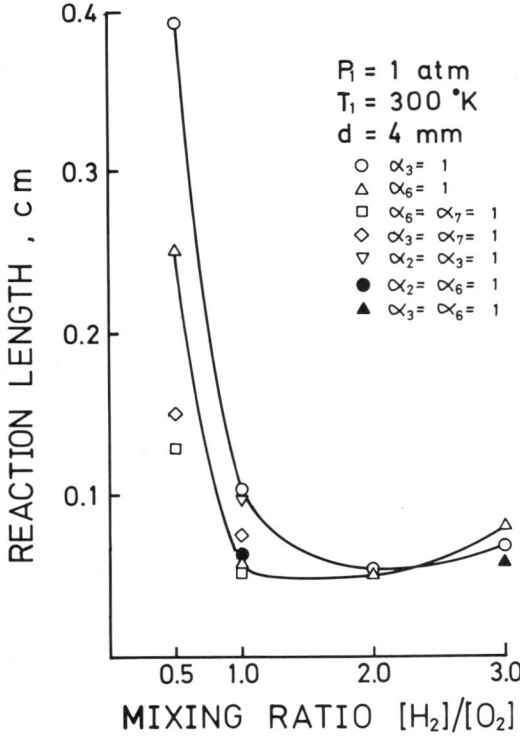

Fig. 6 Lengths of reaction zone as function of mixing ratios for oxyhydrogens at $p_I = 1$ atm and $T_I = 300$ K in tube of 4 mm i.d., calculated under various assumptions on catalytic efficiencies.

Species conservation

$$\frac{d[i]V}{dx} = W_i^+ - W_i^* \quad \text{mole} \cdot \text{cm}^{-3} \cdot \text{s}^{-1} \tag{16}$$

where R_0 is the universal gas constant.

The energy equation can be written in a simple singular form:

$$\frac{dT}{dx} = f(T,M; D)/(1-M^2) \tag{17}$$

which determines the propagation velocity D of the detonation as an eigenvalue by utilizing the generalized Chapman-Jouguet condition:

$$f(T,M; D) = 0 \quad \text{at} \quad M = 1 \tag{18}$$

Results and Discussions

Observed and calculated results are compared in Figs. 3-5. In equimolar oxyhydrogen mixtures, as shown in Fig. 3, the detonation velocities observed in a Pyrex tube agree with the calculated ones, assuming no surface reactions. The deviation starts from virtually zero at $p_1 \cong 1$ atm to about 2% at reduced pressures, presumably due to the increased influence of surface area caused by longer reaction lengths. The observed deviations become far more marked in a gamma-alumina tube, i.e., 9~11% slower than the Pyrex counterparts, indicating a higher surface activity of the material. Steady propagation at $p_1 > 0.6$ atm has been confirmed by the agreement of the velocities between the test section (solid circles) and the upstream (open triangles), while the detonation limit occurs at $p_1 = 0.6$ atm. As illustrated in Fig. 3, the slowdown is successfully explained either by surface reactions or water adsorption. Among others, the best fit is provided by two simultaneous surface reactions among H, O, and OH taking place with each catalytic efficiency unity, as well as water adsorption with its accommodation coefficient 0.3. Since plenty of radical species exist in the reaction zone, as shown in Table 3, loss of energy due to surface recombinations can naturally yield a significant amount of slowdown. Along the same line, the adsorption of the most abundant molecules (H_2O), has provided the slowdown systematically throughout the pressure range under consideration.

In oxygen-rich mixtures $H_2 + 2O_2$, as shown in Fig. 4, the observed velocities in Pyrex deviate from the calculated Chapman-Jouguet values by 2.6%, which is much higher than in equimolar mixtures. This can be attributed to long reaction zones suggested by Fig. 6 and a low T_{C-J} shown in Table 3. The steady deviation, observed only above the detonation limit $p_1 \cong 0.87$ atm, extends even to 17~20%, i.e., 330~370 m/s. The slowdown is explained either by a surface reaction $O + OH \rightarrow HO_2$ between the most abundant radicals confirmed in Table 3, or again by water adsorption assuming the accommodation coefficient 0.3.

The results for hydrogen-rich mixtures $3H_2 + O_2$ are illustrated in Fig. 5. Recombination of hydrogen atoms is more dominant in reducing detonation velocities than any other surface reaction, because H is the most abundant and of much higher energy than OH, which also exists in plenty according to Table 3. Although both a set of surface reactions ($H + H \rightarrow H_2$ and $H + OH \rightarrow H_2O$) and water adsorption provide excellent explanations throughout the pressure range of steady propagation, the best fit is obtained by an adsorption coefficient 0.5 unlike the preceding two cases $H_2 + O_2$ and $H_2 + 2O_2$.

Conclusions

The amount of the observed experimental data and analytically calculated results is limited and far from sufficient. The following discussions are possible, however, in determining the relative importance between surface catalysis and water adsorption as a true mechanism of slowdown.

For different oxyhydrogen mixtures, different combinations of surface reactions are taken into account to explain the observed velocity deficits up to about 20% of lossless Chapman-Jouguet values. However, the reactions which provide good agreement with experiments are always between the most abundant molecules under a given mixing ratio: $O+OH \rightarrow HO_2$, $O+O \rightarrow O_2$, $H+OH \rightarrow H_2O$, and $H+H \rightarrow H_2$ for H_2+O_2 mixtures; $O+OH \rightarrow HO_2$ for H_2+2O_2 mixtures; and $H+H \rightarrow H_2$ for $3H_2+O_2$ mixtures.

In the case of water adsorption formulation, on the other hand, an accommodation coefficient of 0.3 is able to yield good agreement for H_2+O_2 and H_2+2O_2 mixtures at a wide range of initial pressures, while a much larger coefficient of 0.5 is needed for $3H_2+O_2$ mixtures. This inconsistency in adsorption coefficients is a rather unconvincing argument for water adsorption being the mechanism of detonation slowdown, if the mechanism of adsorption is not strongly influenced by mixing ratios. This, however, needs to be studied.

At the present stage, it is more consistent to consider that surface recombination reactions are responsible for the heat loss to the tube wall and the resulting slowdown of detonations.

References

Fujiwara T. (1977) *16th (Intl.) Symposium on Combustion,* p. 1771. The Combustion Institute, Pittsburgh, Pa.

Hair M.L. (1967) *Spectroscopy in Surface Chemistry,* p. 141. Marcel Dekker, New York.

Munday G., Ubbelohde A. R., and Wood I. F. (1968) *Proc. Roy. Soc.* (London) **A306,** 159.

Zeldovich Ia. B. and Kampaneets A. S. (1960) *Theory of Detonation,* p. 145, 187. Academic Press, New York.

Pressure and Wall Heat Transfer behind a Hydrogen/Azide Detonation Wave in Narrow Tubes

C. Paillard,* G. Dupre,* R. Lisbet,* and J. Combourieu*
Université d'Orléans, Orléans, France
and
V.P. Fokeev,† L.G. Gvozdeva,† and T.V. Bazhenova†
Academy of Sciences, Moscow, U.S.S.R.

Predictions of heat fluxes for constant-flow parameters did not agree with observations obtained with heat-transfer gages, except just behind the detonation front or under some special conditions of initial pressure and tube diameter. To characterize the influence of variable-flow parameters on wall heat losses, the pressure profiles behind the wave were measured for HN_3 detonation in narrow tubes (1, 2, 5, and 10 mm i.d.) and at initial pressures of 1-20 Torr. The experimental pressure profiles behind the wave were compared to different theoretical solutions. In all cases, the observed pressure decreased much faster than that predicted by the isentropic solution. Predictions that account for wall heat losses and steady flow of gas downstream were not in accord with observations in any case except just behind the wave front, or for initial conditions such that the velocity deficit was inversely proportional to the diameter. Near detonation limits, and for tubes of very small diameters, the simplifying hypotheses of Edwards et al. (1970) were no longer valid. The trajectories of triple-shock intersections were observed on the soot-coated inner surface of the tube elements. This method has clearly shown the influence of tube diameter on the cell length and the role of the chemical kinetics on the wave structure.

Introduction

THE effect of confinement on the behavior of gaseous detonations has been extensively investigated. In general, this effect causes a small velocity deficit due to the development of a wall boundary layer behind the initiating shock (Fay, 1959). The characteristics of the flow behind the wave are assumed

---

Presented at the 7th ICOGER, Göttingen, Federal Republic of Germany, Aug. 20-24, 1979.
Copyright © American Institute of Aeronautics and Astronautics, Inc., 1981. All rights reserved.
*Centre de Recherches sur la Chimie de la Combustion et des Hautes Températures (CNRS).
†Institute of High Temperatures.

to be close to the Chapman-Jouguet (C-J) state values. When the wave propagates in a narrow tube at low initial pressure in the case of compounds such as gaseous azides, which decompose exothermically, the velocity deficit can reach 50% (Paillard et al., 1974). In such conditions, the wall heat losses may modify the parameters of the gases behind the shock front. Edwards et al. (1970) have analyzed the influence of wall heat transfer on the expansion following a C-J detonation wave in stoichiometric hydrogen-oxygen mixtures at 1 atm initial pressure. Wall heat fluxes measured in tubes larger than 15 mm i.d. were found to be independent of tube diameter, and to agree with predictions obtained from Sichel and David's (1966) hypothesis. A constant friction coefficient, deduced from heat-transfer measurements was used to determine the pressure, temperature, and density profiles behind the detonation wave.

For a gaseous HN_3 detonation wave, which was confined in narrow tubes (i.d. ≤ 10 mm), the experimental heat-transfer rates did not agree with theoretical values and exhibited a strong dependence on tube diameter (Paillard et al., 1979). For the conditions studied, the above calculations of the flow parameter profiles were no longer justified. The present study, therefore, will attempt to determine the pressure evolution behind the wave in detonation tubes of 1-10 mm i.d. Since Gordon's (1949) and Edwards et al.'s (1959) measurements, technological improvements have made possible the adaptation of piezoelectric transducers to very narrow tubes. On the other hand, it is well known that all self-sustained detonation waves exhibit a three-dimensional structure behind the leading shock front through the reaction zone. The interaction of wall and flow inside this zone can exert a noticeable influence on the detonation structure and, consequently, on the gas characteristics and on the detonation limits. To investigate the effect of tube diameter on the structure, the trajectories of triple-point collisions on the inner wall of narrow tubes coated with soot have been observed.

Experimental

Hydrogen azide HN_3 at initial pressures between 1 and 20 Torr was detonated in small diameter tubes with length-to-diameter ratios > 375. The preparation of HN_3 and the experimental system were described by Paillard et al. (1979). The detonation apparatus consisted of four tubes (1, 2, 5, and 10 mm i.d. and each 3.75 m long) connected with a larger tube where ignition was effected. Heat-transfer rates, pressure transients, and soot traces were obtained in separate experiments for each tube at distances X of 1.4, 2.4, and 3.4 m from the ignition chamber. Thermal gages and pressure transducers were mounted flush with the wall and connected to oscilloscopes. Elapsed time between the responses of adjacent stations were used to deduce precise measurements of the wave velocity.

The pressure behind the leading shock was measured with piezoelectric transducers made of lead titanate and lead zirconate. These transducers were calibrated either with a shock wave produced in a shock tube or with a steep rise of pressure produced in a small vessel. Two types of transducers were used:

1) French pressure transducers (LEM 20 H48) fabricated by LEM Society under license from ONERA. Each transducer can be adapted on tubes with 2-10 mm i.d. through a mechanical mounting. The sensitive area is mounted flush with the inner wall. Because of the small diameter (0.8 mm) pressure records were obtained with a good spatial resolution. The short rise time (0.2 μs) is convenient for measurement of the rapid pressure transients behind the detonation wave, and sensitivity of the order of 3-4 Torr per mV permits experiments at low initial pressures.

2) Russian transducers, fabricated in the Soviet Union at the Institute of High Temperatures of Moscow from Gvozdeva's design (1978). These have a sensitive area 1 mm in diameter, a sensitivity varying between 5 and 100 Torr/mV and a rise time of about 1 μs. They are machined to give the proper curvature of the inner wall of each tube. The transducers are cemented on a tube element with Araldite.

The detonation structure was studied by the classic method of soot traces. A tube section 45 mm long was cut in two hemicylinders and modified so that it could be sealed and connected to the detonation tube. A soot layer was deposited on the inner surfaces of the hemicylinders. The section was removed after each experiment and trajectories of triple-shock intersections were observed.

Results

In previous studies, the HN_3 detonation velocities were measured in the four detonation tubes and the C-J parameters calculated for an initial pressure P_I, in the range of 1-25 Torr (Paillard et al., 1973 and 1979).

Heat-Transfer Rate

The present study of wall heat-transfer rates complements the prior work of Paillard et al. (1979). The influence of tube diameter ϕ on experimental heat-transfer rates is shown in Fig. 1. The theoretical values were computed from Mirels' (1958) theory as modified by Sichel et al. (1966) for a C-J detonation. For this evaluation, convection was assumed to be the predominant mode of the heat transfer. The thermal radiation flux (hot gases to the tube wall) was estimated from the Stefan-Boltzmann's law. The temperature, pressure, and composition of the gaseous mixture behind the detonation wave were assumed to be those of the Chapman-Jouguet state. The estimated radiation flux of the gases, produced by a HN_3 detonation in a 10 mm tube at 10 Torr initial pressure, is less than 1.7×10^2 J·m^{-2}·s^{-1}. The radiation losses are thus quite negligible (less than 0.003% of the convective heat-transfer rate).

The best agreement between experiment and theory was obtained with the largest tube and with predictions that were calculated for a laminar regime frozen-gas boundary layer. The discrepancy between experimental and theoretical heat-transfer rates increased as ϕ decreased. The heat flux observed just before extinction of a detonation in the 1 mm tube (open triangles in Fig. 1) suggests a decoupling between the shock front and the reaction zone. The great influence of ϕ on heat-transfer rates emphasizes the importance of the flow parameter gradients that exist behind the wave front in narrow tubes.

Predicted:
―――― Turbulent boundary layer and frozen gas
---------- Laminar boundary layer and frozen gas
—·—· Laminar boundary layer with recombination

Observed:
● ○, $\phi = 10$ mm;   ▼, $\phi = 5$ mm
■ □, $\phi = 2$ mm;   △, $\phi = 1$ mm

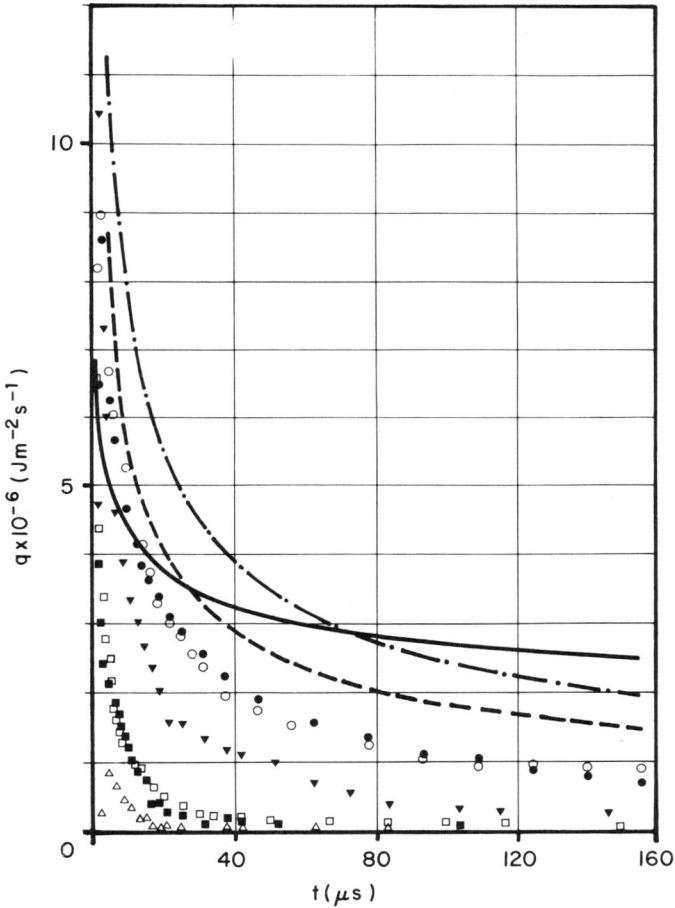

Fig. 1   Effect of tube diameter on wall heat flux ($P_1 = 10$ Torr).

Pressure Gradients Behind the Detonation Wave

For a tube of large diameter, Edwards et al.'s (1970) results suggested that the experimental pressure profile determined at a distance $(X-x)$ behind the wave front should be in good accord with the pressure predicted by the isentropic solution:

$$\frac{P}{P_{\text{C-J}}} = \left\{ 1 - \frac{\gamma}{\gamma-1} \frac{X-x}{X} \right\}^{2\gamma/(\gamma-1)} \tag{1}$$

where P_{C-J} is the Chapman-Jouguet pressure, γ the ratio of specific heats, and X the position of the wave front from the origin.

For a tube of small diameter, the pressure gradient is greater than that predicted from Eq. (1). Edwards et al. (1970) considered both a steady and a nonsteady solution for this case. Cromack (1967) postulated that the flow is steady behind the detonation front for a region whose extent depends on the position of the wave front from the onset of detonation. If dissociation and recombination effects are ignored, and if only the convective heat transfer through the boundary layer is taken into account, the pressure P and temperature T gradients for a steady flow are:

$$\frac{dP}{dx} = -\frac{2\gamma C_f P(T° - T_w)}{\phi T(\gamma \xi - 1)} \tag{2}$$

$$\frac{dT}{dx} = -\frac{2\gamma C_f (T° - T_w)(1 - \xi)}{\phi(\gamma \xi - 1)} \tag{3}$$

where C_f is the friction coefficient, $T°$ and T_w correspond to stagnation and wall temperatures, respectively, ϕ is the tube diameter, and $\xi = P^2/RT\rho_I^2 D_{C-J}^2$, with ρ_I the specific mass of gases at initial conditions, D_{C-J} the C-J detonation velocity in laboratory coordinates, and R the perfect gas constant per unit mass. At the plane $X - x = 0$ at time $t = 0$, $T = T_{C-J}$, and $P = P_{C-J}$. C_f and γ are assumed to be constant over the time interval considered. Equations (2) and (3) were integrated by the Runge-Kutta method for different C_f values.

For a nonsteady flow, an approximate solution, valid only at planes close to the detonation front (Edwards et al., 1970), is

$$\ln \frac{P}{P_{C-J}} = -\left\{ \frac{2\gamma(4\gamma - 3)}{3(\gamma - 1)} \frac{C_f}{\phi}(X - x) + \frac{2(X - x)}{X} \right\} \tag{4}$$

From Eq. (4), the condition for obtaining a steady pressure profile can be deduced

$$X \gg \frac{3(\gamma - 1)}{\gamma(4\gamma - 3)} \frac{\phi}{C_f} \tag{5}$$

For a HN_3 detonation at an initial pressure of 10 Torr, C_f is larger than 3×10^{-3} (cf. 2 below), and the steady-state is attained for X much greater than 100ϕ when the time range is approximately $100 \mu s$.

1) Influence of Pressure Transducer Position on Pressure Profiles for a Given Tube Diameter and Initial Pressure

The observed P/P_{C-J} are deduced from oscillograms, such as the one given in Fig. 2. The pressure profiles obtained for the three transducer positions ($X = 1.4, 2.4, 3.4$ m) on the 10 mm tube are shown in Fig. 3, and coincide only in the first 50 μs, after which the pressure profiles vary with the transducer

HEAT TRANSFER BEHIND A DETONATION WAVE

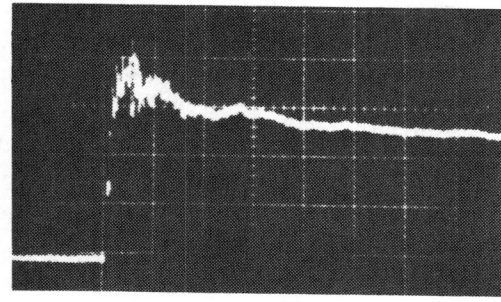

Fig. 2 Typical response of a piezoelectric, transducer to a HN_3 detonation ($P_1 = 20$ Torr, $X = 1.4$ m, $\phi = 5$ mm, sweep rate = 10 μs/division, sensitivity = 50 mV/division).

Observations curve	X, m
———	3.4
– – –	2.4
–·–·–	1.4

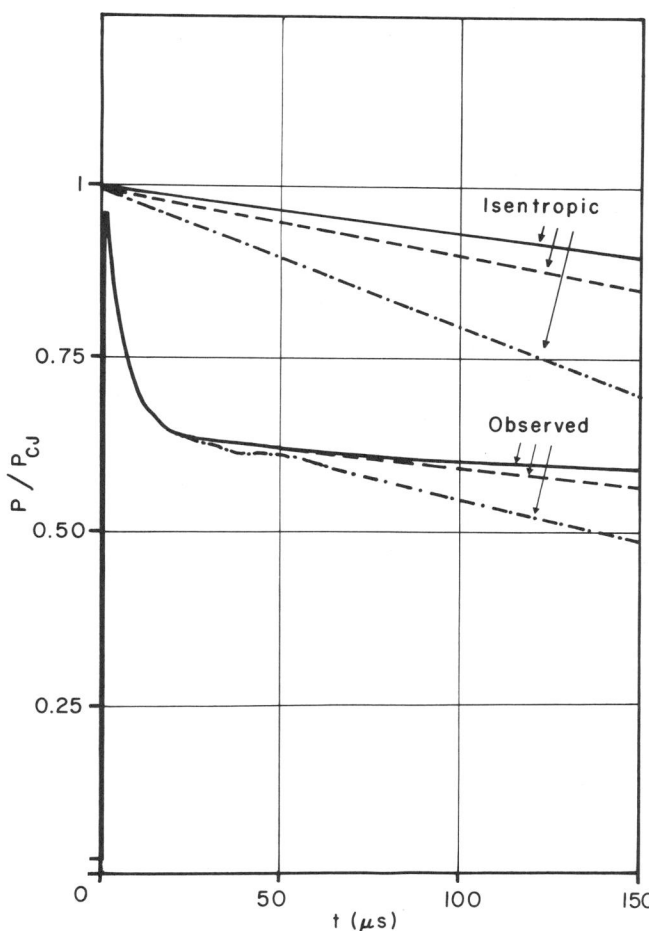

Fig. 3 Variation of pressure profiles with distance from ignition ($P_1 = 20$ Torr, $\phi = 10$ mm).

position. For $X = 3.4$ m, Eq. (5) indicated that the steady flow was reached. As can be seen in Fig. 3, the pressure rose steeply and attained a maximum within 2 μs. The maximum value was close to the C-J pressure but remained much lower than the von Neumann value. The pressure jump was followed first by a rapid decrease for 20 μs, then by a slower decrease. The rate of decay of pressure immediately behind the front was greater than that given by the isentropic solution as shown in Fig. 3.

2) Influence of Tube Diameter on Pressure Profiles for a Given Initial Pressure

The dependence of the pressure profile for large X ($= 3.4$ m) on tube diameter is shown in Fig. 4. In the larger tubes, 10 and 5 mm i.d., the pressure peak is somewhat lower than the C-J value, whereas it is much lower for the tubes of 1 and 2 mm diameter. The pressure attained a maximum value after 1 or 2 μs for the 10 mm tube, 10 μs for the 1 mm tube. For all diameters, two distinct zones are evident: for $X - x$ small, the pressure gradient was large; for

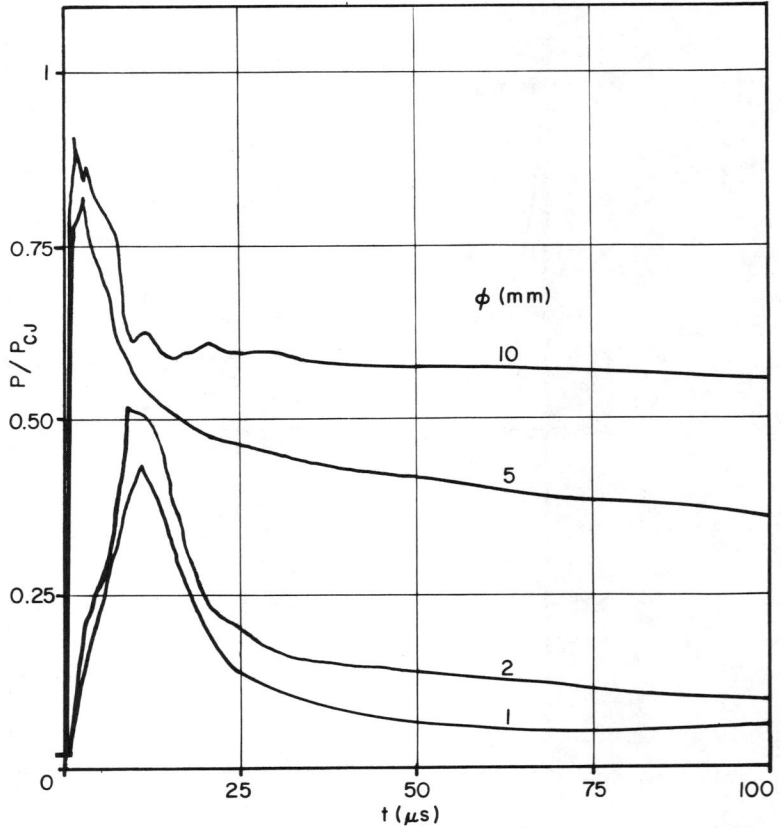

Fig. 4 Influence of tube diameter ϕ on pressure profiles vs time ($P_I = 20$ Torr, $X = 3.4$ m).

$X-x$ large, the pressure gradient was small. The extent of the zone of high-pressure gradient increased as the diameter decreased. In the diameter and initial pressure range considered, a linear relation was not found between the pressure behind the detonation front and the reciprocal diameter, as was observed by Brochet et al. (1969) with much larger tubes and at higher initial pressures.

The observed $P/P_{C\text{-}J}$ profile for $P_I = 10$ Torr, $\phi = 10$ mm, and $X = 3.4$ m is compared in Fig. 5 to the theoretical values calculated for the three solutions considered: isentropic, steady, and nonsteady flow. The pressure predicted by the isentropic solution does not agree at all with the experimental results, even for the widest tube. The steady and nonsteady solutions were computed with

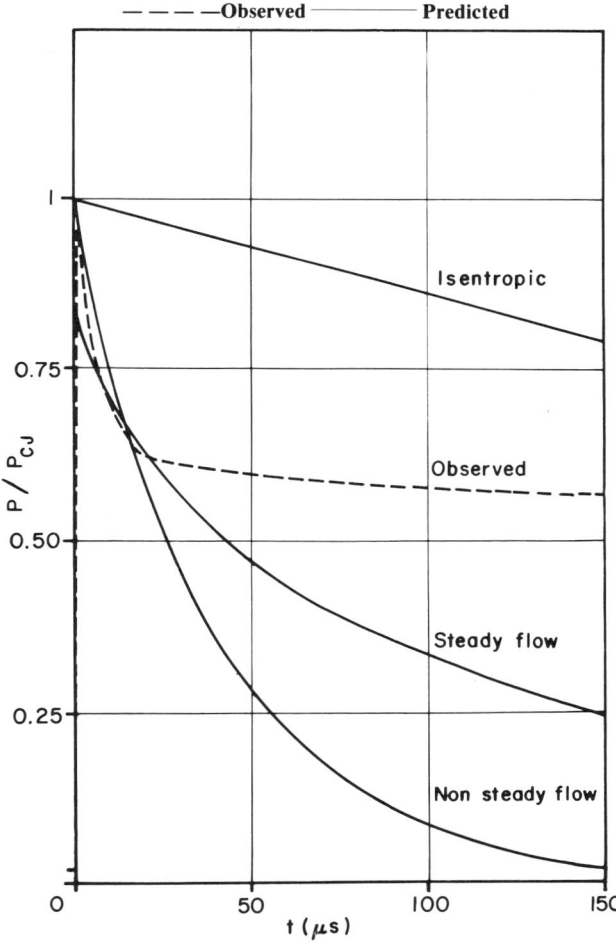

Fig. 5 Comparison between observed pressure profiles and several predictions ($P_I = 10$ Torr, $\phi = 10$ mm, $X = 3.4$ m and $C_f = 1.5 \; 10^{-2}$).

the assumption of a constant friction coefficient C_f equal to 1.5×10^{-2}. Heat-transfer rate measurements just behind the detonation front were used with the hypothesis of a C-J state of gases in the downstream flow to show that C_f exceeded 10^{-2}. For $t < 20$ μs, there is a good agreement between experimental pressure profiles and steady or nonsteady values. On the other hand, for $t > 20$ μs, the discrepancy between theory and experiment is large, especially for the nonsteady flow. The value of C_f used for this estimate is much larger than those usually considered, which range from 3×10^{-3} to 5×10^{-3} (Zeldovich and Kompaneets, 1960; Edwards et al., 1970).

Predicted steady flow pressure profiles, for several values of friction coefficients, at an initial pressure of 10 Torr and for tube diameters of 10, 5, and 2 mm i.d are shown in Figs. 6-8. Computations were made with the assumption of a C-J state immediately behind the front. With the largest tubes, (Figs. 6 and 7), the experimental profiles are bounded by the theoretical curves whose C_f are 3×10^{-3} and 15×10^{-3}. Although the observed pressure near the front is lower than P_{C-J}, the predicted pressure profiles are in satisfactory accord with the experimental results, during the first 20 μs, for $C_f = 15 \times 10^{-3}$. Further downstream, the rate of the pressure decay is smaller

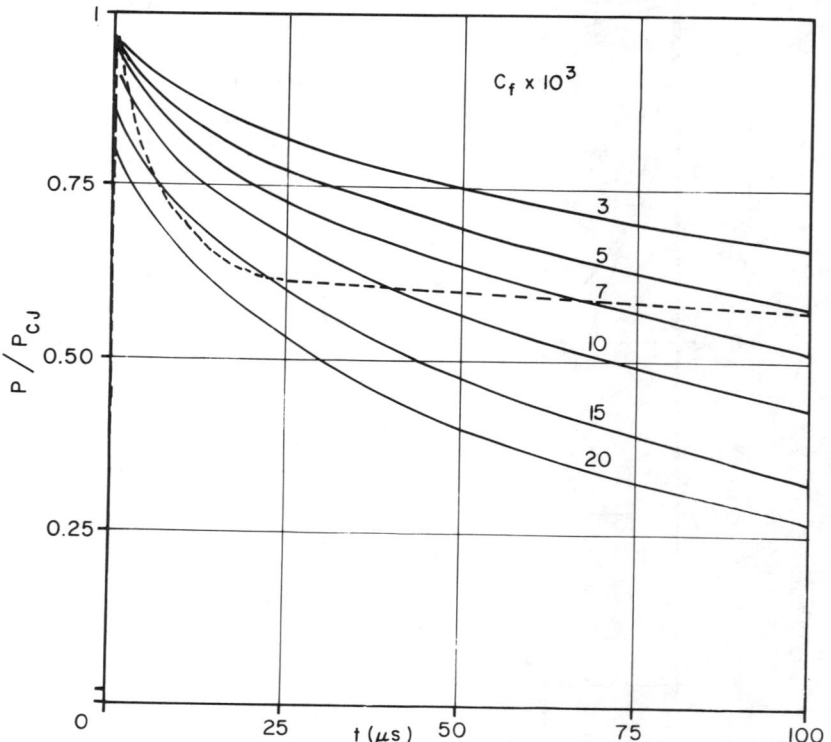

Fig. 6 Influence of friction coefficient (C_f) on pressure profiles predicted from a steady flow solution ($P_1 = 10$ Torr, $\phi = 10$ mm, $X = 3.4$ m). ———— predicted, ———— observed.

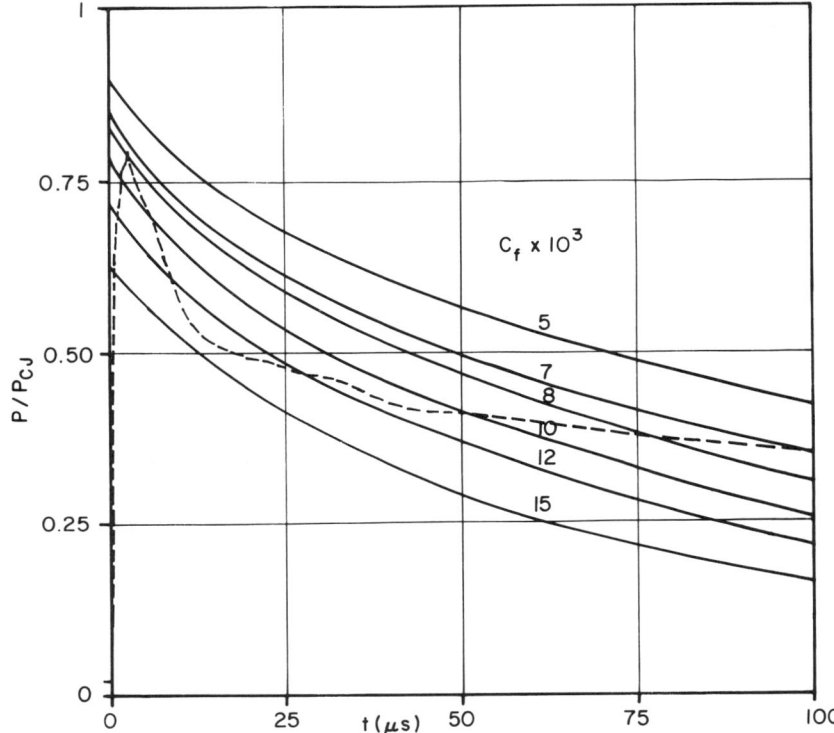

Fig. 7 Influence of friction coefficient C_f on pressure profiles predicted from a steady flow solution ($P_1 = 10$ Torr, $\phi = 5$ mm, $x = 3.4$ m). ———— predicted, — — — — — observed.

than that predicted by the steady flow solution. In this region, the calculations are probably invalid because of the variation of γ with temperature and the evolution of the gas composition in the downstream flow. In the case of the 2 mm tube (Fig. 8), the observed pressure profile deviates appreciably from those recorded for the largest tubes. The pressure peak is much lower and the pressure pulse is much broader. At these conditions the velocity deficit is no longer inversely proportional to the diameter, and reaches 15% (Paillard et al., 1979). The C-J characteristics and the one-dimensional structure of the detonation wave become, therefore, an unacceptable model of the phenomenon.

3) Influence of Initial Pressure on Pressure Profiles for a Given Tube Diameter

Just as the velocity deficit slightly increased with decreased P_1 in the pressure range 2.8-20 Torr for the largest tube $\phi = 10$ mm (Paillard et al., 1979), the observed pressure profiles did not vary greatly, as shown in Fig. 9. At 100 μs, the ratio P/P_{C-J} is included between 0.5 and 0.6. Near the detonation front, the curves coincide then diverge slightly after a time of about

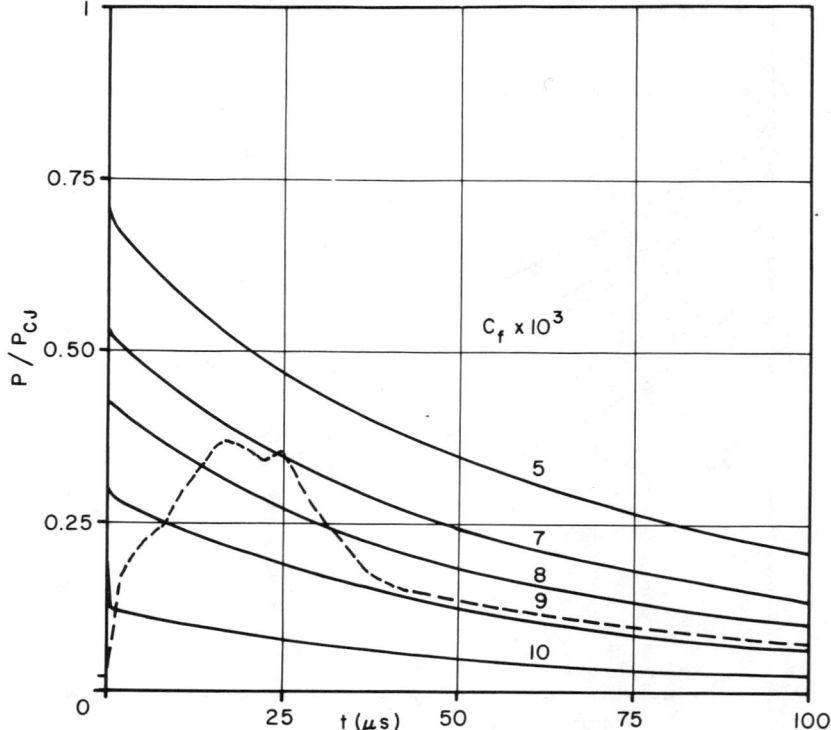

Fig. 8 Influence of friction coefficient (C_f) on pressure profiles predicted from a steady flow solution ($P_1 = 10$ Torr, $\phi = 2$ mm, $X = 3.4$ m). ——— predicted, — — — observed.

15 μs. At about 25 μs, the profiles approached quasiconstant values decreasing as the initial pressure decreased. Similar curves are given in Fig. 10 for tubes of smaller diameters. The influence of P_1 on the pressure profile was slight when ϕ was equal to 5 mm. For the 2 mm tube, the maximum values of P/P_{C-J} were diminished and shifted to longer times with decreased P_1.

Three-Dimensional Structure of the Detonation Wave

It is well-known that all self-sustained low-pressure detonations exhibit a three-dimensional structure in the region between the leading shock and the end of the reaction zone. The detonation front is characterized by a nonplanar shock wave composed of many shocks that are locally convex toward the upstream flow. The collisions of these shocks are triple-shock intersections. Their trajectories are a series of cells on soot-coated walls. The dimensions of these cells depend on the reaction kinetics and the tube geometry (Strehlow, 1968).

The reaction zone in the three-dimensional model cannot be regarded as being isolated from the expansion wave. Thus, heat losses and pressure

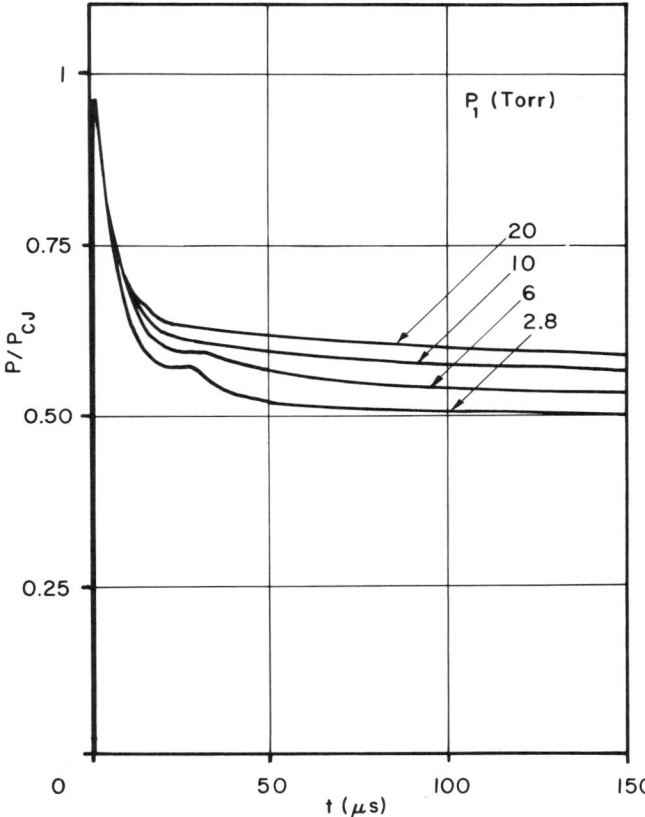

Fig. 9 Influence of initial pressure on observed pressure profiles ($\phi = 10$ mm, $X = 3.4$ m).

gradients that result from confinement may exert a noticeable influence on the detonation structure.

An example of a smoked-foil record obtained with a detonation wave propagating from left to right at 2.6 Torr initial pressure in a 10 mm tube is shown in Fig. 11. The cell length is equal to about 1 cm. In the present work, soot traces cannot be observed for a cell length less than 1 mm. To obtain a clear record, the initial pressure had to be lower than 5 Torr for the 10 mm tube, or lower than 10 Torr for the 5 mm tube. As can be seen in Fig. 11, the cell dimension increased in the propagation direction along the smoked tube element. This effect was more pronounced near the propagation limits, or for narrower tubes as shown in Fig. 12 for $\phi = 5$ mm. Different causes can be given to the cell dimension increase: variation of the tube diameter due to the soot layer thickness, change of wall roughness, physical or chemical processes such as adsorption, chemical reactions, etc. For $P_1 = 3.8$ Torr, the average length of the detonation cells was about 5 mm, and did not depend on distance from the point of ignition. The influence of tube diameter and initial pressure on the

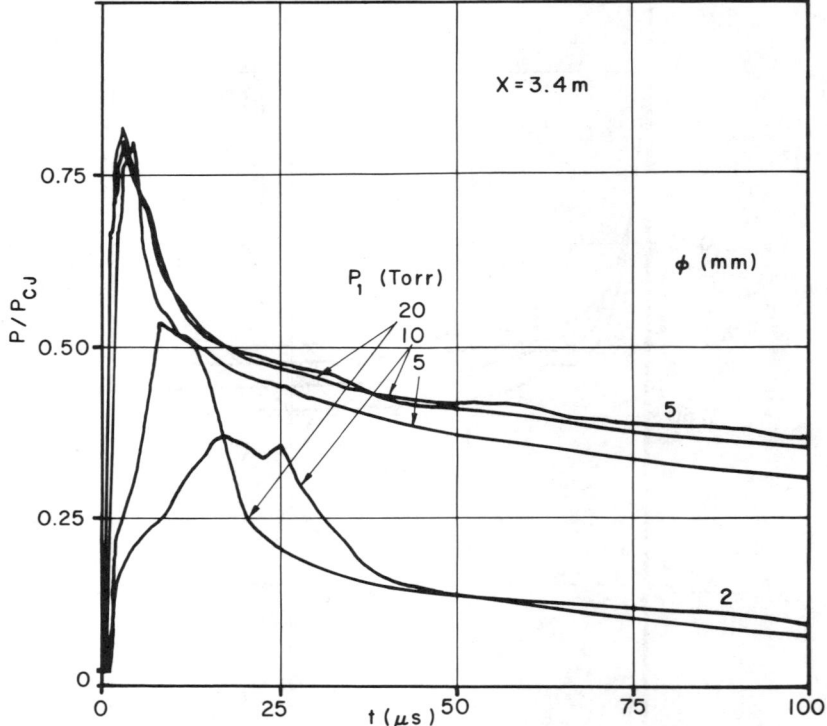

Fig. 10 Influence of initial pressure on observed pressure profiles ($\phi = 5$ and 2 mm, $X = 3.4$ m).

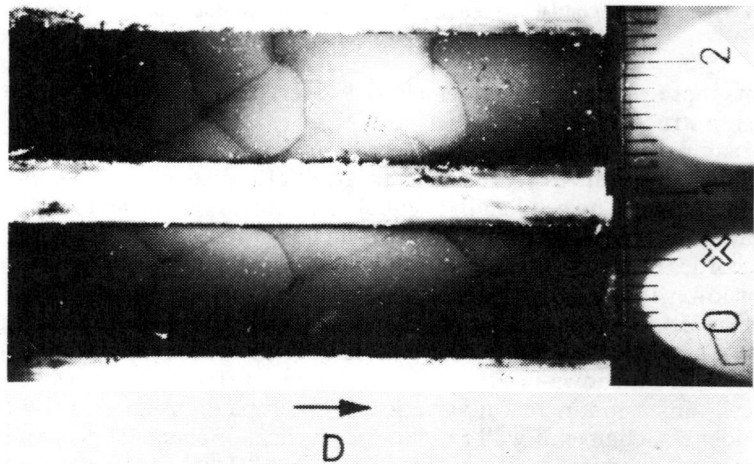

Fig. 11 Photograph of trajectories drawn by the triple point collisions on the inner surface of a tube element coated with soot ($P_1 = 2.6$ Torr, $\phi = 10$ mm, $X = 3.4$ m).

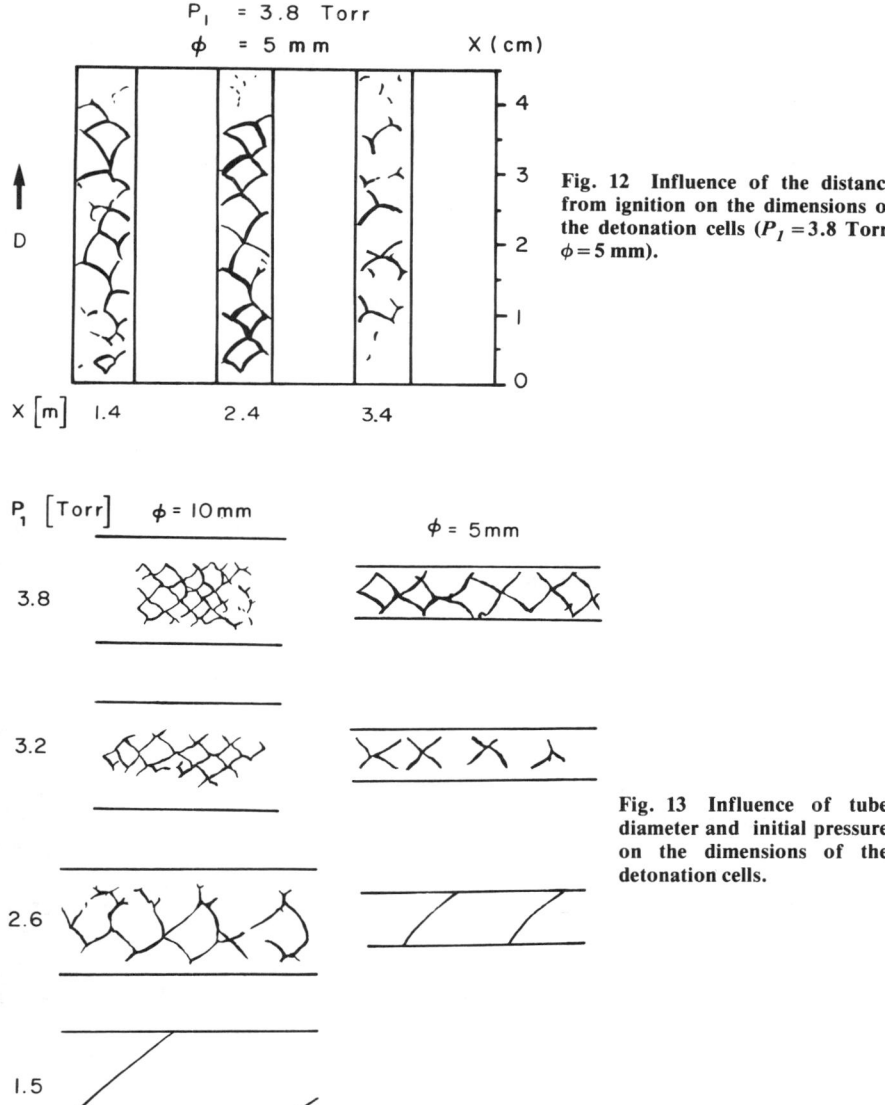

Fig. 12 Influence of the distance from ignition on the dimensions of the detonation cells ($P_1 = 3.8$ Torr, $\phi = 5$ mm).

Fig. 13 Influence of tube diameter and initial pressure on the dimensions of the detonation cells.

average length L of a cell is shown in Fig. 13. L increased substantially at a given initial pressure with decreased ϕ. According to Strehlow et al. (1967), the cell length depends on the nature of gas and on the propagation mode. In the present case, it depended on tube diameter and initial pressure as well. In the initial pressure range and for the diameters studied, the heat losses had a strong effect on the chemical reactions and, therefore, on the detonation structure. The influence of heat losses on the cell dimensions also appears, as shown by Voitsekhovskii et al. (1969), in the planar mode propagation. In that mode, the heat exchange between the reaction zone and the tube wall is more important than in the rectangular mode. Consequently, an increase of the cell dimension is observed when the propagation mode, initially rectangular, becomes planar.

The cell length also increased when initial pressure decreased for a given tube diameter. Near the limit pressure, spin phenomenon appeared at a pressure decreasing as the tube diameter increased. In the case of spinning detonations, soot traces present at the downstream end of the tube element sometimes disappeared when the extinction of detonation was observed. As a general rule, when the cell length exceeded a few millimeters (when it was larger than the sensitive region of the transducer or the thermal gage), the shape of the signals were modified. The response depended on the part of the cell that came into contact with the gage at the wall. For example, near the pressure limits of propagation, or with the smallest tubes, the pressure record was not reproducible during the first microseconds following the wave. For the same reasons, the wall temperature records exhibited different features as shown in the previous paper (Paillard et al., 1979), depending on whether the triple-point collision happened on the thermal gage or not.

Conclusion

The results of the present investigation of HN_3 detonation in narrow tubes can be summarized as follows:

1) The detonation velocity depends on wall heat losses.

2) The wall heat flux $q(t)$ observed behind the wave front was not in accord with that calculated for constant flow parameters. In the diameter and pressure range considered, $q(t)$ varies strongly with tube diameter. This observation can result from flow deviations from the C-J detonation parameters.

3) The observed pressures behind the detonation front are lower than those given by the isentropic solution. When the wall effect remained slight (when the velocity deficit was less than a few percent), the observed pressure profile near the detonation front nearly agreed with that calculated from the steady-state solution given by Edwards et al. (1970). This calculation accounts for friction and heat-transfer effects by means of a friction coefficient that was deduced from wall heat-flux measurements.

4) When the wall effect became important (when the velocity deficit was no longer inversely proportional to the diameter), the pressure near the front was much lower than the C-J pressure. Investigation of the detonation structure in this case has shown that the cell length depended strongly on the tube diameter. The heat losses in the reaction zone appeared to modify the chemical processes.

Acknowledgments

This research was carried out in the framework of the Cooperation treaty between France and U.S.S.R. on High-Temperature Physics.

References

Brochet C., Guerraud C., Manson N., and Veyssiere M. (1969) Variation de la pression des gaz derrière le front des détonations dans des mélanges stricts propane-oxygène. Influence du diamètre des tubes et de la pression initiale. *C.R. Acad. Sci.* (Paris), **268B**, 361-364.

Cromack D.E. (1967) Steady flow analysis of the flow induced in the wake of a disturbance propagating supersonically through an infinite duct. *TR AE 6704*, Rensselaer Polytechnic Institute, Troy, N.Y.

Edwards D.H., Williams G.T., and Breeze J.C. (1959) Pressure and velocity measurements on detonation waves in hydrogen-oxygen mixtures. *J. Fluid Mech.* **6**, 497-517.

Edwards D.H., Brown D.R., Hooper G., and Jones A.T. (1970) The influence of wall heat transfer on the expansion following a C-J detonation wave. *J. Phys. D: Appl. Phys.* **3**, 365-376.

Fay J.A. (1959) Two-dimensional gaseous detonations: Velocity deficit. *Phys. Fluids* **2** (3), 283-289.

Gordon W.E. (1949) Pressure measurements in gaseous detonations by means of piezoelectric gauges. *3rd Symp. (Intl.) on Combustion, Flame and Explosion Phenomena*, pp. 579-586. Williams and Wilkins Co., Baltimore, Md.

Gvozdeva L. G. and Jiline Yu. V. (1978) Pribory i Techn. *Eksperim.* USSR, **5**, 249-250.

Mirels H. (1958) The wall boundary layer behind a moving shock wave. *Symp. on Boundary Layer Research*, Freiburg, Fed. Rep. Germany (Edited by H. Görtler), pp. 283-295. Springer-Verlag, Berlin.

Paillard C., Dupré, G., and Combourieu J. (1973) Etude de la détonation de composés endothermiques gazeux: I. Célérit ès de détonation de l'azoture de chlore dans des tubes cylindriques. Limites de détonation de l'azoture d'hydrogène. *J. Chim. Phys.* **70** (5), 811-818.

Paillard C., Dupré G., and Combourieu J. (1974) Etude de la détonation de composés endothermiques gazeux: II Propagation et conditions critiques d'extinction dans des tubes capillaires de la flamme de décomposition de l'azoture de chlore gazeux pur ou dilué. *J. Chim. Phys.* **71** (2), 175-181.

Paillard C., Dupré G., Lisbet R., Combourieu J., Fokeev V.P., and Gvozdeva L.G. (1979) Study of hydrogen azide detonation and wall heat transfer behind the wave in narrow tubes. *Acta Astronautica* **6** (3-4), 227-242.

Sichel M. and David T.S. (1966) Transfer behind detonations in H_2-O_2 mixtures. *AIAA J.* **4** (6), 1089-1090.

Strehlow R.A., Liaugminas R., Watson R.H., and Eyman J.R. (1967) Transverse wave structure in detonations. *11th Symposium (Intl.) on Combustion*, pp. 683-692 The Combustion, Institute, Pittsburgh, Pa.

Strehlow R.A. (1968) *Fundamentals of Combustion*. International Textbook Co., Scranton, Pa.

Voitsekhovskii B.V., Mitrofanov V.V., and Topchian M.E. (1969) Investigation of the structure of detonation waves in gases. *12th Symposium (Intl.) on Combustion*, pp. 829-837, The Combustion Institute, Pittsburgh, Pa.

Zeldovich Ya.B. and Kompaneets A.S. (1960) *Theory of Detonation*. Academic Press, London.

Pressure Evolution behind Spherical and Hemispherical Detonations in Gases

D. Desbordes* and N. Manson†
Université d'Orléans, Bourges, France
and
J. Brossard‡
Université d'Orleans, Bourges, France

Investigations on the detonation front velocity $D_S(R)$ and the pressure field $p(t)$ behind, in hemispherical (0.38 m i.d.) and spherical charges (1 m i.d.) of stoichiometric C_2H_2-O_2 mixtures diluted with N_2 or A at initial pressure $p_f \leq 1$ bar, were carried out by means of microwave interferometer and quartz pressure gages. These investigations demonstrate that: 1) When the distance R^* (at which the detonation velocity D_S reaches a constant value $D_{S\infty}$ equal to about 98-99% of the computed $D_{C\text{-}J}$ Chapman-Jouguet value) is small ($R^* \sim 0$), then for $R > R^*$: a) about 4 µs after the arrival of the pressure peak p_R, the best agreement between the observed pressure-time profiles with those calculated from the Zeldovitch-Taylor (Z-T) ideal-wave theory is obtained if the calculations are made with a frozen value $\Gamma_{C\text{-}J}$ of the isentropic coefficient, rather than with the equilibrium values (Γe) and b) the peak pressure p_R increases with R and reaches for $R \sim 0.5$ m a value lower than $p_{C\text{-}J}$ by about 3%; and 2) as the initial pressure p_f and the mixture strength approach their detonation limit values, the distance R^* increases. Except at low initial pressure and $R < R^*$, the values of D_S and p_R and the pressure profiles approach those given by the C-J and Z-T theories.

Nomenclature

a = sound velocity
α = nondimensional sound velocity
D = absolute detonation front velocity
p = static pressure
P = nondimensional static pressure
Q = chemical energy per unit mass (at $T = 298$ K)

Presented at the 7th ICOGER, Göttingen, Federal Republic of Germany Aug. 20-24, 1979. Copyright © American Institute of Aeronautics and Astronautics, Inc., 1981. All rights reserved.
*Assistant Professor, Laboratoire d'Energétique et de Détonique.
†Professor, Laboratoire d'Energétique et de Détonique.
‡Professor, Laboratoire de Recherche Universitaire.

r	=	radial distance
R	=	detonation front radius
R_c	=	critical radius
R^*	=	minimum radius R where $D_S = D_{S\infty}$
t	=	time
T	=	temperature
$\mathcal{V}$	=	absolute particle speed
V	=	nondimensional absolute particle speed
W	=	electrical energy release
δ	=	relative difference
Γ	=	isentropic coefficient
λ	=	nondimensional radius
ρ	=	mass density
$\bar{\rho}$	=	nondimensional mass density

Subscripts

C-J	=	Chapman-Jouguet condition
D	=	absolute detonation front velocity
e	=	thermodynamic equilibrium
f	=	fresh gas
m	=	minimal value
M	=	maximal value
p	=	static pressure
R	=	measured detonation front characteristics
S	=	spherical

Introduction

BECAUSE of self-confinement (cf. Brochet et al., 1970), the instantaneously self-sustained spherical detonation is amenable to description with a minimum of simplifying assumptions, as an ideal wave. As shown by Zeldovitch (1942) and Taylor (1950), a special solution of the gasdynamics of the detonation products behind the front predicts the propagation of a spherical detonation with time-independent characteristics. As Friedrichs (1946) observed, such a solution, even though highly interesting because of its simplicity, assumes the validity of physically incompatible infinite compression gradients immediately followed by infinite expansion gradients. Thus, the existence of such detonation seemed to be doubtful. To eliminate the fundamental difficulties due to this incompatibility, the studies performed later tended either to describe the spherical detonation using the strong blast waves theory (Lee, 1972; Oppenheim et al., 1972) or to specify the wave structure (Cheret, 1971; Lee, 1972). Nevertheless, a satisfactory answer to the question of whether the propagation of the spherical detonation front is closely dependent on the flow behind it, as pointed out by Jouguet (1917),§ was not found.

Improvement in measurement techniques, particularly those concerning the spherical detonation front velocity (Brossard et al., 1974; Edwards et al., 1976)

---

§"Rien ne de précisément que ce mouvement restera qu'il faut pour que la célérité soit constante."

permit one to define more accurately the behavior of such detonations. Thus it is known that: 1) the initiation of spherical detonation is not always instantaneous; it is preceded by a predetonation zone of a variable amplitude (Manson and Ferrie, 1953; Struck, 1968; Lee, 1969; Desbordes, 1973; Edwards et al., 1976); and 2) when a self-sustained spherical detonation is observed, its velocity tends asymptotically to a constant value close to (but slightly lower than) that calculated (D_{C-J}) for an ideal plane wave (Desbordes, 1973). The observation of the evolution of the flow parameters behind the spherical detonation front is still very difficult. Because of point symmetry, all the difficulties are due to the existence of very high gradients just behind the front. As far as we know, the only experimental results are the mass density (Struck, 1968) and the pressure (Edwards et al., 1978); but their comparison with theory, because of measurement uncertainty, cannot be considered decisive for the validation of any model.

The purpose of the present work was to measure the pressure and to compare the observed profiles at different distances from the explosion center with those calculated by the Zeldovitch-Taylor model. The results of our calculations, in which a "frozen" value of isentropic coefficient Γ of the reaction products deduced from the usual thermodynamic properties of gaseous mixtures at high temperatures is used, appear to be in excellent agreement with experiments. Our experiments show the influence of the predetonation zone on the pressure profiles and lead to some complementary information on nonstationary characteristics of the front propagation in the zone preceding the constant-velocity regime.

Theoretical Pressure Field

The Zeldovitch and Taylor (Z-T) model of ideal wave assumes that the spherical detonation: 1) is created instantaneously at the center ($t=0$, $R(t=0)=0$) with quasizero energy deposition in a homogeneous and quiet reactive mixture; 2) is a discontinuity formed by the shock and the combustion wave, with the reactive zone of negligible thickness; and 3) propagates with a steady velocity $D=D_{C-J}$, the C-J characteristics being just behind the front, and the burned gases (products) expand isentropically behind it. Thus, the characteristics of products depend on only two variables $r(0 \leq r \leq R(t))$ and t. They may be deduced from conservation equations with boundary conditions at the center ($r=0$) and on the front ($r=R$), if self-similarity is assumed. The similar solution depends only on the single dimensionless variable λ:

$$\lambda = \frac{r}{D_{C-J}t} = \frac{r}{R} \qquad (1)$$

where $R(t)$ is the front coordinate at time t.

The conservation equations and the boundary conditions lead to:

$$\frac{d\ell n\, \mathcal{C}}{d\ell n\lambda} = -\frac{(\Gamma-1)(\lambda-V)V}{\mathcal{C}^2-(\lambda-V)^2}$$

$$\frac{d\ell n\, V}{d\ell n\lambda} = -\frac{2\mathcal{C}^2}{\mathcal{C}^2-(\lambda-V)^2} \qquad (2)$$

Table 1 Computed values of p_{C-J}, Γ_{C-J}, k, and kernel pressures $p(\Gamma_{C-J})$ and $p(\Gamma_e)$ for mixture A, $C_2H_2 + 2.5O_2$ (mixture at $T_f = 293$ K and different pressures p_f)

p_f, bar	P_{C-J}, bar	Γ_{C-J}	k	$p = P(\Gamma_{C-J}) p_{C-J}$, bar ($0 \leq r/R \leq 0.5$)	$p = P(\Gamma_e) p_{C-J}$, bar ($0 \leq r/R \leq 0.5$)
1	34.46	1.1527	$1.41 \cdot 10^{-2}$	10.53	9.90
0.8	27.32	1.1511	$1.41 \cdot 10^{-2}$	8.36	7.86
0.6	20.25	1.1490	$1.39 \cdot 10^{-2}$	6.21	5.83
0.4	13.27	1.1457	$1.33 \cdot 10^{-2}$	4.08	3.83
0.2	6.44	1.1406	$1.26 \cdot 10^{-2}$	1.98	1.87
0.1	3.13	1.1357	$1.17 \cdot 10^{-2}$	0.97	0.91

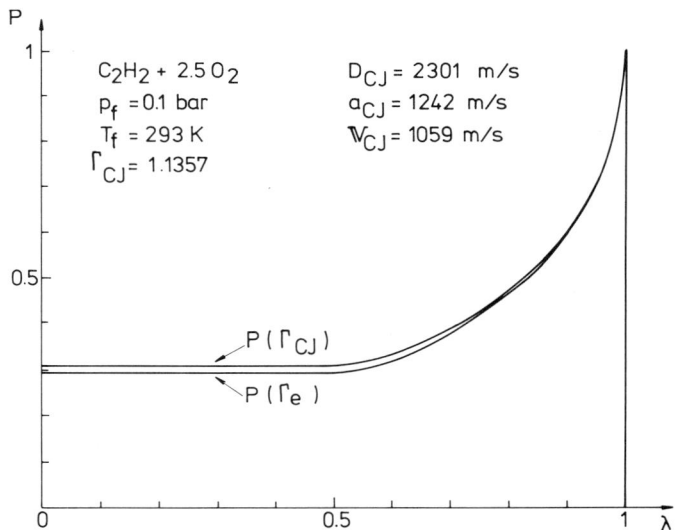

Fig. 1 Variation of the Z-T pressure profile $P = p/p_{C-J}$ vs $\lambda = r/R$ for $\Gamma = \Gamma_{C-J}$ and $\Gamma = \Gamma_e(\lambda)$.

with $V = 0$ for $\lambda = 0$ and the C-J condition for $\lambda = 1$ (Zeldovitch, 1942) where:

$$\alpha = \frac{a}{D_{C-J}}, \text{ the normalized sound velocity} \tag{3a}$$

$$a = \left(\frac{\partial p}{\partial \rho}\right]_s\right)^{1/2} = \left(\Gamma \frac{p}{\rho}\right)^{1/2}, \text{ the sound velocity} \tag{3b}$$

$$\Gamma = \frac{\partial \ln p}{\partial \ln \rho}\bigg]_s, \text{ the isentropic coefficient} \tag{3c}$$

$$V = \frac{\mathcal{V}}{D_{C-J}}, \text{ the normalized particle speed} \tag{4}$$

and $\mathcal{V}$ is the particle speed.

The integration of this set of equations from $\lambda=1$, gives profiles $\alpha(\lambda)$, $V(\lambda)$ as well as the pressure $P(\lambda) = p/p_{C-J}$, and the mass density $\bar{\rho}(\lambda) = \rho/\rho_{C-J}$ referred to their C-J values. Integration can be performed during the isentropic expansion for the assumption that the Γ coefficient is either: constant, $\Gamma = \Gamma_{C-J}$ (frozen value); or equal to its local equilibrium value $\Gamma_e(\lambda)$. Our calculations show that (see Appendix):

$$\Gamma_e(\lambda) = \Gamma_{C-J} + k\ell n P(\lambda) \tag{5}$$

Γ_{C-J} and k being the constants that refer to the mixture with fixed initial conditions (see Table 1 for the values of Γ_{C-J} and k obtained in the case of the $C_2H_2 + 2.5O_2$ mixture at an initial temperature and pressure equal to $T_f = 293$ K and $0.1 \leq p_f \leq 1$ bar, respectively).

Generally, the values of α and V calculated with Γ_e are practically the same as those calculated with Γ_{C-J}. But the behavior of $P(\lambda)$[¶] and $\bar{\rho}(\lambda)$ are different, as shown in Fig. 1 for the case of a $C_2H_2 + 2.5O_2$ mixture. The values of $P(\lambda)$ calculated for $\Gamma = \Gamma_e$ are lower than those calculated for $\Gamma = \Gamma_{C-J}$. Therefore, it appears that: 1) the value $\lambda \sim 0.5$ which delimits the "kernel" does not depend on the choice of Γ, $\alpha(\lambda, \Gamma_e) \sim \alpha(\lambda, \Gamma_{C-J})$; and 2) in the kernel $(V=0, 0 \leq \lambda \leq 0.5)$, $P(\Gamma_{C-J})$ exceeds $P(\Gamma_e)$ by about 6% (see Table 1).

The computed $P(\lambda)$ profiles must be transposed to the t variable for fixed values of distance r_i from the explosion center, because the experimental evolution of the parameter P is given only by $P(t)$ records, where [see Eq. (1)]:

$$t = \frac{r_i}{D_{C-J}\lambda} \tag{6}$$

For $0 \leq t < t_i = r_i/D_{C-J}$, $P(t) = p_f/p_{C-J} = $ constant. When $t_i \leq t < \infty$, $P(t)$ is obtained from the similar solution with t defined by Eq. (6).

The beginning of the spherical detonation propagation is governed essentially by the deposition of external energy W. The detonation is considered as self-sustained provided that the chemical energy $(4/3\pi R^3 \rho_f Q)$, released in a sphere of radius R, can be considered as large compared to the value of W. This happens when R is large compared to the critical radius:

$$R_c = \left(\frac{3W}{4\pi \rho_f Q}\right)^{1/3} \tag{7}$$

where Q is the specific heat of reaction.

Experimental Apparatus

The acetylene-oxygen (diluted with nitrogen or argon) mixtures used were:

$$A - C_2H_2 + 2.5O_2$$
$$B_1 - C_2H_2 + 2.5O_2 + Z_1N_2$$
$$B_2 - C_2H_2 + 2.5O_2 + Z_2A$$

[¶]In the Runge-Kutta method, the integration step was chosen sufficiently small so as not to influence the profile.

the initial temperature and pressures being: for mixture A, $T_f = 293 \pm 2$ K, $p_f \leq 1$ bar; and for mixtures B_1 and B_2, $T_f = 293 \pm 2$ K, $p_f = 1$ bar. The four gases, C_2H_2 (purity 99.5%), O_2, N_2, and A (purity 99.99% for each of them), were mixed in the two experimental vessels and their partial pressures measured as discussed in previous work (Brossard et al., 1974). For mixture A the initial pressure p_f was varied; for mixtures B_1 and B_2 the dilution was varied by the addition of nitrogen and argon, respectively.

The first vessel (Fig. 2) was the spherical one of $R_0 = 500$ mm radius used by Brossard (1970) and Desbordes (1973). The second one (Fig. 3) was a cylinder of radius $R_0 = 190$ mm and height $H_0 = 493$ mm. The ignition device was placed in the center of one end of the cylinder. Either a Teflon lens of a microwave interferometer or a pressure gage support was placed at the other end. Detonations were always ignited by means of a capacitor ($C = 18.8$ μF) discharge of a nominal electrical energy of $W = 235$ J, in exploding U-shaped wires. With such wires set in the center of the spherical vessel, spherical detonation (S) of 0.5 m radius was observed. In the cylindrical vessel, the wire extends over the bottom in its center by only a half-circle of $\phi = 5$ mm and, thus, hemispherical detonations(HS) of 0.19 m radius have been observed.

Velocity Measurements

The ionized front (combustion front) velocity $D_S(R)$ as a function of the radius R was deduced from the Doppler frequencies recorded with a 3 cm wave length microwave interferometer (cf. Brossard et al., 1974). The values $\bar{D}_i$ of the mean velocity were deduced from transit time measurements between the explosion center and the distance r_i of each pressure gage. The asymptotic value $D_{S\infty}$ for the spherical detonation velocity also for each distance r_i the relative difference:

$$\delta_i = \frac{D_{S\infty} - \bar{D}_i}{D_{S\infty}} \tag{8}$$

were determined.

Pressure Measurements

Static pressures were recorded at various distances r_i by piezoelectric quartz gages (Kistler Model 603 B). These gages were chosen over other commercial gages as well as over those made in our laboratory (Desbordes et al., 1978) after consideration of preliminary tests to determine the quality of the transfer function, corrosion resistance, and thermal strength, as well as of the size of the gages. The sensitive surface ($\phi = 5.5$ mm), in contact with a piezoelectric quartz element of smaller dimension, is made of stainless steel. Its resonance frequency is about 500 KHz, the maximum frequency response being limited by the concurrent use of a charge amplifier. The gages were tested in: 1) a shock tube in order to check their sensitivity and response quality (2% of maximum error on the amplitude), and 2) a detonation tube of $\phi = 50$ mm to find the rise time (~ 2.5 μs for the detonation conditions) and to determine any thermal effects on pressure measurements. Thermal effects became important for times greater than 10 μs after the combustion front had crossed the gage, and a silicone grease of 0.1-0.2 mm thickness on sensitive surface of gage was necessary to provide thermal protection.

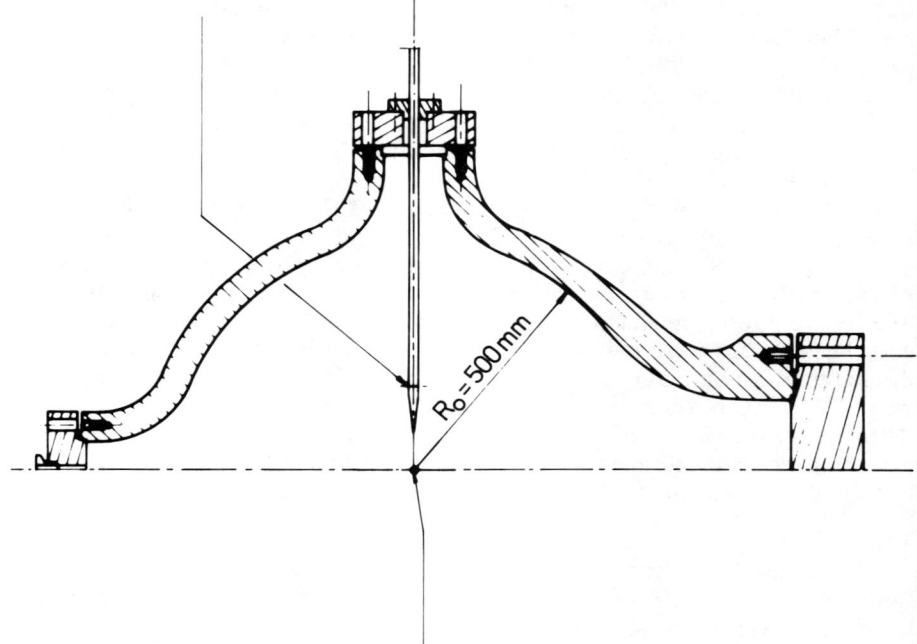

Fig. 2 Spherical vessel (central ignition).

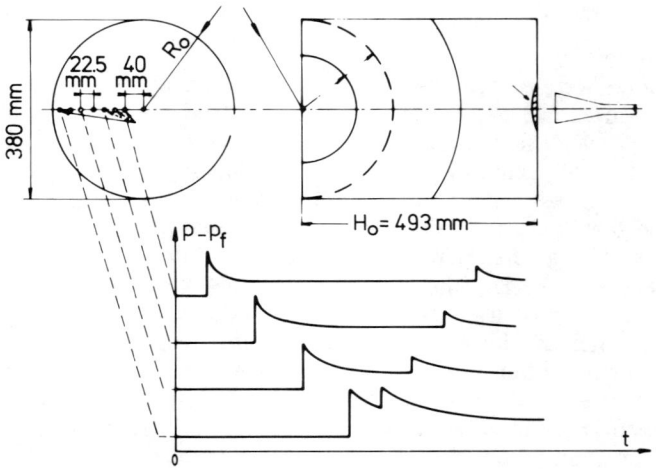

Fig. 3 Cylindrical vessel (hemispherical propagation).

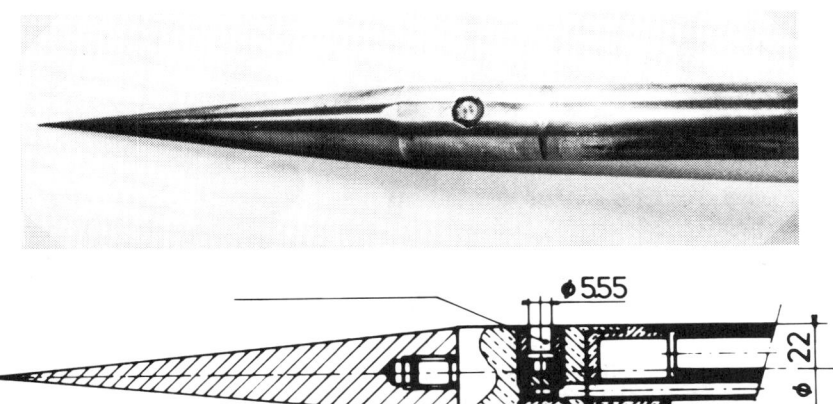

Fig. 4 Pressure gage support.

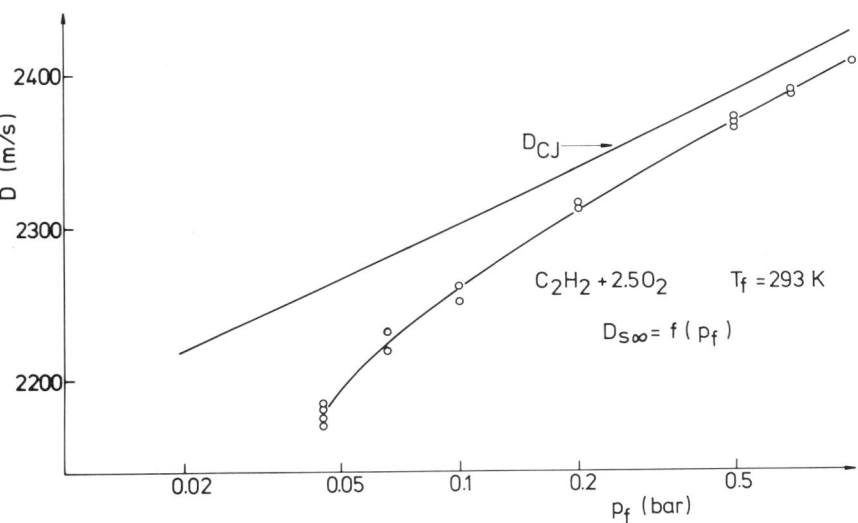

Fig. 5 Variation of detonation velocity $D_{S\infty}$, in mixture $C_2H_2 + 2.5O_2$ at $T_f = 293$ K vs p_f.

The front pressure gages were set on one end of the cylindrical vessel at four of the seven positions r_i. These positions were separated by 22.5 mm from each other and by 40-175 mm from the vessel center. In some experiments, one of the four gages was placed on a water-cooled cylindrical support ($\phi = 22$ mm) having on its end a sharp cone (Fig. 4). Its support axis was normal to the plane wall, and gage could not be placed nearer than 130 mm from the center. It was used in place of the microwave device to compare the pressure records and mean velocities $\bar{D}_i$ with those obtained by gages placed at the bottom of the vessel. In the spherical vessel the distance r_i of the gage on its support varied 130-500 mm from the center. The pressure records $p(t)$ are interesting only between the moment when the detonation front crosses the pressure gage and the moment when the reflected shock wave (from the walls) interacts with the gage.

Results and Discussions

As a general rule, the velocities and the pressure profiles (for $R > 130$ mm), for both hemispherical (*HS*) and spherical (*S*) propagation were quite similar, despite the possible influence of the wall in the case of hemispherical detonations.

Failure of sphericity was noticed only near the detonability limits in mixtures B_1 and B_2. This feature may not be due to the presence of the wall, however, for such phenomena were observed by Struck (1968), Lee et al. (1969), and Kamel and Oppenheim (1973) where walls did not play a role.

Detonability Limits

The minimum value p_{f_m} in mixture A and the maximum dilutions Z_{1M} and Z_{2M} of mixtures B_1 and B_2 beyond which detonations cannot be observed are: $p_{f_m} = 0.046$ bar, $Z_{1M} = 5$, and $Z_{2M} = 14$. The spherical detonation becomes self-sustained when its radius R is larger than the critical radius R_c [Eq. (7)] whose value for mixture A is in the detonability range of $0.046 \leq p_f \leq 1$ bar, $42 \geq R_c \geq 15$ mm; whereas for mixture B_1 for $5 \geq Z_1 \geq 0$, $22 \geq R_c \geq 15$ mm; and for mixture B_2 for $14 \geq Z_2 \geq 0$, $27 \geq R_c \geq 15$ mm.

Detonation Velocities

The relative difference value δ_D between theoretical and experimental velocities (Brossard et al. 1974):

$$\delta_D = \frac{D_{C\text{-}J} - D_{S\infty}}{D_{C\text{-}J}} \qquad (9)$$

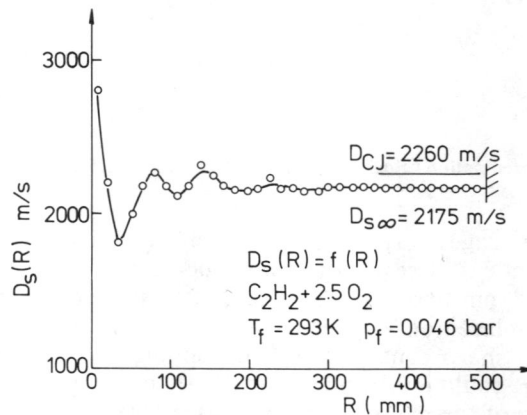

Fig. 6 Variation of ionized front velocity $D_S(R)$ in mixture $C_2H_2 + 2.5O_2$ at $T_f = 293$ K, $p_f = 0.046$ bar, vs R.

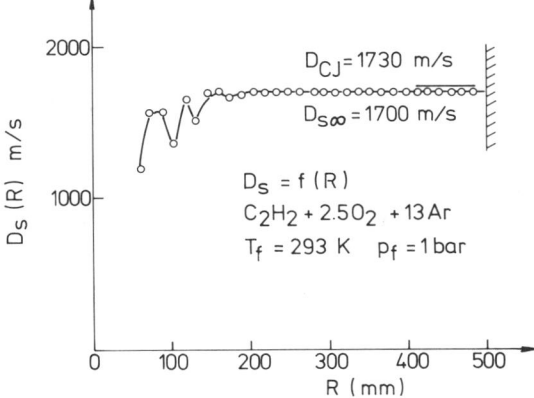

Fig. 7 Variation of ionized front velocity $D_S(R)$ in mixture $C_2H_2 + 2.5O_2 + 13\,A$ at $T_f = 293$ K, $p_f = 1$ bar, vs R.

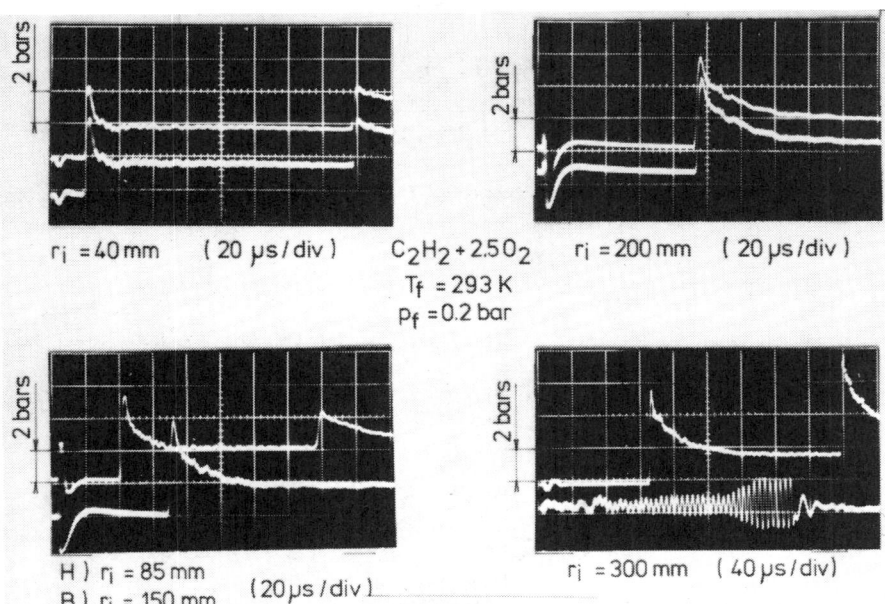

Fig. 8 Typical pressure records for "stable" detonations ($D_S(R) = D_{S\infty}$).

are 1-4% for mixture A (cf. Fig. 5), and 1-2% for mixtures B_1 and B_2. From the variation of the velocity $D_S(R)$ with R as well as the values of δ_i defined by Eq. (8), we deduce, for each reactive mixture, the radius R^* (cf. Brossard et al., 1974) above which the detonation velocity can be considered as steady ($D_S(R) = D_{S\infty}$). When $p_f \geq 0.1$ bar in mixture A, $Z_I \leq 3$ in mixture B_1, and

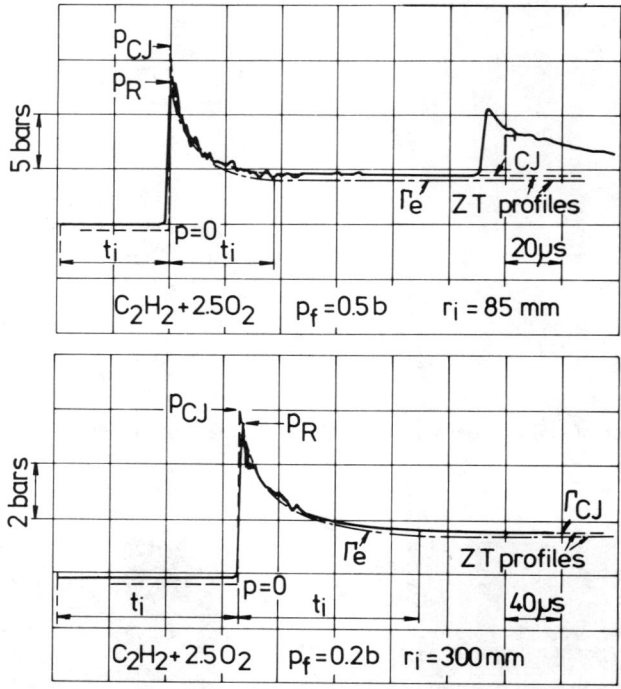

Fig. 9 Variation of observed and calculated (Z-T model) pressure behind detonation front propagating with the $D_{S\infty}$ velocity.

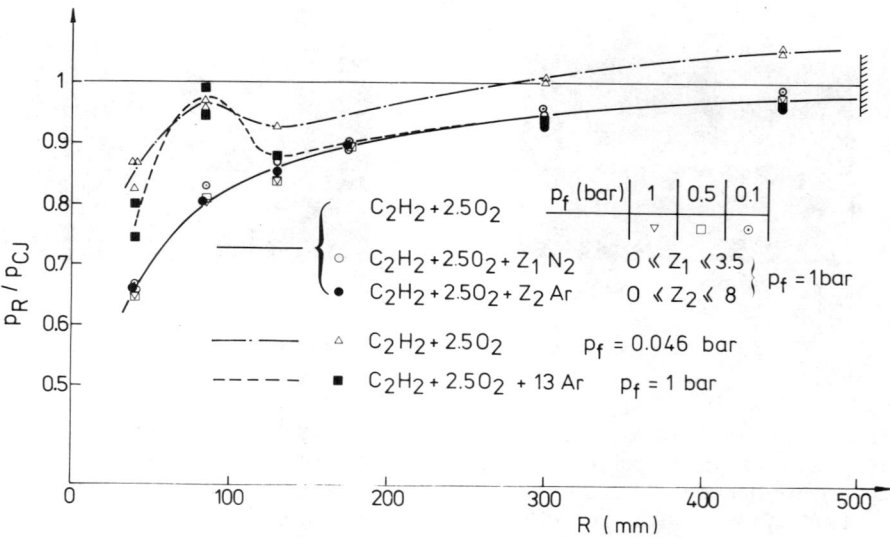

Fig. 10 Variation of ratio p_R/p_{CJ} vs distance from center.

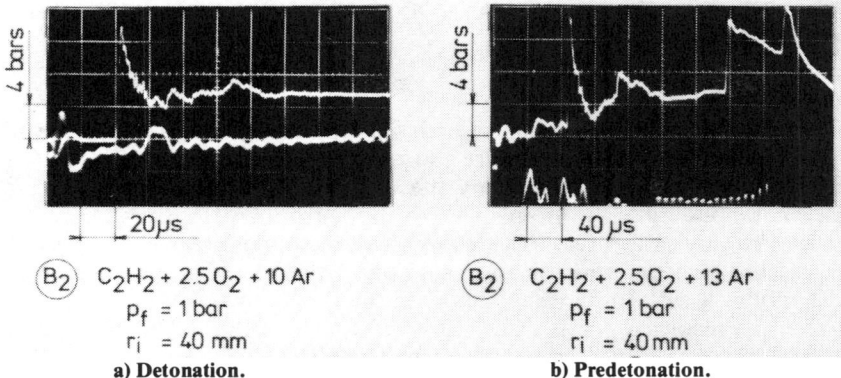

Fig. 11 Examples of pressure records near center for nonsteady velocity detonation at detonability limits.

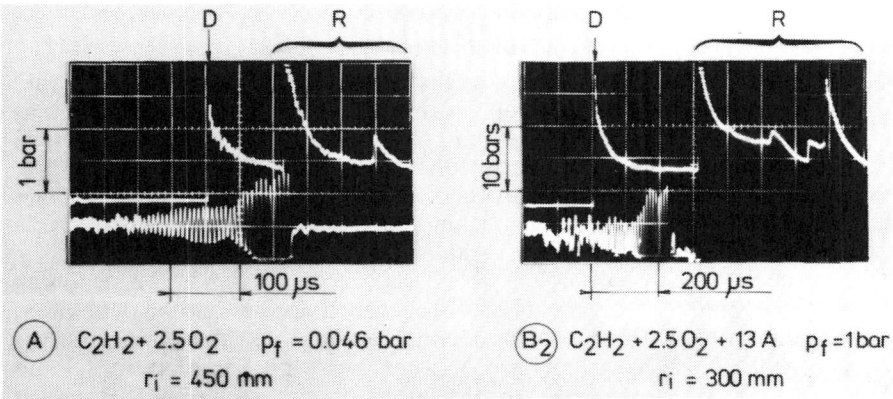

Fig. 12 Examples of pressure records for established detonations ($D_S(R) = D_{S\infty}$) at detonability limits (D = detonation front and R = reflected shock waves).

$Z_2 \leq 8$ in mixture B_2, the practically zero values of δ_i (and thus $R^* \sim 0$) indicate that the $\Delta_D = D_{S\infty} - D_S(R)$ are not significant, and such detonations can be called "stable." For $p_f < 0.1$ bar in mixture A and for $Z_1 > 3$, $Z_2 > 8$ in mixtures B_1 and B_2, values of Δ_D different from zero are observed at the beginning of the propagation (i.e., $R < R^*$) and become increasingly large as the detonability limits are approached. In mixture A, oscillations of the velocity $D_S(R)$ (Fig. 6) around $D_{S\infty}$ appear when p_f decreases and since the δ_i values are equal to zero, the predetonation zone, if it exists, is very small. In mixtures B_1 and B_2 the values of δ_i increase with Z and, near the detonation limits, the predetonation (separated shock and flame fronts) is clearly evident on pressure records for $R < 6\text{-}7$ cm.

A typical variation of the velocity $D_S(R)$ vs R, obtained from Doppler effect records, when the value of R^* is about 200 mm, is reported in Fig. 7.

These phenomena have also been observed by Struck (1968), Brossard et al. (1974), and Edwards et al. (1978).

Pressure Profiles

Examples of pressure profile records in the case of "stable" detonations ($\delta_i = 0$) for different distances r_i are shown in Fig. 8. If these profiles $p(t)$ are compared with those calculated by the Z-T model (the theoretical pressure peak corresponding to the moment at which the front crosses the gage, see Fig. 9), the following salient features are apparent: For times greater than 4 μs after the pressure peak p_R and for any value of r_i, the recorded pressure profiles agree very well with those calculated with the $\Gamma = \Gamma_{C-J}$ value. Differences exist between the recorded profiles and those calculated with the $\Gamma = \Gamma_e$, especially in the kernel. The measured peak pressure amplitude p_R is only a function of the r_i distance. The variation of p_R/p_{C-J} vs r_i given in Fig. 10, shows that: 1) for $40 \leq r_i \leq 500$ mm, this ratio increases with r_i from 0.65 to 0.97, 2) the relative difference

$$\delta_p = \frac{p_{C-J} - p_R}{p_{C-J}} \tag{10}$$

is about 3% for $r_i = 500$ mm, and is of the same order as δ_D (1-1.5%). Within the limits of experimental measurement this agreement may be considered as an indication of the existence of a "quasi C-J" state for the "stable" spherical detonation; and, 3) the large relative difference value $\delta_p \simeq 35\%$ observed at $r_i = 40$ mm arises mainly because of the finite response time of pressure gages and the high pressure gradients just behind the front. These gradients lead to a stronger pressure decrease when r_i is small.

When the detonations are not "stable," i.e., $D_S(R) \neq D_{S\infty}$ at the beginning of the propagation, two cases may be distinguished: in the first the detonation is established at $R < 40$ mm; in the second, $R > 40$ mm. The former occurs when $0.1 \geq p_f \geq 0.046$ bar in mixture A, when $3 < Z_1 \leq 4.5$ in mixture B_1, and when $8 < Z_2 \leq 12$ in mixture B_2. In the latter, the predetonation is observed only in mixtures B_1 and B_2, near the detonability limits. Examples of pressure records when $r_i = 40$ mm are given in Fig. 11. When the detonation is established, the first pressure peak p_R is followed by other peaks of smaller amplitude. Afterward, the pressure falls and is stabilized near the Z-T kernel value. The ratio p_R/p_{C-J} is greater than the value measured at the same position for "stable" detonations. When the pressure gage is set in the predetonation zone, it records a weak shock wave produced by the ignition device, followed by a high-amplitude pressure peak p_R followed itself, as seen previously, by peaks of smaller amplitudes. For greater radii, when velocity $D_S(R)$ tends to $D_{S\infty}$ (cf. Figs. 6 and 7) the larger R is, the more the recorded pressures tend to Z-T profiles (Fig. 12). The ratio p_R/p_{C-J} vs r_i tends in mixtures B_1 and B_2 (cf. Fig. 10, $B_2 - Z_2 = 13$) to the value corresponding to "stable" detonations; and in mixture A, more especially if p_f is small (cf. Fig. 10, A—$p_f = 0.046$ bar), remains greater than the one corresponding to "stable" detonations.

It can be noticed that, at the detonability limits, the values $\delta_D \simeq 2\%$ obtained for mixtures B_1 and B_2, as with "stable" detonations, agree well with the δ_p value of 3% obtained for great radii ($R \sim 500$ mm). But in mixture A, the δ_D value of 4% does not lead to a two-fold δ_p value. The negative value obtained for δ_p (-5%) means that the measured pressure exceeds p_{C-J}.

This latter observation can be explained by the fact that the distance between the C-J plane and the shock front is no longer negligible, and grows as p_f decays. This is contrary to the Z-T model hypothesis. At sufficiently low initial pressure p_f, the three-dimensional structure of the detonation becomes significant** and, therefore, cannot be neglected since the cell length L in the case of mixture A at $p_f = 0.046$ bar is about 5 mm [cf. Strehlow and Engels (1969) for detonations in tubes and Edwards et al. (1978) for spherical detonations]. As was shown by Vassiliev et al. (1972) and Edwards et al. (1976) in tubes, the distance between the C-J plane and the shock front is a function of the cell length L. Thus, the peak pressures p_R measured with our gages for low initial pressure p_f does not correspond to the C-J plane.

Conclusions

The results of our investigations on the propagation of spherical detonations in stoichiometric acetylene-oxygen mixtures at initial pressures $p_f \leq 1$ bar, or diluted by nitrogen or argon at $p_f = 1$ bar, show that:

1) The variation of the ionized front velocity $D_S(R)$ and the pressure field recorded at different distances ($r_i \geq 40$ mm) from the center are, within experimental accuracy, the same for the spherical and hemispherical detonations.

2) When the detonation is "stable" (i.e., $D_S(R) = D_{S\infty}$ for $R > R^* \sim 0$): a) the pressure $p(t)$ for times larger than 4 μs after the detonation has crossed the gage are in good agreement with those given by the Z-T model when the frozen $\Gamma = \Gamma_{C-J}$ value is used; b) far from the center ($R \sim 500$ mm), the relative pressure differences $\delta_p = (p_{C-J} - p_R)/p_{C-J}$ of about 3% might correspond effectively, as demonstrated in classical C-J theory, to the relative velocity difference $\delta_p = (D_{C-J} - D_{S\infty})/D_{C-J}$ of 1-1.5% often observed for spherical detonations; and c) near the center, the large value of δ_p (35% at $r_i = 40$ mm) is mainly due to the finite response time of the pressure gage used. The gage is subject to rapid pressure variations especially if the detonation curvature is large.

3) When the detonation is not "stable" near the center as the detonability limit is approached: a) for small front radii ($R < R^*$) during the formation of detonation and the first instants of its propagation, pressure profiles observed cannot be compared to those of the Z-T model—indeed the pressure decrease behind the first peak of relative amplitude p_R/p_{C-J} larger than that associated with "stable" detonations is nonmonotone and sharper; b) for greater radii (when $R > R^*$) the detonation velocity is steady—the pressure profiles $p(t)$ and the relative difference δ_p tend to the theoretical values in the same way as in "stable" detonations; and c) at low initial pressures, when the three-dimensional structure becomes important, the pressure peaks are systematically greater than those obtained for "stable" detonations.

**This not the case in mixtures B_1 and B_2 at $p_f = 1$ bar, even for such large dilutions as $Z_1 = 5$ and $Z_2 = 14$.

Appendix

The Zeldovitch-Taylor (Z-T) theory was used to determine the evolution of burned gases characteristics behind a spherical Chapman-Jouguet (C-J) detonation. The set of equations (2) were integrated by the Runge-Kutta method, the calculation began at the C-J point at the front. During isentropic expansion the coefficient Γ_e was computed for successive states of thermodynamic equilibrium as follows. First, for each reactive mixture, the composition of the reaction products at equilibrium, enthalpy and entropy at different pressures p and temperatures T were determined with a computer program (Johnson, 1968) which uses the *JANAF Thermochemical Tables* data. Then, for the initial conditions of temperature T_f and pressure p_f, the C-J state characteristics were computed

$$(\mathcal{V}_{C-J}, a_{C-J}, p_{C-J}, \rho_{C-J}, \Gamma_{C-J} \ldots)$$

Along the isentropic $S = S_{C-J}$, values of:

$$\Gamma_e = \frac{\partial \ln p}{\partial \ln \rho} \bigg]_{S_{C-J}} \tag{A1}$$

were shown to be well-represented by the relationship where k is a positive constant:

$$\Gamma_e(p) = \Gamma_{C-J} + k \ln p / p_{C-J} \tag{A2}$$

in the range $0.3 p_{C-J} \leq p \leq p_{C-J}$ and $0.8 T_{C-J} \leq T \leq T_{C-J}$.

Acknowledgments

The authors wish to thank technicians of our laboratory: M. M. Brugier, J.-L. Guerraud, and S. Sec.

References

Brochet C., Brossard J., Manson N., Cheret R., and Verdes G. (1970) A comparison of spherical, cylindrical and plane detonation velocities in some condensed and gaseous explosives. *Proceedings Fifth (Intl.) Symposium on Detonation,* p. 41, Pasadena, Calif.

Brossard J. (1970) Contribution à l'étude des ondes de choc et de combustion dans les gaz. *Thèse d'Etat,* Poitiers, France.

Brossard J., Desbordes D., and Manson N. (1974) Célérité de la détonation sphérique divergente en fonction du rayon. *Acta Astronautica* 1, 873.

Cheret R. (1971) Contribution à l'étude des détonations sphériques divergentes dans les explosifs solides. *Thèse de Doctorat d'Etat,* Poitiers, France.

Desbordes D. (1973) Célérité de propagation des détonations sphériques divergentes dans les mélanges gazeux. *Thèse de 3ème Cycle,* Poitiers, France.

Desbordes D., Manson N. and Brossard J. (1978) Explosion dans l'air de charges sphériques non confinées de mélanges réactifs gazeux. *Acta Astronautica* 5, 1009.

Edwards D. H., Jones A. T., and Phillips D. E. (1976) The location of the Chapman-Jouguet surface in a multiheaded detonation wave. *J. Phys. D: Appl. Phys.* 9, 1331.

Edwards D. H., Hooper G., and Morgan J. M. (1976) An experimental investigation of the direct initiation of spherical detonation. *Acta Astronautica,* **3,** 117.

Edwards D. H., Hooper G., Morgan J. M., and Thomas G. O. (1978) The quasisteady regime in critically initiated detonation waves. *J. Phys. D: Appl. Phys.* **11,** 2103.

Friedrichs K. O. (1946) On the mathematical theory of deflagrations and detonations. NAVORD Rept. 79-46.

Johnson C. (1968) Contribution à l'étude des détonations dans les mélanges hydrogene-oxygene-azote. *Thèse ès Sciences Appliqu*ées, Poitiers, France.

Jouguet E. (1917) *La Mécanique des explosifs.* DOIN, Paris.

Kamel M. M. and Oppenheim A. K. (1973) Photographic laboratory studies of explosions. *Aero Missili e Spazio* **2,** 122.

Lee J. H., Knystautas R., and Bach G. G. (1969) Theory of explosions. MERL Rept. McGill Univ., Montreal, Canada.

Lee J. H. (1972) Gasdynamics of detonations. *Acta Astronautica* **17,** 455.

Manson N. and Ferrie F. (1953) Contribution of study of spherical detonation waves. *Fourth (Intl.) Symposium on Combustion,* p. 486. Williams and Wilkins, Baltimore, Md.

Oppenheim A. K., Kuhl A. L., and Kamel M. M. (1972) On self similar blast waves headed by the Chapman-Jouguet detonation. *J. Fluid Mech.* **55,** 257.

Strehlow R. A. and Engels C. D. (1969) Transverse waves in detonations: II-Structure and spacing in H_2-O_2, C_2H_2-O_2, C_2H_4-O_2 and CH_4-O_2 systems. *AIAA J.* **7,** 492.

Struck W. (1968) Kugelförmige detonationswellen in gasgemischen. *Thèse de Doctorat,* Aachen, Fed. Rep. Germany.

Taylor G. I. (1950) The dynamics of the combustion products behind plane and spherical detonation fronts in explosives. *Proc. Roy. Soc.* (London) **A 200,** 235.

Vasiliev A. A., Gravilenko T. P., and Topchian M. E. (1972) On the Chapman-Jouguet surface in multiheaded gaseous detonations. *Acta Astronautica* **17,** 499.

Zeldovitch Ya. (1942) Distribution de la pression et de la vitesse dans les produits de détonation; cas d'une onde sphérique divergente. *Zh.E.T.P.* (USSR) **12,** 389.

Deflagration Explosion of an Unconfined Fuel Vapor Cloud

Shiro Taki* and Yuji Ogawa†
Fukui University, Fukui, Japan

Unconfined vapor cloud explosions of liquefied fuel gas have been studied in laboratory scale by ignition of a few grams of liquefied petroleum gas (LPG) which have been ejected into air. Although the amount of the fuel is quite small, the subsequent development of the explosion is similar to those of much larger scale. In other words, unconfined fuel vapor cloud explosions may be divided into two stages: a deflagration propagating in premixed gases, followed by a diffusion flame promoted by buoyancy-dominating convection. The observed fireball diameters are in good agreement with the existing fuel mass diameter relationship for much larger fireballs, presumably because the final diameter is controlled by the stoichiometric combustion of spilled LPG. Pressure waves induced by unconfined combustion are numerically simulated using a spherically expanding piston: $V/V_* = \exp[-(c/a+t)^n]$, where V is the time-dependent volume of the piston. The calculated peak overpressure is proportional to $r^{-1.5}$ for rapid expansion, similar to a TNT blast, whereas it becomes proportional to $r^{-1.0}$ for slow expansion.

Introduction

AS liquefied fuel gases have boiling points lower than the ambient temperature, accidental failure of high-pressure tanks or containers results in rapid escape of the fuels into open surroundings, forming large clouds of vapor. If such clouds are ignited by accident, rapid combustion spreads, generating strong compression waves and thermal radiation. Understanding the mechanisms of unconfined fuel vapor cloud explosions may be of importance in the prevention of such accidents. Such experiments are difficult to perform at reasonably large scale. For example, vapor cloud explosions of 3.1-31 kg n-pentane are considerably smaller than real accidents, and yet too

---

Presented at the 7th ICOGER, Göttingen, Federal Republic of Germany, Aug. 20-24, 1979. Copyright © American Institute of Aeronautics and Astronautics, Inc., 1981. All rights reserved.
*Associate Professor, Department of Mechanical Engineering.
†Technician, Department of Mechanical Engineering.

large to handle and measure in detail (Hasegawa and Sato, 1978). In this sense, a small amount of unconfined fuel was utilized in the laboratory by Fay et al. (1977). However, the concentration of the fuel used was above the upper flammability limit, so that only diffusive combustion occurred and compression waves were not.

In the present study, explosions are produced by injecting a small amount of liquefied petroleum gas (LPG) into air. The ignition and subsequent evolution of the explosion of the unconfined vapor cloud are observed by the simultaneous use of direct photographs and pressure recording. Parallel with this, the strength of the compression waves generated by unconfined combustion have been modeled by the solution of the conservation equations for the flow associated with a spherically symmetric expanding piston, and these results are compared with the measurements.

Analysis of Pressure Waves Generated by a Spherical Piston

The generation of pressure waves by combustion is analyzed by assuming that the increase of gas volume due to combustion can be represented by a spherical piston

$$V/V_* = \exp\{-(c/a+t)^n\} \tag{1}$$

where V is the volume of the piston, t time, c and n the parameters characterizing the motion of the piston, a the arbitrary constant, and V_* the final volume of the piston.

The motion of the gas outside the piston is analyzed under the following restrictions: 1) the motion is spherically symmetric, 2) transport phenomena are negligible, and 3) the gas is perfect. Then, the fundamental equations governing the gas are written in the following Lagrangian form:

$$\frac{\partial}{\partial t}(1/\rho) - \frac{\partial}{\partial \xi}(r^2 u) = 0, \qquad \frac{\partial u}{\partial t} + r^2 \frac{\partial p}{\partial \xi} = 0$$

$$\frac{\partial e}{\partial t} + \frac{\partial}{\partial \xi}(r^2 up) = 0, \qquad d\xi = \rho r^2 (dr - u dt), \qquad e = \frac{1}{\gamma - 1}\frac{p}{\rho} + \frac{1}{2}u^2 \tag{2}$$

where ρ, u, p, r, and ξ denote the density, velocity, pressure, distance from the origin, and the Lagrangian coordinate, respectively. The boundary conditions at the piston can be easily derived from Eq. (2)

$$\left\{\frac{\partial}{\partial t}\left(\frac{p}{\rho^\gamma}\right)\right\}_{\xi=0} = 0, \qquad \left\{\frac{\partial p}{\partial \xi}\right\}_{\xi=0} = -\frac{1}{R^2}\frac{d^2 R}{dt^2} \tag{3}$$

where R is the radius of the piston. The gas velocity at the piston is the same as the piston velocity, i.e.,

$$[u]_{\xi=0} = \frac{dR}{dt} \tag{4}$$

The boundary conditions at the other side, i.e., at $\xi = \infty$, are the same as the initial state of the atmosphere. Initial conditions are

$$u = 0, \qquad \rho = \rho_0, \qquad p = p_0 \qquad (5)$$

where the subscript 0 denotes the initial state.

Equations (2) are solved by a second-order accurate, noncentered, two-step finite-difference scheme (MacCormack, 1971)

$$(1/\rho)_i^* = (1/\rho)_i^j - \lambda [\,(r^2 u)_{i+1}^j - (r^2 u)_i^j\,]$$

$$u_i^* = u_i^j - (r^2)_i^j \lambda (p_{i+1}^j - p_i^j)$$

$$e_i^* = e_i^j - \lambda [\,(r^2 u p)_{i+1}^j - (r^2 u p)_i^j\,]$$

$$r_i^* = r_i^j + \Delta t \cdot u_i^j \qquad (6)$$

$$(1/\rho)_i^{j+1} = \tfrac{1}{2} \{ (1/\rho)_i^j + (1/\rho)_i^* - \lambda [\,(r^2 u)_i^* - (r^2 u)_{i-1}^*\,] \}$$

$$u_i^{j+1} = \tfrac{1}{2} [u_i^j + u_i^* - (r^2)_i^j \lambda (p_i^* - p_{i-1}^*)]$$

$$e_i^{j+1} = \tfrac{1}{2} \{ e_i^j + e_i^* - \lambda [\,(r^2 u p)_i^* - (r^2 u p)_{i-1}^*\,] \}$$

$$r_i^{j+1} = r_i^* + (\Delta t/2)(u_i^{j+1} - u_i^j) \qquad (7)$$

where i and j denote a lattice point in ξ and t directions, respectively, and $\lambda = \Delta t / \Delta \xi$. For stability, the condition

$$\frac{\Delta t}{\Delta \xi} \leq 1/r^2 \sqrt{\gamma \rho p} \qquad (8)$$

must be satisfied at every lattice point.

The cumulative calculational error is examined by using energy conservation to compare the total energy increment in the calculating region with the work done by the piston. The difference was found to be less than 2.0% throughout the calculation.

Typical results are presented (Fig. 1) in dimensionless form. The parameters for nondimensionalization are

$$t_* = r_* / \sqrt{p_0/\rho_0}, \qquad V_* = (4/3) \pi r_*^3 \qquad (9)$$

The two assumed typical rates of expansion represented in Fig. 1 are constructed so as to have the half-widths τ of 5.9 t_* and 11.8 t_* providing the identical V_* when integrated over time. The maximum rates of expansion become about 0.15 and 0.075, respectively. When the results are compared for these two different piston motions, it seems that the duration of the pressure waves for case a is almost half of that for case b, whereas the peak over-

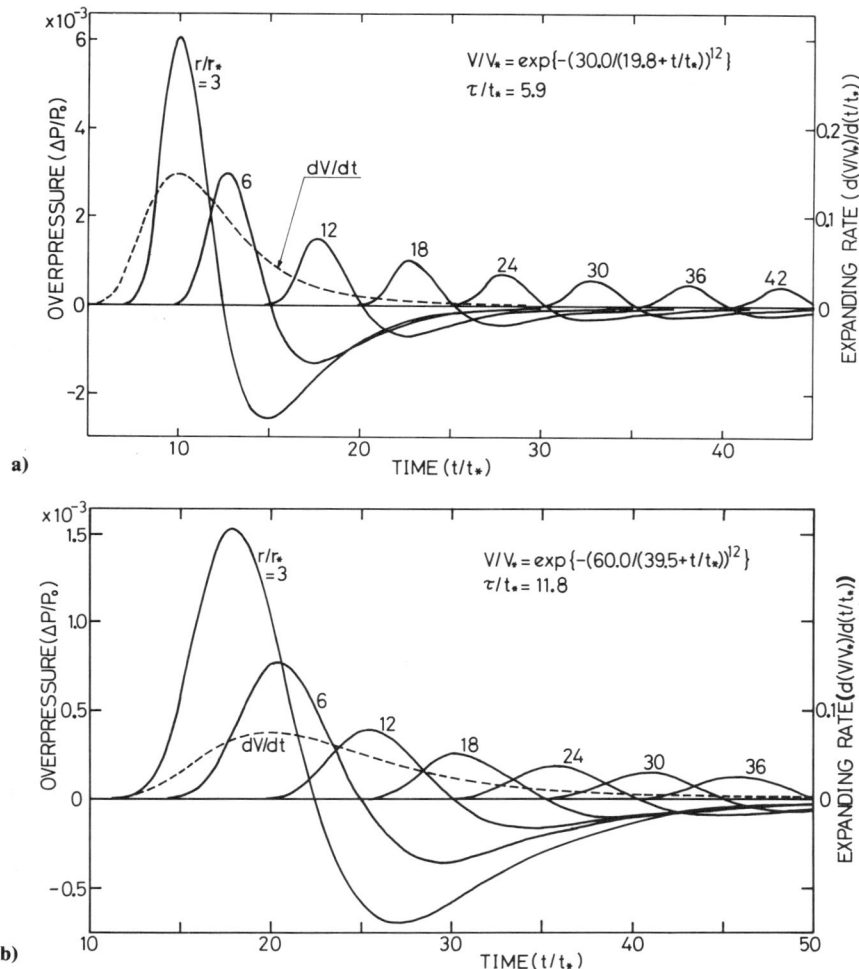

Fig. 1 Calculated overpressure and rate of piston expansion (solid line denotes overpressure, broken line denotes rate of piston expansion).

pressures $(\Delta p/p_0)_{max}$ for case a are about four times that for case b. Interestingly enough, the case a produces peak overpressures twice as strong as the case b. The relation between peak overpressure and the distance is shown in Fig. 2. Note that the peak overpressure generated by spherical piston is related to the dimensionless distance by the equation:

$$\Delta_p/P_0 \simeq \{0.64(t_*/\tau)^2\}(r/r_*)^{-1.0} \tag{10}$$

for the case where the parameter $n = 12.0$ and when $\tau/t_* \geq 3.0$. When $\tau/t_* < 3.0$, the expansion occurs rapidly and the relation approaches that for the

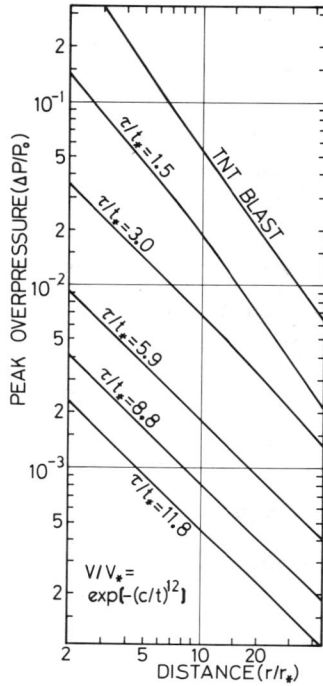

Fig. 2 Calculated peak overpressure, including curve for TNT blast (Strehlow and Baker, 1975).

TNT curve, i.e.,

$$\Delta p/p_0 \simeq \text{const } (r/r_*)^{-1.4 \sim -1.6} \tag{11}$$

It should be noted that our study shows that the transition of dependency on distance remains unchanged over the range of $n = 3\text{-}18$.

Experimental Apparatus

LPG is ejected into the open atmosphere from a fuel ejector. As shown in Fig. 3, the fuel ejector is mainly composed of a cylinder and a piston. The cylinder has a 22 mm i.d. and the volume between the piston and the cellophane is filled with 4-6 cm³ of liquefied fuel. As the piston is pushed up from the bottom by air compressed to 0.5 MPa, the film of cellophane is ruptured and the liquefied fuel is ejected. Ejection produces a fuel droplet-vapor cloud that mixes with the air. The shape of the fuel cloud is observed by a shadowgraph method. The vertical velocity of the fuel at ejection is about 10 m/s; it rapidly decreases to under 2 m/s about 0.15 s later. At 20-200 ms after ejection, the fuel vapor cloud is ignited by an electric spark located 250-1000 mm above the fuel ejector nozzle. The schematic of the system for ignition and time recording is shown in Fig. 4. The condenser (capacity 1μ F) for ignition is charged to 1.5-3.0 kV, and the spark gap is made of tungsten wires of 0.8 mm diam.

EXPLOSION OF AN UNCONFINED FUEL VAPOR CLOUD

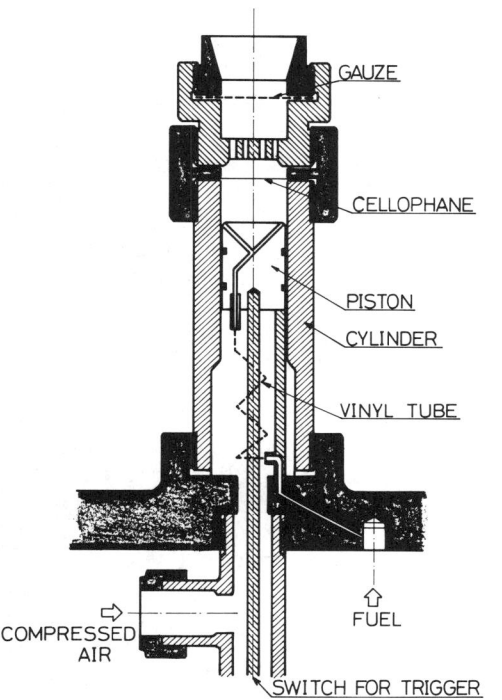

Fig. 3 Details of fuel ejector design.

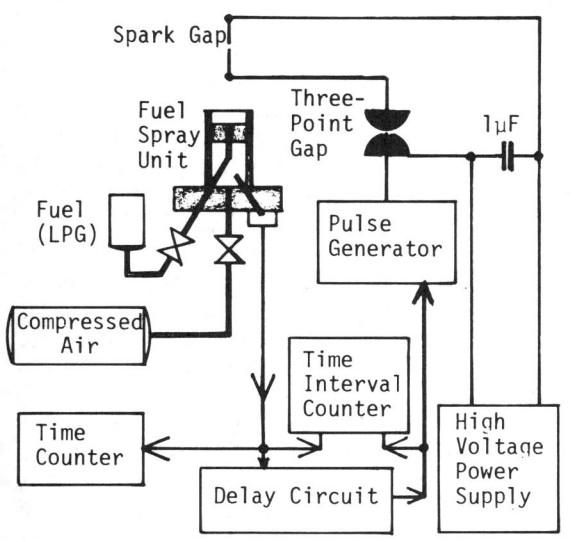

Fig. 4 Schematic of the system for ignition and time recording.

The development of the fireball is recorded by a 16 mm cine camera operated at the framing rate of 64 frames/s. The diameter of the fireball is measured from the film, assuming that the flames are axisymmetric. Pressure waves are observed using piezoelectric pressure transducers specifically designed to measure small pressure amplitudes.

Experimental Results

Ignition of an unconfined fuel vapor cloud can lead to combustion in a fireball mode, when both the timing and position of the ignition source are chosen to lie within a certain range. The influence of the location of the ignitor and the amount of fuel ejected on the appearance of observed fireballs have been investigated. The fuel was LPG, the main component of which is butane. The temperature of the fuel and the room are both about 10 °C. The fuel is ejected by air which was compressed to about 0.4 MPa. The sequence of the combustion process for three typical cases are shown in Fig. 5. In the first case (run 53), shown in Fig. 5a, the spark gap is located near the center of the fuel cloud. Immediately after the ignition a deflagration rapidly propagates in the premixed fuel-rich gas, until it becomes a diffusion flame at about $t = 100$ ms, when the flame becomes the brightest. During this period, the flame shape is governed mainly by the shape of the initial fuel cloud. At later periods, however, for example at $t = 203$ ms, eddies are clearly seen near the bottom of the flame. These are obviously created by the effect of buoyancy or convection, which along with the initial fuel movement, reinforce the flame by introducing more fresh air into the flame. As a result, the flame shape is transformed into a pancake under the action of a large eddy, the core of which is a ring around the vertical axis.

In the second case (run 3), shown in Fig. 5c, the volume of fuel is the same as in the first case, while the spark gap is located close to the bottom of the fuel cloud. The time for the deflagration to propagate through the whole fuel cloud is somewhat longer than the first case, because the upper boundary of the fuel cloud is much farther away from the ignition source. Although the initial shape of the flame is longer in the vertical direction, the final form of the flame becomes similar to the first case. In the third case (run 406), shown in Fig. 5c, the volume of the liquid fuel is increased, but the spark gap location is the same as that of the first case. The flame development is quite similar to the first case. For unknown reasons the duration of combustion is a little shorter than for the previous two cases in spite of a larger amount of the fuel.

From photographic records, the horizontal radii of the illuminous regions are measured and the rate of the change of the horizontal radii are computed. These results are shown in Fig. 6. In every case, the radius of the illuminous region increases rapidly during the first stage, followed by a sudden decrease of its rate. The maxima of the rates of expansion of the combustion region in this first stage are 4.4, 2.3, and 10.4 m/s for the three cases, respectively. Generally, the larger the volume, the greater the maximum rate of expansion in our experiments. After the initial peak, the rates of expansion decrease to below 1 m/s and fluctuate until burnout. In Fig. 6b, an estimated mean radius is also shown. In the first half of the duration, the mean radius is larger than the horizontal one, presumably due to the influence of the shape of the fuel

EXPLOSION OF AN UNCONFINED FUEL VAPOR CLOUD 173

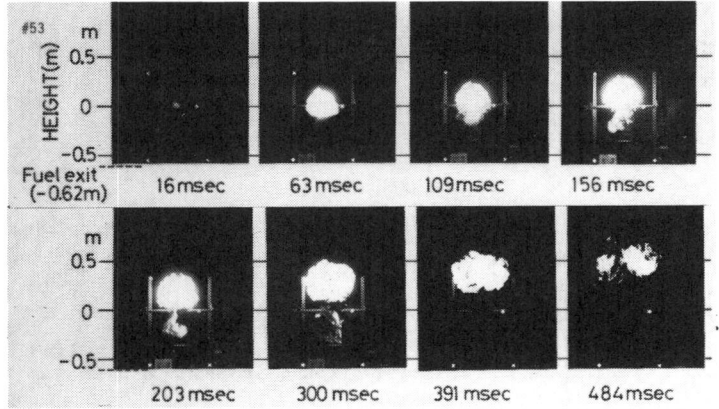
a) Run 53, LPG volume 4 cm^3, spark gap 0.62 m above ejector nozzle.

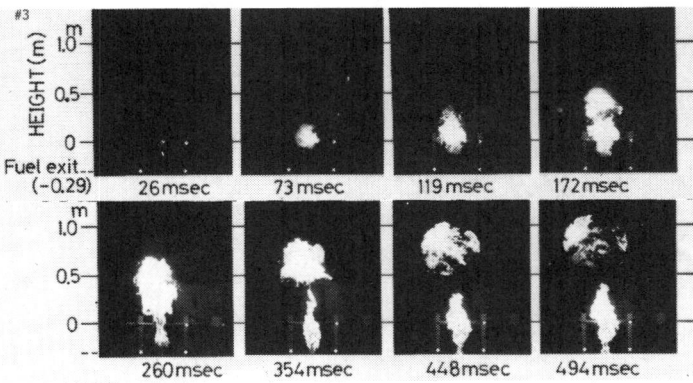
b) Run 3, LPG volume 4 cm^3, spark gap 0.29 m above ejector nozzle.

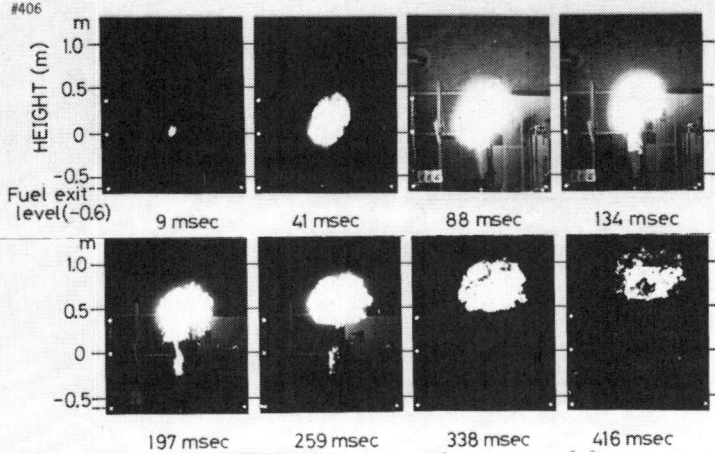
c) Run 406, LPG volume 6 cm^3, spark gap location 0.60 m above ejector nozzle.

Fig. 5 Photographs of combustion development.

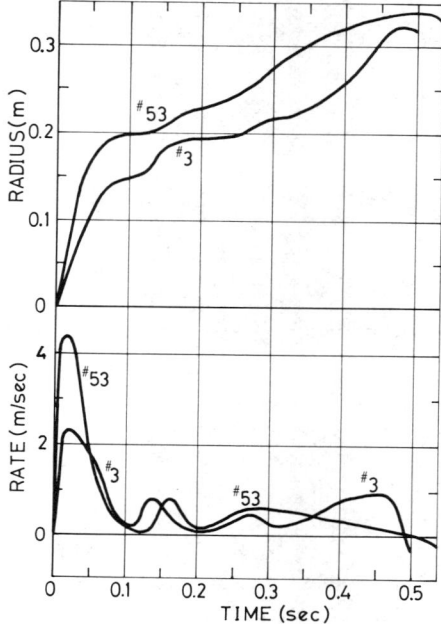

Fig. 6a Radius (horizontal direction) and its rate of change, runs 3 and 53 (fuel volume and spark gap location as given in Fig. 5).

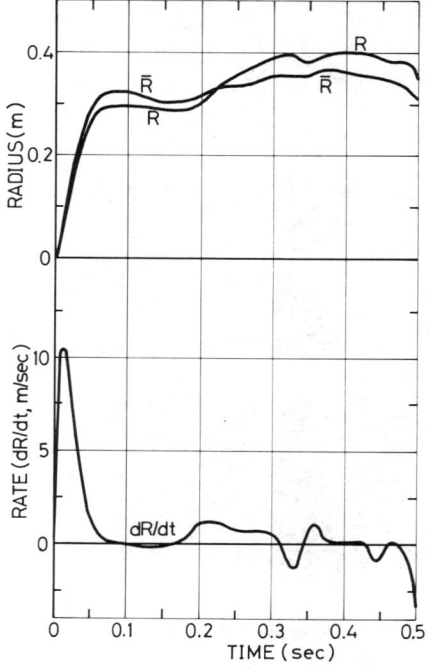

Fig. 6b Change in mean radius $\bar{R}$ and radius (horizontal direction) and rate of change vs time for run 406.

cloud. In the latter half, however, the horizontal radius increases and exceeds the mean one, because the flame shape is changed into a flat pancake by eddies.

In the present observation, the final radius of the fireball is 0.30-0.34 m for the fuel mass 2.4 g, and 0.36-0.40 m for the fuel mass 3.6 g. These values confirm the existing conception that the diameter of a fireball depends mainly on the fuel mass. The diameter of fireballs has been observed by several investigators. The relation between the diameter of a fireball and the fuel mass is shown in Fig. 7. It is seen that for an extremely wide range of fuel mass from below 10^{-4} kg to over 10^6 kg, the data are almost approximated by a single line. This indicates that the fireball size is determined by stoichiometric combustion of similar fuels.

Typical pressure waves, measured using a piezoelectric pressure transducer, are shown in Fig. 8. The behavior of flame developments for run 513 is similar to that for run 406 shown previously. The sensitivity of the pressure transducer previously is estimated to be 0.7 Pa/mV according to the calibration at higher pressures. The pressure peak occurs at about 70 ms after ignition, and slightly before the flame is brightest. The duration of the compression wave is of the order of 30 ms; that is extremely long.

Comparison of Experimental and Calculated Results

When the fuel volume is 6 cm³, for example, the reference values appearing in Eq. (9) are

$$r_* = 0.42 \text{ m}, \quad t_* = 1.48 \text{ ms} \tag{12}$$

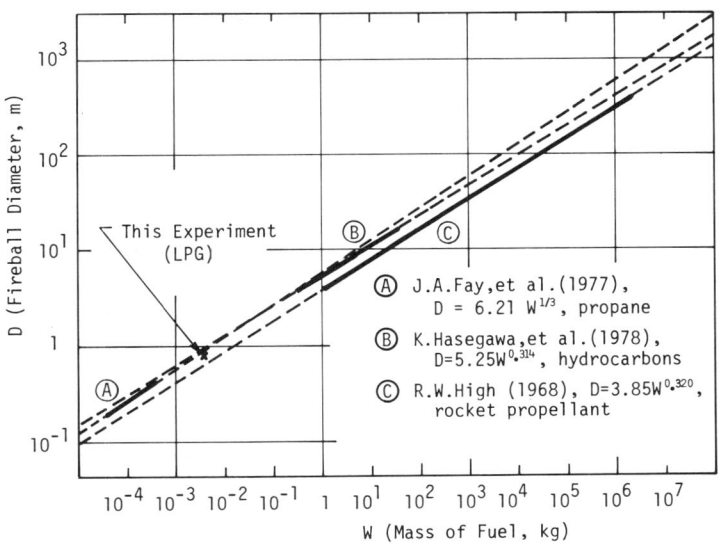

Fig. 7 Correlation for diameter of fireballs (solid lines denote observed results, broken lines denote extrapolations.

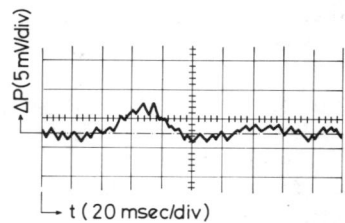

Fig. 8 Typical pressure oscillogram: transducer location 1.4 m from ignition point for run 513, fuel volume 6 cm³, spark gap location 0.6 m above ejector.

The location of pressure measurement shown in Fig. 8 corresponds to $r/r_* = 3.3$. Since the measured pressure peak is about 5 Pa, Eq. (10) yields $\tau/t_* \simeq 60$. It is seen in Fig. 1 that the duration of compression wave is nearly a half of τ, $30\,t_* \simeq 45$ ms, which is in reasonable agreement with the experiment.

From the preceding simple discussion, it seems useful to use dimensionless distance and time. If the fuels are hydrocarbons and the combustion efficiency equals 0.8, the following relations are approximately correct

$$r_* \simeq 2.78 W^{1/3} \text{ m}; \quad t_* \simeq 9.67 W^{1/3} \text{ ms} \tag{13}$$

where W is the fuel mass expressed in kg. The value of r_* nearly coincides with the radius of a fireball (Fig. 7). The strength of compression wave is naturally determined by the dimensionless combustion rate. The most important parameter of the piston model is τ/t_*, the half width of $dV/dt - t$ curve.

Conclusions

1) An experimental technique to produce unconfined fuel vapor cloud explosions on a laboratory scale is established. Although the volume of fuel (LPG) is extremely small (4-6 cm³), the flame development is similar to those of much larger scales. The development of unconfined fuel vapor cloud explosions can be divided into two stages, i.e., a deflagration propagating in premixed gases, followed by a diffusion flame promoted by buoyancy and convection.

2) Pressure waves induced by unconfined combustion are simulated using a spherically expanding piston: The volume increment V of the spherical piston changes with time t, according to $V/V_* = \exp\{-(c/a+t)^n\}$. Calculated results show that the peak overpressure is inversely proportional to the distance from the center. This property does not correspond to that of blast waves of the same dimensionless distance, but does to that of weak blast waves in the far field.

3) The experimental result from the pressure measurement is quantitatively consistent with the result obtained from the spherical piston model.

4) The results of a spherical piston analysis models the fireball diameter for an extremely wide range of fuel mass.

Acknowledgments

The authors express their sincere thanks to T. Fujiwara, Nagoya University, for his helpful advice.

References

Fay J. A. and Lewis D. H. (1977) Unsteady burning of unconfined fuel vapor clouds. *Proceedings of the Sixteenth Symposium (Intl.) on Combustion,* pp. 1397-1405. The Combustion Institute, Pittsburgh, Pa.

Hasegawa K. and Sato K. (1978) Study on the fireball following steam explosion of n-pentane. *Proceedings of the Second Intl. Symposium on Loss Prevention and Safety Promotion in the Process Industries* **VI,** 297.

MacCormack R. W. (1971) Numerical solution of the interaction of a shock wave with a laminar boundary layer. *Lecture Notes in Physics* **11,** 151-163.

Strehlow R. A. and Baker W. E. (1975) The characterization and evaluation of accidental explosions. NASA CR-134779.

Self-Similar Blast Waves Supported by Variable Energy Deposition in the Flowfield

R. H. Guirguis* and A. K. Oppenheim†
University of California, Berkeley, Calif.
and
M. M. Kamel‡
Cairo University, Cairo, Egypt

The problem of blast waves supported by variable energy deposited in the flowfield is formulated. In general, such waves can arise as a consequence of an exothermic process taking place in the flowfield or by absorption of radiation emanating from an outside source. A particular solution is obtained for a self-similar blast wave in which the amount of the deposited energy is directly proportional to local temperature and inversely proportional to time. Such condition can be considered as representative of a special kind of blast wave driven by laser irradiation. As typical of self-similar solutions, the applicability of the numerical results we present is restricted to strong waves, that is, cases where the effect of pressure in the medium into which their fronts propagate is negligible.

Nomenclature

a,b,c = parameters defined in Eq. (7)
D = function defined in Eq. (4)
E = energy deposited in blast wave per unit area for $j=0$, per unit polar and axial length for $j=1$, and per unit steric angle for $j=2$
f = nondimensional particle velocity, u/w
F = reduced coordinate, f/x
g = nondimensional pressure, $p/\rho_a w^2$
h = nondimensional density, ρ/ρ_a
j = geometric factor· 0, 1, or 2 for plane, cylindrical, or spherical symmetry
J = classical energy integral, $\int_0^1 [g/(\gamma-1) + hf^2/2]x^j dx$
p = pressure
P = function defined in Eq. (6)
$\dot{q}$ = rate of energy deposited per unit mass
Q = function defined in Eq. (5)

Presented at the 7th ICOGER, Göttingen, Federal Republic of Germany, Aug. 20-24, 1979. Copyright © American Institute of Aeronautics and Astronautics, Inc., 1980. All rights reserved.
*Research Assistant.
†Professor of Engineering.
‡Associate Professor.

r = radius
R = gas constant per unit mass
t = time
T = temperature
$\mathfrak{I}$ = energy integral, defined in Eq. (19)
u = particle velocity
w = front velocity, dr_n/dt
x = similarity variable, r/r_n
y = front coordinate, $\gamma p_a/p_a w^2$
Z = reduced coordinate, $\gamma g/hx^2$
γ = specific heat ratio
δ_j = factor related to unit angle, $2\pi j + (j-1)(j-2)$
ϵ = wave power index, $d\ell n E/d\ell n t$
λ = decay coefficient, $d\ell n y/d\ell n r_n$
μ = velocity index, $d\ell n r_n/d\ell n t$
ν = exponent in Eq. (10), $(\lambda + \phi)/(j+1)$
ρ = density
ϕ = energy deposition factor, defined in Eq. (9)
Λ = energy partition ratio, defined in Eq. (17)

Subscripts

a = ambient condition
B = singularity $B(P=0, Q=0,$ and $D \neq 0)$
D = singularity $(Q/Z=0$ and $Z=\infty)$
n = front or state immediately behind it

Introduction

ENERGY for blast waves can be delivered in four basic modes. First, it may be deposited instantaneously, as in the classical case of adiabatic point explosions treated by Von Neumann (1963), Sedov (1946), and Taylor (1941). Second, it may be provided by continuous deposition at the inner boundary of the flowfield, as in the cases of piston (Taylor, 1946) or deflagration (Kuhl et al., 1973) driven waves. Third, it may be deposited by chemical or radiative transfer at the front of the flowfield. Self-similar regimes of this type were exposed comprehensively by Barenblatt et al. (1980). Finally, it may be distributed throughout the flowfield, a mode that so far has received just scant attention (Bishimov, et al., 1970).

A judicious choice of the distribution function can provide a proper simulation of the effects of chemical reactions or of the radiative transfer of energy into or out of the flowfield. The latter is of particular interest in the case of blast waves driven by laser irradiation when the front is not totally opaque to the laser beam. To our knowledge this case has not been treated yet despite its potential importance to nuclear fusion.

In this paper the problem of blast waves associated with distributed energy deposition in the flowfield is formulated in terms of a general function governing this process. A solution is obtained for the special case of a self-similar blast wave—a flowfield for which the effects of the atmosphere into which the wave propagates are negligible.

In order to satisfy the condition of self-similarity the quantity $\dot{q}t/T$ (where $\dot{q}$ is the rate of energy deposition per unit mass, t the time, and T the temperature) must not be a function of the front coordinates. For the solutions presented here this quantity was assumed to be constant, simulating in effect a mode of energy deposition in laser-driven blast waves (Raizer, 1974).

A remark about the practical significance of self-similar solutions may be in order. As pointed out by Barenblatt (1979), such solutions are most often viewed best as intermediate asymptotics. They describe usually the flowfield after the process of initiation and before that of its decay, while both of them are essentially nonself-similar in character. This is, incidentally, not the case for the best known classical Taylor-Sedov self-similar solution describing the initial stages of a point explosion—the nonsteady flowfield formed by an instantaneous deposition of a finite amount of energy at a point. The most important feature of self-similar solutions is that they represent a dynamically compatible flow configuration; the structure of a nonsteady flowfield consisting of some prescribed elementary components under conditions satisfying the conservation principles of gasdynamics.

In our case this involves the coexistence of a continuous flowfield (where energy is being deposited) with a shock front (situated at its outer boundary). In order to conform to the self-similarity requirement, the trajectory of the latter must be a power function of the radius. When this is observed experimentally, one may infer from our solution a possible law that then governs the process of energy deposition and the corresponding structure of the flowfield. Such an inference is, of course, not unique, but it has the distinct advantage of simplicity—a feature that is so often encountered in nature that it has to be taken into account, at least as a point of departure for a more complex interpretation, if a good reason for it is thereby indicated. It is in this sense that self-similar solutions have a well-established physical significance.

Formulation

Governing Equations

Following our previous exposition (Oppenheim et al., 1971), the equations governing one-dimensional unsteady flowfields were expressed in terms of the following nondimensional variables

$$x \equiv r/r_n$$

where r is the space coordinate while subscript n denotes conditions at (or more accurately immediately behind) the front,

$$f \equiv u/w$$

where u is the flow velocity while $w \equiv dr_n/dt$,

$$h \equiv \rho/\rho_a$$

where ρ is the density while subscript a denotes ambient conditions and

$$g \equiv p/\rho_a w^2$$

where p is the pressure.

Then by introducing two reduced coordinates

$$F \equiv f/x \quad \text{and} \quad Z \equiv \gamma g/hx^2 \tag{1}$$

where γ is the specific heat ratio, the conservation equations were transformed into the following autonomous set:

$$\frac{dF}{d\ell nx} = -\frac{Q(F,Z)}{D(F,Z)} \tag{2}$$

and

$$\frac{dZ}{d\ell nx} = -\frac{Z}{1-F}\frac{P(F,Z)}{D(F,Z)} \tag{3}$$

where

$$D(F,Z) \equiv Z - (1-F)^2 \tag{4}$$

$$Q(F,Z) \equiv (j+1)(F-b)Z - (a-F)(1-F)F \tag{5}$$

and

$$P(F,Z) \equiv \{(j+1)(\gamma-1)+2\}(c-F)D(F,Z) + (\gamma-1)Q(F,Z) \tag{6}$$

while the parameters a, b, and c controlling the singularities of the above system of ODEs (Oppenheim et al., 1972) are specified as follows:

$$a \equiv \frac{\lambda+2}{2} \quad b \equiv \frac{\lambda+\phi}{(j+1)\gamma} \quad \text{and} \quad c \equiv \frac{\lambda+\phi+2}{(j+1)(\gamma-1)+2} \tag{7}$$

where $j = 0$, 1, or 2 for plane, line, or point symmetrical flowfields, respectively. In the above

$$\lambda \equiv \frac{d\ell n y}{d\ell n r_n} \tag{8}$$

is the so-called decay coefficient, while

$$y \equiv \gamma p_a/\rho_a w^2$$

and

$$\phi \equiv (\gamma-1)(h/g)\dot{q}(r_n/w^3) \tag{9}$$

is the energy deposition factor, $\dot{q}$ expressing the rate at which it is deposited per unit mass.

It should be noted that, in order to satisfy the self-similarity condition, ϕ has to be independent of the front coordinate y while, in order to preserve the autonomous structure of the governing equations, it can be at most a function of F and Z. Here it is taken as a constant. As a consequence of this, one has the added advantage of having the energy equation integrated in closed form, namely

$$\frac{g}{g_n}\left(\frac{h}{h_n}\right)^{-\gamma} = \left[\frac{h}{h_n}x^{j+1}\frac{1-F}{1-F_n}\right]^{-\nu} \quad (10)$$

where $\nu \equiv (\lambda+\phi)/(j+1)$.

Energy Deposition

The total amount of energy contained within the blast wave per unit area, or per unit length and polar angle or per unit steric angle for $j=0, 1,$ or 2, respectively, can be expressed for a perfect gas, as

$$E = \int_0^{r_n} \left(\frac{p}{\gamma-1} + \rho\frac{u^2}{2}\right) r^j dr = \rho_a r_n^{j+1} w^2 J \quad (11)$$

where

$$J \equiv \int_0^1 \left(\frac{g}{\gamma-1} + \frac{hf^2}{2}\right) x^j dx$$

referred to as the energy integral, is a constant for a given self-similar blast wave. Since the internal energy of the ambient atmosphere is zero, it follows from the principle of global energy conservation that

$$\frac{dE}{dt} = \int_0^{r_n} \rho\dot{q} r^j dr = \rho_a r_n^j w^3 \int_0^1 \frac{\phi g}{\gamma-1} x^j dx \quad (12)$$

Differentiating Eq. (11) with respect to time yields

$$\frac{dE}{dt} = \rho_a r_n^j w^3 (j+1-\lambda) J \quad (13)$$

By comparing Eqs. (12) and (13) one obtains the condition

$$\int_0^1 \frac{\phi g}{\gamma-1} x^j dx = (j+1-\lambda) J \quad (14)$$

relating the energy deposition function to the energy distribution within the field. It should be noted that the case of the classical adiabatic point explosion, $\lambda = j+1$ [Eq. (14)], is satisfied when $\phi = 0$, as expected. In terms of the velocity index

$$\mu \equiv \frac{d\ell n r_n}{d\ell n t} = \frac{wt}{r_n} = \frac{2}{\lambda+2} \quad (15)$$

a constant for a given self-similar wave, related to λ as shown above in consequence of Eq. (8), one gets using Eq. (9) and noting that $p/\rho = RT$,

$$\dot{q} = [\phi\mu/(\gamma-1)]/R(T/t) \tag{16}$$

Thus, one can observe that the assumption of $\phi = $ const is tantamount to the condition that the specific power is directly proportional to temperature and inversely proportional to time. From Eq. (14) it then follows that the relative amount by which energy in the flowfield is divided between kinetic and internal is given by

$$\Lambda \equiv \frac{\int_0^1 \frac{hf^2}{2} x^j dx}{\int_0^1 \frac{g}{\gamma-1} x^j dx} = \frac{\phi}{j+1-\lambda} - 1 \tag{17}$$

an energy partition ratio which, as will be shown later, is a constant bounded between 0 and 1.

The variation of blast wave energy with time can be most conveniently expressed in terms of the power index

$$\epsilon \equiv \frac{d\ell n E}{d\ell n t}$$

a constant for a given self-similar wave, since by virtue of Eqs. (11) and (15),

$$\epsilon = (j+3)\mu - 2 \tag{18}$$

On the basis of dimensional analysis, as shown by Barenblatt et al. (1980), the shock trajectory can be expressed in terms of time explicitly as follows:

$$r_n = \left(\frac{E(t)\delta_j}{t^3 \mathfrak{I} \rho_3}\right)^{1/(j+3)} t^\mu \tag{19}$$

The energy integral $\mathfrak{I}$, appearing above is related to J, the classical energy integral, as follows:

$$\mathfrak{I} = \delta_j \mu^2 J \tag{20}$$

where δ_j is the total field angle that is equal to 2, 2π, or 4π for $j = 0$, 1, or 2, respectively.

Solution

Integral Curves

Equations (2) and (3) yield a single first-order differential equation governing the solution in the phase plane of the reduced coordinates F and Z, namely

$$\frac{dZ}{dF} = \frac{Z}{1-F} \frac{P(F,Z)}{Q(F,Z)} \tag{21}$$

Solving the problem for a given value of λ and ϕ is thus reduced to the integration of the above equation, subject to the appropriate boundary conditions at the front specified by the strong shock condition (Oppenheim et al., 1972)

$$F_n = \frac{2}{\gamma+1}, \quad Z_n = \left(\frac{\gamma-1}{2}F_n + 1\right)(1-F_n)$$

On the other side, each integral curve has to culminate at singularity D, representing the conditions at the center, namely

$$F_D = b = \frac{\lambda+\phi}{(j+1)\gamma}, \quad Z_D = \infty \qquad (22)$$

It should be noted that the parameters λ and ϕ are interrelated by Eq. (17). This relation may be regarded as a compatability condition that has to be satisfied by ϕ for any given value of λ. However, Λ is unknown in advance. Thus the evaluation of integral curves for Eq. (21) becomes a boundary value problem, solved by iterating on Λ using estimates obtained from Eq. (17).

Once the integral curves are obtained, the similarity variable x can be found by the quadrature of Eqs. (2) or (3). The gasdynamic parameters are then evaluated from the following relations obtained from Eqs. (1) and (10) with the equation of state expressed by Eq. (26)

$$u/u_n = x(F/F_n) \qquad (23)$$

$$T/T_n = (Z/Z_n)x^2 \qquad (24)$$

$$\frac{\rho}{\rho_n} = \left[\frac{Z}{Z_n}\left(\frac{1-F}{1-F_n}\right)^\nu x^{(j+1)\nu+2}\right]^{1/(\gamma-1-\nu)} \qquad (25)$$

and

$$\frac{p}{p_n} = \frac{\rho}{\rho_n}\frac{T}{T_n} \qquad (26)$$

Limiting Cases

1) Lower Limit: $\lambda = -\infty$, $\Lambda = 0$

As the center is approached, $Z \to \infty$, $F \to F_D = b = 1$, so that a consequence of expression for b, Eq. (7), $\lambda + \phi = (j+1)\gamma$. Hence $\phi = +\infty$ and Eqs. (5) and (6) reduce to

$$Q = -\lambda/2(1-F)F \to \infty$$

and

$$P = (\gamma-1)Q \to \infty$$

while from Eq. (17) it follows that $\Lambda = 0$, i.e., all the energy deposited in the flowfield is transformed into internal energy. Under such circumstances, the governing equation becomes

$$\frac{dZ}{dF} = (\gamma - 1)\frac{Z}{1-F}$$

and its integration yields immediately

$$\frac{Z}{Z_n}\left(\frac{1-F_n}{1-F}\right)^{\gamma-1} \tag{27}$$

The flowfield can be then evaluated by the quadrature of

$$\frac{d\ell nx}{dF} = -\frac{D}{Q} = \frac{2D}{\lambda F(1-F)}$$

yielding

$$\ln\frac{X}{X_n} = \frac{2Z_n(1-F_n)^{\gamma-1}}{\lambda}\int_{F_n}^{F}\frac{dF}{F(1-F)^{\gamma-1}} - \frac{2}{\lambda}\int_{F_n}^{F}\frac{1-F}{F}dF \tag{28}$$

It is apparent from Eq. (28) that at $\lambda = -\infty$, $x = x_n = 1$, except at $F = 1$ when the first term becomes $-\infty$ causing x to drop abruptly to zero. The whole flowfield is therefore concentrated on the phase plane at one point: $F = 1$, $Z = \infty$, while the whole integral curve corresponds to $x = 1$.

Since for the strong shock condition

$$u/u_n = [(\gamma+1)/2]Fx \tag{29}$$

if $F = 1$, then from $x = 0$ to $x = 1$

$$u/u_n = [(\gamma+1)/2]x$$

so that at $x = 1$, u/u_n undergoes a step change from 1 to $(\gamma+1)/2$. On the other hand, Eqs. (24-26) imply, respectively, that at $x = 1$, T/T_n, ρ/ρ_n, and p/p_n all change abruptly from 1 to $+\infty$.

2) Upper Limit: $\lambda = \lambda_0$, $\Lambda = 1$

As λ increases, singularity B (Oppenheim et al., 1972) reaches the point on the phase plane representing the strong shock condition, imposing an upper limit on λ, λ_0. This corresponds to

$$F_B = c \equiv \frac{\lambda+\phi+2}{(j+1)(\gamma-1)+2} = \frac{2}{\gamma+1} \tag{30}$$

and, since B is on the $Q = 0$ locus,

$$Z_B \equiv \frac{(a-F_B)(1-F_B)F_B}{(j+1)(F_B-b)} = \frac{2\gamma(\gamma-1)}{(\gamma+1)^2} \tag{31}$$

Solving Eqs. (30) and (31) for ϕ and λ, one gets

$$\lambda_0 = 2[1 + 2j/(\gamma+1)] \tag{32}$$

For $j=2$ and $\gamma=1.4$, $\lambda_0 = 16/3$. Substituting Eqs. (30) and (32) into Eq. (17), it turns out that $\Lambda = 1$ independently of the values of j and γ, i.e., the energy deposited is shared equally between kinetic and internal.

Since $d\ell nx/dF = \infty$ and $d\ell nx/dZ = -\infty$, the value of x changes abruptly from 1 to 0 at (F_n, Z_n). Substituting $F = F_n = 2/(\gamma+1)$ in Eq. (29), one obtains

$$u/u_n = x$$

while, using Eqs. (24-26), it follows that for $Z = Z_n$.

$$T/T_n = x^2 \tag{33}$$

$$\frac{\rho}{\rho_n} = x^{[(j+1)\nu+2]/(\gamma-1-\nu)} \tag{34}$$

and

$$\frac{p}{p_n} = x^{2+[(j+1)\nu+2]/(\gamma-1-\nu)} \tag{35}$$

where

$$\nu \equiv \frac{\lambda+\theta}{j+1} = \frac{2j}{j+1}\frac{\gamma-1}{\gamma+1}$$

for $j=2$ and $\gamma=1.4$, Eqs. (34) and (35) yield

$$\rho/\rho_n = x^{15} \text{ and } p/p_n = x^{17}$$

The energy integral J can then be evaluated in a closed form, as follows:

$$J = 2 \int_0^1 \frac{g}{\gamma-1} x^j dx = \frac{4}{(\gamma-1)(\gamma+1)} \left[1 \bigg/ \left((j+3) + \frac{(j+1)\nu+2}{\gamma-1-\nu} \right) \right] \tag{36}$$

For $j=2$ and $\gamma=1.4$, $J=0.20833$ and Eq. (20) yields $\mathfrak{I} = 0.19472$.

Results

Parameters of self-similar blast waves, associated with energy deposition throughout the field according to the condition $qt/T=$ const, are, for the case of $j=2$ and $\gamma=1.4$, cited in Table 1 and plotted in Fig. 1.

Table 1 Solution parameters ($j = 2$, $\gamma = 1.4$)

λ	μ	ϵ	F_D	Λ	ϕ	$\phi\mu$	J	$\mathfrak{J}$
$-\infty$	0	-2	1	0	∞	-2	—	—
-10	$-1/4$	-3.25	0.917	0.06553	13.852	-3.463	2.48	1.95
-6	$-1/2$	-4.5	0.897	0.08517	9.767	-4.883	1.812	5.693
-4	-1	-7	0.882	0.10036	7.703	-7.703	1.478	18.58
-2	$\mp\infty$	$\mp\infty$	0.860	0.12236	5.612	$\mp\infty$	1.156	$+\infty$
-1	2	8	0.845	0.13753	4.550	9.100	1.000	50.26
0	1	3	0.827	0.15718	3.472	3.472	0.849	10.67
1	2/3	1.333	0.802	0.18358	2.367	1.578	0.704	3.930
2	1/2	0.50	0.767	0.22125	1.221	0.611	0.566	1.778
3	2/5	0	0.714	—	0	0	0.424	0.852
4	1/3	-0.333	0.622	0.38637	-1.386	-0.462	0.322	0.450
5	2/7	-0.571	0.393	0.67420	-3.384	-0.957	0.228	0.234
16/3	3/11	$-7/11$	0.159	1.00000	$-14/3$	$-14/11$	0.208	0.195

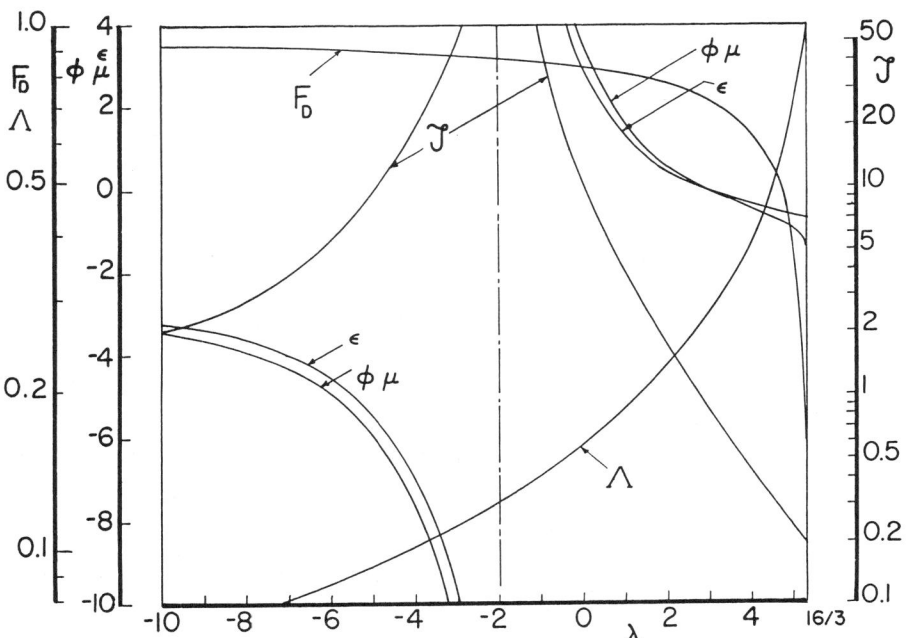

Fig. 1 Solution parameters corresponding to Table 1; $j = 2$, $\gamma = 1.4$.

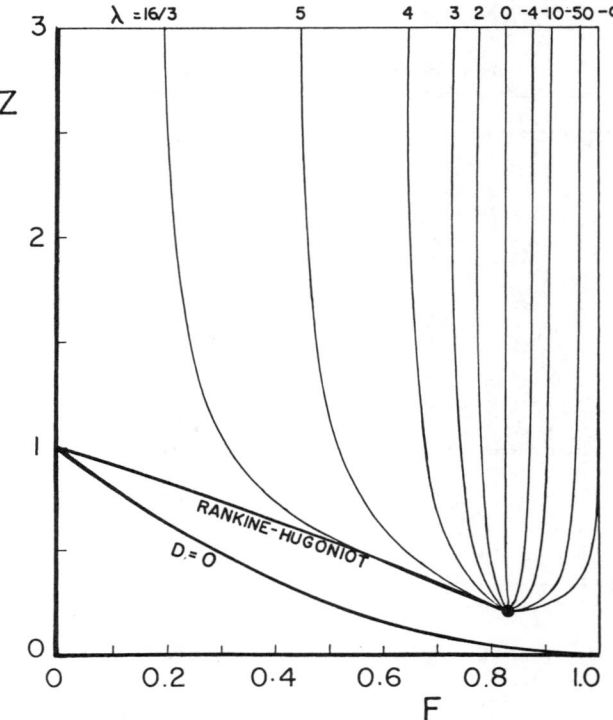

Fig. 2 Integral curves in the phase plane; $j=2$, $\gamma=1.4$.

Those corresponding to negative values of the front velocity modulus, μ (or $\lambda < -2$), according to which $r_n \sim t^\mu$, are considered admissible upon the understanding that the front trajectory is situated in the second quadrant of the time-space domain, i.e., its time scale is negative (viz. Oppenheim et al., 1971). For positive values of μ down to $\mu=0.4$ (or $-2<\lambda<3$), as well as for its negative values, energy is deposited while for $\mu<0.4$ (or $\lambda<3$) down to $\mu=0.273$ (or $\lambda=16/3$) it is withdrawn.

The solutions are bounded between-two limits. The lower limit of $\lambda=-\infty$ corresonds to a logarithmic trajectory (Oppenheim et al., 1972). It is to be regarded as a limit of negative μ's only since between $\mu=0$ and $\mu=3/11$ there is a gap, i.e., a region devoid of any solution. The upper limit of $\lambda=16/3$ corresponds to the maximum energy which can be withdrawn from the field due to the fact that pressure and temperature at the center are reduced to zero. It is of interest to note that the energy partition ratio Λ at this limit becomes independent of j and γ.

Integral curves for $j=2$ and $\gamma=1.4$ are displayed on the phase plane in Fig. 2. Each integral curve starts from the strong shock condition and ends at the center of symmetry represented by the singularity D for which $Z=\infty$. Plotted on the graph also is the sonic line $D=0$. All the integral curves are above this line, pointing out that all the flowfields under consideration are subsonic.

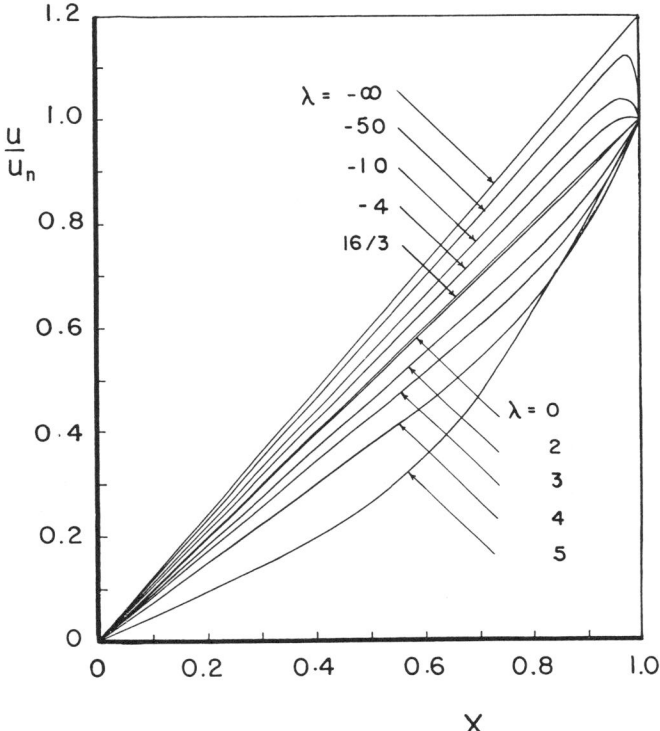

Fig. 3 Particle velocity profiles corresponding to Fig. 2.

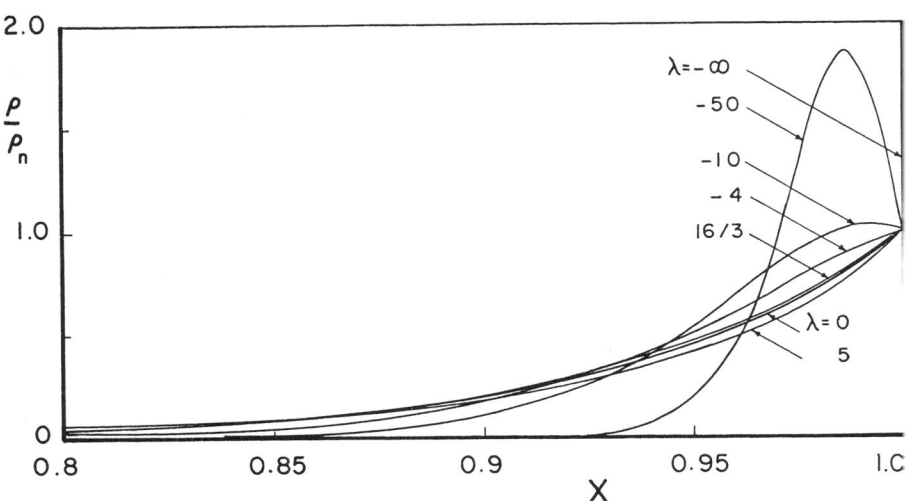

Fig. 4 Density profiles corresponding to Fig. 2.

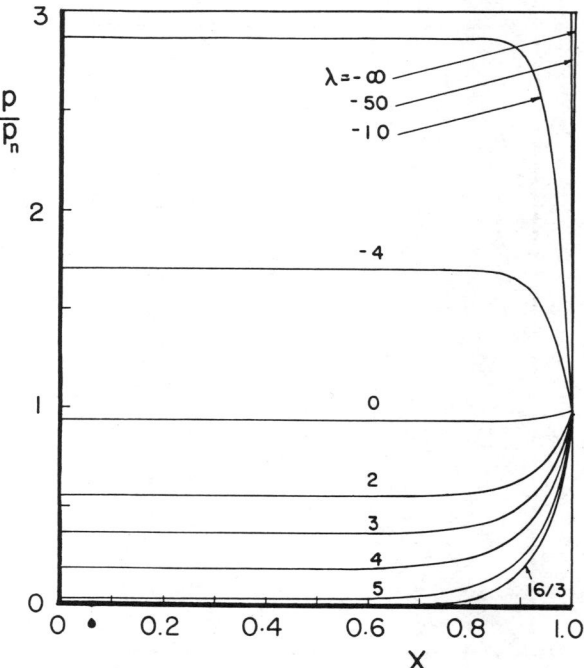

Fig. 5 Pressure profiles corresponding to Fig. 2.

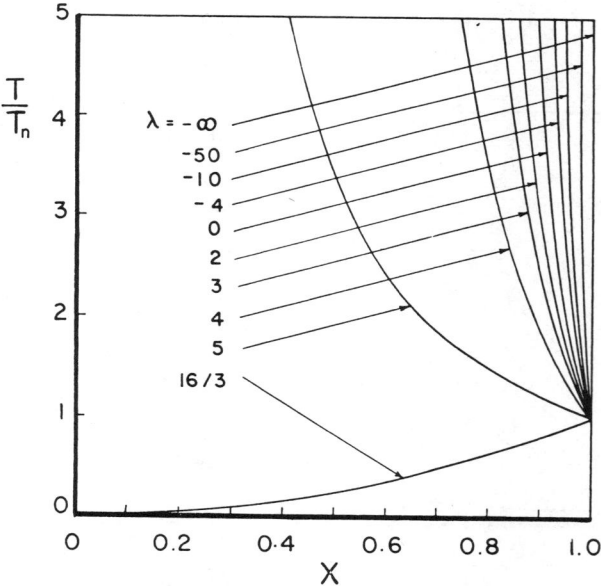

Fig. 6 Temperature profiles corresponding to Fig. 2.

The corresponding profiles of the gasdynamic field parameters u/u_n, ρ/ρ_n, p/p_n, and T/T_n are presented for a set of representative values of λ in Figs. 3-6.

Summary

In summary, we have presented a method of approach to the analysis of self-similar blast waves associated with energy deposition (or withdrawal) throughout the flowfield. As an illustration, a comprehensive solution has been provided for a class of waves for which the rate of energy deposited per unit mass is directly proportional to local temperature and inversely proportional to time. The solution consists of a set of integral curves on the phase plane of reduced blast wave coordinates and the corresponding set of space profiles of gasdynamic parameters: particle velocity, density, pressure, and temperature.

It should be noted that the solution we present constitutes just a representative case of self-similar blast waves with energy deposited in the field. It is, in fact, the simplest case, corresponding to energy deposition factor ϕ assumed constant. Other solutions could be generated in a similar way for variable ϕ, provided that its variation could be described entirely as a function of the reduced coordinates F and Z. In other cases, even though for the sake of self-similarity ϕ is still a function of just the field and space coordinates, the governing equations lose their autonomous character. Under such circumstances, our method does not offer any advantage since the solution is then governed by two rather than one differential equation, making it impossible to decouple the evaluation of integral curves from the determination of the profiles of gasdynamic parameters.

Acknowledgments

This work was supported by the U.S. Army Research Office under Grant DAAG29-77-G-0064 and the Department of Energy under Contract W-7405-ENG-48.

References

Barenblatt G. I. (1979) *Similarity, Self-Similarity, and Intermediate Asymptotics,* Vol. XVII, p. 218. Consultants Bureau, New York and London.

Barenblatt G. I., Guirguis R. H., Kamel M. M., Kuhl A. L., Oppenheim A. K., and Zeldovich Ya. B. (1980) Self-similar explosion waves of variable energy at the front. *J. Fluid Mech.* **99** (4), 841-858.

Bishimov E., Korobeinikov V. P., and Levin V. A. (1970) Strong explosion in combustible gaseous mixture. *Astronautica Acta* **15**, 267-274.

Kuhl A. L., Kamel M. M., and Oppenheim A. K. (1973) Pressure waves generated by steady flames. *14th Symposium (Intl.) on Combustion,* pp. 1201-1215. The Combustion Institute, Pittsburgh, Pa.

Oppenheim A. K., Lundstrom E. A., Kuhl A. L., and Kamel M. M. (1971) A systematic exposition of the conservation equations for blast waves. *J. Appl. Mech.* **38**, 783-794.

Oppenheim A. K., Kuhl A. L., Lundstrom E. A., and Kamel M. M. (1972) A parametric study of self-similar blast waves. *J. Fluid Mech.* **52** (4), 657-682.

Raizer Yu. P. (1974) *Lasernaya Iskra* (The Laser Spark). Izd. Nauka, Moscow (in Russian).

Sedov L. I. (1946) Propagation of intense blast waves. *Prikl. Mat. i Mekh.* **10**(2), 241-250.

Taylor G. I. (1941) The formation of a blast wave by a very intense explosion. British Rept. RC-210 (see also 1950, *Proc. Roy. Soc.* (London) **A201**, 175-176.

Taylor G. I. (1946) The air wave surrounding an expanding sphere. *Proc. Roy. Soc.* (London) **A186**, 273-292.

Von Neumann J. (1963) The point source solution. *N.D.R.C.*, Div. B. Rept. AM-9, classified report written in 1941 and published in *John Von Neumann Collected Works* (Ed. A. H. Taub) **XI**, 219-237. Pergamon Press, New York.

Chemical Kinetics in LNG Detonations

Charles K. Westbrook* and Leonard C. Haselman*
*Lawrence Livermore National Laboratory,
University of California, Livermore, Calif.*

The problem of detonability of vaporized mixtures of liquified natural gas and air is addressed, using a characteristic time analysis. Separate numerical models are used to treat the evolution of the blast wave produced by a charge of high explosive and the chemical ignition delay of the fuel-air mixture. These models are combined with experimental data to predict the amount of high explosive required to initiate a detonation of a stoichiometric mixture of methane and air, giving an estimate of 50-100 kg of high explosive in spherical geometry. The effects of minor constituents such as ethane and propane on methane-air detonability are examined, and the mechanism by which these minor constituents kinetically sensitize the fuel is discussed.

Introduction

IN the event of a large-scale spill of liquified natural gas (LNG) or other liquified energy fuel, the liquid fuel will rapidly vaporize and mix with air. As the fuel mixes with air, some portion of the fuel-air mixture may become flammable and/or detonable, depending on a variety of factors that are functions of the spill, spill site, the fuel itself, and many other parameters. In the present work, factors influencing the possibility of detonation of the fuel-air mixtures are examined by means of computer modeling techniques. The type of modeling approach described here is intended to be used in coordination with experimental programs. The models must be validated by means of comparison with experimental data, after which they can be used to assist in the analysis of those experiments and to extrapolate to conditions that are difficult or expensive to achieve experimentally. Model predictions must periodically be verified by means of further experiments. The primary goal of modeling complex systems such as gaseous detonations is to provide additional diagnostic tools to aid in the interpretation of given experiments and to substantially reduce the cost and time requirements of a research program.

Presented by the 7th ICOGER, Göttingen, Federal Republic of Germany, Aug. 20-24, 1979. Copyright © American Institute of Aeronautics and Astronautics, Inc., 1980. All rights reserved.
*Physicist.

In addition, model predictions can often indicate potentially fruitful areas for further experimental research or point out potential dangers.

Gaseous Detonations

Perhaps the most dangerous hazard that can result from LNG spill is the possibility of an atmospheric gaseous detonation. Detonations can be produced either by transition from deflagration or by direct initiation from a blast wave. In either case there are quite restrictive conditions that must be satisfied if the detonation is to propagate. The shock wave associated with a detonation compresses and heats a mixture of unreacted gases very rapidly. In the absence of chemical reactions this shock wave would gradually weaken, decaying into a simple compressional sound wave. It is possible to define a characteristic shock wave decay time, in the absence of reaction, as the time required for the shock pressure to fall from one value to some other value. If the shocked gas is reactive, then once the shock wave has compressed and heated the gas, chemical reactions will begin to take place. At the end of a chemical ignition delay period, rapid energy release again heats the mixture. This exothermic process generates the rarefaction that pursues and reinforces the detonation, counteracting the gradual decay of the shock wave. Therefore, a useful measure of the stability of a detonation wave can be derived by comparing the characteristic shock wave decay time with the chemical induction time. If the chemical time scale is longer than the shock decay time, the detonation will weaken, decaying into a sound wave preceding a conventional deflagration. On the other hand, if the chemical time scale is shorter than or comparable to the shock wave time scale, the detonation will be stable and continue to propagate.

Lee (1977) has shown that the minimum high explosive charge required to initiate detonation is strongly dependent on geometrical factors and for spherical configurations this blast energy would depend on the cube of the chemical induction time. Recently Westbrook (1979) has developed a kinetic model for the oxidation of methane, ethane, and air mixtures. This model makes it possible to calculate chemical induction times with a precision and generality not previously possible. The evolution of high explosive blast waves is computed using a numerical model developed specifically for such conditions (Wilkins, 1969). The detonation stability and direct initiation processes thus have been split conceptually into a fluid mechanical model dealing with the blast wave, and a chemical kinetic model dealing with the induction times. We will describe these two submodels and show how they have been combined to analyze certain detonation phenomena.

Chemical Ignition Delay

A great deal of work has been done in recent years on the ignition of methane in shock tubes and some studies of the shock tube ignition of ethane and higher alkanes have also appeared. A shock wave is propagated through a reactive gas sample, rapidly raising its density, temperature, and pressure to relatively high values. These postshock conditions are similar to those that are produced in detonation shock fronts. Under these postshock conditions the fuel first breaks apart into smaller fragment chemical species. This ignition or induction phase, during which the gas temperature and pressure are nearly constant, is followed by a very rapid oxidation phase during which these fragments react to form final products, with water and carbon dioxide being the most significant. The duration of the ignition phase is much longer than the oxidation phase. In a typical case, the combined reaction time was 250 μs, with the final oxidation phase taking less than 1 μs. The dominance of the ignition period is an important feature of the chemical evolution of these systems.

A detailed reaction mechanism describing the chemical kinetic evolution of mixtures of methane and ethane has been presented by Westbrook (1979) and is given in Table 1. This model reproduces experimental data reported by Burcat et al. (1971) for the ignition delay of $CH_4/O_2/Ar$ and $C_2H_6/O_2/Ar$ mixtures. These results are summarized in Fig. 1, in which the logarithm of the induction time is plotted as a function of reciprocal temperature. The upper solid line represents the experimental data for methane and the lower solid line shows the experimental data for ethane. Computed induction times are in-

Table 1 Methane-ethane oxidation mechanism [reaction rates in $cm^3 \cdot mole \cdot s \cdot kcal$ units, $k = AT^n \exp(-E_a/RT)$]

	Reaction		log A	n	E_a	Reference
1	$CH_4 + M$	$\rightarrow CH_3 + H + M$	17.1	0	88.4	Hartig et al. (1971)
2	$CH_4 + H$	$\rightarrow CH_3 + H_2$	14.1	0	11.9	Baldwin et al. (1970)
3	$CH_4 + OH$	$\rightarrow CH_3 + H_2O$	3.5	3.08	2.0	Zellner and Steinert (1976)
4	$CH_4 + O$	$\rightarrow CH_3 + OH$	13.2	0	9.2	Herron (1969)
5	$CH_4 + HO_2$	$\rightarrow CH_3 + H_2O_2$	13.3	0	18.0	Skinner et al. (1972)
6	CH_3HO_2	$\rightarrow CH_3O + OH$	13.2	0	0.0	Colket (1975)
7	$CH_3 + OH$	$\rightarrow CH_2O + H_2$	12.6	0	0.0	Fenimore (1969)
8	$CH_3 + O$	$\rightarrow CH_2O + H$	14.1	0	2.0	Peeters and Mahnen (1973)
9	$CH_3 + O_2$	$\rightarrow CH_3O + O$	13.4	0	29.0	Brabbs and Brokaw (1975)
10	$CH_2O + CH_3$	$\rightarrow CH_4 + HCO$	10.0	0.5	6.0	Tunder et al. (1967)
11	$CH_3 + HCO$	$\rightarrow CH_4 + CO$	11.5	0.5	0.0	Tunder et al. (1967)
12	$CH_3 + HO_2$	$\rightarrow CH_4 + O_2$	12.0	0	0.4	Skinner et al. (1972)
13	$CH_3O + M$	$\rightarrow CH_2O + H + M$	13.7	0	21.0	Brabbs and Brokaw (1975)
14	$CH_3O + O_2$	$\rightarrow CH_2O + HO_2$	12.0	0	6.0	Engleman (1976)
15	$CH_2O + M$	$\rightarrow HCO + H + M$	16.7	0	72.0	Schecker and Jost (1969)
16	$CH_2O + OH$	$\rightarrow HCO + H_2O$	14.7	0	6.3	Bowman (1975)
17	$CH_2O + H$	$\rightarrow HCO + H_2$	12.6	0	3.8	Westenberg and deHaas (1972a)
18	$CH_2O + O$	$\rightarrow HCO + OH$	13.7	0	4.6	Bowman (1975)
19	$CH_2O + HO_2$	$\rightarrow HCO + H_2O_2$	12.0	0	8.0	Lloyd (1974)
20	$HCO + OH$	$\rightarrow CO + H_2O$	14.0	0	0.0	Bowman (1970)
21	$HCO + M$	$\rightarrow H + CO + M$	14.2	0	19.0	Westbrook et al. (1977)
22	$HCO + H$	$\rightarrow CO + H_2$	14.3	0	0.0	Niki et al. (1969)
23	$HCO + O$	$\rightarrow CO + OH$	14.0	0	0.0	Westenberg and deHaas (1972b)
24	$HCO + HO_2$	$\rightarrow CH_2O + O_2$	14.0	0	3.0	Baldwin and Walker (1973)
25	$HCO + O_2$	$\rightarrow CO_2 + HO_2$	12.5	0	7.0	Westbrook et al. (1977)
26	$CO + OH$	$\rightarrow CO_2 + H$	7.1	1.3	-0.8	Baulch and Drysdale (1974)

(Table 1 continued on next page)

Table 1 (cont) Methane-ethane oxidation mechanism [reaction rates in $cm^3 \cdot mole \cdot s \cdot kcal$ units, $k = AT^n \exp(-E_a/RT)$]

	Reaction		log A	n	E_a	Reference
27	$CO + HO_2$	$\to CO_2 + OH$	14.0	0	23.0	Baldwin et al. (1970)
28	$CO + O + M$	$\to CO_2 + M$	15.8	0	4.1	Simonaitis and Heicklen (1972)
29	$CO_2 + O$	$\to CO + O_2$	12.4	0	43.8	Gardiner et al. (1971)
30	$H + O_2$	$\to O + OH$	14.3	0	16.8	Baulch et al. (1973a)
31	$H_2 + O$	$\to H + OH$	10.3	1	8.9	Baulch et al. (1973b)
32	$H_2O + O$	$\to OH + OH$	13.5	0	18.4	Baulch et al. (1973b)
33	$H_2O + H$	$\to H_2 + OH$	14.0	0	20.3	Baulch et al. (1973b)
34	$H_2O_2 + OH$	$\to H_2O + HO_2$	13.0	0	1.8	Baulch et al. (1973b)
35	$H_2O + M$	$\to H + OH + M$	16.3	0	105.1	Baulch et al. (1973b)
36	$H + O_2 + M$	$\to HO_2 + M$	15.2	0	-1.0	Baulch et al. (1973b)
37	$HO_2 + O$	$\to OH + O_2$	13.7	0	1.0	Lloyd (1974)
38	$HO_2 + H$	$\to OH + OH$	14.4	0	1.9	Baulch et al. (1973b)
39	$HO_2 + H$	$\to H_2 + O_2$	13.4	0	0.7	Baulch et al. (1973b)
40	$HO_2 + OH$	$\to H_2O + O_2$	13.7	0	1.0	Lloyd (1974)
41	$H_2O_2 + O_2$	$\to HO_2 + HO_2$	13.6	0	42.6	Lloyd (1974)
42	$H_2O_2 + M$	$\to OH + OH + M$	17.1	0	45.5	Baulch et al. (1973b)
43	$H_2O_2 + H$	$\to HO_2 + H_2$	12.2	0	3.8	Baulch et al. (1973b)
44	$O + H + M$	$\to OH + M$	16.0	0	0.0	Moretti (1965)
45	$O_2 + M$	$\to O + O + M$	15.7	0	115.0	Jenkins et al. (1967)
46	$H_2 + M$	$\to H + H + M$	14.3	0	96.0	Baulch et al. (1973b)
47	C_2H_6	$\to CH_3 + CH_3$	19.4	-1	88.3	Pacey (1973)
48	$C_2H_6 + CH_3$	$\to C_2H_5 + CH_4$	-0.3	4	8.3	Clark and Dove (1973)
49	$C_2H_6 + H$	$\to C_2H_5 + H_2$	2.7	3.5	5.2	Clark and Dove (1973)
50	$C_2H_6 + OH$	$\to C_2H_5 + H_2O$	13.8	0	2.4	Greiner (1973)
51	$C_2H_6 + O$	$\to C_2H_5 + OH$	13.4	0	6.4	Herron and Huie (1973)
52	C_2H_5	$\to C_2H_4 + H$	13.6	0	38.0	Lin and Back (1966)
53	$C_2H_5 + O_2$	$\to C_2H_4\ HO_2$	12.0	0	5.0	Cooke and Williams (1971)
54	$C_2H_5 + C_2H_3$	$\to C_2H_4 + C_2H_4$	17.5	0	35.6	Benson and Haugen (1967)
55	$C_2H_4 + O$	$\to CH_3 + HCO$	13.0	0	1.1	Davis et al. (1972
56	$C_2H_4 + M$	$\to C_2H_3 + H + M$	14.0	0	98.2	Just et al. (1977)
57	$C_2H_4 + H$	$\to C_2H_3 + H_2$	13.8	0	6.0	Benson and Haugen (1967)
58	$C_2H_4 + OH$	$\to C_2H_3 + H_2O$	14.0	0	3.5	Baldwin et al. (1966)
59	$C_2H_4 + O$	$\to CH_2O + CH_2$	13.4	0	5.0	Peeters and Mahnen (1973)
60	$C_2H_3 + M$	$\to C_2H_2 + H + M$	16.5	0	40.5	Benson and Haugen (1967)
61	$C_2H_2 + M$	$\to C_2H + H + M$	14.0	0	114.0	Jachimowski (1977)
62	$C_2H_2 + O_2$	$\to HCO + HCO$	12.6	0	28.0	Gardiner and Walker (1968)
63	$C_2H_2 + H$	$\to C_2H + H_2$	14.3	0	19.0	Browne et al. (1969)
64	$C_2H_2 + OH$	$\to C_2H + H_2O$	12.8	0	7.0	Vandooren and Van Tiggelen (1977)
65	$C_2H_2 + O$	$\to C_2H + OH$	15.5	-0.6	17.0	Browne et al. (1969)
66	$C_2H_2 + O$	$\to CH_2 + CO$	13.8	0	4.0	Vandooren and Van Tiggelen (1977)
67	$C_2H + O_2$	$\to HCO + CO$	13.0	0	7.0	Browne et al. (1969)
68	$C_2H + O$	$\to CO + CH$	13.7	0	0.0	Browne et al. (1969)
69	$CH_2 + O_2$	$\to HCO + OH$	14.0	0	3.7	Benson and Haugen (1967)
70	$CH_2 + O$	$\to CH + OH$	11.3	0.68	25.0	Mayer et al. (1967)
71	$CH_2 + H$	$\to CH + H_2$	11.4	0.67	25.7	Mayer et al. (1967)
72	$CH_2 + OH$	$\to CH + H_2O$	11.4	0.67	25.7	Peeters and Vinckier (1975)
73	$CH + O_2$	$\to CO + OH$	11.1	0.67	25.7	Peeters and Vinckier (1975)
74	$CH + O_2$	$\to HCO + O$	13.0	0	0.0	Jachimowski (1977)
75	$CH_3OH + M$	$\to CH_3 + OH + M$	18.3	0	80.0	Westbrook and Dryer (1979)

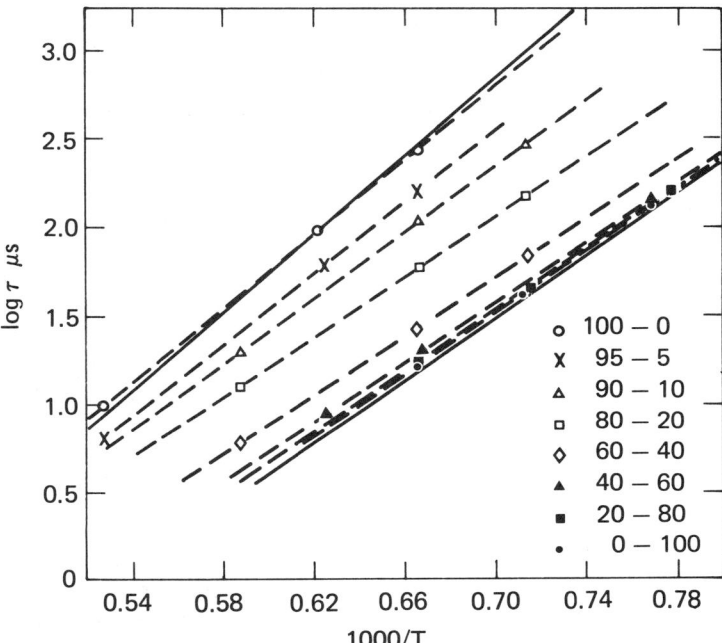

Fig. 1 Logarithm of the chemical induction time (μs) at different temperatures: dashed lines represent computed cases, key gives CH_4-C_2H_6 percentages, respectively; upper solid line shows experimental data for CH_4 and lower solid line for C_2H_6.

dicated as dashed lines, with the key indicating the relative fuel fractions of methane and ethane, respectively. From Fig. 1, it is clear that the model reproduces very well the data of Burcat et al. for the pure fuels.

With the mechanism validated at both ends of this compositional spectrum, the model was then used to investigate the evolution of mixtures of methane and ethane, combined first with stoichiometric amounts of oxygen. Particular attention was directed toward the compositional range that is closest to that encountered in normally occurring LNG, with aproximately 90% CH_4 and 10% C_2H_6. While the kinetic mechanism is not yet able to deal with propane or higher alkane species, there is both experimental and theoretical evidence to suggest that as far as kinetic sensitization and induction delay are concerned, propane and ethane behave quite similarly.

This study of the induction period of methane-ethane mixtures demonstrated several very significant points. First, the addition of quite small amounts of ethane (5-10%) to methane very sharply reduced the induction time of the composite fuel relative to that of pure methane as shown in Fig. 1. This reduction is also illustrated in Fig. 2, in which the induction time at several postshock temperatures is plotted as a function of fuel composition. For example, when ethane is 5% of the fuel, the induction time is roughly half that for pure methane. This reduction in induction time by a factor of two would correspond to a reduction in the critical energy for direct initiation of a detonation by a factor of eight (Lee, 1977). This effect is quite large and illustrates dramatically the need for detailed chemical kinetic analysis of these systems. In an important sense, the chemical behavior of LNG, at least as far as its detonability is concerned, appears to be dominated by minor constituents such as ethane.

This work was able to determine the detailed chemical mechanism for the fuel sensitization process. Methane itself is difficult to detonate, due primarily to its very long chemical induction time. The CH_4 molecule is unusually stable, with the hydrogen atoms bound tightly to the carbon atom. In addition, when a hydrogen atom is abstracted, the resulting methyl radicals (CH_3) are even more difficult to consume. Rather than being oxidized directly, methyl radicals combine together to form ethane ($CH_3 + CH_3 \rightarrow C_2H_6$). Much of methane consumption thus proceeds through ethane. The hydrogen atoms in the ethane molecule are more easily abstracted than in methane, and the consumption of ethane is much more rapid than methane. When ethane is present initially, more hydrogen atoms are available, and these hydrogen atoms initiate the chain branching reactions that rapidly consume the available fuel. The kinetic process, by which small amounts of ethane can dominate the consumption of methane and dramatically reduce the induction times, not only explains all of the experimental data but also demonstrates conclusively the inadequacy of so-called thermal sensitization mechanisms. Crossley et al. (1972) reached the same conclusion as to the inadequacy of the thermal sensitization mechanism, based on purely experimental results. The consumption of the two fuels occurs simultaneously, and it is through the free radical chain

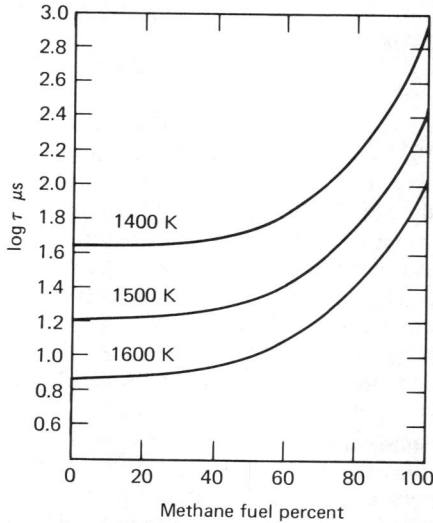

Fig. 2 Logarithm of chemical induction time (μs) at selected postshock temperatures showing the effect of fuel composition; remaining fuel is ethane.

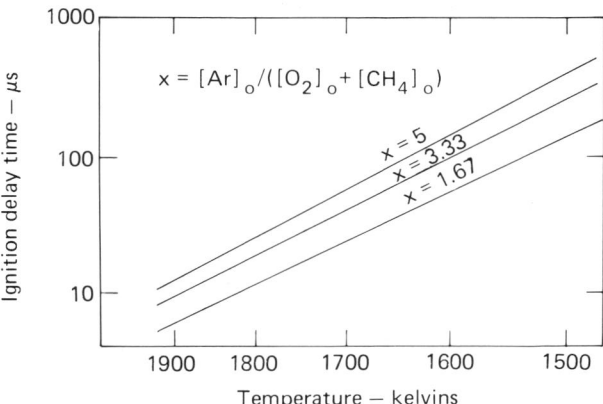

Fig. 3   Ignition delay times as functions of temperature for different amounts of argon dilution.

branching reactions that the coupling occurs, not through a sequential release of heat.

It was also shown that the same degree of kinetic sensitization occurs for fuel-air mixtures that are not stoichiometric. This conclusion is significant since wide ranges in local fuel-air equivalence ratio would be expected in an actual LNG spill. In addition, the presence of water vapor in the air was found to have a negligible effect on the computed induction delay times.

In addition to the earlier kinetic modeling work already described, we have carried out another series of shock tube ignition delay time calculations to examine the effects of changes in the amount of inert diluent that is present along with the fuel and oxygen. For the sake of illustration we consider here the results for stoichiometric methane-oxygen mixtures with different amounts of dilution; results for fuel mixtures of methane and ethane and for nonstoichiometric mixtures are very similar. The experimental work of Burcat et al. used argon as the diluent, and the mole fraction of argon was five times that of the oxygen. This was done in order to better approximate the heat capacity of air than if greater dilution were used. In the new series of calculations, the amount of argon diluent was varied by $\pm 50\%$, with ignition delay times computed for each composition at a variety of initial postshock temperatures. The ignition delay times for some of these mixtures are plotted in Fig. 3 as functions of reciprocal temperature. The effective activation energy appears to increase slightly with increasing argon dilution. The computed results at 1500 and 1900 K are summarized in Fig. 4 as functions of dilution. The ignition delay time can be seen to be proportional to the ratio $[Ar]_0/([O_2]_0 + [CH_4]_0)$ with the constant of proportionality changing with temperature.

Relating Shock Decay Times to Induction Times

A considerable amount of experimental information is available on the

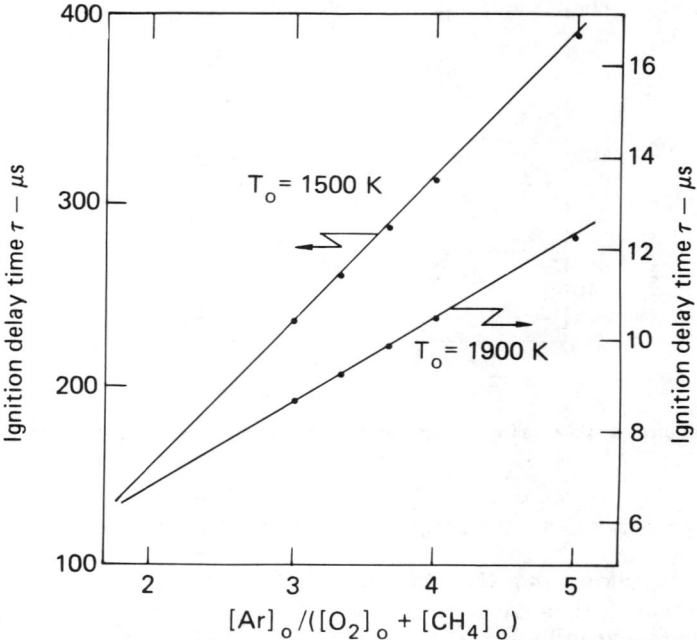

Fig. 4 Ignition delay times as functions of argon dilution evaluated for $T_0 = 1500$ and 1900 K.

detonability of fuels that are either pure methane or primarily methane in oxygen and in air. These experiments have been carried out under nearly unconfined, atmospheric conditions and with carefully defined amounts of fuel and oxidizer. In one series of experiments, Bull and co-workers (1976) used stoichiometric mixtures of methane and oxygen, diluted with varying amounts of nitrogen. In each case they determined the minimum amount of high explosive required to initiate a steady detonation in an unconfined spherical configuration. One goal of their study was to use results at low nitrogen concentrations, where the experiments were simpler to perform, to extrapolate to conditions with large amounts of nitrogen (as in normal air) where the experiments could not be carried out. In the second series of experiments reported by Bull et al. (1977, 1979), critical masses of high explosive were determined for various mixtures of methane and ethane in air. These data, shown in Fig. 5, display the same rapid sensitization of methane by ethane described earlier. Again the data were extrapolated to the limit of pure methane in air, with both extrapolations indicating that approximately 22 kg of high explosive would be required to detonate a methane-air mixture.

Comparisons were made with these two sets of experiments in a series of model calculations. For each mixture selected a one-dimensional finite-difference hydrodynamic numerical model was used to calculate the evolution of the time-dependent shock wave produced by spherical charges of high

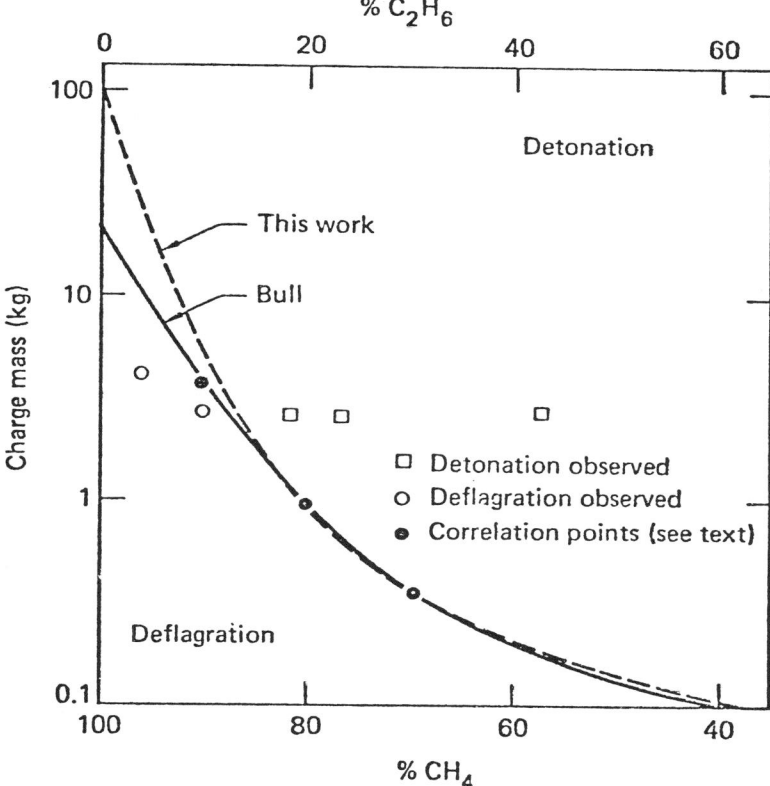

Fig. 5 **Detonation thresholds as functions of composition.**

explosive for charge masses ranging 0.01-22 kg. The shock decay time was defined somewhat arbitrarily as the time required for the shock to decay from 20 to 10 bars. This pressure range was selected primarily because it encompasses the Chapman-Jouguet pressure for a detonation in atmospheric LNG-air of approximately 16 bars. The result of these calculations was a relation between charge mass and shock decay time and is shown in Fig. 6. The shock decay time was found to vary as the cube root of the charge mass, as would be expected from point blast theory treatments of spherical shock front decay. At the same time, chemical induction times were calculated for each mixture. In these kinetics calculations, the initial postshock density was held fixed at 6×10^{-3} g/cm^3 and the initial postshock temperature was varied over a wide range. The results of the induction time calculations for the CH_4-C_2H_6 mixtures in air are plotted in Fig. 7. By equating the chemical induction time with the shock decay time, a correlation was established between the critical mass of high explosive and the initial postshock temperature of the reactive gas mixture. This precedure is illustrated by the three large dots in Figs. 5-7,

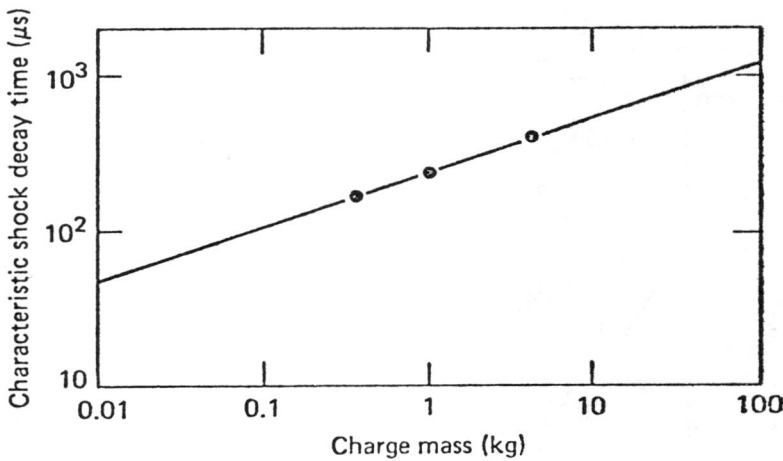

Fig. 6 Shock decay times (20-10 bars) for varying high explosive charge masses.

showing the effect of fuel composition on critical high explosive mass characteristic shock decay time and induction time. Let us illustrate this argument explicitly. From Fig. 5, experimental results show that a mixture of 70% CH_4-30% C_2H_6 corresponds to an HE charge mass of about 360 g. The shock decay calculations plotted in Fig. 6 show that the shock decay time for 360 g charge is about 130μs. Finally, the kinetics results in Fig. 7 show that 130μs induction time intersects the 70% CH_4-30% C_2H_6 line at about 1230 K. All of these indicated values in Figs. 5-7 are marked by a heavy dot. As a result, we have identified a mixture of 70% CH_4-30% C_2H_6 with a postshock temperature of 1230 K and an initiator mass of 360 g. The same procedure for the other heavy dots connects these quantities (80% CH_4-20% C_2H_6; 1 kg HE, 1250 K, 240μs) and (90% CH_4-10% C_2H_6; 3.7 kg HE, 1280 K, 390μs). Note that these temperatures are all very similar. An entirely analogous procedure was carried out for the experimental data in which the nitrogen content varied. Finally the characteristic temperature correlations were extrapolated in both cases to estimate the relevant induction time for methane-air. This procedure is illustrated in Fig. 7 by the three sets of dashed lines leading from the large dots to the line representing methane air. The dashed lines are intended to represent reasonable upper and lower limits for the extrapolation and the central dashed line shows a straight line extrapolation. Translating these intersection points back into critical charge masses gives limits of 24 kg and 100 kg, while the straight line extrapolation give a value of 55 kg, as shown in Fig 7. The corresponding values based on the other set of experimental data (not shown) give a range of 50-100 kg of high explosive.

The choice of a characteristic pressure decay time based on a decrease of the shock pressure from 20 to 10 bars is of course somewhat arbitrary. An interesting correlation can be derived in a somewhat different fashion from that described above, requiring only that the shock decay time depend on the cube

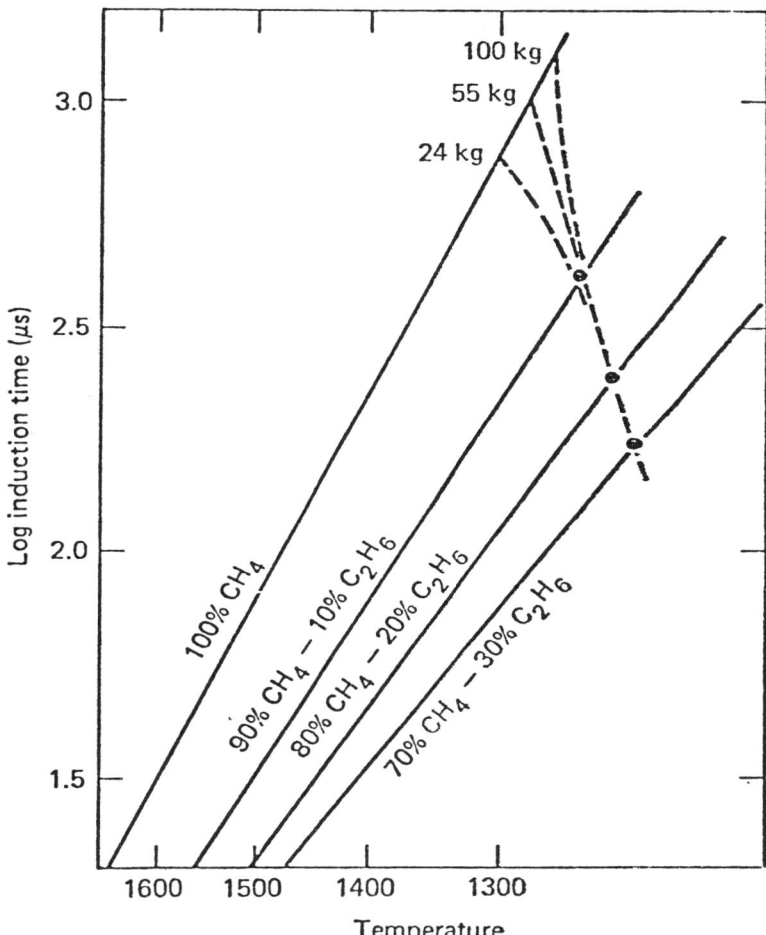

Fig. 7 Logarithm of chemical induction time (μs) for various fuel mixtures showing the points used to extrapolate to pure methane.

root of the charge mass. The fuel mixture with 70% CH_4- 30% C_2H_6, at an initial postshock temperature of 1400 K, was chosen for reference. From Fig. 7, the induction time is 33 μs. From Fig. 5, the experimental data show that the critical high explosive mass for detonation initiation for this mixture is 360 g. These two points, together with the assumption of a cube root dependence of pressure decay time on charge mass, allow us to use the other data of Fig. 7, all at 1400 K initial temperature, to predict the dependence of critical charge mass on composition. For an initial temperature of 1400 K, the computed chemical induction time from Fig. 7 are: 80% CH_4-20% C_2H_6, 44 μs; 90% CH_4-10% C_2H_6, 75 μs; and 100% CH_4, 218 μs. If we assume that these induction times are equal to pressure decay times, we can express the charge mass in terms of the characteristic time τ

$$\tau \alpha m^{1/3} \qquad \tau/\tau_0 = (m/m_0)^{1/3} \qquad m = m_0(\tau/\tau_0)^3$$

Here m_0 is the 360 g critical charge mass observed for 70% CH_4-30% C_2H_6 and τ_0 is the computed induction time of 33 μs for this mixture. Insertion of the aforementioned times (44, 75, and 218 μs) into this equation, gives values for m of 0.85, 4.2, and 106 kg, respectively. The result of this process is illustrated by the dashed curve in Fig. 5 which shows a remarkable agreement between predicted and experimental results. Therefore, the extrapolation of this predicted curve to pure methane in air gives a high explosive mass of 106 kg. Data points from Lind (1978), quoted by Boni et al. (1978), are also indicated in Fig. 5 and are consistent with the predicted curve. The use of a constant initial postshock temperature reflects the fact that the Chapman-Jouguet conditions for methane and ethane are very similar, so that the postshock conditions which lead to a propagating detonation should be the same for all mixtures of fuels. The temperature of 1400 K was used because it provided the best fit to the portion of Fig. 5 for which experimental data are available.

The extrapolations carried out in this type of analysis have used temperature as a variable because this dependence is rather weak. The predicted critical charge mass for pure methane in the air is quite sensitive to the extrapolation method, so that the range quoted of 24-106 kg must be regarded as somewhat uncertain. However, there appears to be no way to justify a predicted charge mass of more than 1000 kg suggested by Boni et al. (1978). Boni et al. used a coupled fluid mechanics-chemical reaction model in their study, while the present study has decoupled the two processes, but it seems unlikely that the large differences in predictions for the methane-air could be attributed solely to this one factor. In addition, the chemical reaction model used in the present work is considerably more sophisticated and more reliable than that used by Boni et al. In this light, no explanation can be given for the disagreements between the present results and those of Boni et al.

It must be emphasized again that while there is considerable theoretical interest in the initiation of a detonation in methane-air, there is some reason to question how relevant that situation is to practical LNG safety. Since LNG contains appreciable amounts of minor chemical species which have been determined, both experimentally and in our modeling studies, to significantly modify its chemical behavior, predictions of LNG detonability made on the basis of studies of pure methane can be seriously misleading. As noted earlier, with only 5% of the fuel consisting of ethane, the induction time is half that of pure methane. This translates into a reduction of a factor of eight in the amount of high explosive needed to detonate such a mixture, and with a typical LNG composition of 90% CH_4 and 10% ethane, propane, and other species, the critical mass is even smaller. In addition, the process of differential boiloff, in which the more volatile component methane evaporates first, will mean that the composition of the LNG vapor resulting from a typical spill will be progressively richer in these minor constituents. This effect would further reduce the required high explosive mass to initiate a detonation.

The kinetic modeling presented here, as well as that discussed earlier by Westbrook, suggests that several types of fuel modification might be used to increase the chemical induction time of LNG-air mixtures, thereby reducing the detonability of the mixtures. If some additive were included in the fuel which could serve as a means of capturing H atoms, the chain branching of these systems would be sharply reduced. Similarily, if the amounts of minor constituents such as ethane could be removed or at least significantly reduced, the results in Fig. 1 indicate that the induction time would again be sharply increased. In addition, a chemically inert diluent species could be added to the LNG. Although in the computations described earlier argon was the diluent, any such inert diluent would produce similar results. The need to be able to liquefy the diluent at about the same temperature as the LNG would place a restriction on the choice of diluent. Any of these processes could significantly enhance the safety of handling and using the fuel, so long as the process being used had no other effect detrimental to the safety or end use of the fuel.

Conclusion

It is important to consider that typical LNG is composed not only of methane, but that approximately 10% of LNG is made up of ethane, propane, and other species. The induction time calculations described here show that this 10% makes a great deal of difference in the induction time and, therefore, in the detonability of LNG. Studies that have not or do not take this composition into account may not be applicable to the question of the detonability of LNG vapor. These impurities or minor constituents play a major role in determining the induction time and detonability of LNG. The purely kinetic model described here has been validated by comparison with experimental data and can be reliably applied to other sets of conditions that have not received experimental attention. This was done to examine the possible effects of the presence of water vapor and effects of inert dilution and fuel stoichiometry, in addition to variations in fuel composition. Finally, the characteristic time analysis described was used to correlate available experimental data on unconfined detonations. Extrapolations were made to estimate that a high explosive mass of 24-106 kg would be required to detonate a stoichiometric methane-air spherical cloud.

Acknowledgment

This work was performed under the auspices of the U.S. Department of Energy by the Lawrence Livermore Laboratory under Contract W-7405- Eng - 48.

This report was prepared as an account of work sponsored by the United States Government. Neither the United States nor the United States Department of Energy, nor any of their employees, nor any of their contractors, subcontractors, or their employees, makes any warranty, express or implied, or assumes any legal liability or responsibility for the accuracy, completeness or usefulness of any information, apparatus, product or process disclosed, or represents that its use would not infringe privately owned rights.

Reference to a company or product name does not imply approval or recommendation of the product by the University of California or the U.S. Department of Energy to the exclusion of others that may be suitable.

References

Boni A., Wilson C. W., Chapman M., and Cook J. L. (1978) A study of detonation in methane/air clouds. *Acta Astronautica* **5**, 1153-1169.

Boni A. A., Su F.Y., and Wilson C. W. (1978) Numerical simulation of unsteady combustion and detonation phenomena. AIAA/SAE 14th Joint Propulsion Conference, Las Vegas, Nev. Paper 78-947.

Bull D. C., Elsworth J. E., Hooper G., and Quinn C.P. (1976) A study of spherical detonations in mixtures of methane and oxygen diluted with nitrogen. *J. Phy. D* **9**, 991-2000.

Bull D. C., Elsworth J. E., and Hooper G. (1977) Initiation of spherical detonation in hydrocarbon/air mixtures. *Acta Astronautica* **5**, 997-1008.

Bull D.C., Elsworth J. E., and Hooper G. (1979) Susceptibility of methane-ethane mixtures to gaseous detonation in air. *Combustion and Flame* **34**, 327-330.

Burcat A., Scheller K., and Lifshitz A. (1971) Shock-tube investigation of comparative ignition delay times for C_1-C_5 alkanes. *Combustion and Flame* **16**, 29-33.

Crossley R. W., Dorko E. A., Scheller K., and Burcat A. (1972) The effect of higher alkanes on the ignition of methane-argon mixtures in shock waves. *Combustion and Flame,* **19**, 373-378.

Lee J. H. (1977) "Initiation of gaseous detonation. *Ann. Rev. of Phys. Chem.* **28**, 75-104.

Lind C. D. (1978) Private communication.

Westbrook C. K. (1979) An analytical study of the shock tube ignition of mixtures of methane and ethane. *Combustion Sci. and Technol.* **20**, 5-17.

Wilkins M. (1969) Calculation of elastic-plastic flow. University of California, Lawrence Livermore Lab. Rept. UCRL-7322, Livermore, Calif.

II. Liquid and Solid Phase Phenomena

Molecular Dynamics of Shock and Detonation Phenomena in Condensed Matter

John R. Hardy*
University of Nebraska, Lincoln, Neb.
and
Arnold M. Karo† and Franklin E. Walker‡
*Lawrence Livermore National Laboratory,
University of California, Livermore, Calif.*

We have studied, using computer molecular dynamics, the behavior of a number of shock-loaded two-dimensional systems. Initially, we studied the behavior of lattices containing voids when an exterior wall was shock loaded by an impacting plate. We did this first for a model system bonded by Morse potentials typical of real systems. We then proceeded to study two more realistic systems: in the first case, the effect of aluminum plate impact on a model face-centered cubic nitric oxide lattice containing a void, and in the second case, a similar situation where the lattice was a model face-centered cubic sulfur nitride lattice. In all three cases we considered the situation when the system was hot, the atoms being given small initial random motions, and when it was cold, i.e., no initial random motion. In all three systems we found that the shocks had violent dynamic and static structure on the atomic scale that led to the spall of highly directed fragments into the void. These produced highly localized disruption of the opposite wall, leading to the spall of a jet-like stream of fragments from its outer face. This last behavior was absent for the nitric oxide lattice, for which something closer to fracture was produced in the second wall. This general pattern of behavior was observed for both cold and hot lattices. Our second topic of study was the effect of shock loading by impact on a metastable lattice that could undergo an exothermic phase transition to a configuration of lower energy. We found that shock loading induced such a transition immediately behind the shock front; thus both phase transition and shock front propagated with the same velocity. Our final study was a microscopic examination of the "Hugoniot" relationship between shock and particle velocities. We were able to show that our studies lead to good agreement with experimental measurements.

Presented at the 7th ICOGER, Göttingen, Federal Republic of Germany, Aug. 20-24, 1979.
Copyright © American Institute of Aeronautics and Astronautics, Inc., 1980. All rights reserved.
*Professor of Physics, Behlen Laboratory of Physics.
†Senior Scientist.
‡Deputy Program Director, Non-Nuclear Ordnance Program.

Introduction

SOME two years ago we presented at the sixth of these colloquia the first results of our molecular dynamics simulations of the shock loading of two-dimensional crystal lattices (Karo et al., 1978). Specifically, we discussed the behavior of two distinct lattices, both consisting of one type of atom and both having interatomic potentials which acted between first and second neighbor atoms. However, while the potentials were the same for both types of neighbor for a given lattice, they were different in form for the two lattices. The first lattice had Morse potential interactions so that bond breaking, and thus lattice disruption, were endothermic processes; while the second lattice was bonded by potentials which were such that, although the atoms were located at local minima in their normal lattice positions, bond breaking and lattice disruption were net-exothermic processes. For both systems the shock was generated through impact loading by offsetting a small section of the lattice and bringing it in with an appropriately chosen velocity to strike the main lattice

The primary objective of these studies, together with our earlier work on solitary structures in one-dimensional systems (Hardy and Karo, 1977), was to demonstrate the manner in which a shock preserves its integrity and its sharp detailed structure on an atomic scale, both factors leading to the highly efficient rupture of bonds when a shock arrives at a free surface.

For the Morse-bonded system this resulted in the spall of a microscopic fragment of the main lattice, to which was imparted the bulk of the energy of the initial impact. As a result, subsequent to the spall, the main lattice was essentially quiescent, while the spalled fragment, in addition to having a high overall translational kinetic energy, became severely distorted and partially disrupted, both indications of a high degree of internal agitation.

In the case of the system bonded by the net-exothermic potential, while the initial history of the spall was similar to that for the Morse system, the subsequent evolution was very different. First the spalled fragment blew itself apart, then the reflexive disturbance in the main lattice, generated by the initial spall to the right, produced further spall at the left-hand end of the lattice. There was thus initiated a spall sequence, both to the right and to the left, that ultimately led to complete disruption of the whole lattice into rapidly moving individual atoms.

Subsequently, we have much refined and extended our molecular dynamics computer codes so that we are now able to simulate the behavior of more complex arrays of atoms in which we are free to specify arbitrarily the potential parameters for each bond, the mass parameter for each atom, and the crystallographic configuration of the system to be studied. Using these codes we have studied shock propagation in: 1) monatomic two-dimensional lattices containing point defects (Karo and Hardy, 1978); 2) diatomic lattices (Hardy and Karo, 1978); 3) a polyatomic lattice designed to model a two-dimensional assembly of PETN (pentaerythritol tetranitrate) molecules, the potential paramenters being adjusted to values appropriate to the various bonds in that molecule (Karo et al., 1979a); and 4) a variety of model lattices containing linear defects and voids (Karo et al., 1979b). In parallel with these calculations, we have carried out thermal transport and accommodation studies which have been reported recently (Buckingham, 1979).

The aim of the present paper is to present studies which represent a partial synthesis of the type of study described above as 3) and 4); specifically, we report results obtained when we shock load two lattices, each containing a void, and bonded by realistic potentials. For purposes of comparison we also present results for a shock-loaded "model" lattice containing a void that has the same initial geometrical configuration as our lattices bonded by realistic potentials. We also report related studies on a metastable lattice that can undergo an exothermic rearrangement (phase transition) to a lower energy configuration. Finally, we present results demonstrating the derivation of the "Hugoniot" relationship between particle and shock velocities by molecular dynamics.

Method of Study (Molecular Dynamics)

The technique of molecular dynamics involves the numerical solution by computer of Newton's equations of motion for all the atoms in our assembly. As a result the coordinates and velocities of the particles are obtained as functions of time. It is thus possible to generate a movie which shows the time evolution of the system by obtaining a sequence of displays of the atomic configurations at successive time intervals of equal length (typically 0.025-0.10 in units of the inverse molecular frequency of the system).

For the purpose of presenting our results in this paper, we shall follow our earlier procedure (Karo et al., 1978) of showing selected stills from the relevant movies chosen so as to display the critical stages in the evolution of each system.

Specific Studies

Lattices Containing Voids

The Model System

The initial configuration for this study is shown in Fig 1. The impacting plate contains 5 columns and 7 rows of atoms and it strikes a lattice containing 10 columns and 33 rows of atoms. Offset 20 units to the right is another lattice containing 8 columns and 33 rows of atoms; the "void" is represented by the space between the two lattices. Strictly speaking, in order to represent a true void we should have additional rows of atoms uniting the two lattices at top and bottom. However, as will be seen later, the major activity is concentrated in the region delimited by the vertical dimensions of the incoming plate and we expect that little, if any, qualitative difference in this activity would be produced by uniting the two lattices. Moreover, following the motion of atoms in the bridging regions, were they present, would obviously increase the amount of computation time required, probably in a prohibitive fashion if the computation could no longer be carried out using only the fast memory storage. This kind of "computational strategy" is typical of molecular dynamics simulations: the cardinal rule is never to expend machine time monitoring the motions of atoms whose history is, in some sense, "irrelevant" to the main purpose of the study.

As was stated in the Introduction this is a model system: to be specific, this means that all the atomic masses are the same and identical Morse potentials

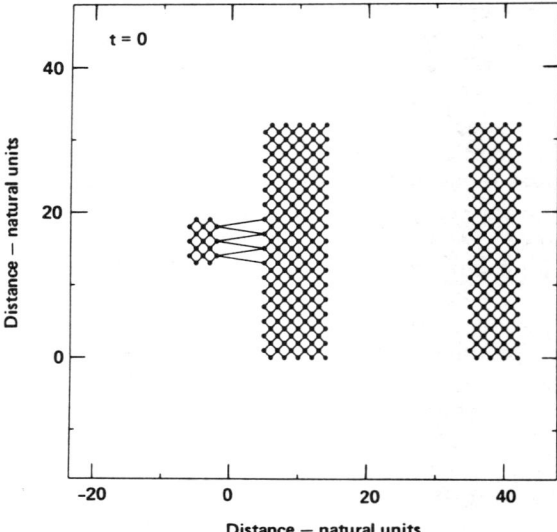

Fig. 1 Initial configuration for all impact-loaded void studies.

are used for all bonds (both first and second neighbor) within each subunit (i.e., the plate and the two lattices), but the parameters may differ for the various subunits. The mass and potential parameters have values that we have found to be reasonable for typical materials in earlier studies. The relative values are chosen to be such that the plate and the first lattice have identical natural frequencies, while the second lattice has a somewhat lower natural frequency (approximately 60%). The motivation for this last choice is to insure that fragments spalled out of the first lattice approach the second lattice at approximately the same Mach number (measured relative to the sound speed in that lattice) as the plate had with respect to the first lattice.

In the runs to be described, the initial offset of the plate was chosen to be sufficiently large for the plate-lattice interaction to be effectively zero, and the plate was brought in with a velocity of 1.4 units normal to the surface of the first lattice. Two runs were made with different initial conditions. For the first run all atoms were initially at rest, apart from the uniform motion of the plate; for the second run all the atoms were given small initial, randomly chosen, horizontal and vertical velocity components, the horizontal components for the plate being additional to the uniform horizontal motion. In this way the effect of finite temperature was simulated. The upper and lower bounds on the random components were such that the mean random kinetic energy per atom was on the same order, relative to the bond dissociation energy, as that of an atom in a typical solid at room temperature.

Figure 2 shows appropriately chosen stills from the movie history for the initially quiescent or "cold" system (i.e., the one with no initial random motion), while Fig. 3 shows a similar sequence of stills from the history of the system with initial random motion, the "hot" system.

MOLECULAR DYNAMICS OF SHOCKS AND DETONATION 213

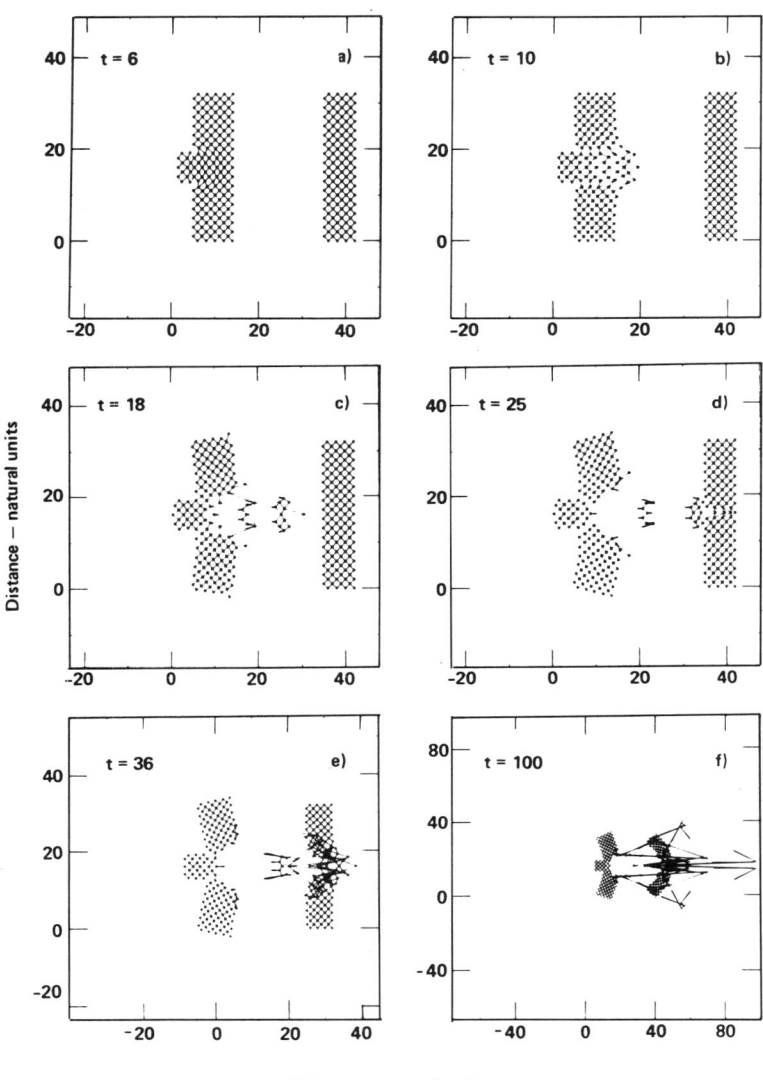

Fig. 2 Configurations of the impact-loaded model void at zero temperature: a) after a shock has been launched in the first wall; b) after spall has commenced from the inner face of the first wall; c) as the fragment spalled from the first wall transits the void; d) as the spalled fragment strikes the inner face of the second wall; e) as spall commences from the outer face of the second wall; and f) in the final state. The corresponding time t is shown in each figure.

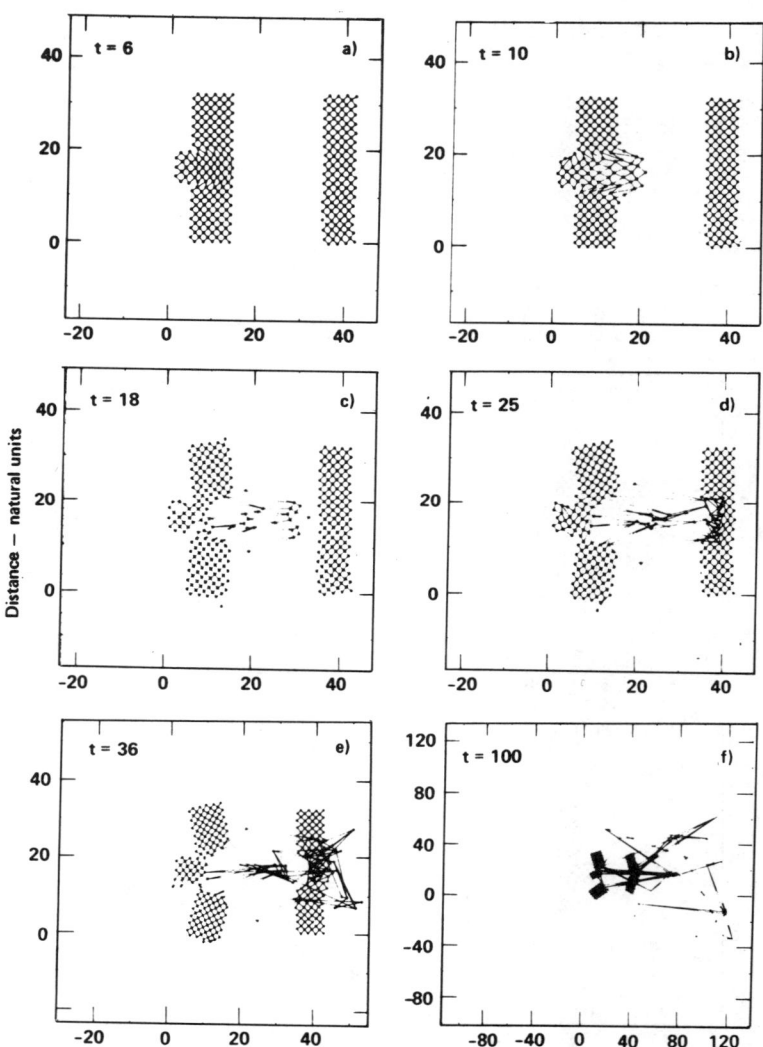

Fig. 3 Configurations of the impact-loaded model void at finite temperature (sequence as defined in Fig. 2).

For the cold system one can see clearly that the bulk of plate energy is imparted to a central section of the first lattice whose width is the same as that of the plate. Within this region one can see propagating a clearly defined, almost planar shock front (Fig. 2a) which, when it reaches the right-hand side, proceeds to spall out this central section of the lattice (Fig. 2b). Subsequently, the spalled fragment proceeds across the void, partially disrupting during the transit (Fig. 2c). On impact with the second lattice, a further shock is generated with a somewhat curved profile (Fig. 2d). When this reaches the far side of this lattice, further spall occurs (Fig. 2e) but in a more "focussed" manner, in the sense that the primary activity is concentrated in the vicinity of the central row of atoms. Thus we get a small cluster of rapidly moving atoms being driven off with a strong tendency to separate so that, ultimately (Fig. 2f), we have a jet of fast-moving isolated atoms and diatomic and polyatomic fragments.

The history of the hot system, as depicted in the corresponding sequence of stills (Fig. 3) is, broadly speaking, very similar. Again we see the ejection of a small rapidly moving fragment from the first lattice, which traverses the void, becomes somewhat disrupted in the process, and then punches through the second lattice, finally driving out rapidly moving atomic fragments from the far side. The major difference is the loss of mirror symmetry between the upper and lower halves of the system owing to the asymmetric nature of the initial conditions brought about by the presence of the randomly chosen velocity components.

In both histories there appear certain unphysical features during the later stages when atoms have undergone large relative displacements. Specifically, pairs of atoms pass through one another without scattering. The reason for this is that atoms interact only if they are first and second neighbors in the *initial* configuration. However, atoms at adjacent *surfaces* (e.g., the inner two surfaces of the void) are connected by the same bonds that they would have if there were no gap between the surfaces. Specifically, in the case of the void the intersurface bonds retained are those that would be present if the two lattices were united by a uniform horizontal displacement of the second lattice with respect to the first. A similar prescription applies to bonds between the impacting plate and the first lattice.

The absence of interactions between other pairs of atoms is less serious than it may seem, since one can safely argue that the amount of activity (spall, bond rupture, etc.) we observe represents a *lower bound* on what would occur if all interactions were included. Thus the true history of the system under study will be even more violent than our results demonstrate.

Realistic Systems: Aluminum Plate and Void in a
Model Face-Centered Cubic Nitric Oxide Lattice

The initial spatial configuration for this system is identical with that for the model system. However, the mass and Morse potential parameters for the plate have values appropriate to aluminum, while the two lattices are now diatomic systems consisting of alternating columns of nitrogen and oxygen atoms, with appropriate masses, bonded by Morse potentials whose parameters are chosen to represent N-O bonds between first neighbors and either N-N or O-O bonds between second neighbors. The appropriate potential paramenters for both this study and the succeeding study were taken from Herzberg (1950).

The geometrical arrangement is such that the first columns of both the right- and left-hand lattices consist of oxygen atoms, while the last columns consist of nitrogen atoms.

Once again, as we did for the model system, we studied the history of this system for two different initial conditions. First, the aluminum plate was brought in with a uniform horizontal velocity to strike the first nitric oxide lattice, both the plate and the two nitric oxide lattices being quiescent (i.e., no initial random motion was present), and then the run was repeated with initial random motion present in all three components of the system. In the latter case a slight variation on the procedure for the model system was employed: the range of random velocites employed for the two lattices differed from that for the plate. The reason for this was that we felt it best to study a hot system in which the mean thermal energy per bond was approximately the same fraction of the bond dissociation energy for all bonds in the system.

The history of the cold lattice is illustrated in Fig. 4 by stills which have been chosen so that they represent approximately the same stages in its temporal development as those illustrated for the model system. There are some interesting differences between the two, particularly after the initial spall from the first lattice. Nonetheless, it should be noted that there is a strong similarity in the earlier stages; specifically, the vertical extent of the shock remains comparable with the width of the incident plate. However, as the shock approaches the right-hand edge of the first lattice one can see that the front has become somewhat bowed, and the subsequent spall pattern reflects this. Thus, we see in Fig. 4c that the spalled material is moving mainly as two large fragments, and the ultimate effect of their impact on the second lattice (Fig. 4f) is to produce a central fracture rather than the jet of rapidly moving fragments spalled out to the right, that we observed for the model system. In spite of this general overall trend, it can be seen that some rapidly moving single atom fragments are driven off during the first spall (see Fig. 4c).

In Fig. 5 we show corresponding stills taken from the movie history of the hot system. Qualitatively these show a very similar behavior to that of the cold system: the first spall is again composed primarily of two relatively large subunits, and the effect of the second impact is to produce a central fracture of the second lattice. The main effect of the initial random motion is again manifested as a progressive loss of vertical symmetry. It also has the result of producing more unphysical effects, due to the failure of close neighbor atoms to "sense" one another, in the later stages of the history (this is also true of the model system). As before, we argue that this simply implies that the history we have obtained represents a lower bound on the true level of activity; the inclusion of all significant interactions at all stages is likely to produce more bond rupture and spall.

Realistic Systems: Aluminum Plate and Void in a Model Face-Centered Cubic Sulfur Nitride Lattice

The runs made for this system are identical to those made for the nitric oxide model lattice containing a void except for the following changes: 1) the oxygen

MOLECULAR DYNAMICS OF SHOCKS AND DETONATION 217

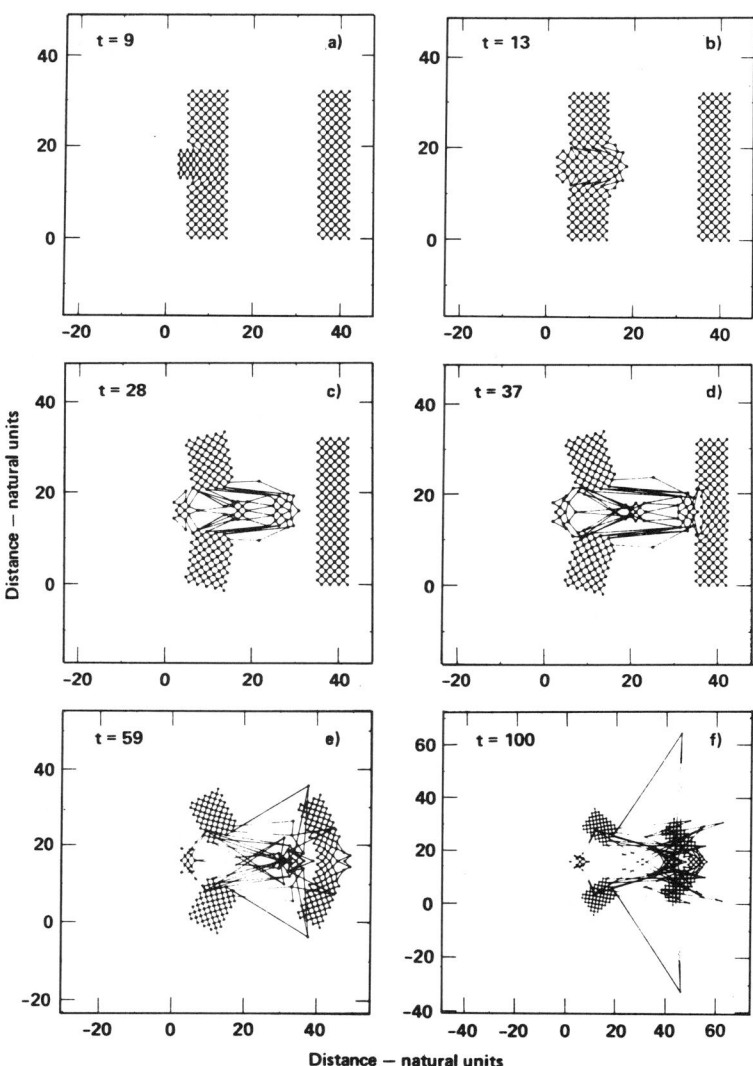

Fig. 4 Configurations of the impact-loaded nitric oxide void at zero temperature (sequence as defined in Fig. 2).

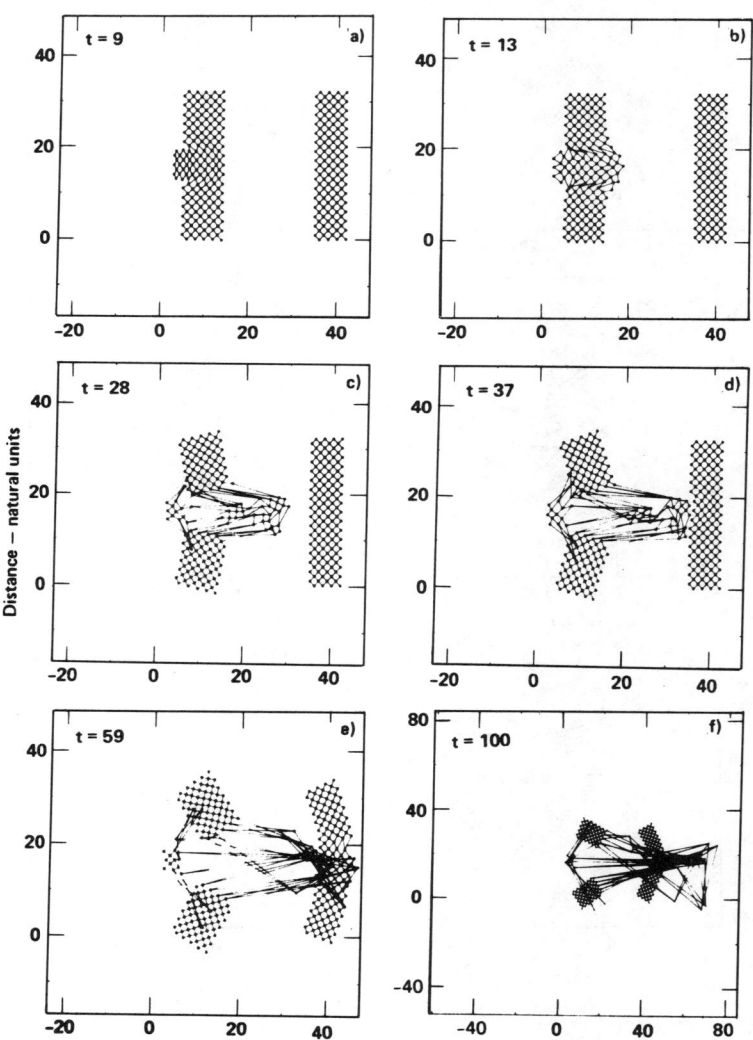

Fig. 5 Configurations of the impact-loaded nitric oxide void at finite temperature (sequence as defined in Fig. 2).

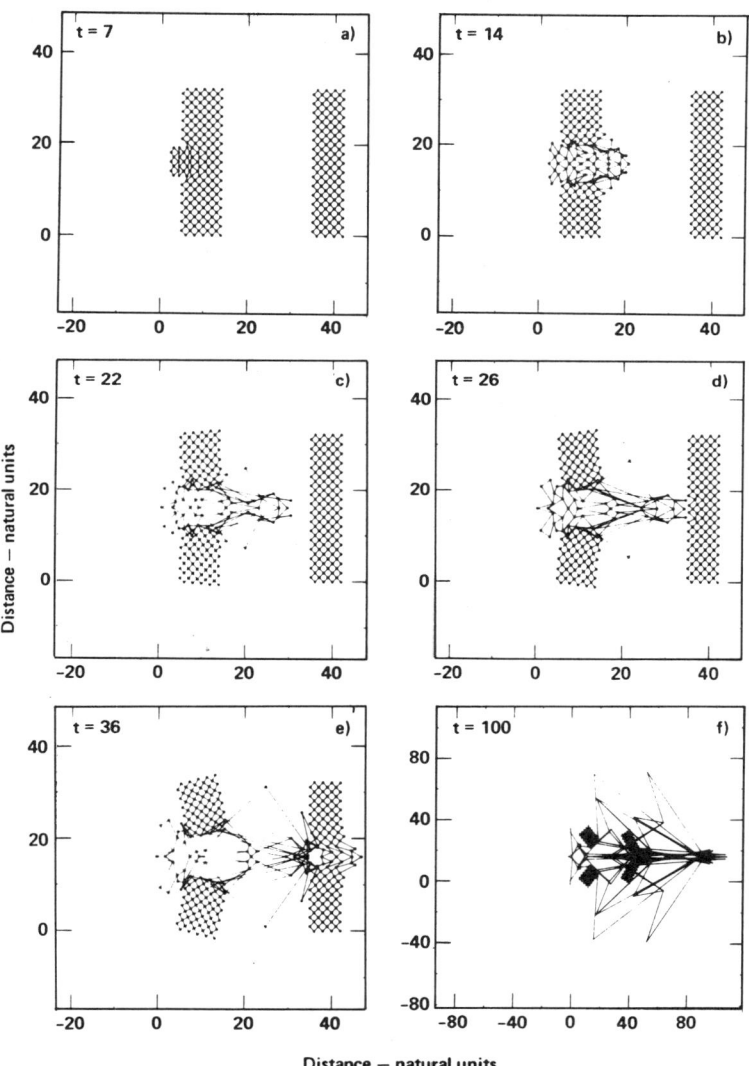

Fig. 6 Configurations of the impact-loaded sulfur nitride void at zero temperature (sequence as defined in Fig. 2).

atoms were replaced by sulfur atoms and the masses and bond potential parameters were appropriately modified; 2) the plate velocity was increased by 20%, which was done to insure that the appropriate atoms of the first spalled fragment sense the left surface of the second lattice (when a lower plate velocity was used, the distortion of the fragment was such that this was not the case); and 3) for the hot system the same range of random velocities was chosen initially for *all* the atoms, which reflects the better match of the aluminum and sulfur nitride lattices that is primarily due to the smaller disparity of the aluminum and sulfur masses.

The stills presented as Fig. 6 show the history of the quiescent system. They have again been chosen to parallel as closely as possible the same stages in the history of the model system and thus do not occur at the same times. It is evident that the qualitative behavior is much closer to that of the model system. Thus we get initial spall and disintegration of the spalled fragment, which subsequently drives a spike-like disruption through the center of the second lattice as its fragments strike in succession.

The stills shown as Fig. 7 show the same stages in the history of the initially hot system. These demonstrate once more that the qualitative behavior is not strongly affected; vertical symmetry is again lost and more unphysical events occur, primarily because of this loss, but the general nature of both spalls is very similar in both hot and cold systems.

In this section we have described a sequence of simulations of shock-loaded voids. *In all cases* we find that the initial shock energy is strongly localized on an atomic scale and is transmitted through the first void wall in a highly nonergodic fashion. This leads in turn to the spall of atomic fragments into the void which are moving in a highly nonrandom, nonergodic fashion, and thus traverse the void to produce very localized and specific bond rupture in its second wall. This occurs irrespective of whether or not the system being studied is initially cold or hot. These results demonstrate that our initial general conclusions for simple model systems and geometries (Karo et al., 1978) remain generally valid for systems bonded by realistic interatomic potentials and having more complex geometries. Specifically, shock loading leads to the propagation of highly nonergodic energetic disturbances through these systems that have the capacity to rupture individual bonds when they encounter free surfaces, and thus to produce spall at the atomic level.

Shock-Induced Phase Transition in a Model System

The initial configuration used in this study is shown in Fig. 8a. The arrangement is very simple; we have a plate of five columns offset to the left by 40 units with respect to the lattice and initally moving toward it with a velocity of 10 units (for this problem our scaling is such that 10 units are equivalent to 1 unit for our earlier models). The lattice is rectangular with the same vertical dimensions as the plate, and both are cold. The basic difference between this simulation and the others described in this paper lies in the nature of the interatomic potential employed. Elsewhere we have used various Morse potentials, each of which has only a single minimum. The new potential has two minima, the outer lying well above the inner one. The potential parameters are adjusted to be such that both first and second neighbor atoms

MOLECULAR DYNAMICS OF SHOCKS AND DETONATION

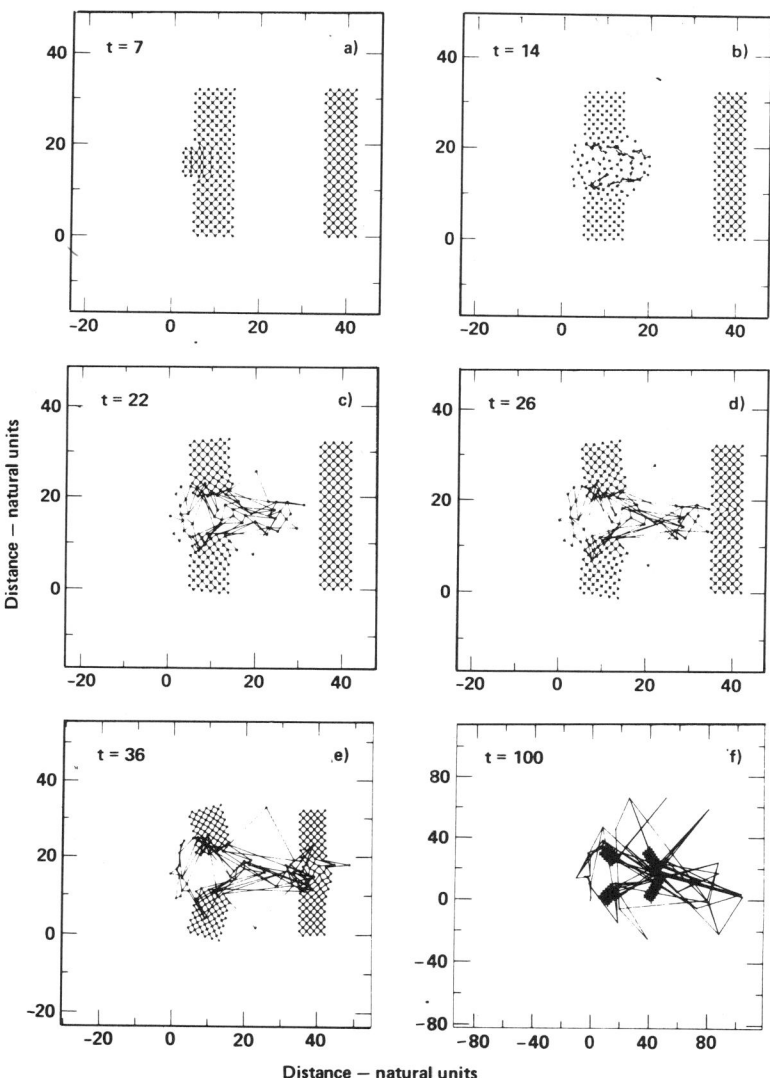

Fig. 7 Configurations of the impact-loaded sulfur nitride void at finite temperature (sequence as defined in Fig. 2).

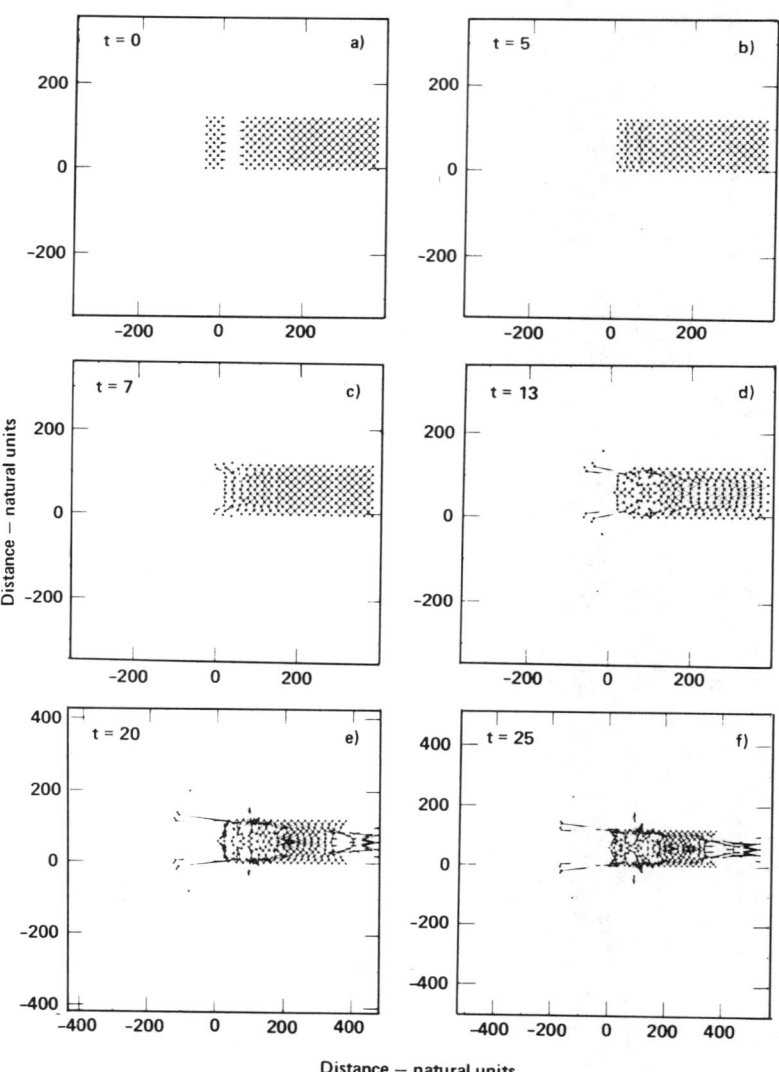

Fig. 8 Configurations of the impact-loaded metastable lattice: a) in the initial state; b) shortly after impact; c) well after impact but well before the shock has reached the right-hand surface; d) just as the shock reaches the right-hand surface; e) after spall to the right has developed; f) after right-hand spall is well advanced. In all cases the corresponding time t in the history is shown in the figure.

are located at the outer minima in the initial configuration. The plate and the lattice are almost identical. The potential parameters are the same for all bonds in the both of them. All the atomic masses are identical except for the upper and lower two rows of the lattice. These atoms are given very large masses in order to "clad" the lattice and prevent early loss of energy to transverse motion.

Figure 8 shows a sequence of stills from the movie history of this system. In Fig. 8b we see the system shortly after impact, and one can see shock fronts propagating forward through the lattice and backward through the plate. Figure 8c shows the situation at a slightly later time, when one can see more clearly what is happening. At the front, the first neighbor atoms are falling into the lower potential well, and we thus have behind the front a lattice that has restructured exothermically into a new phase. We therefore have a shock-induced phase transition propagating with the shock velocity. Figure 8d shows the situation just as the shock reaches the right-hand end of the lattice and immediately before the commencement of right-hand spall. Figure 8e shows the situation after spall has developed. Finally, Fig. 8f shows the situation after spall is well advanced. These three last figures also show clearly that considerable activity is developing within the body of the lattice due to the energy released in the transition. As time proceeds one can see form the movie that the system ultimately "burns itself up" and largely escapes the cladding walls which are only sensed by their first and second neighbor rows. (One can see this beginning in Figs. 8e and 8f.)

This simulation demonstrates a shock front producing a succession of highly coherent effects each of which takes place during a time comparable with the inverse of the natural (atomic) frequency of the system involved. Once again this represents highly nonergodic motion of the atoms in the front.

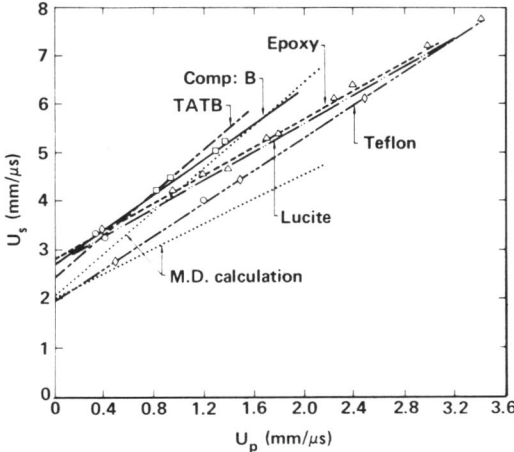

Fig. 9 Comparison of our theoretical "Hugoniot" plots of shock velocity vs particle velocity, computed for two different lattice models, with various experimental data.

Hugoniot Relation between Shock and Particle Velocities

Experimentally it is observed that the shock velocities in many materials can be fitted to what is often called the "Hugoniot" (Davison and Graham, 1979)

$$U_s = a + bU_p$$

where U_s is the shock velocity, U_p the particle velocity behind the shock, and a and b constants for a given material. Tsai (1971) in his early studies of shocks using molecular dynamics found that his computational results appeared to follow such a relation rather closely and, very recently, Klimenko and Dremin (1978) in a molecular dynamics simulation of strong shocks in liquid argon found similar behavior.

We have also frequently tested this result for our model lattices by making a succession of studies of shock transit time (as measured by the onset of spall) as a function of the velocity of the incoming plate used to initiate the shock. The results for two representative model systems are shown in Fig. 9 along with corresponding experimental data for various materials. As would be expected, the Hugoniot calculated for a somewhat stiffer lattice has a greater slope than that associated with the softer lattice. The two model potentials bracket a wide range of realistic potentials, and it is evident that the results obtained from our simulations of the microscopic dynamics agree remarkably well with those obtained by macroscopic measurements.

Conclusions

Our specific findings for the various markedly different situations that we have studied during the present work have already been presented in the appropriate sections of this paper and will not be repeated. In this section we only wish to re-emphasize our main thesis, which we first enunciated at the sixth of these colloquia (Karo et al., 1978), which is further substantiated by the present work and by our work during the intervening period (Karo and Hardy, 1978; Hardy and Karo, 1978; Karo et al., 1979a, 1979b). This thesis asserts that no true understanding of the physical and chemical effects of shocks in energetic materials is possible unless they are studied on a microscopic or atomic scale. The basic reasons for this are that shock dimensions are comparable with interatomic spacings and shock rise times are comparable with the natural periods of atomic vibrations in condensed matter. In these circumstances, the equations of continuum mechanics, equilibrium thermodynamics, and equilibrium statistical mechanics cannot possibly describe the situation in the shock front, because *the implicit conditions for their validity are explicitly violated by these two features.*

Our present studies have also demonstrated, once more, the crucial role played by free surfaces in converting shock energy into individual particle kinetic energy by bond rupture.

These general findings, together with the specific conclusions discussed in the relevant sections, therefore lead us to assert again that, as far as shock-initiated detonation is concerned, a *full* understanding of the process can only come from studies in the spirit of those we have presented here and elsewhere. Both shock dynamics and chemistry must be studied at the atomic level; only by this approach will it be possible to elucidate the true path, or paths, along which a given explosive system is driven by mechanical shock loading from an initial unreacted state to some mixture of the final decomposition products resulting from the highly exothermic reaction.

Acknowledgments

We should again like to thank W. Cunningham for his invaluable programming and computational assistance.

References

Buckingham A. C. (1979) Modeling additive and hostile particulate influences in gun combustion turbulent erosion. *Proceedings of the 1979 CPIA-JANNAF Combustion Conference*. Chemical Propulsion Information Agency, Johns Hopkins Univ. Applied Physics Lab., Laurel, Md.

Davison L. and Graham R. A. (1979) Shock compression of solids. *Phys. Rept.* **55**, 255-379.

Hardy J. R. and Karo A. M. (1978) Theoretical studies of the soliton-like behavior of shocks in one-dimensional systems. *Proceedings Conference (Intl.) on Lattice Dynamics*, pp. 163-166. Flammarion Press, Paris.

Hardy J. R. and Karo A. M. (1978) Theoretical studies of shock dynamics in two-dimensional structures, II: Diatomic lattices. *Bull. Amer. Phys. Soc.* **23**, 252.

Herzberg G. (1950) *Molecular Structure and Molecular Spectra*, pp. 501-581, Table 39. D. Van Nostrand Co., Inc., New York.

Karo A. M., Hardy J. R., and Walker F. E. (1978) Theoretical studies of shock-initiated detonation. *Acta Astronautica*, **5**, 1041-1050. See also Karo A. M. and Hardy J. R. (1977). Molecular dynamics of shock initiated detonations. *Intl. J. Quantum Chem.* **XII**, Sup. 1, 333-343.

Karo, A. M. and Hardy, J. R. (1978) Theoretical studies of shock dynamics in two-dimensional structures. I. Interaction of shocks with monatomic impurities. *Bull. Amer. Phys. Soc.* **23**, 36.

Karo A. M., Walker F. E., and Hardy J. R. (1979a) Shock dynamics in solids: prelude to chemical reaction? ACS/JCS Chemical Congress, Honolulu, Hawaii, Univ. of California Rept. UCRL-81887.

Karo A. M., Walker F. E., and Hardy J. R. (1979b) Theoretical studies of shock dynamics in two-dimensional structures. III. Point imperfections and grain boundaries. *Bull. Amer. Phys. Soc.* **24**, 726.

Klimenko V. Y. and Dremin A. N. (1978) Structure of a shock wavefront in a liquid. *Detonatsiya*, pp. 79-84. Chernogolovka, Akad. Nauk, U.S.S.R.

Tsai D. H. (1971) An atomistic theory of shock compression of a perfect crystalline solid. National Bureau of Standards Special Pub. 326, pp. 105-123.

TNT Explosions in a Hard Vacuum

Allen L. Kuhl*
R & D Associates, Marina Del Rey, Calif.
and
Michael R. Seizew†
TRW Defense and Space Systems, Redondo Beach, Calif.

The evolution of spherical and planar TNT explosions in a hard vacuum were successfully calculated with a Lagrangian finite-difference scheme. Solution accuracy was verified by comparing the front expansion velocity to the theoretical value. Solution stability near the vacuum boundary was a result of the staggered mesh formulation with thermodynamics at cell centers and kinematics at cell boundaries. Presented herein is the flowfield distribution at various stages of expansion for a spherical TNT explosion in a vacuum. This analysis has shown that the asymptotic density and dynamic pressure profiles depend on the mode of energy release assumed (instantaneous vs self-similar Chapman-Jouguet detonation), the equation of state used for the detonation products [Jones-Wilkins-Lee (JWL) model vs constant gamma], and the flow geometry (spherical vs planar). The solution is dominated by the inward propagating rarefaction wave which accelerates the flow to very near the inertial limit (velocity being proportional to distance), thus rendering the rarefaction wave reflected from the center quite ineffective.

Introduction

CONSIDERED here is the detonation and subsequent expansion of solid explosives in a vacuum. The charge starts with an initial density of about 1.6 g/cm^3. The detonation wave burns through the explosive releasing chemical energy which creates pressures of hundreds of kbars and sets up a hydrodynamic motion. When the detonation wave reaches the edge of the charge, the vacuum boundary condition sends a centered simple rarefaction into the charge accelerating the detonation products outward. Along

---

Presented at the 7th ICOGER, Göttingen, Federal Republic of Germany, Aug. 20-24, 1979. Copyright © American Institute of Aeronautics and Astronautics, Inc., 1980. All rights reserved.
*Senior Research Specialist.
†Head, Energy Dynamics Section.

characteristics $dr/dt = u+a$, the particle velocity u and sound speed a are related by

$$d(u+\frac{2}{\gamma-1}a) = \frac{ua}{u+a}j\frac{dt}{r} \qquad (1)$$

where γ denotes the specific heat ratio, and the geometric index j equals 0, 1, and 2 for planar, cylindrical, and spherically symmetric flow, respectively. On the vacuum boundary $a=0$, so that for all geometries the head of the rarefaction wave, i.e., the leading edge of the detonation products, moves at the constant escape velocity of

$$u_{max} = \text{const} = u_n + 2a_n/(\gamma-1) \qquad (2)$$

which can be evaluated from the initial conditions of the problem (for example, subscript n denotes the conditions behind the detonation wave). At this material front the temperature is zero, thus all the initial chemical energy has gone into kinetic energy of the flow. When the tail of the rarefaction reflects from the charge center a nonsimple wave is created. In general, this flowfield can only be solved exactly by numerical calculations, either by finite-difference techniques or by the method of characteristics. Approximate analytical solutions for the escape of the detonation products from planar and spherical charges can be found in the treatise by Stanyukovich (1960), with certain limitations on initial conditions and γ. Courant and Friedrichs (1948) also consider expansions of this type.

When the charge has expanded many diameters the thermal pressure becomes negligibly small and cannot influence the momentum of the flow that is now moving at many km/s. The flow becomes inertial with the particle velocity being proportional to the radius $u/u_n = r/r_n$ (where $r_n = u_{max} \cdot t$). Sedov (1959) has investigated these inertial flows in general. Brode and Enstrom (1972) and others (Keller, 1956; Stuart, 1965; and Greenspan and Butler, 1962) have explored similarity solutions of this type, usually for explosions in a low-density atmosphere. Mirels and Mullen (1963) have derived approximate solutions based upon certain assumptions on the asymptotic shape of the density profile. As we shall demonstrate, this asymptotic form is not unique but depends on the initial conditions of the problem.

The objective of this work was to determine the evolution of the flowfield for a spherical TNT explosion in a hard vacuum under the continuum assumption. Since γ varies greatly during the expansion, a real equation of state was used for the detonation products. Also of interest was the asymptotic form this solution reaches in the inertial limit. This solution is necessary for calculating blast damage from solid explosives in space (e.g., blast effects on spacecraft), and is also pertinent to barium cloud release experiments (Kaplan and Kurt, 1960; and Davidson, 1972). Finally, through global checks on mass and energy, scaling relations are derived to extend the solutions to arbitrary distances.

Analysis

Finite-Difference Equations

The method of analysis used here was a standard finite-difference algorithm‡ to the one-dimensional Lagrangian equations of gasdynamics. The charge is divided into cells with mass,

$$m_{i-½} = \rho^0_{i-½}\left[(r^0_i)^{j+1} - (r^0_{i-1})^{j+1}\right] \tag{3}$$

and initial cell interface positions r^0_i. Thermodynamic quantities of density ρ, pressure p, and internal energy e are defined at cell centers denoted by subscript $i-½$; kinematic properties such as position and velocity are defined at cell interfaces denoted by subscript i. Superscripts indicate time levels.

First, the momentums are changed according to pressure forces, thus updating the interface velocities:

$$u_i^{\ell+½} = u_i^{\ell-½} - \frac{\Delta t^\ell \cdot (p^\ell_{i+½} - p^\ell_{i-½})}{\rho_i^\ell \cdot \Delta r_i^\ell} \tag{4}$$

where

$$\rho_i^\ell \cdot \Delta r_i^\ell = [\rho^\ell_{i+½}(r^\ell_{i+1} - r^\ell_i) + \rho^\ell_{i-½}(r^\ell_i - r^\ell_{i-1})]/2$$

Since this problem does not involve compressions, an artificial viscosity term is not required. Time steps Δt^ℓ are limited by the Courant condition. Next, the interface positions are advanced based on the new velocities:

$$r_i^{\ell+1} = r_i^\ell + \Delta t^{\ell+½} \cdot u_i^{\ell+½} \tag{5}$$

Then the cell densities are changed, consistent with the new interface positions:

$$\rho^{\ell+1}_{i+½} = m_{i-½}/[(r_i^{\ell+1})^{j+1} - (r_{i-1}^{\ell+1})^{j+1}] \tag{6}$$

Finally, the cell internal energies are modified according to the pressure work:

$$e^{\ell+1}_{i-½} = e^\ell_{i-½} + (p^\ell_{i-½} + p^{\ell+1}_{i-½}) \cdot 2 \cdot \frac{(\rho^{\ell+1}_{\theta-\infty} - \rho^\ell_{i-½})}{(\rho^\ell_{i-½} + \rho^\ell_{i-½})^2} \tag{7}$$

In this problem the flow is homentropic since the constant velocity detonation induced the same entropy throughout the field, and since we are neglecting any afterburning as this requires additional oxygen which is absent in this case. As an alternative to the energy Eq. (7), an isentropic relation could have been used between pressure and density.

Equation of State

The Jones-Wilkins-Lee (JWL) equation of state was used as the primary

---

‡A modified version of the WONDY code described in Laurence and Mason (1971).

equation of state for the detonation products. This gives the pressure as a function of density and internal energy in the following form:

$$p = \omega \rho e + A \cdot \left[1 - \frac{\omega}{R_1(\rho_a/\rho)}\right] \cdot e^{-R_1 \rho_a/\rho} + B \cdot \left[1 - \frac{\omega}{R_2(\rho_a/\rho)}\right] \cdot e^{-R_2 \rho_a/\rho} \quad (8)$$

According to Dobratz (1974), the JWL parameters for TNT are:

$$A = 3.738 \text{ Mbars} \qquad R_1 = 4.15$$
$$B = 0.03747 \text{ Mbars} \qquad R_2 = 0.90$$
$$\rho_a = 1.63 \text{ g/cm}^3 \qquad \omega = 0.35$$

Some calculations were also performed with constant gamma equation of state:

$$p = (\Gamma - 1)\rho e \quad (9)$$

with $\Gamma = 1.35$ or $\Gamma = 2.70$.

Initial Conditions

Two types of initial conditions were considered. First, the charge was assumed to detonate at the center of symmetry, with an ideal constant velocity Chapman-Jouguet detonation progressing out to the charge radius r_0. Initial conditions were taken to be the self-similar flowfield distribution behind a strong Chapman-Jouguet detonation shown in nondimensional form in Table 1. The conditions at the detonation front are related to the heat release q and the detonation velocity w_n, by the following:

$$\rho_n = \rho_a(\Gamma+1)/\Gamma \quad e_n = q + \frac{1}{2(\Gamma+1)^2} w_n^2 \quad p_n = \frac{1}{\Gamma+1}\rho_a w_n^2 \quad u_n = \frac{1}{\Gamma+1} w_n \quad (10)$$

For TNT ($\rho_a = 1.63$ g/cm^3, $q = 3.68 \times 10^{10}$ erg/g, $w_n = 6.93$ km/s, $\Gamma = 2.727$), the detonation conditions become:

$$\rho_n = 2.228 \text{ g/cm}^3 \qquad e_n = 5.40 \times 10^{10} \text{ erg/g}$$
$$p_n = 210 \text{ Kbars} \qquad u_n = 1.859 \text{ km/s}$$

Table 1 Flowfields behind planar and spherical self-similar detonations, $\Gamma = 2.70$

	Planar, $j=0$				Spherical, $j=2$			
x	u/u_n	p/p_n	ρ/ρ_n	e/e_n	x	u/u_n	p/p_n	ρ/ρ_n	e/e_n
0.00	0.000	0.300	0.640	0.469	0.00	0.000	0.230	0.580	0.396
					0.459	0.000	0.230	0.580	0.396
0.50	0.000	0.300	0.640	0.469	0.50	0.023	0.237	0.587	0.404
0.55	0.100	0.347	0.675	0.513	0.55	0.063	0.254	0.602	0.422
0.60	0.200	0.397	0.710	0.559	0.60	0.111	0.277	0.621	0.445
0.65	0.300	0.453	0.746	0.607	0.65	0.164	0.304	0.643	0.472
0.70	0.400	0.514	0.781	0.657	0.70	0.222	0.336	0.668	0.503
0.75	0.500	0.580	0.817	0.709	0.75	0.286	0.375	0.695	0.539
0.80	0.600	0.652	0.853	0.764	0.80	0.356	0.422	0.726	0.581
0.85	0.700	0.729	0.889	0.820	0.85	0.435	0.480	0.762	0.630
0.90	0.800	0.813	0.926	0.878	0.90	0.531	0.557	0.805	0.691
0.95	0.900	0.903	0.963	0.938	0.95	0.658	0.668	0.861	0.775
1.00	1.000	1.000	1.000	1.000	1.00	1.000	1.000	1.000	1.000

In some of the calculations, the Stanyukovich instantaneous detonation model was used as the initial conditions. In this case, the energy is released instantaneously throughout the entire charge, so that no motion or density change occurs. Initial conditions are constant in this case:

$$\rho = \rho_a = 1.63 \text{ g/cm}^3 \qquad e = q = 3.68 \times 10^{10} \text{ erg/g}$$
$$p = p(\rho, e) \qquad u = 0$$

for $0 \leq r \leq r_0$.

Boundary Conditions

Symmetry conditions were used at the charge center. At the vacuum boundary, the pressure, mass, and internal energy were set equal to zero in the ghost cell outside the charge, while the interface velocity, being a calculated quantity, remained unconstrained. The latter, when compared with the escape speed, served as a check on the solution accuracy. This Lagrangian staggered mesh (with thermodynamics at the cell centers and kinematics at the cell interfaces) is ideally suited for applying this vacuum boundary condition. Early attempts at calculating this problem with an Eulerian space-centered difference scheme met with failure. The vacuum boundary leaked mass due to numerical diffusion and the front velocity then increased unboundedly.

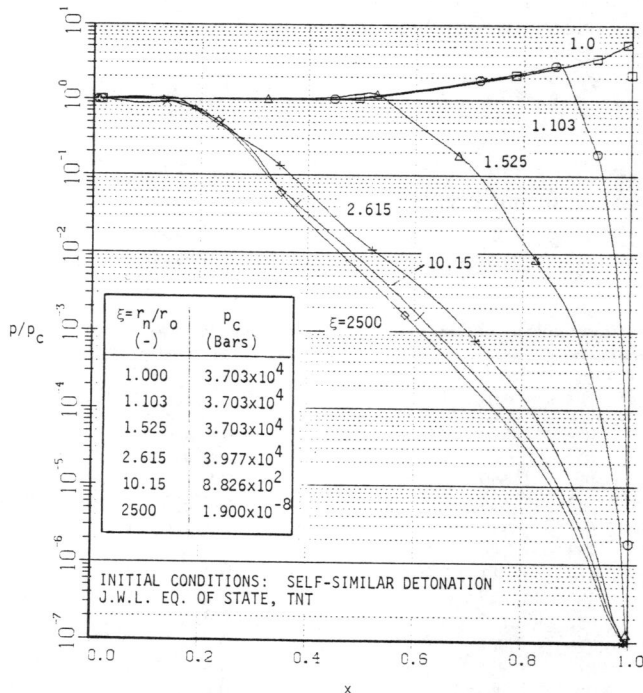

Fig. 1 Evolution of the pressure distribution for a spherical TNT explosion in a vacuum.

Discussion of Results

The primary calculation analyzed the evolution of a spherical explosion in a vacuum starting from the self-similar detonation conditions for TNT; the JWL equation of state was used for the detonation products. The resulting flowfield distribution at various stages of the expansion are shown in Figs. 1-5. In order to compare the solution at different times and see the variation in the flowfield distribution, the results are presented in nondimensional form. Positions were divided by front location at that time ($x = r_i/r_n$ where r_n equals the leading edge of the detonation products), thermodynamic variables were

Table 2 Nondimensionalizing parameters for Figs. 1-5 vs ξ

$\xi = r_n/r_0$	p_c, bars	ρ_c, kg/m^3	e_c, J/kg	u_n, km/s	$\tfrac{1}{2}\rho_c u_n^2$, bars
1.000	3.703×10^4	1.294×10^3	2.208×10^6	1.859	2.234×10^4
1.103	3.703×10^4	1.294×10^3	2.208×10^6	8.800	5.009×10^5
1.525	3.703×10^4	1.294×10^3	2.208×10^6	9.005	5.245×10^5
2.615	3.977×10^4	1.321×10^3	2.270×10^6	9.089	5.458×10^5
10.15	8.826×10^2	6.591×10^1	3.826×10^5	9.140	2.753×10^4
2500	1.900×10^{-8}	4.553×10^{-3}	1.192×10^3	9.153	1.907×10^{-3}

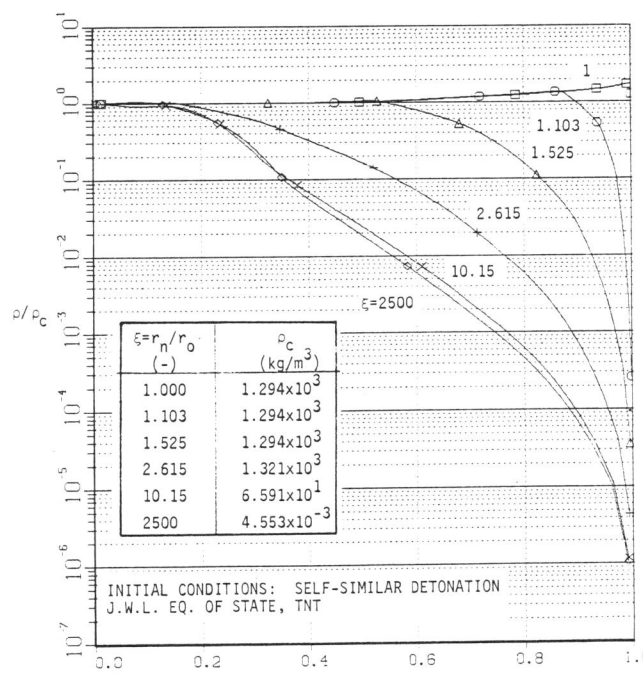

Fig. 2 Evolution of the density distribution for a spherical TNT explosion in a vacuum.

divided by their instantaneous values at the center (denoted by subscript c) and velocities were divided by the instantaneous front velocity u_n. Curves are labeled by $\xi = r_n/r_0$ which denotes the front position nondimensionalized by the initial charge radius r_0. Table 2 gives the nondimensionalizing parameters ξ for each figure.

The flowfield distributions show the progression of the rarefaction wave into the detonation products. The wave diagram of Fig. 6 indicates that the rarefaction reaches the center at a nondimensional time $\tau = 2.6$, where $\tau \equiv tu_n/r_0$. After a front expansion of ten charge radii the velocity is almost linear with distance, indicating that the flow is nearly inertial. With further expansion from $\xi = 10$ to $\xi = 2500$, the flowfields changed only imperceptibly. It may be concluded that this flow is indeed dominated by the initial rarefaction wave. The second rarefaction created after reflection from the center is relatively unimportant because the flow is very near the inertial limit and, therefore, the rarefaction can no longer influence the flow that is moving at many km/s.

Note that flowfields corresponding to solid explosive detonations in the atmosphere are bounded by an air shock. The detonation products overexpand to high velocities, incompatible with (i.e., much higher than) gas velocities just behind the main shock. This situation is resolved by an inward facing secondary shock that eventually implodes at the center (Brode, 1959). Thus,

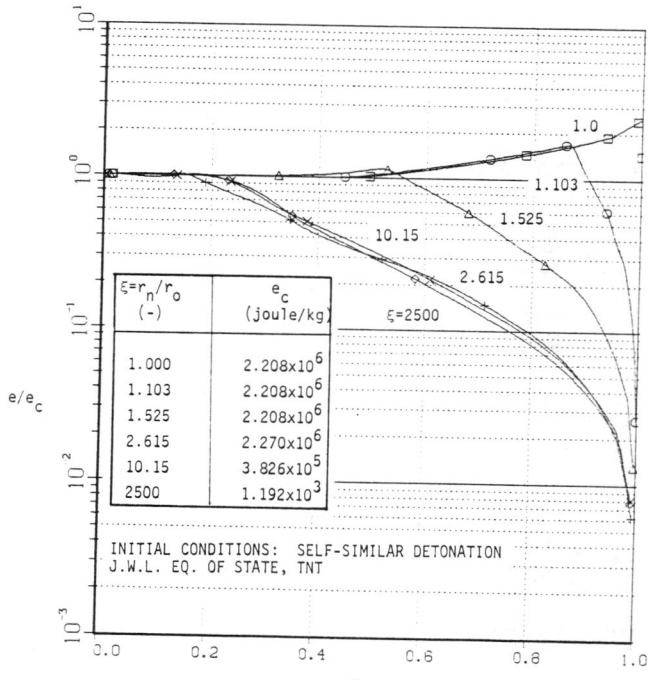

Fig. 3 Evolution of the internal energy distribution for a spherical TNT explosion in a vacuum.

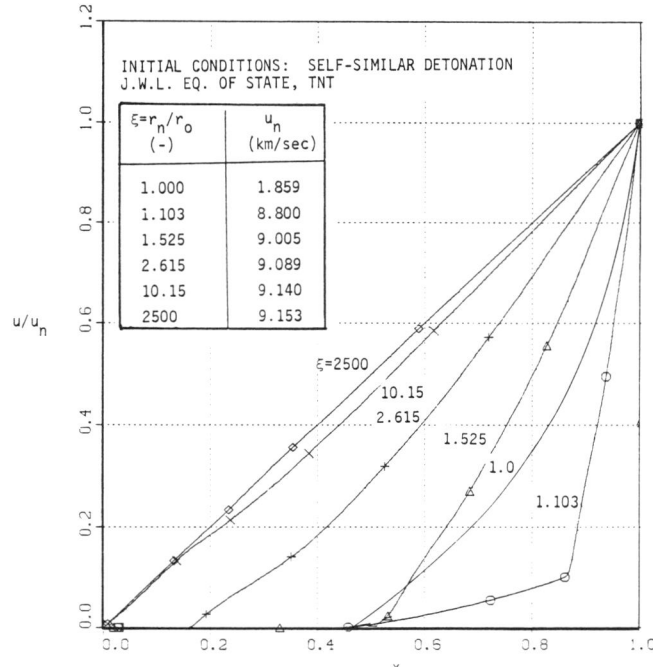

Fig. 4 Evolution of the velocity distribution for a spherical TNT explosion in a vacuum.

both the main shock and the imploding shock are a gasdynamic reaction to the atmospheric pressure. For explosions in a vacuum both these shocks are absent.

The accuracy of the present calculations was verified first by checking global mass and energy conservation. Mass was conserved exactly, by formulation of the problem. Global energy was conserved to better than 1% for the entire calculation. The front velocity proved to be a more sensitive check on the solution accuracy. For initial conditions of a self-similar detonation, the theoretical front velocity is

$$u_{max} = w_n/(\Gamma+1) + 2\sqrt{e_n\Gamma/(\Gamma-1)} \tag{11}$$

while for the instantaneous detonation case the front velocity should be:

$$u_{max} = 2\sqrt{q\Gamma/(\Gamma-1)} \tag{12}$$

Calculated front velocities are compared with the theoretical values in Table 3. For the primary calculation (case 1) the calculated front velocity was within 0.3% of the theoretical value. To obtain this accuracy very fine zoning was required near the front ($\Delta r_n^0/r^0 = 10^{-6}$) with coarser zoning ($\Delta r_I^0/r^0 = 0.05$) near the center.

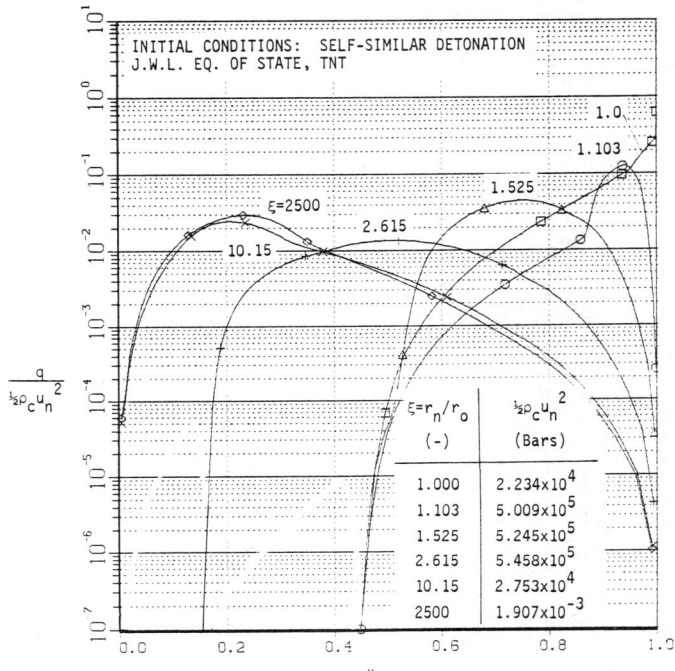

Fig. 5 Evolution of the dynamic pressure distribution for a spherical TNT explosion in a vacuum.

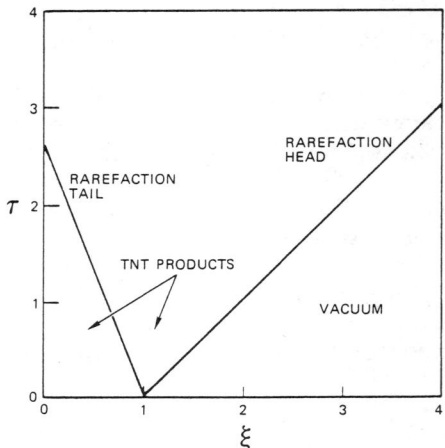

Fig. 6 Time-distance diagram for a spherical TNT explosion in a vacuum.

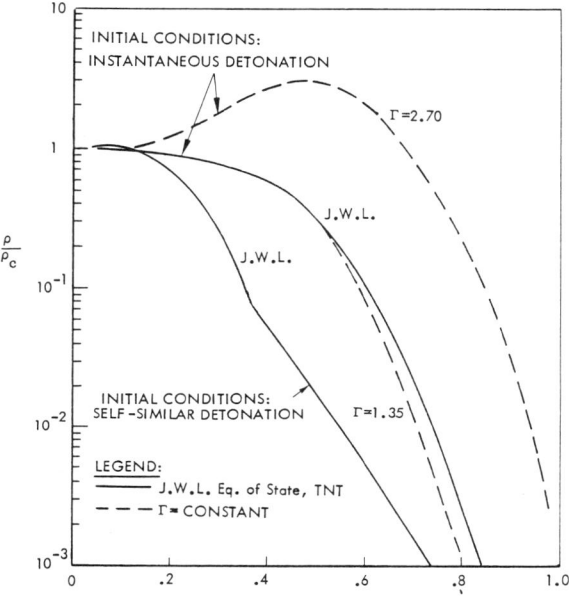

Fig. 7 Asymptotic form of the density distribution for spherical explosions in a vacuum.

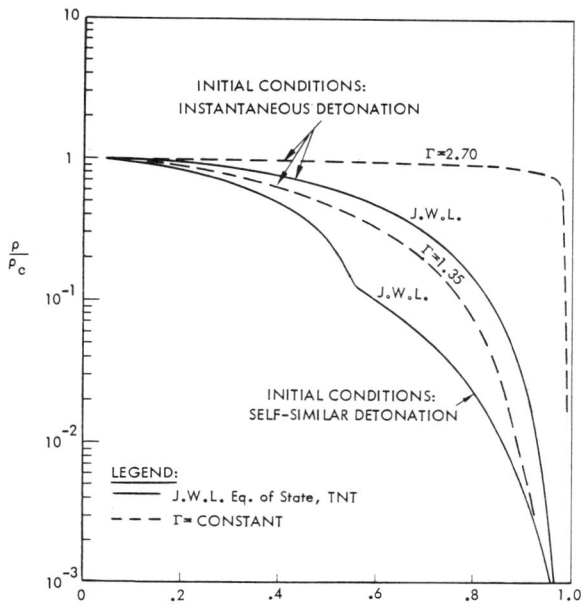

Fig. 8 Asymptotic form of the density distribution for planar explosions in a vacuum.

Table 3 Comparison of theoretical and calculated front velocities

Case	Conditions	u_n ($\xi = 2500$), km/s	u_{max}, km/s	$u_n/u_{max} \times 100$, %
	Spherical			
1	SSD + JWL[b]	9.153	9.124[a]	100.3
2	ID + JWL[b]	6.018	7.535[a]	79.9
3	ID + $\Gamma = 2.7$	4.639	4.835	95.9
4	Planar			
5	SSD + JWL	9.532	9.124[a]	104.5
6	ID + JWL	6.505	7.535[a]	86.3
7	ID + $\Gamma = 2.7$	4.686	4.835	96.9
8	ID + $\Gamma = 1.35$	6.832	7.535	90.7

[a] Based on $\Gamma = 1.35$.
[b] SSD = self-similar Chapman-Jouguet detonation; ID = instantaneous detonation.

How the problem formulation assumptions (namely, the initial conditions, the equation of state, and the geometry) influence the asymptotic flowfield distribution is now considered. Additional calculations were performed for the spherical geometry starting from the instantaneous detonation conditions for TNT, some using the full JWL equation of state and some using $\Gamma = $ const. Similar calculations were performed for the planar geometry. The resulting asymptotic density and dynamic pressure profiles are shown in Figs. 7-10.

From these figures we make the following observations: 1) Holding the equation-of-state fixed, the profiles for the instantaneous detonation (ID) initial conditions are fuller than the profiles from the self-similar detonation (SSD) initial conditions. This is a consequence of the fact that the SSD initial conditions already contain a rarefraction. 2) Holding the initial conditions fixed at say the ID conditions, the results based on assuming $\Gamma = 1.35$ are essentially identical to the results for the JWL equation of state. This is, in some sense, a measure of how rapidly the variable gamma approaches the constant value of 1.35. Utilizing a solid explosives gamma ($\Gamma = 2.70$), however, changes even the qualitative nature of the solution giving a density peak at $x \simeq 0.5$ in the spherical case, and a very flat or constant density profile in the planar case. Thus, results for a large constant value of Γ are not a close approximation to the results from a complete equation of state. Caution is therefore advised when applying approximate analytic solutions such as those of Stanyukovich (1960), if they are based on a large Γ near three. 3) Holding both the equation of state and the initial conditions fixed, we find the density profiles for the planar case are fuller than the spherical case—a consequence of the spherical divergence creating a stronger rarefaction than the planar case.

To extend these solutions to arbitrary distances, the density at center of the wave ρ_c is required. Global mass conservation provides an analytic relationship between ρ_c and the front position:

$$(\rho_c/\rho^0)\xi^{j+1} = 1/[(j+1)J_I] \tag{13}$$

where J_I denotes the nondimensional mass integral:

$$J_I \equiv \int_0^1 h(x)x^j dx \tag{14}$$

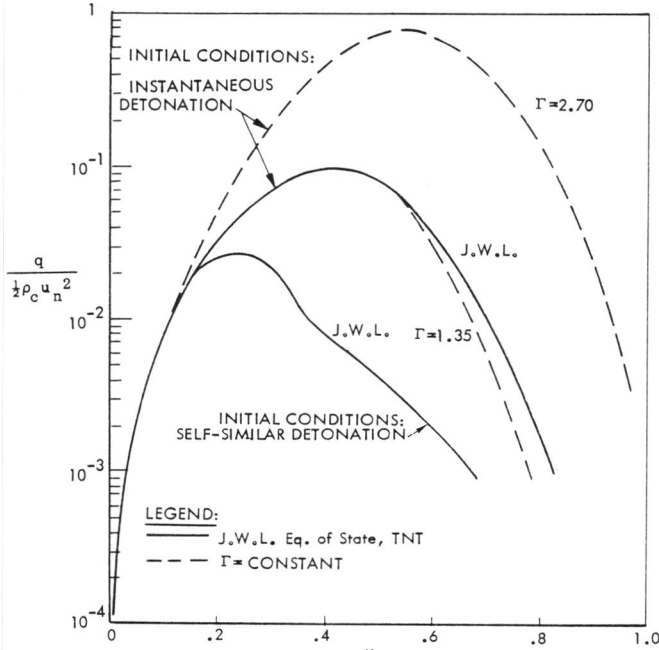

Fig. 9 Asymptotic form of the dynamic pressure distribution for spherical explosions in a vacuum.

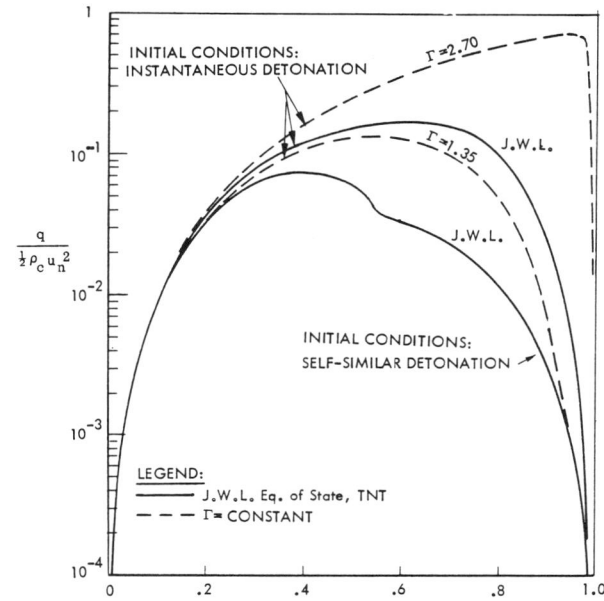

Fig. 10 Asymptotic form of the dynamic pressure distribution for planar explosions in a vacuum.

(see the Appendix for the derivation). Since the density profile $h(x)$ approaches a constant form for ξ large, J_1 is asymptotically constant, and Eq. (13) provides the required relation. Figure 11 presents the center density ρ_c vs the front position from the numerical calculations for the spherical geometry cases. These results indeed verify that $\rho_c \alpha 1/(r_n{}^{j+1})$.

These calculations have been checked by comparison with test data from Jack and Armendt (1965) who performed some experiments with 56.8 g spherical charges of pentolite detonated in a blast chamber at low atmospheric pressures (0.1 mm Hg). Stagnation pressure measurements were made on a flat

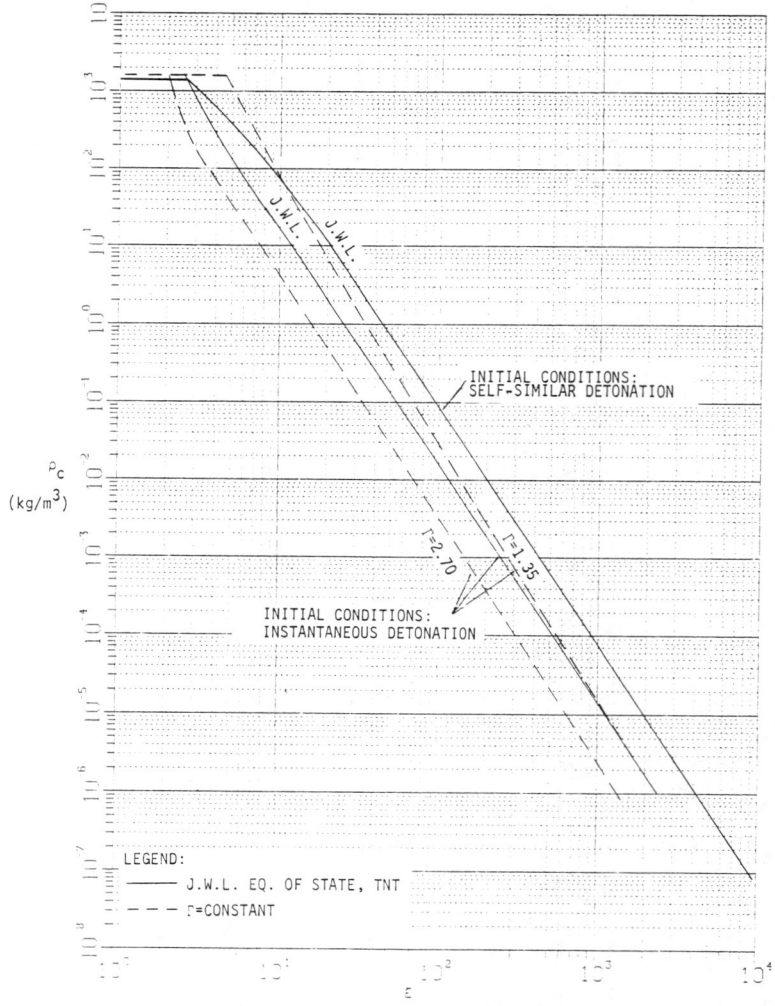

Fig. 11 Comparisons of the density at the center of the spherical explosion vs nondimensional front position ξ.

A) EXPERIMENT CONFIGURATION:

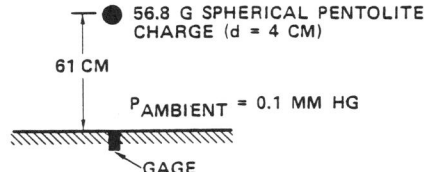

B) PRESSURE MEASUREMENT

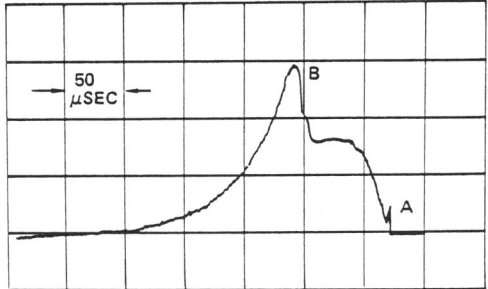

Fig. 12   Reflected pressure measurement of a spherical detonation in a partial vacuum (from Jack and Armendt, 1965).

surface 0.61 m below the charge ($\xi \simeq 30$). An example of an experimental pressure history is shown in Fig. 12. Two peaks are evident: The first peak, labeled A, corresponds to the reflection of the air shock created by the combustion products pushing on the residual atmosphere. The second peak, labeled B, which occurs after about half the positive phase duration, corresponds to the reflection of the peak dynamic pressure pulse far behind the leading edge of the detonation products. This figure qualitatively confirms our predicted dynamic pressure distributions (unfortunately the gage calibration was inaccurate at low ambient pressures so a pressure scale was not included for this figure in the aforementioned reference).

The present calculations represent a continuum solution to the problem of a TNT explosion in a vacuum. During the later stages of the explosion, the gas mean free path can approach or exceed characteristic dimensions of the expanding cloud, and the flow will transition into the free molecular flow regime. Under such circumstances the density does not decrease monotonically from the center of the cloud to the edge, but acquires a second peak in density at $x \simeq 0.5$-0.9. This can be inferred from observations of barium cloud release experiments that show a ring structure with bright emission near $x \simeq 0.8$ indicating an accumulation of mass there, and much less emission near the center (Davidson, 1972). Further evidence of the free molecular flow nature of the late stages of the gas cloud expansion can be inferred from the work of

Trinks and Schilf (1980). Clearly, the solutions presented here apply to the continuum regime only.

Summary and Conclusions

The evolution of spherical and planar TNT explosions in a hard vacuum were successfully calculated with a Lagrangian finite-difference scheme. Solutions accuracy was verified by comparing the front expansion velocity to the theoretical value. Solution stability near the vacuum boundary was a result of the staggered mesh formulation with thermodynamics at cell centers and kinematics at cell boundaries. This analysis has shown that the asymptotic flowfield distribution depends on the initial conditions, the equation of state used, and the flow geometry. The solution is dominated by the inward propagating rarefaction wave that accelerates the flow to very near the inertial limit, thus rendering the reflected rarefaction wave quite ineffective.

Appendix: Global Integrals for Explosions in a Vacuum

Consider first the initial explosives mass M^0 and energy E^0:

$$M^0 = [\sigma_j/(j+1)]\rho^0(r^0)^{j+1} \tag{A1}$$

$$E^0 = M^0 q = [\sigma_j/(j+1)]\rho^0(r^0)^{j+1}q \tag{A2}$$

where $\sigma_j = 1$, 2π, and 4π for $j = 0$, 1, or 2. Global conservation implies that the mass and energy within the wave equals the initial mass and energy in the explosive:

$$M^0 = \int_0^{r_n} \rho r^j \sigma_j dr = \sigma_j \rho_c r_n^{j+1} \cdot J_1(\xi) \tag{A3}$$

$$E^0 = \int_0^{r_n} (e + \tfrac{1}{2}u^2)\rho r^j \sigma_j dr = \sigma_j \tfrac{1}{2}\rho_c r_n^{j+1} u_n^2 \cdot J_3(\xi) \tag{A4}$$

where $J_1(\xi)$ and $J_3(\xi)$ denote the nondimensional mass and energy integrals:

$$J_1(\xi) = \int_0^1 h x^j dx \tag{A5}$$

$$J_3(\xi) = \int_0^1 (c \cdot g + f^2) h x^j dx$$

$$= \int_0^1 h x^{j+2} dx \text{ for } \xi \to \infty \tag{A6}$$

and where

$$f(x) = u/u_n \; (=x \text{ inertial limit}), \quad g(x) = e/e_c, \quad h(x) = \rho/\rho_c,$$

$$c = e_c/(\tfrac{1}{2}u_n^2) \; (\to 0 \text{ as } \xi \to \infty), \quad \text{and } g = \xi_n/r^0$$

Combining Eqs. (A1-A4) we find

$$(\rho_c/\rho^0)\xi^{j+1} = 1/(j+1)J_1 \qquad (A7)$$

$$u_n^2/q = 2(J_1/J_3) \qquad (A8)$$

Note J_1 and J_3 each approach a constant value for large ξ, since the flowfield distribution approaches a constant in the inertial limit.

Thus, global mass conservation provides a relationship [Eq. (A7)] between the density at the center of the wave and the front position. Global energy conservation [Eq. (A8)] provides a check on the front velocity, or alternately, a way to evaluate the total kinetic energy of the flow.

Acknowledgments

The authors are indebted to S. Fink of TRW who performed the computer programing.

References

Brode H.L. (1959) Blast wave from a spherical charge. *Phys. Fluids,* **II**, 217-229.
Brode H.L. and Enstrom J.E. (1972) Analysis of gas expansion in a rarefied atmosphere. *Phys. Fluids,* **15**, 1913-1917.
Courant R. and Friedrichs K.O. (1948) *Supersonic Flow and Shock Waves,* pp. 191-195. Interscience Publishers, New York.
Davidson R.E. (1972) Ring structure of a neutral gas cloud studied in a one-dimensional expansion into space. NASA TN D-6760.
Dobratz B.M. (1974) Properties of chemical explosives and simulants. Univ. of California, Lawrence Livermore Lab., Rept. UCRL-51319, Rev. 1.
Greenspan H.P. and Butler D.S. (1962) On the expansion of a gas into vacuum. *J. Fluid Mech.* **13**, 101-119.
Jack W.H. and Armendt B.F. (1965) Measurements of normally reflected shock parameters from explosive charges under simulated high altitude conditions. U.S. Army Ballistic Res. Lab., Rept. 1280, pp. 1-69.
Kaplan S.A. and Kurt V.G. (1960) Expansion of a sodium cloud in interplanetary space. *Soviet Astronomy* **AJ 4**, 508-514.
Keller J.B. (1965) Spherical, cylindrical, and one dimensional gas flow. *Quart. of Appl. Math.* **14**, 171-184.
Laurence R.J. and Mason D.S. (1971) WONDY IV-A computer program for one dimensional wave propagation with rezoning. Sandia Rept. SC-RR-710284.
Mirels H. and Mullen J.F. (1963) Expansion of gas clouds and hypersonic jets bounded by a vacuum. *AIAA J.* **1**, 596-602.
Sedov, L.I. (1959) *Similarity and Dimensional Methods in Mechanics,* pp. 271-281. Academic Press, New York.
Stanyukovich D.P. (1960) *Unsteady Motion of Continuous Media,* pp. 147-163, 498-506. Pergamon Press, New York.
Stuart G.W. (1965) Explosions in a rarefied atmosphere. *Phys. Fluids* **8**, 603-606.
Trinks H. and Schilf N. (1981) Gasdynamic investigations of lead azide/lead styphnate detonation processes in vacuum by multichannel mass spectrometry. *Gasdynamics of Detonations and Explosions, Progress in Astronautics and Aeronautics,* Vol. 75 (Eds. J. R. Bowen, N. Manson, A. K. Oppenheim, and R. I. Soloukhin). AIAA, New York.

Gasdynamic Investigations of Lead Azide/Lead Styphnate Detonation Processes in Vacuum by Multichannel Mass Spectrometry

H. Trinks* and N. Schilf*
Hochschule der Bundeswehr, Hamburg, Federal Republic of Germany

For the determination of the gasdynamics and the kinetics of fast reactions, samples of solid high explosives are ignited in a large vacuum chamber. During the fast expansion of the blast wave and subsequent motion of the decomposition products (and intermediates) through the vacuum, both gaseous and condensed reaction products remain in a frozen state. The expansion process in the vacuum and the reaction intermediates and products are observed by short-time photography, multichannel mass spectrometry, and condensate analysis. The magnetic mass spectrometer, installed in the vacuum chamber, measures four different gaseous reaction products (in the range 1-250 amu) simultaneously during the measuring time (10-500 μs) with a time resolution of 2 μs. In experiments with lead azide/lead styphnate samples, the pseudohalogenide N_6, N_2, N_3, O, O_2, CO_2, CO, C_2O, and NO were detected. Near the exploding surface lead condenses, and the resulting droplets (0.1-100 μm diam) have velocities up to 6 km/s.

Introduction

EXPERIMENTAL investigations of the reactions that occur during the detonation of solid explosives are difficult because the pressure and the temperature in the reaction zone is high and the reaction time is very short, only a few microseconds.

A reaction process is characterized by the initial state z_1, the chemical kinetic mechanism of the reaction and the reaction path(s) dictated by the mechanism, and by the final state z_2 a long time after the reaction. Usually the primary state z_1 is known, e.g., by the structure formula of the explosive material just before ignition. The final state z_2 can be determined experimentally, e.g., by mass spectrometric analysis and pressure measurements.

Presented at the 7th ICOGER, Göttingen, Federal Republic of Germany, Aug. 20-24. 1979.
Copyright © American Institute of Aeronautics and Astronautics, Inc., 1980. All rights reserved.
*Professor, Department of Electrical Engineering.

However, the knowledge of the states z_1 and z_2 is insufficient for the description and the understanding of the entire reaction. There are numerous possibilities concerning different reaction paths from z_1 to z_2. Theoretically exact computations of explosive reactions are often inaccurate because of the lack of information on essential reaction process rates.

For the investigation of the detonative decomposition of high explosives, the reaction process can be studied in a large vacuum chamber (Schilf, 1979). The reaction products' produced in the reaction zone, are expanding just after the reaction into the vacuum. As a consequence, interactions of the primary reaction products not only with the surrounding gas, but also the reaction products themselves, are suppressed. As the primary state of the reaction products is essentially frozen during the reaction products' motion through the vacuum chamber, it is possible to measure these reaction products—even if they are unstable or reactive, such as atomic oxygen.

A shadow short-time photograph (Fig. 1) of the expanding blast waves of the detonating lead azide sample, in air under normal pressure and in high vacuum, indicates the freezing of the state of reaction products at vacuum conditions.

Experimental Setup

The experimental setup (shown schematically in Fig. 2) consists of: the high-vacuum chamber, the detonative sample ignited in the vacuum, and the detector probe located on the axis of the chamber for sampling the reaction products.

The gaseous reaction products are analyzed in the following manner by the specially constructed four-channel mass spectrometer (Schilf and Trinks, 1981) shown schematically in Fig. 3.

A small fraction of the gaseous reaction products in the blast wave, flowing over the skimmer housing, passes through the skimmer pinhole and into the ion source. A beam of ions is formed by electron impact, and a portion of the beam is focused on the entrance slit of each of four independent double-focusing magnetic mass spectrometers. Each consists of a cylindrical condenser as energy separator, a magnetic field separator, and an electron multiplier. As the magnetic field strength is adjusted, ions of different mass-to-charge ratio (m/e) are detected.

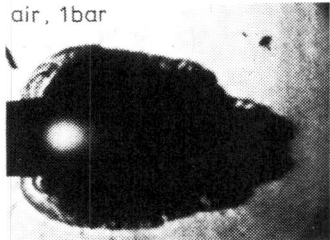
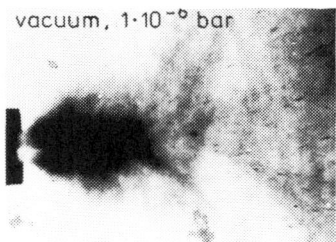

Fig. 1 Shadow short-time photography of blast waves, produced by detonating a lead azide sample in air ($p = 1$ bar) and in vacuum ($p = 10^{-6}$ bar).

The ion beam is amplified by a secondary electron multiplier and the multiplier output is recorded on a storage oscilloscope. The mass analyzer measures four different ions in the mass range of 1-250 amu simultaneously with a time resolution of 2 μs.

The mass spectra deduced from the responses of one of the four channels at various magnetic field strengths are shown in Fig. 4. Typical mass spectrometric results from two of the channels, one adjusted for the mass-to-charge ratio of atomic oxygen, and the other for molecular oxygen, are shown in Fig. 5. The blast wave was produced by the detonation of lead styphnate. The detonation is initiated by an exploding wire. The oscilloscope is triggered by the signal from a photomultiplier which senses the emitted light from the detonation. At the moment $t_0 = 0$ the blast wave starts at the point of the detonation and moves to the mass spectrometer skimmer (located at $a = 1.45$

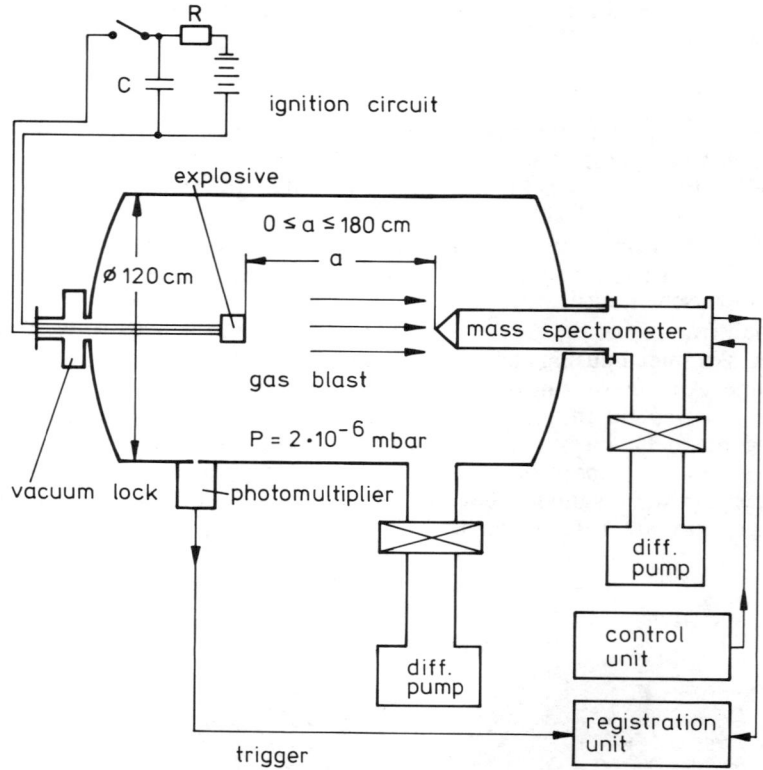

Fig. 2 Schematic view of the experimental apparatus.

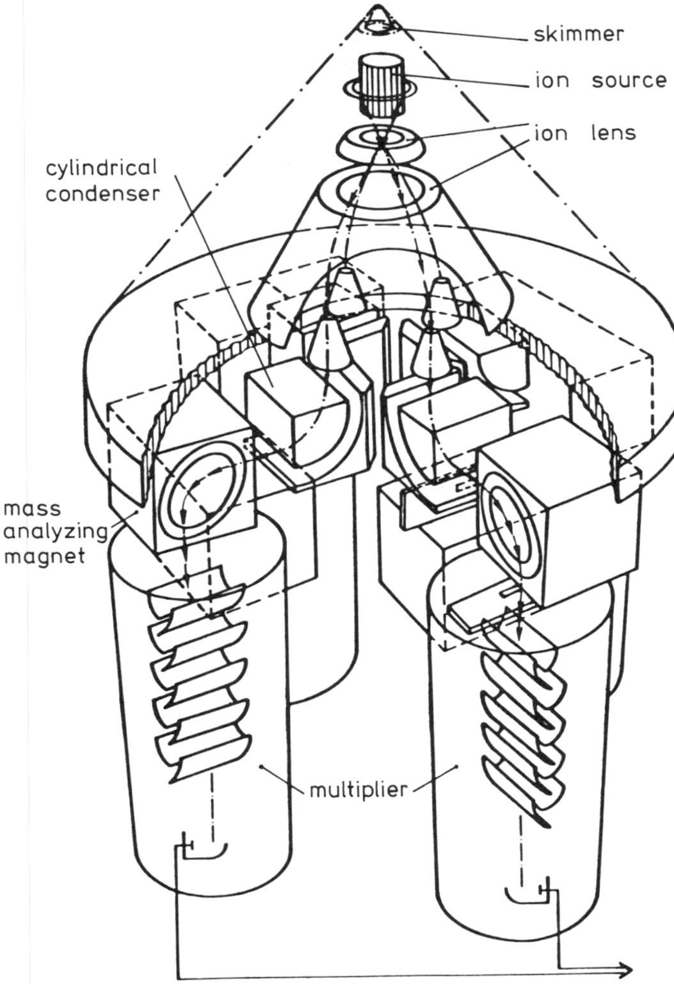

INVESTIGATIONS OF DETONATION PROCESSES

Fig. 3 Schematic of the mass spectrometer.

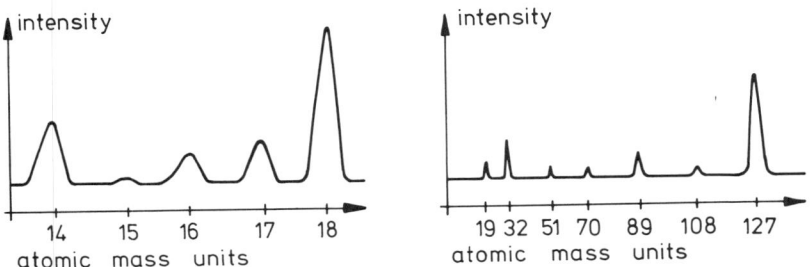

Fig. 4 Typical mass spectra obtained by one channel with adjustment for high and low atomic mass units.

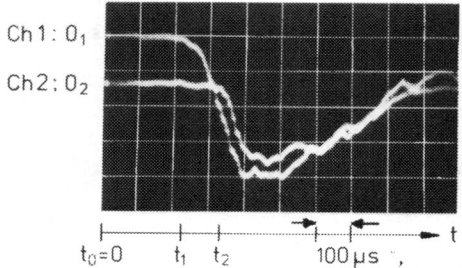

Fig. 5 Typical mass spectrometric signals measured in the blast wave: channel 1 set for atomic oxygen; channel 2 set for molecular oxygen.

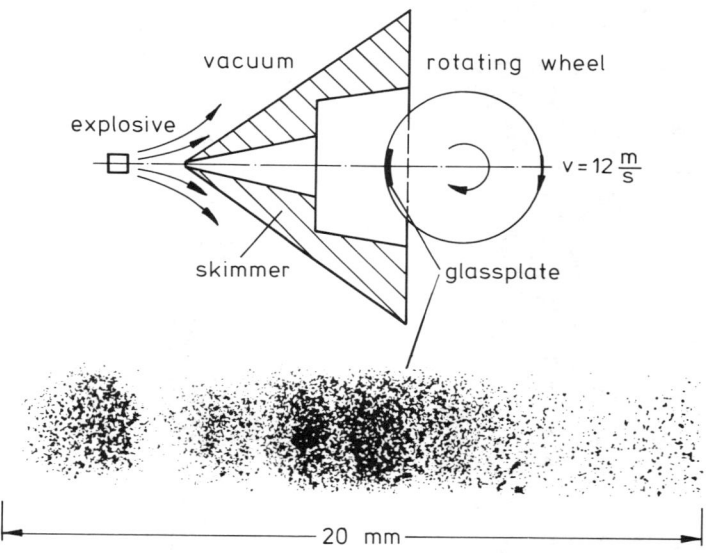

Fig. 6 Schematic of the device for the condensate detection and typical condensate prints observed by microscopy.

m). After the time of flight t_1, the reaction product atomic oxygen O_1 is detected by the mass spectrometer. Subsequently at the time t_2 the molecular oxygen O_2 in the blast wave is detected by the mass spectrometer. The maximum velocity ($v_{\max,1} = a/t$) for atomic oxygen is 6.3 km/s and that ($v_{\max,2} = a/t_2$) for molecular oxygen is 4.4 km/s. These results indicate that the quantitative detection of unstable and reactive reaction products, such as atomic oxygen, is possible by this mass analyzer. (The results shown in Fig. 5 demonstrate that the atomic oxygen O_1 is, indeed, originated in the detonation process and has not been formed by electron impact of molecular oxygen in the ion source or the like. In this case, simultaneously with the O_1 signal the O_2 signal would have to be observed.)

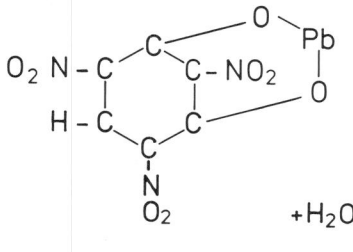
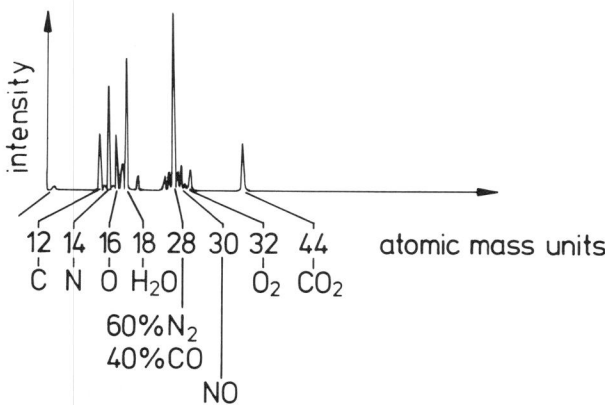

Fig. 7 Structural formula of lead azide and lead styphnate and the mass spectrum of the final state of detonative decomposition.

Condensates are detected by a special arrangement shown schematically in Fig. 6. The condensates in the blast waves impinge on a rotating wheel located behind a skimmer. The condensate prints on the wheel are interpreted by microscopic techniques and condensate analysis techniques, such as secondary electron energy-loss spectroscopy. These methods provide data on condensate velocity distribution, particle size distribution, and the chemical composition.

With this system data interpretation is facilitated by conducting a large number of experiments with identical samples of explosive material. During gas analysis, experiments should be repeated to confirm results and also the settings of the four channels of the mass analyzer should be varied to avoid bias.

Results

The system has been used to investigate the detonative decomposition of small samples of a mixture of lead azide (140 mg) and lead styphnate (60 mg)

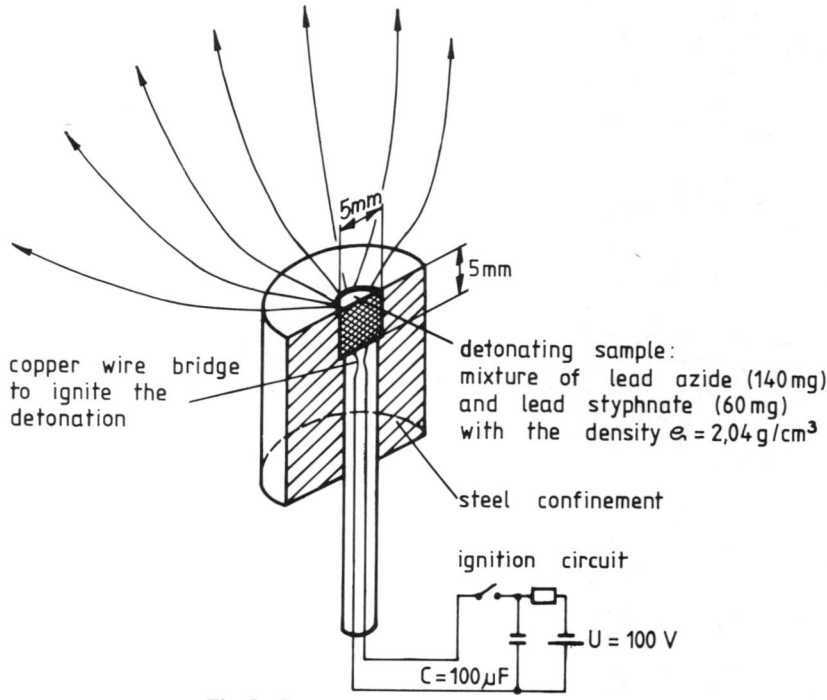

Fig. 8 Description of the explosive system.

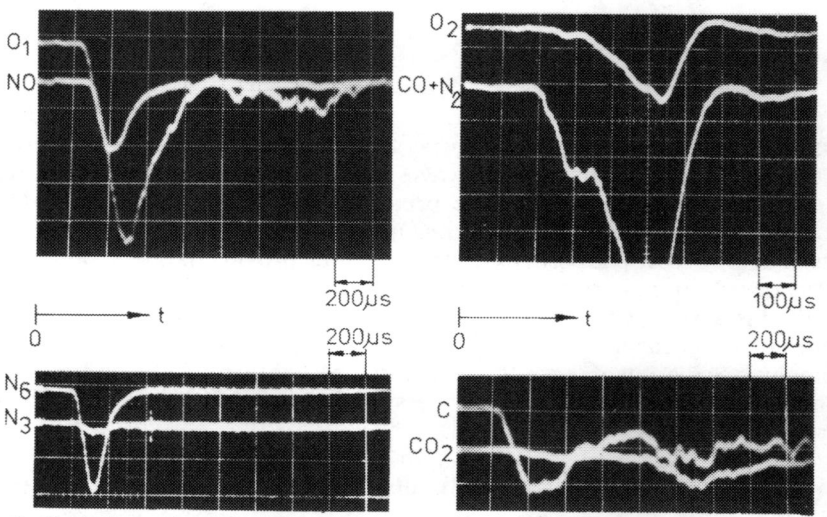

Fig. 9 Typical experimental results obtained by four-channel mass spectrometry in the blast waves of lead azide/lead styphnate with $a = 1.45$ m.

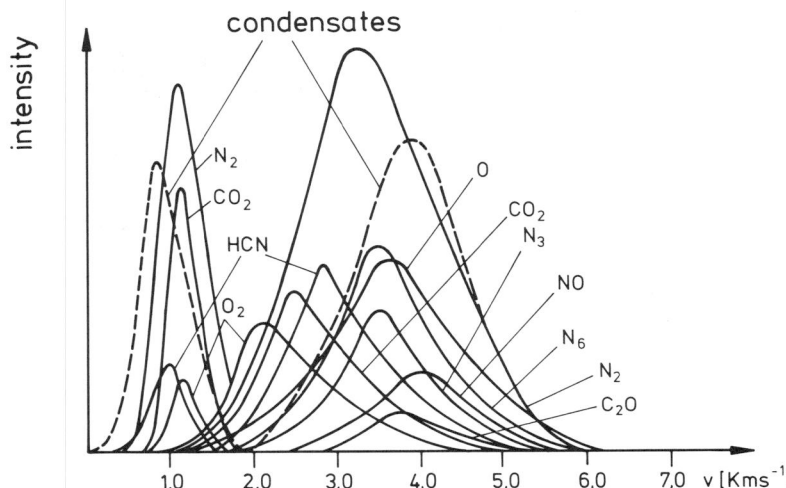

Fig. 10 Velocity distributions of the reaction products in the blast waves.

(see Figs. 7 and 8). Using identical samples, 180 experiments were made. Typical experimental results are shown in Fig. 9.

The velocity distribution of all gaseous reaction products found in the blast waves are deduced from such signals (see Fig. 10). (A large number of reactive or radical products has been found. These products seem, indeed, to be originated in the detonation process and not by electron impact in the ion source or the like. This assertion is confirmed by results such as shown in Fig. 5 and by the fact that some of the found reactive or unstable products, e. g., N_6, can hardly be formed by electron impact with the observed high intensity.) Although the reaction products N, N_2, N_3, and N_6 were detected, there was no evidence of either N_4 or N_5. The detonative decomposition of lead azide appears to be accompanied by the formation of pseudohalogenide N_6 structures, which decompose to N_3 and finally N_2 molecules. The existence of N_6 structures during the detonative decomposition of lead azide had been found earlier (Clusius and Schumacher, 1958; Aulinger and Trinks, 1972). Gaseous lead or lead compounds were not detected. The velocity distributions of the blast-wave reaction products has bimodal character: mostly unstable or reactive reaction products, such as N_3, N_6, O_1, fall into the higher velocity distribution range, while only stable products fall into the lower velocity distribution. This effect is postulated as caused by the burst into the vacuum (Gruschka and Wecken, 1971). The reaction zone, accompanied by a shock wave, travels with the detonation velocity within the detonating sample. When the shock wave reaches the boundary to the vacuum, the outer part of the reaction zone escapes with a velocity up to 7.0 km/s into the vacuum. The primary originated reaction products expand into the vacuum immediately. Secondary reactions are strongly suppressed during this process. As a consequence, a number of unstable and radical reaction products have been found in the fast part of the blast wave.

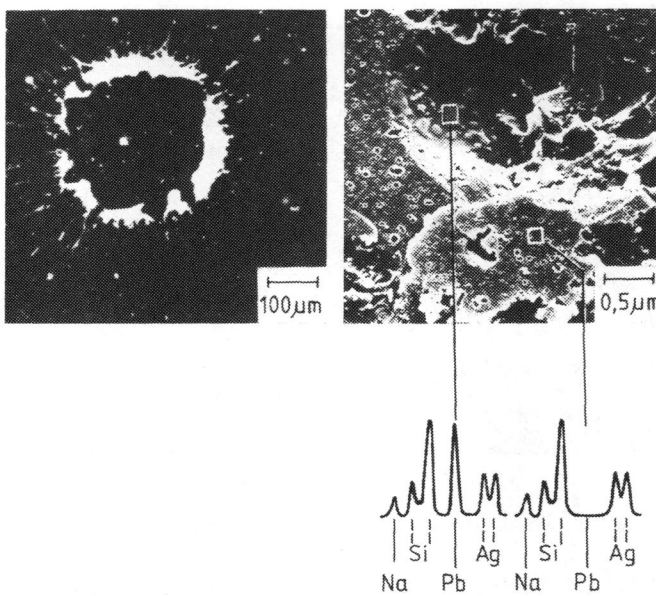

Fig. 11 Photography of the plate and analysis of the condensates. Electron energy-loss spectra from the bottom of a crater produced by a condensate droplet are shown and compared to the undamaged glass plate.

The shock wave, reaching the boundary to the vacuum, not only gives rise to this fast precursor. The shock wave is partly reflected and moves back into the decomposed zone. The more slowly expanding reaction products in this zone are interacting while unstable and reactive products are destroyed. In this way, during this second and slow part of the expanding blast waves mostly stable and nonreactive products are found.

The stable reaction products (HCN, N_2, O_2, and CO_2) are originated in the primary reaction. For this reason these products are observed in the low- and high-velocity products of the detonation—each product having its own velocity distribution.

The condensate reaction products also appear to have a bimodal velocity distribution (see Figs. 6 and 10). The condensates are analyzed by microscopy and secondary electron-loss spectroscopy. In Fig. 11 parts of the surface of the glass plates are shown. On the one hand, a large droplet is perceptible which burst at the surface, and on the other hand, a small crater can be seen, broken by a very fast condenseous particle. The electron energy-loss spectrum shows that this particle consists of lead. The other detected mass peaks, e.g., silicon and silver, are based on impurities in the glass surface. This is shown by the spectrum taken at the undamaged glass surface near the crater. The experimental results show that most of the detected condensates consist of lead droplets with a size of 0.1-100 μm. All lead atoms appear to condense immediately or soon after formation during the detonation process, and the

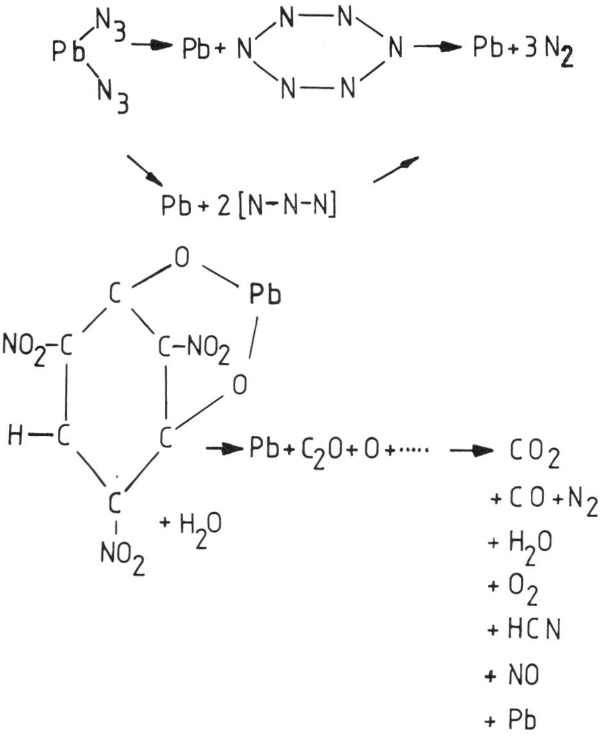

Fig. 12 Schmatic conception of detonative decomposition of lead azide and lead styphnate. Shown is the primary state before the reaction is started, the medium state observed as some reaction products expand in a frozen state, and the final stable state of all reaction products a long time after the reaction. (Lead condenses near the reaction zone and escapes as droplets).

resulting particles expand together with the gaseous reaction products. Apart from the condenseous lead droplets one should expect solid carbon from the detonated lead styphnate. The detection of solid carbon, however, has not been carried out yet; further investigations are continuing.

Figure 12 shows schematically a conception of the detonative decomposition of lead azide as well as lead styphnate, derived from all experimental results.

Conclusions

The experimental investigation of detonation and combustion processes of solid explosives in a vacuum chamber is an effective method to freeze the intermediates and products of the postreaction state and to extend the original, extremely short measuring time.

With multichannel mass spectrometry detailed information concerning composition of the detonation products can be obtained, including the dependence on the expanding process. The mass spectrometric data suggest that unstable, reactive products can also be observed. To demonstrate the effectiveness of the technique, the detonative decomposition of lead azide/lead styphnate was studied. Of particular interest are the absence of N_4 and N_5 and the nature of the condensation products.

Further investigations are in progress, especially to elucidate the mechanism of formation of N_3 and N_6 structures and the condensation processes of lead immediately after the detonative reaction.

References

Aulinger F. and Trinks H. (1972) Massenspetrometrische untersuchungen an schwaden detonierender sprengstoffe. German-French Res. Inst., Saint Louis, France, ISL 30/72.
Clusius K, and Schumacher H. (1958) Langsame und explosive zersetzung von metallaziden. *Helvetica Chimica Acta* **41**, 2264-2268.
Gruschka H.D. and Wecken F. (1971) *Gasdynamic Theory of Detonation*. Gordon and Beach Science Publishers, New York.
Schilf N. (1979) Massenspetrometrische analyse an detonationsschwaden im vakuum. Dissertation, RWTH-Aachen.
Schilf N. and Trinks H. (1979) Four channel magnetic mass spectrometer for high dynamic analysis in detonation processes. *Intl. J. Mass Spectrom. Ion Phys.* **37** (1), 123-126.

The Effect of the Shock-Wave Front on the Origin of Reaction

A. N. Dremin* and V.Yu. Klimenko†
Academy of Sciences, Chernogolovka, U.S.S.R.

A considerable body of data has been accumulated thus far which characterizes the decomposition of condensed explosives in shock and detonation waves. The data indicate that the kinetics of decomposition of condensed explosives is extremely rapid. To account for this observation it has been proposed that one of the possible reasons is the specific action of the shock front. It is possible that shock action can lead to a situation in which the explosive reaction is not excited simultaneously upon passage of the shock but one in which the explosives decompose partially in the shock front itself. Model results on the real shock-front structure in fluids and solids have been obtained by the molecular dynamic method. The highly nonequilibrium state of the molecules in the shock front suggests the validity of the assumption of partial explosive decomposition in the shock front.

Introduction

AN examination of the available data on the detonation of solid and liquid explosives has led one of the present authors to the conclusion that the associated decomposition reaction has a heterogeneous character and that it originates and develops in individual centers (Dremin, 1978). This conclusion was based on the following data and arguments.

It had been previously shown that, with identical detonation pressures, the time required for reaction in porous charges with different liquid fillers is equal to the time of reaction in charges without fillers (Dremin et al., 1970). But when the pressures are equal, the shock-front temperature in the first case must be lower, and the reaction time (provided it develops homogeneously) longer than in the second case. This discrepancy implied that in powder and pressed charges the explosive detonation decomposition mechanism was not homogeneous.

It was later discovered that for each solid explosive charge there is a certain shock compression pressure P^* above which the decomposition reaction progress is analogous to that in liquid explosives (Dremin and Schvedov,

---

Presented at the 7th ICOGER, Göttingen, Federal Republic of Germany, Aug. 20-24, 1979.
Copyright © American Institute of Aeronautics and Astronautics, Inc., 1981. All rights reserved.
*Professor, Chief of the Laboratory, Institute of Chemical Physics.
†Doctor of Science, Junior Research Worker, Institute of Chemical Physics.

1976). Since P^* is lower than the detonation pressure, this conclusion is also valid for detonation waves in solid explosives. Moreover, it was shown that the reaction time in liquid trotyl on detonation is equal to the reaction time in pressed trotyl if the initial densities were the same (Schvedov and Dremin, 1975). Direct experimental evidence has been obtained concerning heterogeneous initiation of reaction in liquid explosives under the effect of shock waves of different strength (Walker and Wasley, 1974; Vorobiev and Trofimov, 1978).

As the reaction originates in individual centers, the question arises naturally as to the underlying cause for the appearance of such centers. One possible reason may be a specific action of the shock front. The present paper contains the results of calculations of the shock-wave front structure in liquid and solid argon. It is as yet impossible to carry out analogous calculations for complex multiatomic molecules of explosives. However, on the basis of the results obtained for simple monoatomic argon molecules, it is possible to develop qualitative ideas about the behavior of the complex polyatomic molecules in the shock front (Karo et al., 1978).

Calculation Method

The continuum mechanics approach to the study of the structure of strong shock waves is not applicable, since the shock-front width is comparable with the size of the molecules (Zeldovitch and Rayzer, 1966). To comprehend the

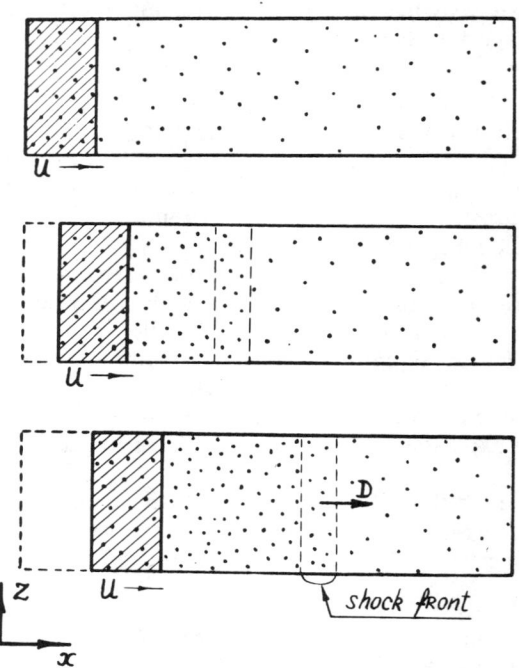

Fig. 1 Formation of shock wave in counting cell.

nature of the processes occurring in a shock front, the molecular structure of the substance must be considered. For gases of low initial pressure the Boltzmann kinetic equation provides a suitable model. There are several approximate methods for the solution of Boltzmann's equation; the use of these methods has led to quite adequate investigations of the shock-wave structure (Mott-Smith, 1951; Hicks et al., 1972; Juck et al., 1973; Genich et al., 1977). The alternative approach to the study of shock-wave structure in gases is the direct numerical simulation of the shock wave by the Monte-Carlo method (Bird, 1967). The most interesting results obtained by application of this method are reported by Bird (1970).

The problem is more complex for the case of modeling the structure of shock waves in dense gases or condensed phases. These systems have high interaction parameters and are not amenable to theoretical treatment. Reliable analytical methods are not available as yet for liquids. However, recent progress in computing technology has stimulated the advent of rather promising methods of mathematical simulation of systems which have high interaction parameters. These are the Monte-Carlo method and the method of molecular dynamics.

To the best of the authors' knowledge, there are no published papers concerned with application of the Monte-Carlo method to the study of shock-wave structure in condensed substances. Some attempts to use the molecular dynamic method for simulate shock-wave propagation in solids have already been undertaken (Paskin and Dienes, 1972; Tsai and McDonald, 1973; Paskin et al., 1977). For the reason given subsequently, these efforts have not led to elucidation of the shock-wave structure in solids. As to the structure of shock waves in liquids, no publications are known to us.

To investigate the structure of a strong shock wave in liquids, solids, and dense gases, the molecular dynamic method for nonequilibrium systems has been developed and is the subject of this paper. The approach is based on the standard molecular dynamic method for equilibrium systems (Alder and Wainwright, 1960; Rahman, 1964; Verlet, 1967) and consists in the real substance being modeled by some finite number of particles (approximately 100-1000) placed in a certain volume. To eliminate the wall effects, periodic boundary conditions are employed. The equations of motion of the particles are integrated numerically to provide the instantaneous position and velocities for all particles at each moment of time.

When the molecular-dynamic approach is used to model liquid argon whose atomic interactions are approximated by the Lennard-Jones potential

$$\Psi(r) = 4\epsilon[(\sigma/r)^{12} - (\sigma/r)^6]$$

where $\epsilon = 119.8$ K and $\sigma = 3.405$ Å. The model results agree well with experimental observations, not only for the thermodynamic properties (temperature, pressure, heat capacity, and compressibility) but also for the transport properties (diffusion, viscosity, and thermal conductivity) (Lagarcov and Sergeev, 1978). Moreover, satisfactory agreement is found for a wide range for the state of the liquid from the triple to the critical point. The molecular dynamic method also effectively describes the argon solid-liquid phase transition (Evseev and Frenkel, 1973).

For the calculations performed in the present work, a volume (counting cell) was isolated which had the form of a parallelepiped with the side ratio 1/1/3 and which was located with its longer side along the x axis (Fig. 1). The cell contained 2592 particles. The differential equations of motion

$$\frac{d^2 r_i}{dt^2} = \frac{1}{m}\sum_{j\neq i} F(r_{ij})$$

where r_i is the position of the ith particle, m its mass, and $F(r_{ij})$ the force of interaction between the ith and jth particles. This equation was solved numerically according to the following differential scheme (Verlet, 1976):

$$r_i(t+h) = 2r_i(t) - r_i(t-h) + \frac{h^2}{m}\sum_{j\neq i} F(r_{ij})$$

The integration step h was taken equal to 10^{-14} s. In this case the fluctuations of the total energy of the system are small and do not exceed 0.01%. In the calculations, the Lennard-Jones potential was terminated at $r_{max} = 3\sigma$, i.e., if spaced further apart than 3σ, the interaction of particles was neglected.

First, all the particles were placed in the face-centered lattice nodes. A molecular chaos characteristic of the real fluids was specified by the Monte-Carlo method; the random-number generator determined the initial directions and velocities of all the particles. The system was given a period of time (100-200 steps) to evolve to the equilibrium state, during which time the redistribution of the kinetic and potential energies of the system occurred. The resulting equilibrium system reveals rather well the structural and kinetic properties of the liquid. Namely, the radial distribution function has spikes, and the particles drift along the system and exchange their sites.

For generation of the shock wave, the particles in a subvolume of the counting cell near $x=0$ were given an initial velocity u in the direction of the x axis (Fig. 1). This compressed the substance, and a compression wave propagated along the counting cell. The formation of three zones was observed in the counting cell: the zone of noncompressed substance, the zone of compressed substance, and an intermediate zone in between the two, which is called the shock-wave front and in which the process of compression occurs. After some time, the process becomes stationary, and the intermediate zone velocity D was taken as that of the shock-wave front. In this way the shock-wave velocity is determined for any specified initial velocity of the subvolume. The calculations verify the Hugoniot equation for shock waves; the adiabatic relations calculated for liquid argon are in rather good agreement with those found experimentally (Fig. 2).

The principal objective of the present work was to investigate the structure of the intermediate zone and to determine the molecular processes which take place within the zone. For this purpose, knowledge of the behavior along the shock-wave propagation path of such fundamental characteristics of the substance as the density, pressure, and temperature was required. These quantities were calculated by the following technique. A transverse analytical layer is displaced in the counting cell along the direction of the shock-wave

EFFECT OF SHOCK-WAVE FRONT ON ORIGIN OF REACTION 257

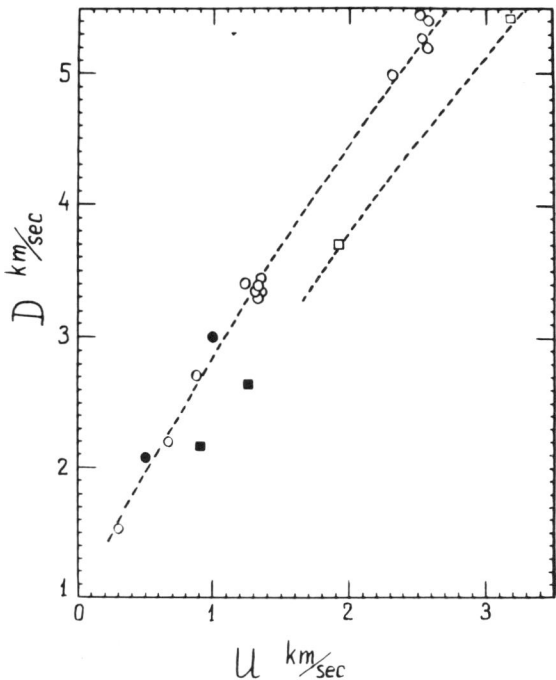

Fig. 2 Experimental (Van Thiel and Alder, 1966) and calculated D-u adiabatic curves of liquid argon (circles—experimental, shaded circles—calculated, data at $\rho_0 = 1.4$ g/cm^3, $T_0 = 88$ K; squares—experimental, shaded squares—calculated, data at $\rho_0 = 0.92$ g/cm^3, $T_0 = 148$ K).

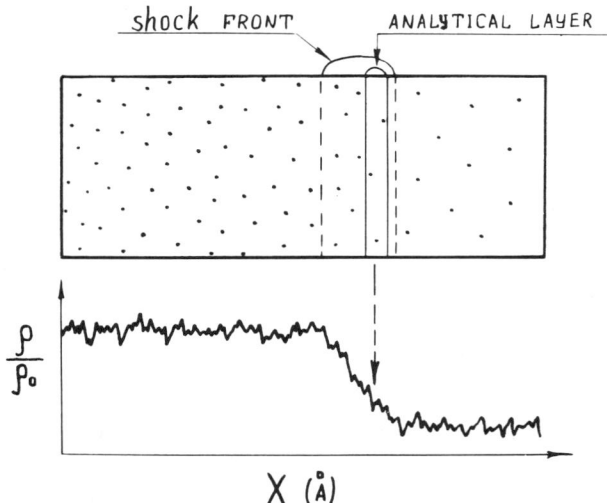

Fig. 3 Profile calculation method.

propagation by a very small step (0.2 Å) (Fig. 3) and in this layer the required properties are calculated. The density is determined by the formula:

$$\rho = \frac{Nm}{V}$$

where V is the analytical layer volume and N the number of particles inside the layer. The mass velocity of the particles in the x direction is calculated by the formula:

$$u_p = \frac{1}{N}\sum_i^N v_{xi}$$

where v_{xi} is the x component of the velocity of the ith particle. The temperature components T_z (across the shock-wave path) and T_x (along the shock-wave path) are determined by the formulas:

$$T_z = \frac{2}{Nk}\sum_i^N \frac{mv_{zi}^2}{2}$$

$$T_x = \frac{2}{Nk}\sum_i^N \frac{m(v_{xi}-u_p)^2}{2}$$

where k is the Boltzmann constant. The pressure is calculated by the formula:

$$P = \frac{2}{3V}\sum_i^N \frac{mv_i^2}{2} - \frac{1}{6V}\sum_i\sum_{i\neq j} r_{ij}\frac{\partial \Psi}{\partial r_{ij}}$$

The foregoing calculation technique has an essential advantage over those used earlier (Paskin and Dienes, 1972; Tsai and McDonald, 1973). In the paper by Paskin and Dienes (1972), a property, say temperature, was calculated for each crystal plane perpendicular to the shock-wave path, while in the paper by Tsai and McDonald (1973) a property was considered as an average of five crystal planes. As a result, discrete profiles were obtained which consisted of far-removed points and, therefore, only three or four points represented a narrow shock front which did not allow a conclusion as to its structure. With the method suggested in the present paper, continuous profiles of density, temperature, and other characteristics of the shock front are obtainable for either solid, liquid, or gaseous states, while the methods suggested by Paskin et al. (1972) and Tsai et al. (1973) are applicable only for crystals.

At first, the application of the analytical layer concept led to some difficulties. On the one hand the layer should be as thin as possible for the profile to be accurately determined; and, on the other hand, it cannot be too thin, since the layer should contain a sufficient number of particles for the statistical estimation of the macroscopic properties of the substance. The width of the intermediate zone, as is shown later, is about 10 Å, while the analytical layer was taken to be 3 Å wide. Since in a 3 Å layer only about 100 particles can be

EFFECT OF SHOCK-WAVE FRONT ON ORIGIN OF REACTION

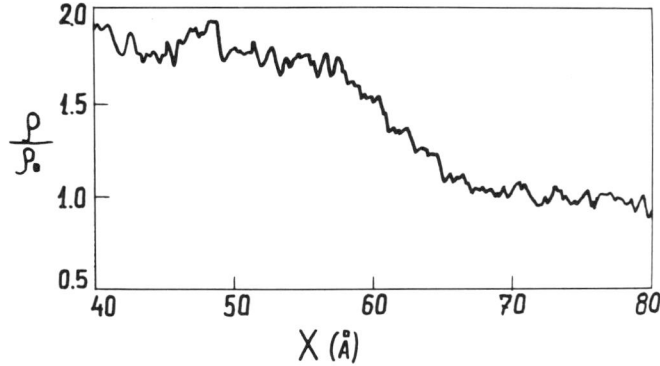

Fig. 4   Instantaneous shock-wave profile of density in liquid argon.

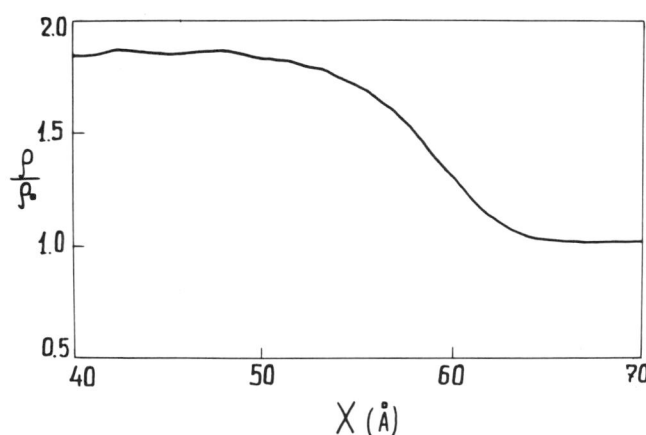

Fig. 5   The statistically averaged shock-wave profile of density in liquid argon.

counted on the average, it is impossible to take a thinner layer, as there would be insufficient particles. Even for a 3 Å layer, high fluctuations in the density and temperature profiles are observed (Fig. 4) and to a considerable degree mask the behavior of these quantities in the shock front. As a consequence, investigation of the shock-wave front structure is stymied.

The above difficulty has been removed with an ensemble method in which the desired property is calculated as the average of some ensembles of the systems. The ensemble of even three to five systems is sufficient to generate rather smooth profiles which can be analyzed. As this approach is, however, inefficient in terms of computer time, an alternative approach has been employed. Since the shock wave moves with constant velocity D, the shock profile is independent of time; and the ergodic hypothesis can be invoked to determine the profile, i.e., ensemble averaging can be replaced by time averaging. About 100 profiles have been thus averaged for various time

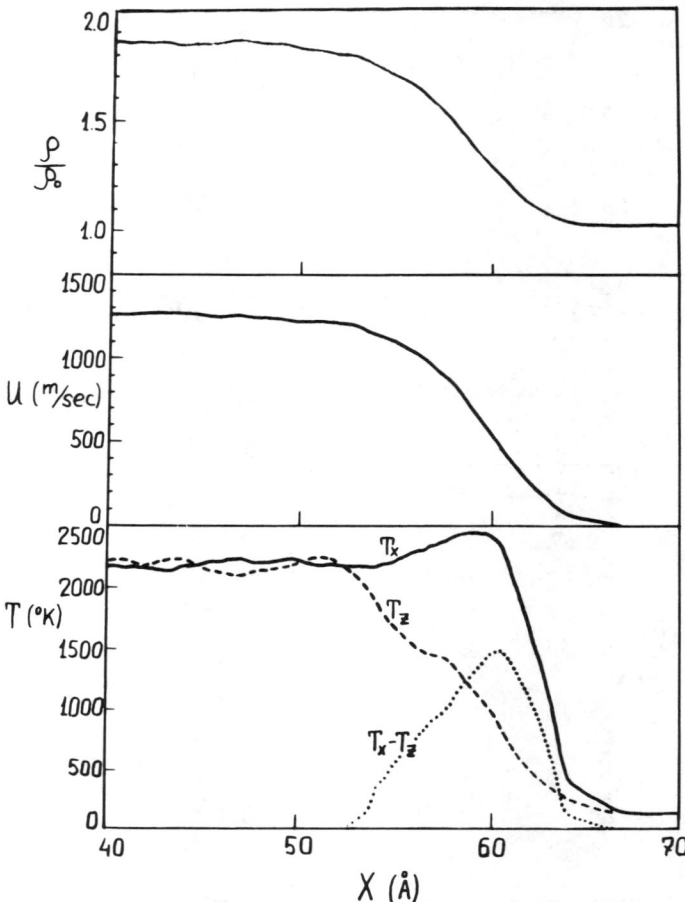

Fig. 6 Density, particle velocity, and temperature components T_x, T_z for shock wave ($u = 1250$ m/s, $D = 2650 \pm 50$ m/s, $P = 30.5 \pm 0.6$ kbar) in liquid argon ($\rho_0 = 0.92$ g/cm^3, $T_0 = 148$ K).

moments, and smooth profiles of density, temperature, etc. (Fig. 5), were obtained.

Calculation and Discussion

The front structure of a strong shock wave in liquid argon has been investigated in a wide range of conditions. The calculated results for liquid argon in the vicinity of the critical point ($\rho_0 = 0.92$ g/cm^3, $T_0 = 148$ K) and for dense liquid in the vicinity of the triple point are shown in Figs. 6 and 7, respectively. It is evident that the substance is compressed very rapidly since the shock front width is 10-15 Å. These results are in good agreement with the experimental estimates obtained by Kormer (1968) in which the shock wave front width in liquid benzol and glycerine did not exceed 50 Å.

EFFECT OF SHOCK-WAVE FRONT ON ORIGIN OF REACTION   261

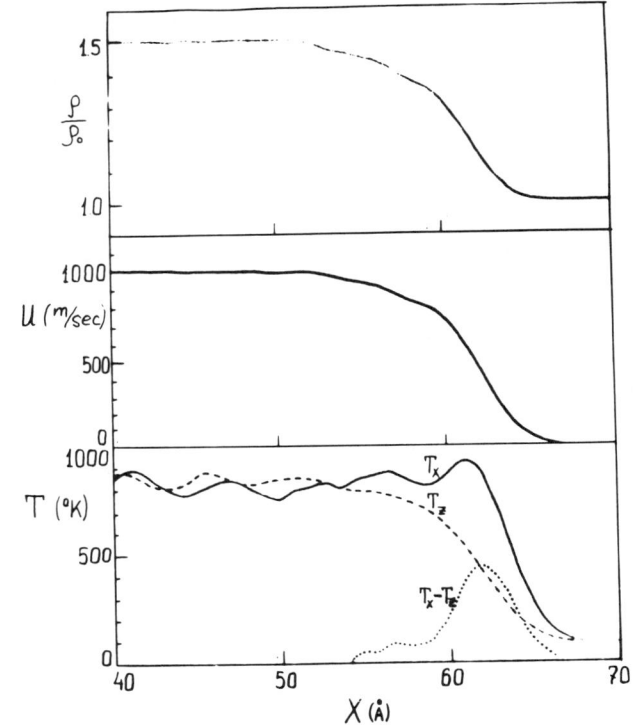

Fig. 7 Density, particle velocity, and temperature components T_x, T_z for shock wave ($u = 1000$ m/s, $D = 3200 \pm 50$ m/s, $P = 44.8 \pm 0.7$ kbar) in liquid argon ($\rho_0 = 1.4$ g/cm^3, $T_0 = 88$ K).

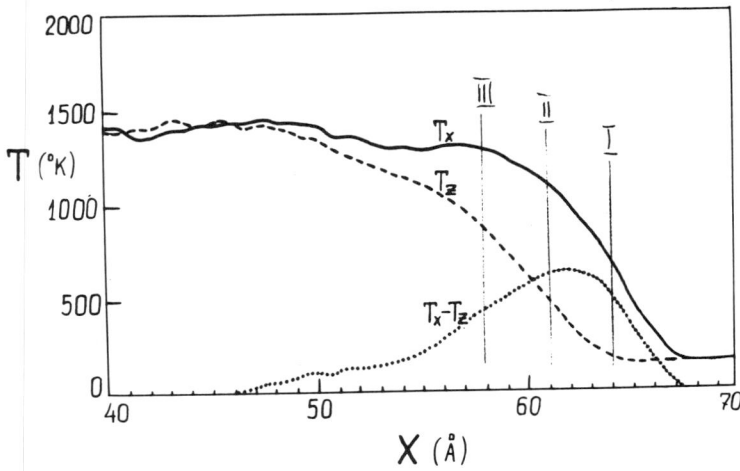

Fig. 8 Temperature components T_x, T_z for shock wave ($u = 900$ m/s, $D = 2160 \pm 50$ m/s, $P = 17.9 \pm 0.4$ kbar) in liquid argon ($\rho_0 = 0.92$ g/cm^3, $T_0 = 148$ K).

The transition of a substance in the shock wave into a compressed and heated state occurs in very short time intervals (from 10^{-12} to 10^{-13} s), which are commensurate with the time of translational relaxation. Therefore, there is no equilibrium by translational degrees of freedom within the front. The translational temperature component T_x increases faster and more intensively than the transverse temperature component T_z.

Calculations of weaker shock waves show that with a decrease in shock-wave intensity, the front structure does not change substantially, although the overheating of the translational temperature component disappears. (See Figs. 8 and 9.)

The temperature, or rather the mean kinetic energy of the particles, does not adequately determine the state of the substance in the shock front, since for one and the same mean kinetic energy of the molecules their velocity distribution can be quite different; while the velocity distribution, or rather the relative velocity distribution of the molecular pairs, directly affects the kinetics of chemical reaction. The distribution functions in three cross sections (I, II, and III in Fig. 8) of the shock-wave front are presented in Fig. 10. It is evident that the distribution function is a Maxwellian one ahead of the wave front and after it. Within the shock-wave front, the function is deformed and differs markedly from an equilibrium one. For illustration, the Maxwellian function is compared in Fig. 11 with the real distribution function over one of the front segments. One can see that the difference is notable.

Figure 12 presents predicted and experimental adiabatic curves for solid argon. The calculated points are somewhat above the experimental ones. This discrepancy is probably due to the calculations having been performed for an ideal monocrystal of argon (the shock wave propagated in 1,0,0 directions), while the experiments were performed with real solid argon which is a polycrystal with different defects.

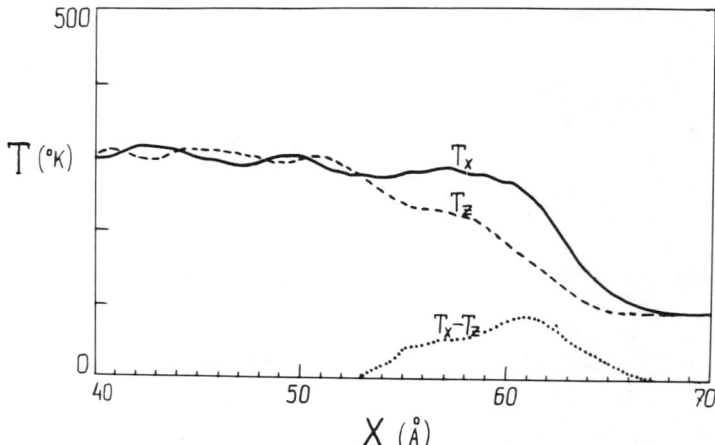

Fig. 9 Temperature components T_x, T_z for shock wave ($u = 500$ m/s, $D = 2080 \pm 50$ m/s, $P = 14.6 \pm 0.4$ kbar) in liquid argon ($\rho_0 = 1.4$ g/cm^3, $T_0 = 88$ K).

EFFECT OF SHOCK-WAVE FRONT ON ORIGIN OF REACTION

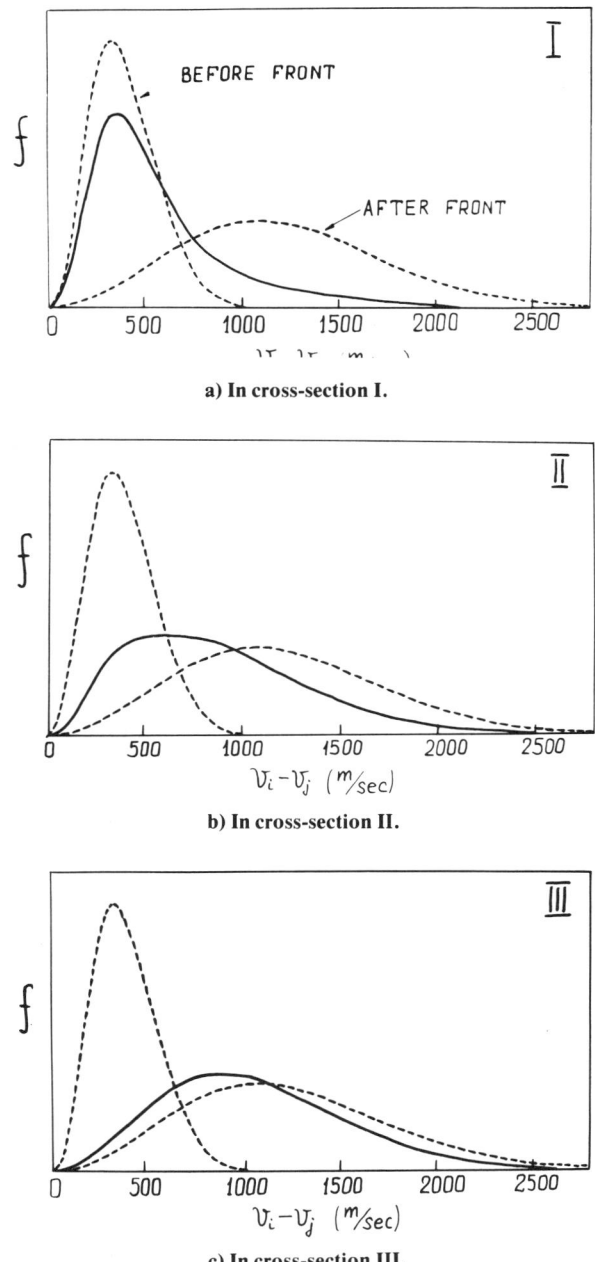

a) In cross-section I.

b) In cross-section II.

c) In cross-section III.

Fig. 10 Distribution function of molecule pairs to relative velocities in several cross sections (see Fig. 8) of shock wave ($u = 900$ m/s, $D = 2160 \pm 50$ m/s) in liquid argon ($\rho_0 = 0.92$ g/cm^3, $T_0 = 148$ K).

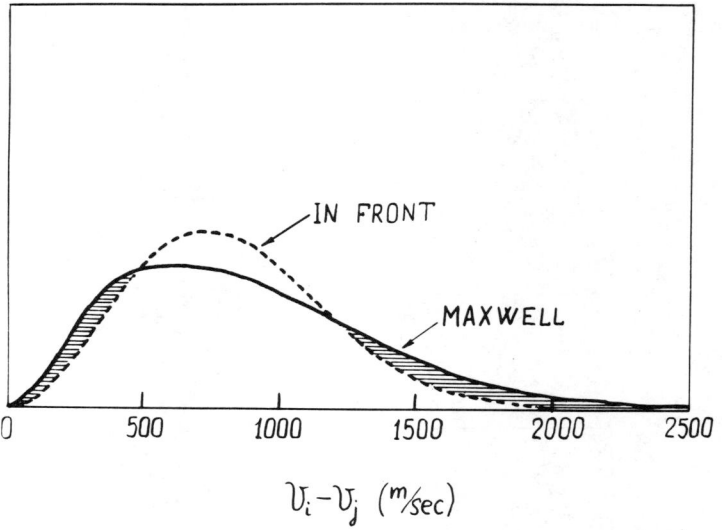

Fig. 11 Comparison of molecule pairs distribution function to relative velocities in shock wave (dashed line) (see Fig. 8, cross-section II) with Maxwell distribution function (solid line).

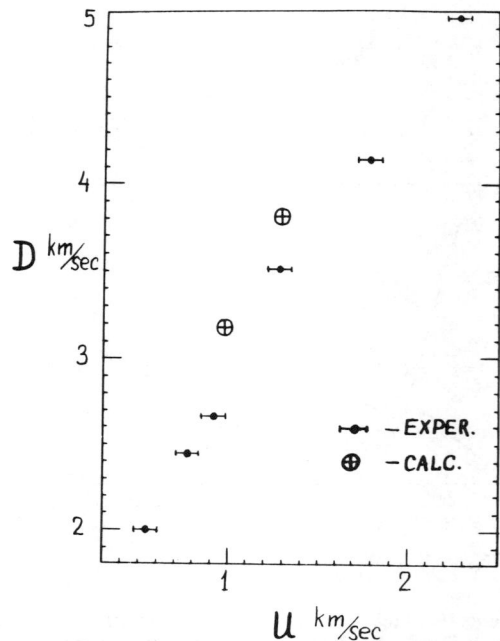

Fig. 12 Experimental (Dick and Warnes, 1970) and calculated D-u adiabats of solid argon.

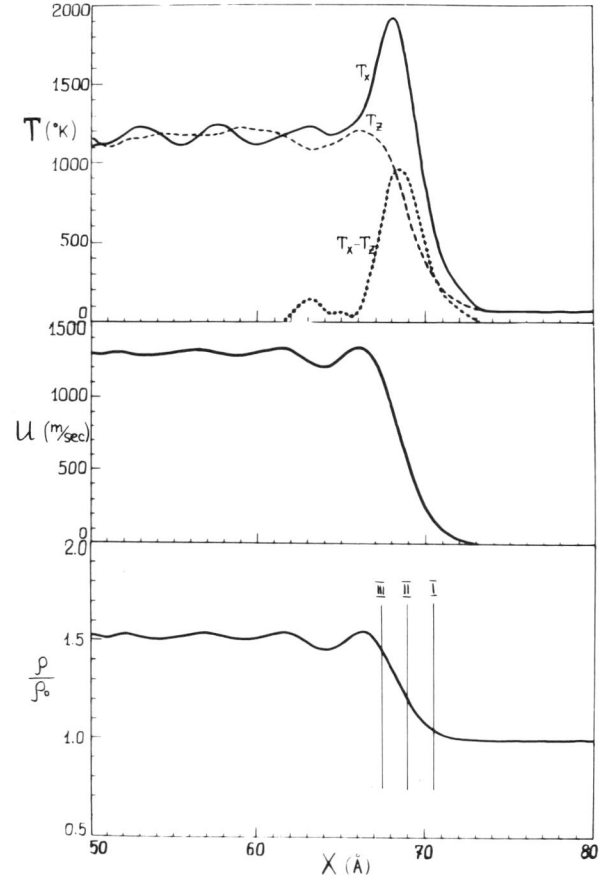

Fig. 13 Density, particle velocity, and temperature components T_x, T_z for shock wave ($u = 1300$ m/s, $D = 3840 \pm 50$ m/s, $P = 82.4 \pm 1.1$ kbar) in solid argon ($\rho_0 = 1.65$ g/cm^3, $T_0 = 75$ K).

The shock-wave profiles in solid argon are presented in Figs. 13 and 14. One can see that the shock-wave structure in a crystal differs markedly from that in a liquid. First, the shock-wave front width in solid argon is about half of the front width in liquid argon. Second, a large overheating of the T_x temperature component is observed in solid argon (it is twice as large in the front than behind it). Third, stationary waves are characteristic of the density and particle velocity profiles behind the shock-wave front. The relative velocity distribution function of molecular pairs within the shock front exhibits an even stronger deviation from the Maxwellian distribution than that predicted for liquid argon shocks (for example, see the function at Sec. II in Fig. 15).

Thus, in strong shock waves in liquids and solids the transition from one equilibrium state (ahead of the front) into the other equilibrium state (behind the front) takes place through a strong nonequilibrium zone of the front.

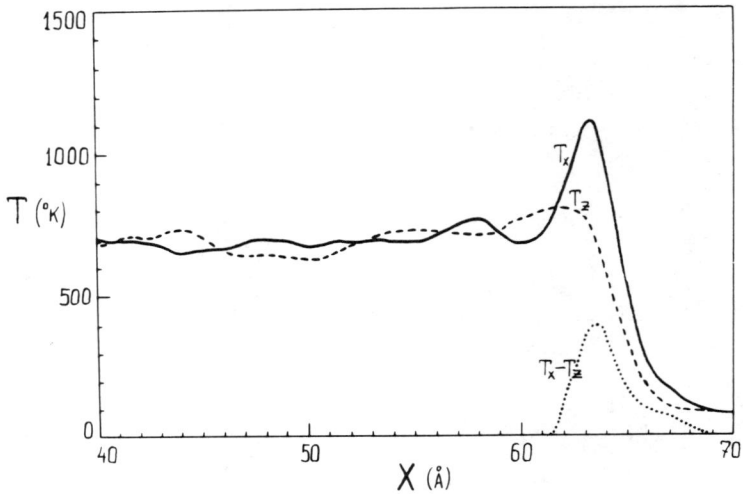

Fig. 14 Temperature components T_x, T_z for the shock wave ($u = 1000$ m/s, $D = 3200 \pm 50$ m/s, $P = 52.8 \pm 0.8$ kbar) in solid argon ($\rho_0 = 1.65$ g/cm^3, $T_0 = 75$ K).

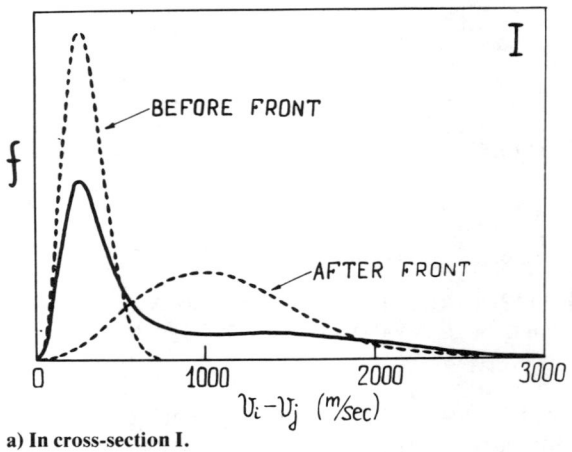

a) In cross-section I.

Fig. 15 Distribution function of molecule pairs to reactive velocities in several cross sections (see Fig. 13) of shock wave ($u = 1300$ m/s, $D = 3840 \pm 50$ m/s) (continued on next page).

Because of the absence of thermodynamic equilibrium, the concept of temperature is actually inappropriate to the shock front.

The transition of complex molecules, such as the molecules of explosives, through the nonequilibrium shock front can cause the destruction of a portion of these molecules and the formation of active particles. Obviously, the appearance of active particles gives rise to the start of reaction in individual centers. Thus, the heterogeneous origin of reaction in liquid and solid ex-

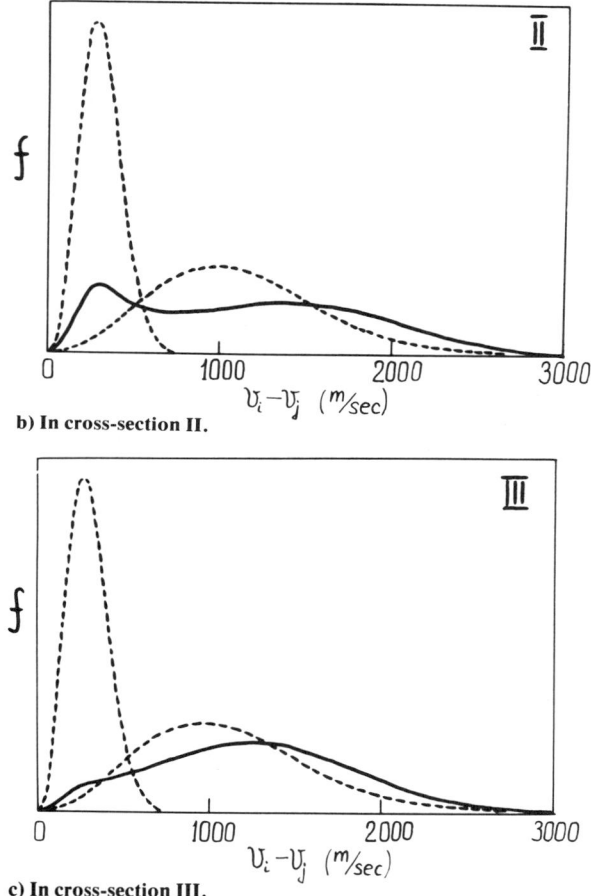

b) In cross-section II.

c) In cross-section III.

Fig. 15 (cont.) Distribution function of molecule pairs to reactive velocities in several cross sections (see Fig. 13) of shock wave ($u = 1300$ m/s, $D = 3840 \pm 50$ m/s).

plosives can, in fact, be conditioned by the specific action of the detonation wave's shock front.

References

Alder B. J. and Wainwright T. E. (1960) Studies in molecular dynamics, II: Behaviour of a small number of elastic spheres. *J. Chem. Phys.* **33**, 1439-1451.

Bird G. A. (1967) The velocity distribution function within a shock wave. *J. Fluid Mech.* **30**, 479-487.

Bird G. A. (1970) Aspects of the structure of strong shock waves. *Phys. Fluids* **13**, 1172-1177.

Bird G. A. (1970) Direct simulation and the Boltzmann equation. *Phys. Fluids* **13**, 2676-2681.

Dick R. D. and Warnes R. N. (1970) Shock compression of solid argon. *J. Chem. Phys.* **53**, 1648-1651.

Dremin A. N. (1978) On condensed explosives detonation decomposition mechanism. *Symp. Int. Sur Le Comportement Des Milieux Denses Sous Hautes Pressions Dynamiques,* pp. 175-182 (Ed. by Commissariat a l'Energie Atomique), Centre d'Etudes Nucléaires de Saclay, Paris.
Dremin A. N., Savrov S. D., Trofimov V. S., and Schvedov K. K. (1970) *Detonation Waves in Condensed Media,* pp. 171. Nauka, Moscow.
Dremin A. N. and Schvedov K. K. (1976) On shock wave explosive decomposition. in *Sixth Symposium (Intl.) on Detonation,* ACR-221, pp. 29-35. ONR, Department of the Navy, Arlington, Va.
Evseev A. M. and Frenkel M. Ya. (1973) Molecular-mechanic simulation of Ar in the region of the liquid-crystal phase transition. *Zh. Fiz. Khimia* **47**, 1333-1334.
Genich A. P., Kasparov G. G., Manelis G. B., and Panov N. V. (1977) On peculiarities of gas heating by strong shock waves. *Chemical Physics of Combustion and Explosion Processes: Detonation,* pp. 39-42. Chernogolovka, U.S.S.R.
Hicks B.L., Yen S. M., and Reilly B. J. (1972) The internal structure of shock waves. *J. Fluid Mech.* **53**, 85-111.
Juck V. J., Rikov V. A., and Schakhov E. M. (1973) Kinetic models and the problem of shock wave structure. *Izv. Acad. Sci. SSSR, Mekhanika Jidkosti i Gasa* **N4**, 135-141.
Karo A. M., Hardy J. R., and Walker F. E. (1978) Theoretical studies of shock-initiated detonations. *Acta Astronautica* **5**, 1041-1050.
Kormer S. B. (1968) Optical study of shock-compressed condensed dielectrics. *Uspekhi Fiz. Nauk* **94**, 641-687.
Lagarkov A. N. and Sergeev V. M. (1978) Molecular dynamics method in statistical physics. *Uspekhi Fiz. Nauk* **125**, 409-448.
Mott-Smith H. M. (1951) The solution of the Boltzmann equation for a shock wave. *Phys. Rev.* **82**, 885-892.
Paskin A. and Dienes G. J. (1972) Molecular dynamic simulations of shock waves in a three-dimensional solid. *J. Appl. Phys.* **43**, 1605-1610.
Paskin A., Gohar A. and Dienes G.J. (1977) Simulations of shock waves in solids. *J. Phys. C* **10**, L563-L566.
Rahman A. (1964) Correlations in the motion of atoms in liquid argon. *Phys. Rev.* **136A**, 405-411.
Schvedov K. K. and Dremin A. N. (1975) Effect of aggregate state and structure of charges on TNT decomposition. *4th All-Union Symposium on Combustion and Explosion,* pp. 440-446. Nauka, Moscow
Tsai D. H. and McDonald K. A. (1973) Second sound in a solid under shock compression. *J. Phys. C* **6**, L171-L175.
Van Thiel M. and Alder B. J. (1966) Shock compression of argon. *J. Chem. Phys.* **44**, 1056-1065.
Verlet L. (1967) Computer "experiments" on classical fluids, I: Thermodynamic properties of Lennard-Jones molecules. *Phys. Rev.* **159**, 98-103.
Vorobjev A. A. and Trofimov V. S. (1978) Study of light scattering in shocked nytromethane. *Fiz. Gorenia i Vzriva* **N3**, 152-154.
Walker F. E. and Wasley R. J. (1974) Initiation patterns produced in explosives by low pressure, long-duration shock waves. *Combustion and Flame* **22**, 53-58.
Zeldovitch Ya. B. and Rayzer Yu. P. (1966) *Physics of Shock Waves and High Temperature Hydrodynamic Phenomena,* Nauka, Moscow.

Single-Shock Curve Buildup and a Hydrodynamic $p_i^N t_f$ Criterion for Initiation of Detonation

M. Cowperthwaite*
SRI International, Menlo Park, Calif.

A similarity solution modeling the one-dimensional reactive flow observed in condensed explosives shocked initially into the 20 kbar region is presented. The solution is used to define an ideal shock initiation process in condensed explosives when the incident shock strength is increased in experiments performed with similar rear-boundary conditions. Properties of this ideal initiation process are derived to provide a theoretical basis for 1) the single-shock trajectory for buildup to detonation postulated by Lindstrom, 2) the linear relationship between the log of the initial pressure p_i and the log of the distance to detonation δ_i formulated by Ramsay and Popolato, and 3) the empirical $p_i^2 t$ condition for detonation postulated by Walker and Wasley. Flows initiated at different pressures are compared to show that a single-shock trajectory and a $p_i^N t_f$ detonation criterion are imposed on the ideal initiation process by the similarity constraint when the hydrodynamics of the buildup process do not change as the incident shock strength is increased. In this case, the $p_i^N t_f$ criterion is a critical condition for initiation that expresses a hydrodynamic scaling law, in terms of the shock parameter N and the time of formation t_f of a reactive rarefaction fan on the rear boundary, rather than a critical energy condition or the condition for a thermal explosion formulated by Hayes.

Introduction

THE similarity solution to the partial-differential equations governing one-dimensional shock-induced reactive flow described in this paper is based on a set of Lagrange pressure histories recorded in initiating PBX9404 shocked initially into the 20 kbar region (Weingart et al., 1978). It is used here to define an ideal shock-initiation process in condensed explosives, and to establish the

---

Presented at the 7th ICOGER, Göttingen, Federal Republic of Germany, Aug. 20-24, 1979. Copyright © American Institute of Aeronautics and Astronautics, Inc., 1981. All rights reserved.
*Senior Chemical Physicist.

significance of some empirical relationships postulated about the shock-initiation process for engineering purposes. The empirical relationships considered are the single-curve buildup to detonation postulated by Lindstrom (1966), the linear relationship between the log of the initial shock pressure p_i and the log of the distance of run to detonation δ_i formulated by Ramsay and Popolato (1965), and the $p_i^2 t$ detonation criterion postulated by Walker and Wasley (1969). Flows initiated with increasing incident shock strength are examined to establish a well-defined set of conditions that the ideal initiation process must satisfy in order to exhibit a single-shock trajectory, and admit a detonation criterion of the form $p_i^N t_i^t$. This criterion is shown to express a hydrodynamic scaling law rather than a critical energy condition (Walker and Wasley, 1969) or a condition for thermal explosion as formulated by Hayes (1976).

Similarity Solution Based on Initiating PBX9404

The similarity solution based on the Lagrange pressure histories recorded in initiating PBX9404 is presented here without derivation because it was constructed in detail in another paper (Cowperthwaite, 1978). Note that the reactive flow is treated as a Zeldovich-von Neuman-Doering (Z-N-D) wave (see Zeldovich, 1940; von Neumann, 1972; and Doering, 1943) and that the explosive and its products are assumed to be polytropic (Cowperthwaite and Rosenberg, 1977) to make the energy equation more tractable.

Let t, τ, v, u, p, and λ denote time, Lagrangian time, specific volume, particle velocity, pressure, and extent of reaction, respectively, and let the subscripts H and i denote, respectively, the quantities immediately behind the shock and the initial shocked condition. Then our similarity solution can be written in terms of the similarity parameter

$$\eta = (1 - t/\alpha)/(1 - \tau/\alpha) \tag{1}$$

as

$$v = v_H V(\eta, m) \tag{2}$$

$$u = u_i (1 - \tau/\alpha)^{-n} U(\eta, m) \tag{3}$$

$$p = p_i (1 - \tau/\alpha)^{-2n} P(\eta, m) \tag{4}$$

$$\lambda = (1 - \tau/\alpha)^{-2n} \Lambda(\eta, m) \tag{5}$$

where α is a characteristic time, n and m are parameters, and τ lies in the range $0 \leq \tau < \alpha$.

The similarity parameter $\eta = 1$ along the shock path where $t = \tau$, and the equation for the shock path in the time-Lagrange distance h plane is

$$(1 - h/\beta) = (1 - \tau/\alpha)^{1-n} \tag{6}$$

where the characteristic distance β is related to the characteristic time α and the initial shock velocity D_i by the expression

$$\beta = \alpha D_i / (1 - n)$$

The functions $V(\eta,m)$, $U(\eta,m)$, $P(\eta,m)$, and $\Lambda(\eta,m)$ are related by the ordinary differential equations obtained by transforming the equations of motion into (τ,η) space using Eqs. (1-5). They satisfy the conditions $V(1,m) = U(1,m) = P(1,m) = 1$, $\Lambda(1,m) = 0$ at the shock front where $\eta = 1$.

The pressure field was chosen to satisfy the equation

$$d^2 P/d\eta^2 = m(\eta - \bar{a}) \tag{7}$$

with $\bar{a}$ a parameter satisfying the condition $0 < \bar{a} < 1$, to account for the fact that the Lagrangian pressure profiles recorded in PBX9404 exhibit a pressure peak and a point of inflection. (The pressure histories are Lagrangian because the pressure gages equilibrate with and follow the shock-induced hydrodynamic flow.) Explicit expressions for P, U, and V were obtained by integrating Eq. (7) to obtain P, and then integrating the momentum and continuity equations to obtain the corresponding expressions for U and V. It is clear from Eqs. (2), (3), and (6) that the parameter n governs the shock, and that $n > 0$, $n < 0$, and $n = 0$, respectively, for accelerating, decelerating, and constant velocity shocks. The parameter m was introduced to allow for compressive flows behind the shock with similar Lagrangian pressure histories but different types of particle velocity and specific volume histories.

The similarity solution provides a theoretical description of a type of shock-initiation process observed in condensed explosives. Transition to detonation occurs in this type of initiation because the lead shock is strengthened by a compressive reactive pulse from the rear. The onset of detonation occurs in the neighborhood of the singular point where $t = \tau \approx \alpha$. Initiating flows with the same shock path and the ensuing compressive wave exhibiting peaks in the Lagrange pressure histories, but with different particle velocity and specific volume histories, are described by solutions with the same value of n but different values of m. With $n = 0.3$ and the polytropic index $k = 2$, for example, the solution with $m = 36$ exhibits maxima in the Lagrange particle velocity histories, the particle velocity/Lagrange distance profiles, and models the flow observed in PBX9404 (Weingart et al., 1978) and cast TNT (Cowperthwaite and Rosenberg, 1976; Kanel and Dremin, 1977). With $m = 18$, however, the solution models the growing step shock with zero pressure gradient discussed by Kennedy (1973) and Hayes (1976). With $m = 12$, the solution models the accelerating shock with positive pressure and particle velocity gradients obtained in numerical studies by Bernier et al. (1963) and Mader and Forest (1976) using Forest Fire.

Ideal Shock-Initiation Process in Condensed Explosives

We consider a series of shock-initiation experiments involving similar rear-boundary conditions to produce increasing values of the initial shock pressure p_i. In this series of experiments we assume that buildup to detonation does not occur when the initial shock pressure is less than a threshold value p_H^T. Similarly, we assume that transition to detonation always occurs, and for practical purposes occurs instantaneously, when the shock pressure is greater than or equal to a critical value p_H^C. Consequently, detonation will be observed if the shock pressure builds up to p_H^C when $p_H^T \leq p_i < p_H^C$, and will always be observed when $p_i \geq p_H^C$. Different explosives will be characterized by different values of p_H^T and p_H^C. The ideal shock-initiation process formulated in this paper is based on the hypothesis that Eqs. (1-7) describe the flows observed in the series of shock-initiation experiments.

In using Eqs. (1-7) to describe the initiation process, we denote shock-initiation experiments performed with the initial shock pressure p_i by a set of values of the parameters (p_i, α, n, m). We assign a characteristic time $\alpha(p_i)$ and a set of order pairs $[n_j(p_i), m_j(p_i)]$ to the class of flows initiated at p_i. A class of flows must be considered at each initial shock pressure to account for the dependence of the initiation process on the rear-boundary conditions. When $p_H^T \leq p_i < p_H^C$, buildup to detonation occurs for accelerating shocks $n_j(p_i) > 0$ but not for decelerating shocks $n_j(p_i) < 0$. For notational convenience when $p_i = p_H^T$, we set

$$\alpha(p_H^T) = \hat{\alpha}, \qquad n_j(p_H^T) = \hat{n}_j, \qquad m_j(p_H^T) = \hat{m}_j$$

and denote time by $\hat{t}$ and $\hat{\tau}$. The equation for the shock pressure in an experiment initiated at p_H^T is then obtained as

$$p_H(\hat{\tau}/\hat{\alpha}) = p_H^T (1 - \hat{\tau}/\hat{\alpha})^{-2\hat{n}_j} \qquad (8)$$

by setting $\eta = 1$ in Eq. (4). When $\hat{n}_j > 0$ (accelerating shocks), transition to detonation will occur when the shock pressure attains the critical pressure after a time $\hat{\tau}_f^C$. The set of these times $\{\hat{\tau}_f^C\}$ defines the times to detonation in the experiments performed at the threshold pressure. We write $\hat{\tau}_f^C = (1 - \hat{\epsilon}_j)\hat{\alpha}$ with $\hat{\epsilon} \approx 0$ to denote that detonation occurs in the neighborhood of the singular point $(\hat{t} = \hat{\alpha}, \hat{\tau} = \hat{\alpha})$. It follows from Eq. (8) that these times to detonation are determined by the values of the parameter $\hat{n}_j$ through the condition $p_H^C = p_H^T(\hat{\epsilon}_j)^{-2\hat{n}_j}$ when p_H^C and p_H^T are known. For the solution based on PBX9404 with $\hat{n}_j = 0.3$, the time to detonation is $\hat{\tau}_f^C = 0.99\hat{\alpha}$ when $p_H^T = 20$ kbar and $p_H^C = 317$ kbar. It is also clear that similar relationships can be written in terms of $\alpha(p_i)$ and $n_j(p_i)$ when the initial pressure is p_i.

We are now in a position to define initiation formally as a collection of maps from (p_i) to (p_i, α, n, m) space. The class of flows in experiments initiated at p_i is represented by a map that associates a set of points $[\alpha(p_i), n_j(p_i), m_j(p_i)]$ in (α, n, m) space with a point in (p_i) space. The shock-initiation process is defined as a set of such maps I_{ij}:

$$[p_H^T \leq p_i < p_H^C, \quad I_{ij}^* \alpha = \alpha(p_i), \quad I_{ij}^* n = n_j(p_i), \quad I_{ij}^* m = m_j(p_i)]$$

which assigns a class of initiating flows to each shock-initiation experiment with initial shock pressure p_i in the range $p_H^T \leq p_i < p_H^C$. Characterization of the initiation process in a given explosive involves finding the functional dependence of the parameters α, n_j, and m_j on the initial shock pressure p_i as p_i is increased from p_H^T to p_H^C in a series of shock-initiation experiments.

For this paper, we define an ideal shock initiation process by the condition that the map from (p_i) to (p_i, α, n, m) space is one-to-one. This type of

SHOCK INITIATION OF DETONATION   273

initiation is ideal in the sense that as the initial shock pressure p_i is increased from p_H^T to p_H^C in our initiation experiments, the same buildup to detonation will be observed in all the experiments performed with the same value of p_i. However, in general, different types of buildup will be observed in experiments performed with different values of p_i. We denote the collection of maps defining this ideal initiation process by I_i:

$$[p_H^T \le p_i < p_H^C, \quad I_i^*\alpha = \alpha(p_i), \quad I_i^*n = n(p_i), \quad I_i^*m = m(p_i)]$$

set $\alpha(p_i) = \alpha_i$, $n(p_i) = n_i$, and $m(p_i) = m_i$ for simplicity, and retain the notation introduced earlier to characterize initiation at the threshold pressure. The solution for buildup to detonation in the experiment with the initial shock pressure p_i is represented by the quadruple $(p_i, \alpha_i, n_i, m_i)$. Buildup to detonation, initiated at p_H^T, is thus characterized by the set $(p_H^T, \hat\alpha, \hat n, \hat m)$ with a time to detonation $\hat\tau^C = (1 - \hat\epsilon)\hat\alpha$, where $\hat\epsilon$ is defined by the condition $p_H^C = p_H^T(\hat\epsilon)^{-2\hat n}$. The value of α_i falls from $\hat\alpha$ to $(1 - \hat\epsilon)\hat\alpha$ as the value of p_i is increased from p_H^T to p_H^C in the shock-initiation experiments. We also denote times in the experiments initiated at p_i by t_i and τ_i, and define the corresponding times to detonation as $\tau_i^C = 1 - \epsilon_i)\alpha_i$ so that the range of τ_i is $0 \le \tau_i \le (1 - \epsilon_i)\alpha_i$ and ϵ_i satisfied the condition $p_H^C = p_H^T(\epsilon_i)^{-2n_i}$. The shock trajectory $\hat t = \hat\tau$ can be considered as the locus of initial states in our series of shock-initiation experiments because the pressure increases along this trajectory from p_H^T to p_H^C. The shock trajectory for buildup to detonation from the threshold pressure, with $\hat\epsilon = 0.01$, is represented by TC in Fig. 1. At T the shock pressure is p_H^T, and at C (where transition to detonation occurs) it is $p_H^C = p_H^T(0.01)^{-2\hat n}$. The point I represents the initial state in experiments performed with the initial pressure $p_i = p_H^T(0.5)^{-2\hat n}$, and the corresponding shock trajectory in the buildup phase is represented by IC until the shock pressure attains the critical value p_H^C.

For convenience in describing the ideal initiation process, we base the time scale on $\hat\alpha$ and change the characteristic time from $\hat\alpha$ to α_i in order to compare flow conditions in an experiment initiated at p_i with those attained in an experiment initiated at p_H^T. The explicit form of the solution for initiation at the threshold pressure $(p_H^T, \hat\alpha, \hat n, \hat m)$ is obtained by setting $p_i = p_H^T$, $\alpha = \hat\alpha$, $n = \hat n$, $m = \hat m$, $\tau = \hat\tau$, $t = \hat t$, $\beta = \hat\beta$, and $\eta = \hat\eta$ in Eqs. (1-6) with

$$\hat\eta = (1 - \hat t/\hat\alpha)/(1 - \hat\tau/\hat\alpha) \tag{9}$$

Use is made of the fact that this solution $(p_H^T, \hat\alpha, \hat n, \hat m)$ is invariant under the transformation $\varphi(\hat t \to t_i, \hat\tau \to \tau_i, \hat h \to h_i, \hat\alpha \to \alpha_i, \hat\beta \to \beta_i)$, defined by the equations

$$\hat t = t_i + \hat\alpha - \alpha_i \tag{10}$$

$$\hat\tau = \tau_i + \hat\alpha - \alpha_i \tag{11}$$

$$\hat h = h_i + \hat\beta - \beta_i \tag{12}$$

A key concept regarding scaling is that the solution $(p_H^T, \hat\alpha, \hat\eta, \hat m)$ is invariant under φ. This is demonstrated as follows. The equations for the shock path are

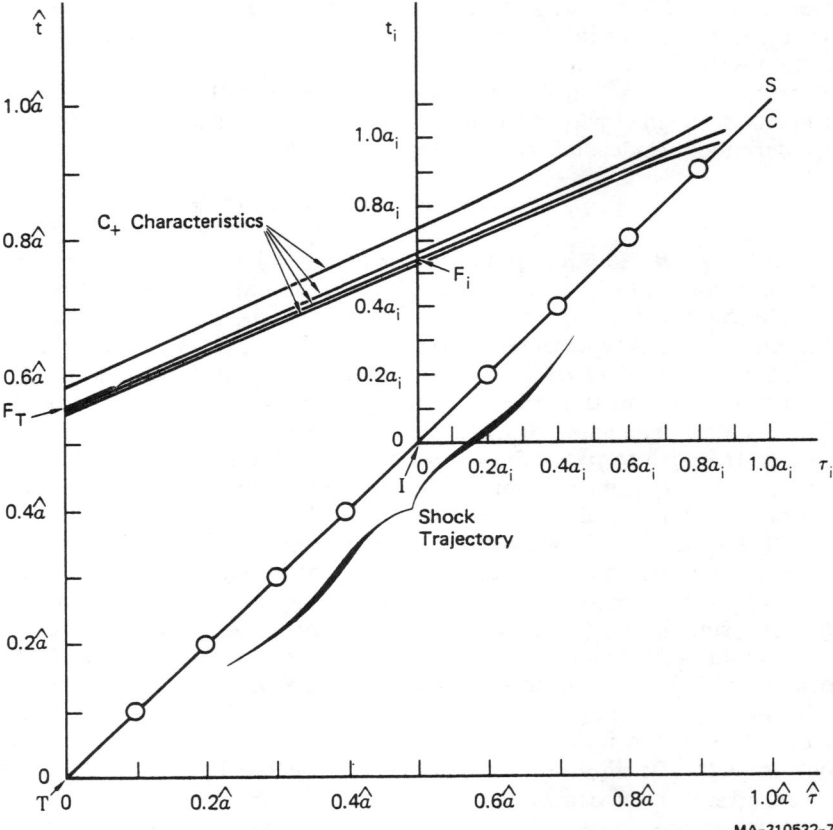

Fig. 1 Plot of shock path and C_+ characteristics in the time-Lagrange time plane for $\hat{n} = 0.3$ and $\hat{m} = 36$.

obviously invariant under φ because $\varphi^*(\hat{t}-\hat{\tau}) = t_i - \tau_i = 0$. In other words, Eqs. (10) and (11) translate the origin of the time coordinates along the shock trajectory $\hat{t} = \hat{\tau}$. For each value of p_i in the range $p_H^T < p_i < p_H^C$, the origin of the (t_i, τ_i) coordinate system is located at the point $(\hat{t} = \hat{\alpha} - \alpha_i,\ \hat{\tau} = \hat{\alpha} - \alpha_i)$ in the $(\hat{t}, \hat{\tau})$ plane because $\hat{t} = \hat{\tau} = \hat{\alpha} - \alpha_i$ when $t_i = \tau_i = 0$. The explicit relationship between α_i and p_i is obtained as

$$p_i = p_H^T (\alpha_i / \hat{\alpha})^{-2\hat{n}} \tag{13}$$

by setting $\hat{\tau} = \hat{\alpha} - \alpha_i$ in Eq. (8). As p_i varies from p_H^T to p_H^C in Fig. 1, for example, the origin of the time axes moves from T to C along TC, and when $p_i = p_H^T (0.5)^{-2\hat{n}}$ the origin of the (t_i, τ_i) coordinate system is located at I. The following relationships among the characteristic distances and times

$$\beta_i / \hat{\beta} = (\alpha_i / \hat{\alpha})(p_i / p_H^T)^{1/2} = (\alpha_i / \hat{\alpha})^{1-\hat{n}} \tag{14}$$

are required to verify that the equation for the shock trajectory in Lagrangian distance-time coordinates is invariant under φ. The equation

$$(1-h_i/\beta_i) = (1-\tau_i/\alpha_i)^{1-\hat{n}} \tag{15}$$

is readily obtained by substituting Eqs. (11) and (12) into the equation

$$(1-\hat{h}/\hat{\beta}) = (1-\hat{\tau}/\hat{\alpha})^{1-\hat{n}} \tag{16}$$

and making use of Eq. (14). We need consider only the equation for the pressure field

$$p = p_H^T(1-\hat{\tau}/\hat{\alpha})^{-2\hat{n}} P(\hat{\eta},\hat{m}) \tag{17}$$

to verify that the solution $(p_H^T, \hat{\alpha}, \hat{n}, \hat{m})$ is invariant under φ. It is convenient to write the shock pressure as

$$p_H(\hat{\tau}/\hat{\alpha}) = p_H^T(1-\hat{\tau}/\hat{\alpha})^{-2\hat{n}} \tag{18}$$

Then Eq. (17) is invariant under φ when $\varphi^* p_H(\hat{\tau}/\hat{\alpha}) = p_H(\tau_i/\alpha_i)$ and $\varphi^* P(\hat{\eta},\hat{m}) = P(\eta_i,\hat{m})$. We first substitute Eqs. (10) and (11) into Eq. (9) to show that the similarity parameter is invariant under φ

$$\varphi^*\hat{\eta} = (1-t_i/\alpha_i)/(1-\tau_i/\alpha_i) = \eta_i \tag{19}$$

It follows that $\varphi^* P(\hat{\eta},\hat{m}) = P(\eta_i,\hat{m})$. We next substitute Eq. (11) into Eq. (18) to obtain

$$\varphi^* p_H(\hat{\tau}/\hat{\alpha}) = p_H^T(\alpha_i/\hat{\alpha})^{-2\hat{n}}(1-\tau_i/\alpha_i)^{-2\hat{n}} \tag{20}$$

and make use of Eq. (13) to obtain the invariance condition for the shock pressure $\varphi^* p_H(\hat{\tau}/\hat{\alpha}) = p_H(\tau_i/\alpha_i)$.

We have shown that the pressure field is invariant under φ, and it is clear from Eqs. (1-5) that the other flow variables are also invariant under φ. Equations for buildup to detonation from the pressure p_i in the flow initiated at p_H^T are thus readily obtained by setting $p_H^T \to p_i$, $\hat{\tau} \to \tau_i$, $\hat{t} \to t_i$, $\hat{h} \to h_i$, $\hat{\alpha} \to \alpha_i$, and $\hat{\beta} \to \beta_i$ in the equations for buildup from p_H^T with

$$\alpha_i = \hat{\alpha}(p_i/p_H^T)^{-\frac{1}{2}\hat{n}} \tag{21}$$

from Eq. (13) and

$$\beta_i = \hat{\beta}(p_i/p_H^T)^{(\hat{n}-1)/2\hat{n}} \tag{22}$$

from Eq. (14). We will denote the solution for buildup to detonation from p_i in the flow initiated at p_H^T by $(p_i, \alpha_i, \hat{n}, \hat{m})$ and set

$$\varphi^*(p_H^T, \hat{\alpha}, \hat{n}, \hat{m}) = (p_i, \alpha_i, \hat{n}, \hat{m}) \tag{23}$$

to show that this solution is obtained from the solution for buildup to detonation from p_H^T by changing the time scale.

We can now compare the flow $(p_i, \alpha_i, \hat{n}, \hat{m})$ in the experiment with initial pressure p_H^T to the flow $(p_i, \alpha_i, n_i, m_i)$ in the experiment with initial pressure p_i and establish a set of well-defined conditions for single-curve buildup and the onset of detonation.

Single-Curve Buildup

The condition our ideal shock-initiation process must satisfy to exhibit the single-curve buildup postulated by Lindstrom (1966) is obtained by comparing Eq. (15) for the shock trajectory in the flow $(p_i, \alpha_i, \hat{n}, \hat{m})$ initiated at p_H^T with the equation

$$(1 - h_i/\beta_i) = (1 - \tau_i/\alpha_i)^{1-n_i} \tag{24}$$

for the shock trajectory in the flow $(p_i, \alpha_i, n_i, m_i)$ initiated at p_i. It is clear from Eqs. (15) and (24) that the set of shock trajectories observed in our series of shock-initiation experiments will fall into a single curve when $n_i = \hat{n}$. We accordingly denote the collection of maps defining our ideal initiation process with single-curve buildup by

$$I_i': [p_H^T \leq p_i < p_H^C, \quad I_i'^* \alpha = \hat{\alpha}(p_i/p_H^T)^{-\frac{1}{2}\hat{n}}, \quad I_i'^* n = \hat{n}, \quad I_i'^* m = m_i]$$

although it is not realistic to assume that the condition $n_i = \hat{n}$ will be satisfied exactly. However, we will use the condition $n_i = \hat{n}$ to derive the relationship between the initial shock pressure and the distance of run to detonation δ_i, because the relationship formulated by Ramsay and Popolato (1965) is obviously associated with single-curve buildup.

Log δ_i-Log p_i Relationship

When $n_i = \hat{n}$, the combination of Eqs. (20) and (24) gives the following relationship between shock pressure and distance

$$p_H(h_i/\beta_i) = p_i(1 - h_i/\beta_i)^{-2\hat{n}/(1-\hat{n})} \tag{25}$$

for our ideal initiation process. We set $(1-\hat{n})/2\hat{n} = \hat{N}$ for convenience. The equation for distance to detonation δ_i in a flow initiated at p_i is obtained as

$$\delta_i = \beta_i \left[1 - (p_i/p_H^C)^{\hat{N}} \right] \tag{26}$$

by inverting Eq. (25) and setting $p_H = p_H^C$. The relationship between δ_i and p_i is obtained as

$$\delta_i = \hat{\beta} \left[(p_H^T/p_i)^{\hat{N}} - (p_H^T/p_H^C)^{\hat{N}} \right] \tag{27}$$

by combining Eqs. (22) and (26). When $p_i = p_H^T$ the distance to detonation $\hat{\delta} = \hat{\beta}[1 - (p_H^T/p_H^C)^{\hat{N}}]$ and when $p_i = p_H^C$ it is zero. Taking logarithms of both sides of Eq. (27) and differentiating with respect to p_i gives the following equation for the slope of the log δ_i vs log p_i curve

$$\mathrm{d}\log\delta_i/\mathrm{d}\log p_i = -\hat{N}/[1-(p_i/p_H^C)^{\hat{N}}] \qquad (28)$$

and it follows that the plot of log δ_i vs log p_i will be approximately a straight line with a slope equal to $-\hat{N} = -(1-\hat{n})/2\hat{n}$ when $p_i \ll p_H^C$. A nondimensional plot of log $(\delta_i/\hat{\beta})$ vs $\log(p_i/p_H^T)$ is shown in Fig. 2 for the solution modeling PBX9404 with $n = 0.3$.

We can now establish the conditions our ideal shock-initiation process must satisfy to exhibit a $p_i^N t_i^l$ detonation criterion.

Hydrodynamic $p_i^N t_i^l$ Detonation Criterion

We use the condition that detonation occurs when the shock pressure attains the critical pressure p_H^C to formulate a hydrodynamic detonation criterion for our ideal shock-initiation process. Note that this criterion does not apply to shocks that first decay and then build up to detonation; only shocks that continuously accelerate to detonation are considered. In this case, the criterion for detonation in an experiment performed at p_i is that the rear boundary supports the flow until the shock attains the critical pressure. The rear-boundary condition must therefore be maintained long enough to generate the forward-facing C_+ characteristic that intersects the shock trajectory at the point where $p_H = p_H^C$. We denote this critical characteristic by C_{+i}^C and denote the time it emanates from the rear boundary by T_i^C. When $p_i = p_H^T$ we set $C_{+i}^C = \hat{C}_+^C$ and $T_i^C = \hat{T}^C$. The time T_i^C is the critical time for initiation of detonation because detonation will always occur in the experiment performed at p_i when the rear-boundary condition is maintained for a time greater than or equal to T_i^C. T_i^C can be determined in principle by integrating C_{+i}^C backward from the shock trajectory to the rear boundary, but we use a different procedure to estimate the critical time here. We make use of the fact that for practical purposes C_{+i}^C and the first C_+ characteristic that does not intersect the shock trajectory emanate from the rear boundary at the same time, $t_i^l \approx T_i^C$. In other words, these characteristics lie in a reactive rarefaction fan centered at the point $(\tau_i/\alpha_i = 0, t_i^l/\alpha_i)$ on the rear boundary. We formulate the detonation criterion in terms of t_i^l because t_i^l is defined simply by the condition that the second derivative of the C_+ characteristic emanating from the rear boundary is zero.

The differential equation for the C_+ characteristics in the flow (p_H^T, $\hat{\alpha}$, $\hat{n}$, $\hat{m}$) can be written as

$$\frac{\mathrm{d}\hat{t}}{\mathrm{d}\hat{\tau}} = \left(\frac{(k-1)}{2k}\frac{V(\hat{\eta},\hat{m})}{P(\hat{\eta},\hat{m})}\right)^{1/2} = C(\hat{\eta},\hat{m}) \qquad (29)$$

The equation for the second derivative

$$\frac{\mathrm{d}^2\hat{t}}{\mathrm{d}\hat{\tau}^2} = -\frac{1}{(\hat{\alpha}-\hat{\tau})}\frac{\mathrm{d}C}{\mathrm{d}\hat{\eta}}\left(\frac{\mathrm{d}\hat{t}}{\mathrm{d}\hat{\tau}} - \hat{\eta}\right) \qquad (30)$$

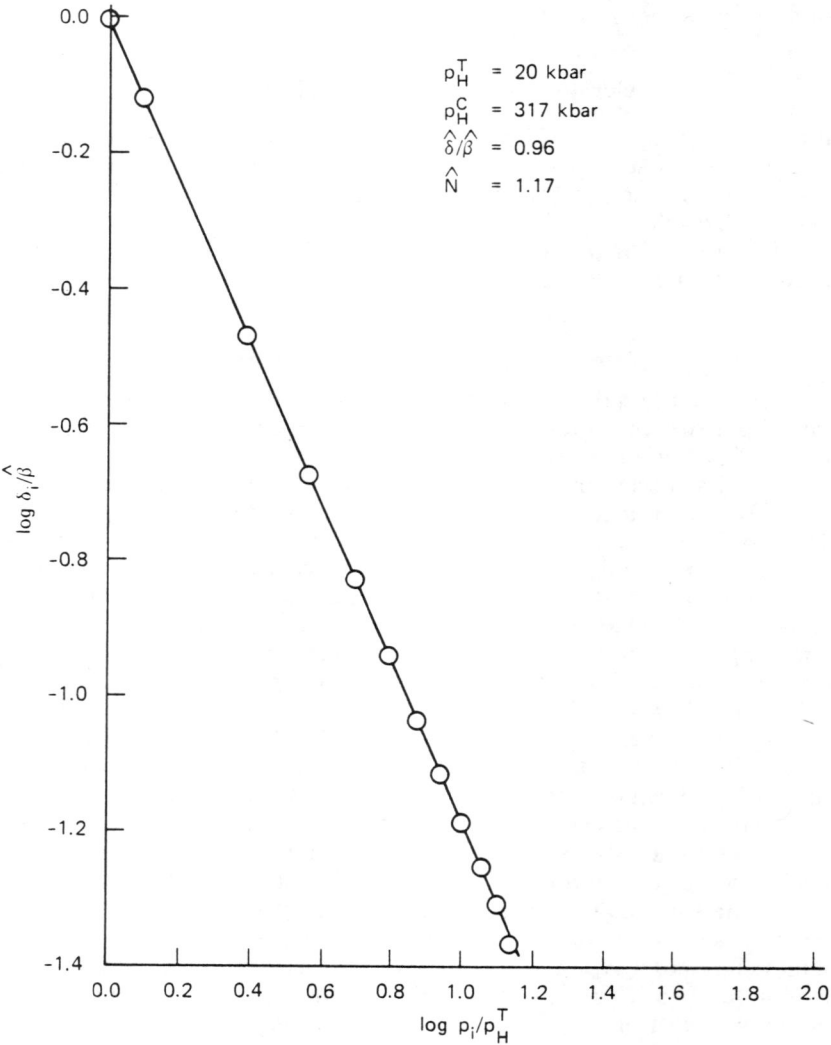

Fig. 2 Nondimensional plot of log distance to detonation vs log initial pressure for $\hat{n} = 0.3$.

obtained by differentiating Eq. (29) shows that $d^2\hat{t}/d\hat{\tau}^2 = 0$ when $\hat{\eta} = \hat{\eta}^f$ satisfies the equation $C(\hat{\eta}^f, \hat{m}) = \hat{\eta}^f$. The time $\hat{t}^f$ is defined by the equation $1 - \hat{t}^f/\hat{\alpha} = \hat{\eta}^f = C(\hat{\eta}^f, \hat{m})$. The center of the rarefaction fan when $\hat{\eta}^f = C(\hat{\eta}^f, \hat{m})$ for the flow modeling the initiation of PBX9404 with $\hat{n} = 0.3$, $\hat{m} = 36$, and $k = 2$ is shown in Fig. 1 as F_T. The C_+ characteristics emanating from $\hat{\tau}/\hat{\alpha} = 0$ in the neighborhood of F_T were constructed by integrating Eq. (29) backward. It follows from our previous discussion that the times t_i^f in experiments performed at p_i are defined by the equation $1 - t_i^f/\alpha_i = \eta_i^f = C(\eta_i^f, m_i)$. A detonation criterion for our ideal initiation process is an expression that relates values of t_i^f in experiments performed at different initial pressures.

We restrict our consideration to the class of flow ($p_H^T \le p_i < p_H^C$, α_i, $\hat{n}$, m_i) that exhibit a single-shock trajectory. Clearly, the condition $\eta_i^f = C(\eta_i^f, m_i)$ defining t_i^f will, in general, be satisfied by different values of η_i, because m_i depends on the initial pressure. For conceptual purposes we eliminate this complication by further restricting our discussion to initiation satisfying the condition $m_i = \hat{m}$. The condition $\eta_i^f = \hat{\eta}^f$ can then be used to formulate a hydrodynamic detonation criterion for this well-defined class of flows. When $n_i = \hat{n}$ and $m_i = \hat{m}$, we can relate the particle velocity fields in the experiments performed at p_H^T and p_i by the equation

$$\varphi^* u(\hat{\tau}/\hat{\alpha}, \hat{\eta}) = u(\tau_i/\alpha_i, \eta_i) \tag{31}$$

The particle velocity along $\tau_i = 0$ in the experiment performed at p_i is the same as the particle velocity along $\hat{\tau} = \hat{\alpha} - \alpha_i$ in the experiment performed at p_H^T; the buildup to detonation in the experiment initiated at p_i is the same as the buildup to detonation in the experiment initiated at p_H^T after the shock attains the pressure p_i. The flow initiated at the threshold pressure thus defines all the flows initiated in our series of shock-initiation experiments as the initial shock pressure is increased from p_H^T to p_H^C, and, moreover, all these flows exhibit similar hydrodynamic features.

Our detonation criterion when $n_i = \hat{n}$ and $m_i = \hat{m}$ expresses the hydrodynamic condition that $d^2 t_i/d\tau_i^2 = 0$ along the line $\hat{\eta} = \hat{\eta}^f$. Accordingly, we set $\eta_i^f = \hat{\eta}^f$ and find that the times the rear-boundary conditions must be maintained in the experiments initiated at p_i and p_H^T are related by the equation,

$$t_i^f/\alpha_i = \hat{t}^f/\hat{\alpha} \tag{32}$$

For the flow with $\hat{n} = 0.3$ and $\hat{m} = 36$ shown in Fig. 1, for example, $d^2 t_i/d\tau_i^2 = 0$ along the line $F_T F_i S$ and Eq. (32) expresses the condition that the triangles TSF_T and ISF_i are similar. Eliminating the characteristic times from Eq. (32) with Eq. (21) gives the hydrodynamic detonation criterion in terms of the initial shock pressure as

$$p_i^{\frac{1}{2}\hat{n}} t_i^f = (p_H^T)^{\frac{1}{2}\hat{n}} \hat{t}^f \tag{33}$$

and we can write this as a $p_i^N t_i^f$ criterion by setting $N = \frac{1}{2}\hat{n}$. Equation (33) reduces to the $p_i^2 t$ criterion formulated by Walker and Wasley (1969) when $2n = 0.5$, but it clearly expresses a hydrodynamic scaling condition rather than a critical energy condition. It is important to realize that all other times associated with significant features of the flow are also related by an equation similar to Eq. (33) when $n_i = \hat{n}$ and $m_i = \hat{m}$, because each significant flow feature propagates along a line of constant η in the flow initiated at p_H^T. The equation for the reaction times t_i^R, for example, is obtained simply by replacing the f superscripts in Eq. (33) with R superscripts.

Although $t_i^f \approx T_i^C$ when $n_i = \hat{n}$ and $m_i = \hat{m}$, it is more correct to write the detonation criterion in terms of T_i^C as

$$p_i^{\frac{1}{2}\hat{n}} T_i^C < (p_H^T)^{\frac{1}{2}\hat{n}} \hat{t}^f$$

because the critical characteristic $\hat{C}_+^C$ lies slightly below the line $\hat{\eta} = \hat{\eta}^J$. When $m_i = m(p_i)$ and $dm_i/dp_i < 0$, the hydrodynamic mechanism of initiation changes as the initial pressure is increased in our series of initiation experiments. Initiation of detonation will occur if the rear boundary is maintained for a time defined by Eq. (33) when the initial pressure is p_i, but this time is not the critical time for detonation. The critical time for initiation is less than that given by Eq. (33) because $dt_i^f/dp_i < 0$ when $dm_i/dp_i < 0$ and $n_i = \hat{n}$.

Results and Conclusions

The similarity solution based on a series of Lagrange pressure histories recorded in initiating PBX9404 was used to define an ideal shock-initiation process in condensed explosives. In this process, initiation of detonation does not occur when the initial shock pressure p_i is less than a threshold pressure p_H^T, but always occurs when the shock pressure builds up to a critical pressure p_H^C. Time scales were chosen so that the flow in an experiment initiated at p_i is governed by two parameters n_i and m_i, which in general depend on p_i. At p_H^T the values of n_i and m_i were set equal to $\hat{n}$ and $\hat{m}$ for notational convenience. A well-defined set of conditions was established for the ideal shock-initiation process to exhibit single-curve buildup, a linear log (initial pressure) vs log (distance of run to detonation relationship), and a $p_i^N t_i^f$ detonation criterion.

The parameter n_i governs the shock trajectory, and the ratio n_i/m_i governs the hydrodynamic flow behind the shock in the buildup to detonation. The ideal initiation process exhibits a single-shock trajectory when $n_i = \hat{n}$ and $dn_i/dp_i = 0$, and admits a $p_i^N t_i^f$ detonation criterion when $n_i = \hat{n}$, $m_i = \hat{m}$, and $dm_i/dp_i = 0$, with $N = \frac{1}{2}\hat{n}$. A single-shock buildup curve and a linear relationship between $\log(p_i)$ vs $\log(\delta_i)$ are good approximations for the ideal initiation process only when n_i is a weak function of p_i. The $p_i^N t_i^f$ relationship expresses a hydrodynamic scaling law rather than a critical energy condition or a thermal explosion condition. The conditions $n_i = \hat{n}$ and $m_i = \hat{m}$ are satisfied when the hydrodynamics of the buildup process do not change as the initial pressure is increased in a series of initiation experiments performed with similar rear-boundary conditions. In this case, flows initiated at different initial pressures are restricted by the similarity constraint, and the $p_i^N t_i^f$ detonation criterion is also a critical condition for the initiation of detonation. When $dm_i/dp_i < 0$, the $p_i^N t_i^f$ criterion is not a critical condition for initiation because $dt_i^f/dp_i < 0$, and detonation will occur in experiments initiated at p_i when the rear-boundary conditions are maintained for times less than t_i^f. The $p_i^N t_i^f$ detonation criterion can thus be used only as a critical condition for initiation when n_i and m_i are weak functions of the initial shock pressure, and when the hydrodynamic flowfields in the buildup to detonation do not change significantly as the initial shock pressure is increased in a series of initiation experiments performed with similar rear-boundary conditions.

Acknowledgments

The author thanks J. T. Rosenberg for reading parts of the manuscript, and B. Y. Lew for programming routines to make plots of the similarity solutions and to calculate the characteristics. This work was sponsored by SRI International.

References

Bernier H., Vidard A., and Prouteau F. (1963) Initiation of detonation by impact. *Proceedings of the Intl. Conference on Sensitivity and Hazards of Explosives.* Explosives Research and Development Establishment, London.

Cowperthwaite M. (1978) Model solution for the shock initiation of condensed explosives. Paper presented at Symposium H. D. P., Behavior of Dense Media under High Dynamic Pressures. Paris.

Cowperthwaite M. and Rosenberg J. T. (1976) A multiple Lagrange gage study of the shock initiation process in cast, TNT. *Proceedings of Sixth Symposium (Intl.) on Detonation,* p. 786. ONR ACR-221.

Cowperthwaite M. and Rosenberg J. T. (1977) Characterization of initiation and detonation by Lagrange gage technology. SRI Final Report, ERDA Contract EY-76-C-03-115, Project Agreement 115, July.

Doering W. (1943) On detonation processes in gases. *Ann. Physik* **43**, 421.

Hayes D. B. (1976) A $P^n t$ detonation criterion from thermal explosives theory. *Proceedings of Sixth Symposium (Intl.) on Detonation,* p. 76. ONR ACR-221.

Kanel G. I. and Dremin A. N. (1977) Decomposition of cast trotyl in shock waves. *Combustion, Explosion, and Shock Waves* **13**(1), 71.

Kennedy J. E. (1973) Pressure field in a shock-compressed high explosive. *Fourteenth Symposium (Intl.) on Combustion,* p. 1251. The Combustion Institute, Pittsburg, Pa.

Lindstrom I. E. (1966) Plane shock initiation of an RDX plastic-bonded explosive. *J. Appl. Phys.* **37**, 4873.

Mader C. L. and Forest C. A. (1976) Two dimensional homogeneous and heterogeneous detonation wave propagation. Los Alamos Scientific Laboratory Rept. LA-6259, June.

Ramsay J. B. and Popolato A. (1965) Analysis of shock wave and initiation data for solid explosives. *Proceedings of Fourth Symposium (Intl.) on Detonation,* p. 233. ONR ACR-126.

von Neumann J. (1942) *Theory of detonation waves. OSRD* 549.

Walker F. E. and Wasley R. J. (1969) Critical energy for the shock initiation of heterogeneous explosvies *Explosivestoffe* **17** (1), 9.

Weingart, R. et al., (1978) Manganin stress gages in reacting high explosive environment. Paper presented at Symposium H. D. P., Behaviour of Dense Media under High Dynamic Pressures. Paris.

Zeldovich, Ia. B. (1940) On the theory of the propagation of detonation in gaseous systems. *Sov. Phys. JETP* **10**, 542.

Induction Delay and Detonation Failure Diameter of Nitromethane Mixtures

H. N. Presles* and C. Brochet†
Université de Poitiers, Poitiers, France

The aim of this work is to determine the influence of additives (acetone, chloroform, bromoform) on critical conditions for the propagation of detonation waves in unconfined homogeneous liquid mixtures with nitromethane. Particular attention is focused on the induction delay value to be used in the calculation of the detonation failure diameter for these mixtures. To measure the induction delay, we developed a method based on an electrical effect and we have shown that it is applicable to very short induction times. Experimental results show that the detonation failure diameter of the nitromethane mixtures increases with additive concentration and depends on the nature of the additive, of which, in accordance with previous results obtained for mixtures with the same additives, acetone has the strongest influence.

Introduction

A PRACTICAL method to vary the detonation characteristics of an explosive in a large field is to decrease its specific energy by dilution with an inert additive. Change in detonation characteristics must lead to variation of its critical conditions for propagation which we tried to determine for mixtures of nitromethane (NM) diluted with chloroform, bromoform or acetone. These critical conditions correspond to the smallest explosive charge diameter for which a steady detonation exists, and, for this reason, are collectively known by the term critical diameter. Trying to calculate this critical diameter, Khariton (1947) proposed that a detonation ceases to propagate when inward lateral rarefaction waves reach the charge's cylindrical axis before completion of those chemical reactions that sustain the detonation. Other proposed models were based on the same idea (Jones, 1947; Eyring et al., 1949; Wood and Kirkwood, 1954; Evans, 1962), but each depends on unknown data (such as temperature and thickness of a detonation induction zone) that greatly limits

---

Presented at the 7th ICOGER, Göttingen, Federal Republic of Germany, Aug. 20-24, 1979. Copyright © American Institute of Aeronautics and Astronautics, Inc., 1980. All rights reserved.
*Chargé de Recherche, Laboratoire d'Energetique et de Détonique.
†Maitre de Recherche, Laboratoire d'Energetique et de Détonique.

their application. Moreover, the critical diameter of a liquid explosive depends on the nature of the confinement.

The model of Dremin and Trofimov (1964) is also based on the Khariton concept and is independent of the confinement, since it was established to calculate the critical diameter (detonation failure diameter) of an unconfined liquid explosive. Moreover, this model relies on data which in principle are known and for some explosives [NM: Enig and Petrone (1970), Tarver et al. (1976); liquid TNT: Dremin and Trofimov (1964); dinitroethane: Tarver et al. (1976)] produces good agreement between calculated and measured values. For all these reasons, we undertook our study within the framework of Dremin and Trofimov model.

Detonation Failure Diameter of Mixtures

The phenomenological model of Dremin and Trofimov has been derived in detail by Enig and Petrone (1970), whose notation is used in the following. When a detonation wave encounters a confinement discontinuity (Fig. 1), the resulting rarefaction wave may cause a lateral failure of the detonation by quenching of the reaction. In the analysis, a quenching wave front propagation in the reaction zone with a constant speed v perpendicular to the direction of detonation propagation is assumed.

In a coordinate system at rest with point A (the failure point) the unreacted explosive in the initial state 0 moves with a speed $D > D_0$ with D_0 the normal detonation velocity. The states (e.g., temperature) of zones 0, 1, 2, and 3 are

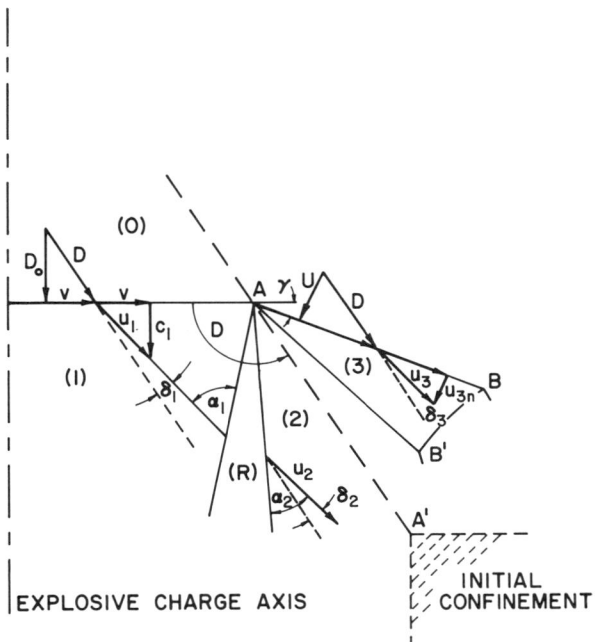

Fig. 1 Dremin and Trofimov model.

denoted by the same number as the indices. The detonation products (zone 1) after expansion in zone R move with velocity u_2 in zone 2 and support the propagation of an oblique shock wave AB. The unreacted explosive (0) which crosses this shock wave moves in zone 3 with a velocity u_3 parallel to u_2 at the pressure P_3. If the compressed explosive remains in zone 3 (its residence time is limited by a rear expansion wave BB') for a time equal to its induction delay

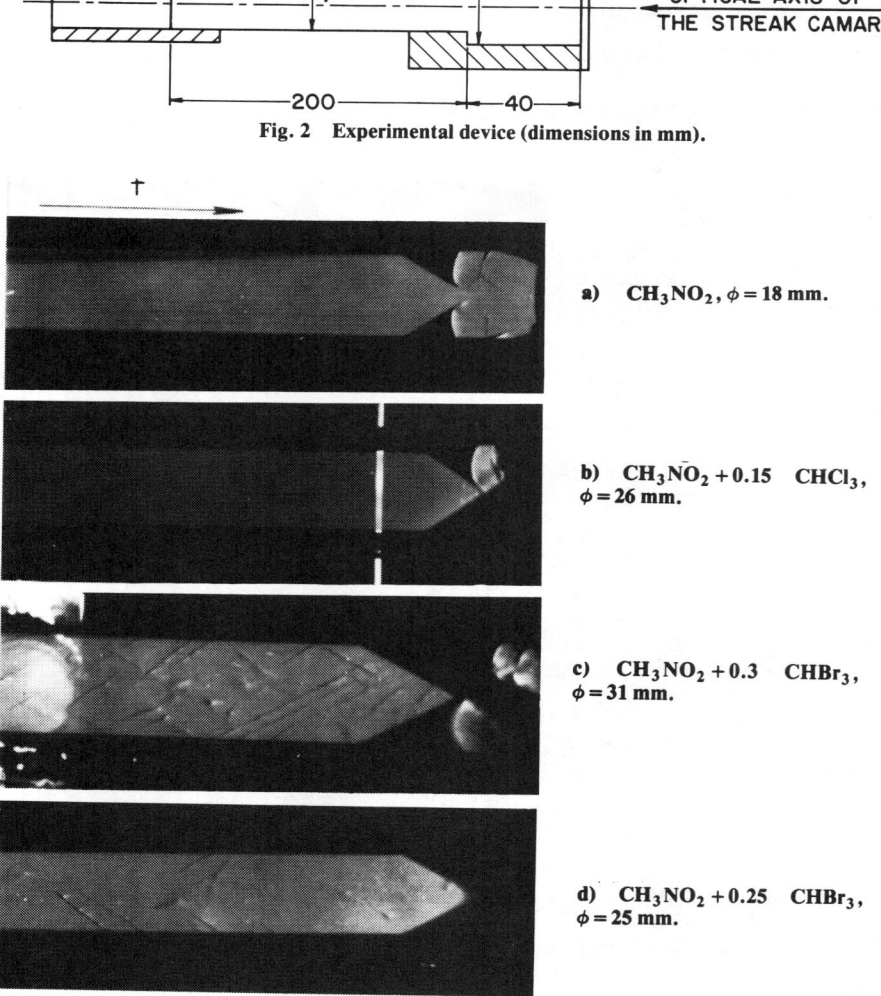

Fig. 2 Experimental device (dimensions in mm).

a) CH_3NO_2, $\phi = 18$ mm.

b) $CH_3NO_2 + 0.15$ $CHCl_3$, $\phi = 26$ mm.

c) $CH_3NO_2 + 0.3$ $CHBr_3$, $\phi = 31$ mm.

d) $CH_3NO_2 + 0.25$ $CHBr_3$, $\phi = 25$ mm.

Fig. 3 Streak camera records of detonation failure phenomena.

τ_3, a detonation is initiated and travels with velocity D_3. If this detonation is initiated at B and reaches A at the instant A arrives at the charge axis, detonation failure occurs. For these conditions, the detonation failure diameter ϕ_f is given by the relation

$$\phi_f = 2u_3 v \tau_3 \left[\frac{1}{D_3 - u_3} + \frac{1}{u_3 - c_3} \right] \qquad (1)$$

where c_3 is the sound speed of the compressed explosive in zone 3.

The apparatus used to determine the detonation failure diameter of mixtures is composed of two tubes (Fig. 2). The detonation propagates first in the tube of diameter ϕ and then emerges into the tube of larger diameter. Four typical streak camera records of the detonation propagation in such a confinement are given on Fig. 3:

1. The detonation was initiated in zone 3, reached the initial detonation prior to its lateral failure and the resulting detonation then propagated in the entire explosive charge.

2. The detonation initiated behind the oblique shock wave reached the charge axis at the same time as the initial detonation disappeared, and failure occurred. This case corresponds exactly to the detonation failure diameter definition given by Dremin and Trofimov and the failure diameter ϕ_f is defined by the tube diameter ϕ.

3. The diameter was smaller than the detonation failure diameter of the explosive. Thus, detonation with velocity D_3 reached the charge axis some time after the complete failure of the initial detonation.

4. The oblique shock did not initiate a detonation, because the diameter ϕ is very much smaller than the failure diameter of the explosive. For a diameter ϕ near to or smaller than the failure diameter, the response of a detonation in compressed explosives to a symmetrical confinement discontinuity was asymmetric. Detonation failure diameters were determined to within 5 or 6% for NM-chloroform and NM-bromoform mixtures at an initial temperature $T_0 = 288 \pm 2$ K. (The failure diameter of NM-acetone mixtures detonation was determined by Dremin et al., 1963.) Typical experimental results for the effects of the additive concentration‡ are given on Fig. 4. They show that the detonation failure diameter of NM-bromoform mixtures is, for a constant concentration, 15-20% higher than those of NM-chloroform mixtures and that the failure diameter increases slowly in comparison with that of NM-acetone mixtures. The detonation failure diameter of NM (15 mm) is slightly smaller than the values given by Dremin et al. (1963) (18 mm) and by Tarver et al. (1976) (16.8 mm).

To calculate the detonation failure diameter of an explosive, the flow (pressure, velocity, direction) in zone 3 must be determined so that a_3, τ_3, and D_3 may be calculated. The determination of the data necessary for this purpose has been the subject of previous works (Brochet et al., 1974; Presles and

---

‡The mixtures composition is defined by the parameter x such that x mole of additive is associated with 1 mole of NM: $0 \leq x \leq 0.4$ if the additive is chloroform or bromoform and $0 \leq x \leq 0.2$ if it is acetone.

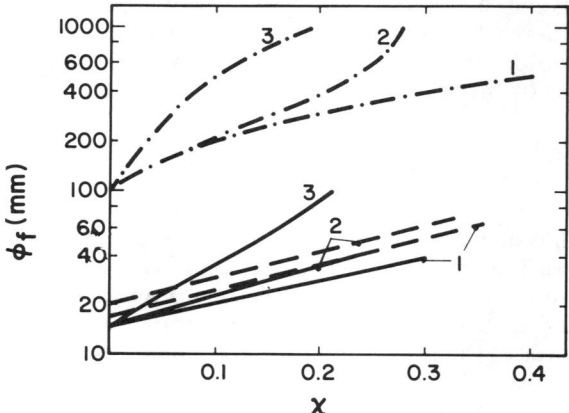

Fig. 4 Failure diameter of mixtures: 1) $CH_3NO_2 + \chi CHCl_3$, 2) $CH_3NO_2 + \chi CHBr_3$, and 3) $CH_3NO_2 + \chi C_3H_6O$; experimental ———; calculated from Eq. (1) with induction delay determined: by dielectric effects — · — · — and from Eq. (3) — — —.

Brochet, 1976; Kusakabe and Fujiwara, 1976; Presles, 1978, 1979). Measurements of the transverse velocity v of the quenching rarefaction wave, already published by Presles and Brochet (1976), were extended to mixtures where $0 \leq x \leq 0.4$. The v values (given in the Table 1) were deduced from the curve $v(x)$ representative of experimental results for each mixture and were used in the calculations.

The flow characteristics around the point A also are given in the Table 1. For constant concentration the pressure P_3 is practically the same for NM-chloroform and NM-bromoform mixtures but is smaller by about 10% for NM-acetone mixtures. The precision on P_3 is estimated to be about 8% from the precision of the different data. The sound speed a_3 in the compressed mixtures was calculated (Presles, 1979) with the method of Enig and Petrone (1965). Two methods were used to evaluate τ_3: one is based on direct measurements of induction delay, and the other on calculation of the delay from the streak camera records performed to determine the detonation failure diameter of the mixtures.

Induction Delay of NM and Mixtures

Among the different methods to determine the detonation induction delay of a liquid explosive, we used the technique proposed by Travis (1965) who showed that electrical signals are generated by initiation processes occurring in dielectric liquid explosives.

In our experiments NM or mixtures were contained in a parallel plane capacitor configuration (Fig. 5). Plate A is made of Plexiglass covered with a thin aluminum film deposited by vacuum evaporation. To minimize the effect of shock-wave curvature experimental requirements were made on a central volume defined by the aluminum plate B (8 mm diam). Plates A and B

Table 1 Flow characteristics for a failure wave

Mixtures	x	v, mm/μs	ν, deg	$P_2 = P_3$, GPa	$\delta_2 = \delta_3$, deg	α_1, deg	α_2, deg	u_2, mm/μs	c_2, mm/μs	ρ_2, g/cm^3	U, mm/μs	u_3, mm/μs	γ, deg
CH_3NO_2 + $xCHCl_3$	0	3.8	121	8.35	11.5	50.5	40	6.39	4.11	1.36	4.36	6.5	22.4
	0.02	3.79	121	8.4	12.1	50.6	40	6.39	4.11	1.36	4.35	6.5	22.6
	0.05	3.76	121	7.9	11.8	51	39.8	6.38	4.08	1.37	4.21	6.49	23.6
	0.1	3.72	121	7.9	11.8	50.6	39.8	6.27	4.01	1.40	4.17	6.37	23.2
	0.15	3.67	121	7.57	11.7	50.8	39.7	6.19	3.96	1.41	4.05	6.3	23.6
	0.2	3.6	121	7.35	11.6	51.1	39.6	6.13	3.91	1.43	3.97	6.23	24.1
	0.3	3.43	120	7.2	11.6	51.5	40	5.89	3.79	1.47	3.88	5.99	24
	0.4	3.25	120	7.15	11.9	52	40.3	5.68	3.68	1.5	3.83	5.77	24
CH_3NO_2 + $xCHBr_3$	0.02	3.76	121.3	8.2	11.7	50.6	40	6.39	4.06	1.41	4.24	6.43	22.8
	0.05	3.69	121.7	8.35	12	49.9	40	6.1	3.92	1.51	4.16	6.20	21.9
	0.1	3.58	122	7.55	11.6	49.9	39.4	5.91	3.76	1.62	3.84	6	22.8
	0.15	3.47	122.4	7.65	11.3	49.7	39.5	5.7	3.6	1.74	3.73	5.79	22.4
	0.2	3.35	122.7	7.3	11.3	49.5	39.5	5.47	3.48	1.84	3.57	5.55	22.1
	0.3	3.13	122.3	6.05	11	51	39	5.28	3.32	1.95	3.19	5.34	24.8
	0.4	2.91	121.8	5.7	10.9	51.6	39.3	4.97	3.15	2.07	3.03	5.03	24.9
CH_3NO_2 + xC_3H_6O	0.1	3.7	122	7.1	12	49.6	39.6	6.08	3.88	1.31	4.09	6.18	22
	0.2	3.6	122.5	6.5	11.5	49.2	40.1	5.82	3.76	1.32	4.06	5.92	20.2

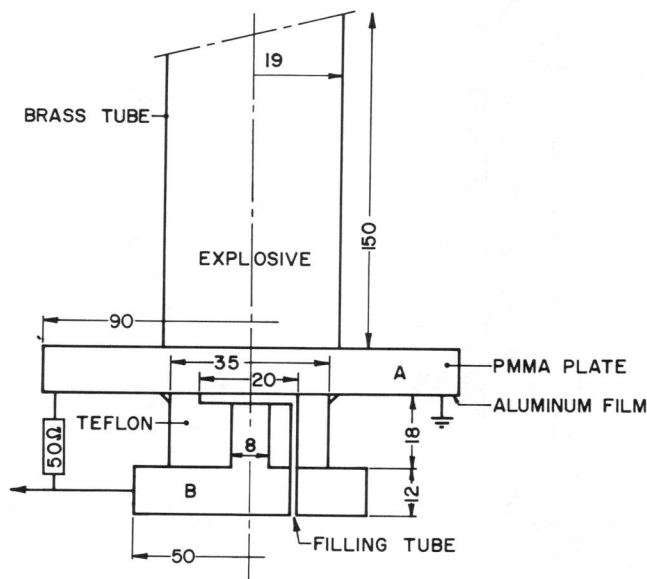

Fig. 5 Apparatus for detonation induction delay measurements (dimensions in mm).

(separation distance X_0 of 2.5 mm) are connected through a 50 Ω resistor to an oscilloscope (Tektronix 7844) of about 3 ns rise time. The absence of a guard ring does not cause any error in the induction delay measurements. A guard ring is absolutely necessary to analyze electrical polarization effects (de Icaza Herrera et al., 1978). The shock-wave generator composed of Astrolite or Pentrite explosive, in contact with the plate A, was calibrated previously (Presles, 1979). The experiments were performed at an initial temperature of $T_0 = 288 \pm 2$ K. A typical oscillographic record of electrical effects resulting from the initiation process in NM (Fig. 6) is composed of four characteristic impulses.

Impulse 1 resulted from NM polarization during the shock-wave propagation.

Impulse 2, which occurred at time τ, indicated an increase of electrical charges on the condenser plates. This impulse was caused by a supplementary polarization (with respect to the polarization induced by the shock wave) which appeared with the initiation of a detonation wave in the compressed NM. Because of the high electrical conductivity in the reaction zone (Hayes, 1967), the detonation wave played the role of capacitor plate A.

Impulse 3 at time t*, the only negative signal resulted from a rapid decrease of electrical charges on the capacitor plates which occurred when the detonation wave overtook the preceding shock wave.

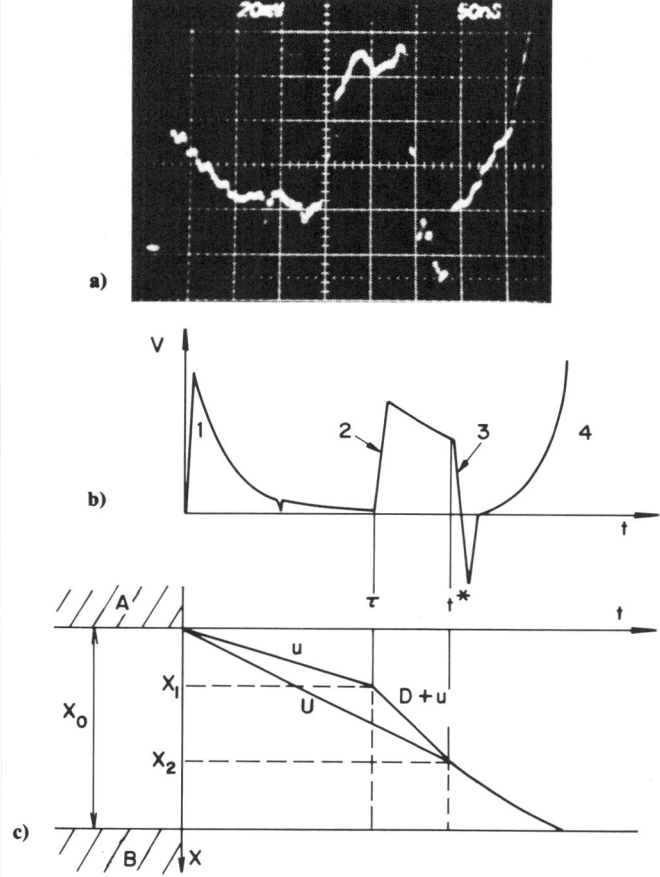

Fig. 6 a) Oscillograph of nitromethane detonation initiation electrical effects; **b)** analysis of oscillograph; and **c)** detonation initiation in liquid explosives (Chaiken, 1958).

From time t^* the detonation was the only source of an electrical polarization effect. Because of the electrode spacing (detonation plate B), the electrical field increased and caused impulse 4 to rise rapidly.

The profile of the shock wave induced by our generator was taken into account and induction times were determined to be smaller than 300 ns. Results (Table 2) show that the NM induction delay, at a shock-wave pressure of about 11-12 GPa, is about 0.1-0.2 μs, which agrees with the extrapolation of results obtained by Hardesty (1976) and de Longueville et al. (1976). A linear fit of our results and those analyzed by Chaiken (1978) is shown in Fig. 7. The constants for the induction delay

$$\tau = AT^2 \exp E/RT$$

were found to be

$$A = 5.32 \times 10^{-2} \mu s/(K)^2 \text{ and } E = 28 \text{ kcal/mole}$$

Since uncertainty on the pressure and temperature§ of the compressed explosive in zone 3 is about 8%, the uncertainty in induction delay is about 100%; and the uncertainty in the detonation failure diameter calculated value is also large.

From Fig. 6c the detonation velocity D in the compressed explosive is given by the relation

$$D = (U-u)/(1-\tau/t^*) \qquad (2)$$

which depends, except for the presence of ratio τ/t^*, only on the induced shock characteristics (U and u). The temperature T was calculated as noted previously. The ratio τ/t^* from Table 2 is, to within 5%, equal to the mean value 0.675. The detonation velocities obtained with this relation and this value of τ/t^* are in good agreement with those obtained from the relation of Campbell et al. (1961). The electrical impulses produced by the detonation initiation processes in mixtures (Fig. 8) are similar to those obtained with NM results (Table 3) and are reproducible; the mean value of the ratio τ/t^* (0.671) is very near that found for NM.

Since the induction delay for the different mixtures are, within experimental uncertainty, represented by the NM curve $\tau(P)$ which is deduced from the relation $\tau(T)$ at an initial temperature of $T_0 = 288$ K (see Fig. 9), the additives used appear to have no effect on NM induction delay as contrasted to additives such as amines studied by Walker (1977).

---

§The temperature T of the compressed NM or mixtures was calculated by the method of Walsh and Christian (1955).

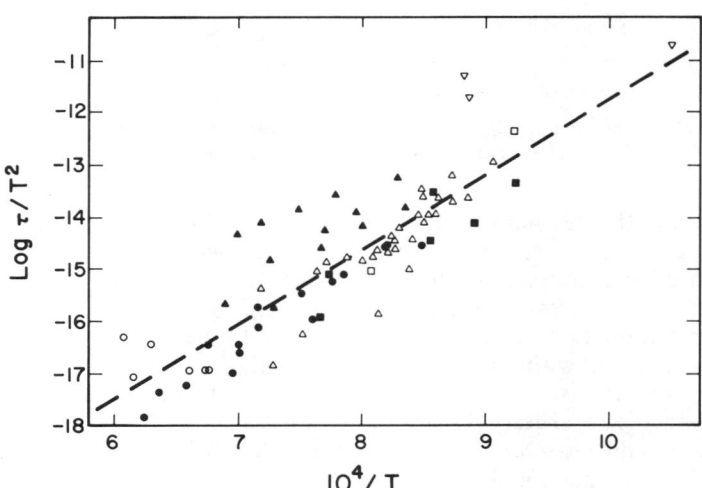

Fig. 7 Induction delays for nitromethane: ● Chaiken (1958), △ Berke et al. (1970), □ Campbell et al. (1971), + Voskoboinikov et al. (1966), ▽ Walker and Wasley (1970), ■ Hardesty (1976), and ○ this investigation.

These results show that detonation delays of the mixtures can be taken as those of NM at the same pressures between 11 and 14 GPa. We believe that this result is still true for smaller pressures.

Dremin and Trofimov (1964) have presented the relation

$$\tau_3 = \tau^* \left[1 - \frac{U - u_{3n}}{D_3} \right] \left[1 + \frac{U}{v} \sin \gamma \right]^{-1} \qquad (3)$$

for evaluation of the induction delay τ_3 from a knowledge of the parameters associated with failure diameter. From the streak camera record of the

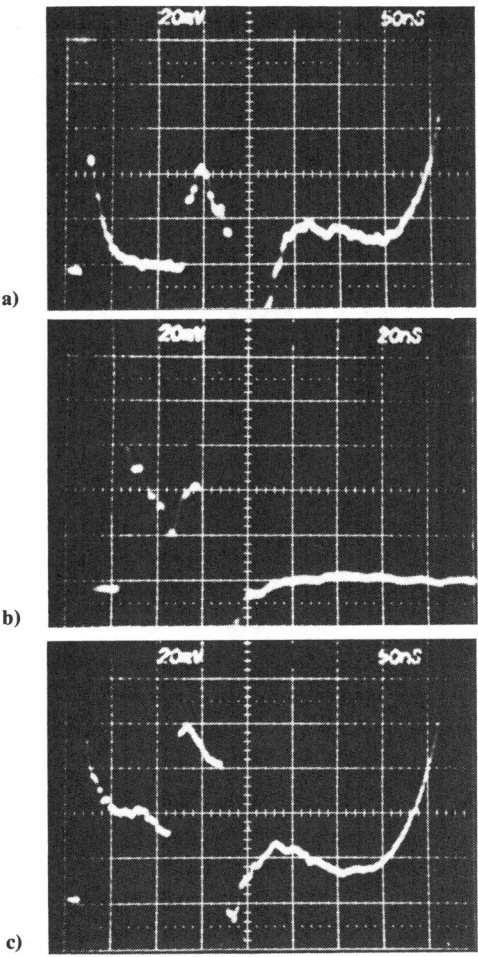

Fig. 8 Detonation initiation electric effects for nitromethane mixtures: a) $CH_3NO_2 + 0.3$ $CHCl_3$, b) $CH_3NO_2 + 0.3$ $CHBr_3$, and c) $CH_3NO_2 + 0.2C_3H_6O$; ordinates 20 mV/cm; abscissas a) and c) are 50 ns/cm, b) 20 ns/cm.

Table 2 Induction delay for nitromethane

T_0, K	P, GPa	T, K	τ, μs	t^*, μs	τ/t^*
288 ± 2	11.8	1624	0.100	0.140	0.714
288 ± 2	12	1652	0.230	0.340	0.676
288 ± 2	11	1514	0.100	0.140	0.714
288 ± 2	11.6	1589	0.182	0.275	0.662
298 ± 2	11.6	1484	0.090	0.140	0.643
298 ± 2	11.6	1484	0.090	0.140	0.643

Table 3 Induction delay for nitromethane with additives

Mixtures	χ	P, GPa	τ, μs	t^*, μs	τ/t^*
$CH_3NO_2 + \chi CHCl_3$	0.05	11.2	0.1	0.15	0.666
	0.1	12	0.07	0.09	0.677
	0.1	11.4	0.187	0.28	0.663
	0.1	11.3	0.095	0.14	0.678
	0.2	11.9	0.12	0.18	0.677
	0.3	11.7	0.11	0.15	0.733
	0.3	11.8	0.11	0.165	0.672
	0.3	11.5	0.12	0.19	0.631
	0.3	12.7	0.06	0.08	0.75
	0.3	12.7	0.06	0.08	0.75
	0.3	12.4	0.07	0.11	0.636
	0.4	12	0.08	0.12	0.655
	0.5	12.2	007	0.09	0.777
	0.5	11.7	0.06	0.08	0.75
	0.5	11.7	0.07	0.1	0.7
$CH_3NO_3 + \chi CHBr_3$	0.1	12.8	0.105	0.149	0.692
	0.1	12.6	0.102	0.158	0.645
	0.2	13.3	0.021	0.029	0.720
	0.3	13.5	<0.04		
	0.3	13.9	0.017	0.024	0.708
	0.3	13.4	0.024	0.036	0.666
	0.4	14	<0.04		
	0.4	13.7	0.018	0.028	0.642
	0.4	13.5	<0.04		
$CH_3NO_2 + \chi C_3H_6O$	0.2	11.6	0.1	0.157	0.637

detonation failure, τ^* is defined along a line parallel to the time axis between the time when the detonation initiated in zone 3 reaches the shock wave AB and the time when the failure wave crosses the line. The values of τ_3 obtained for NM and mixtures (Table 4) at a constant pressure from streak records are smaller by about one order of magnitude than those deduced with direct measurements of electrical signals. This shows that induction delay depends greatly on experimental conditions.

Discussion and Conclusion

The estimates of the induction delay vary greatly with the method of determination and for the same explosive give very different values of the detonation failure diameter as calculated from the theory of Dremin and

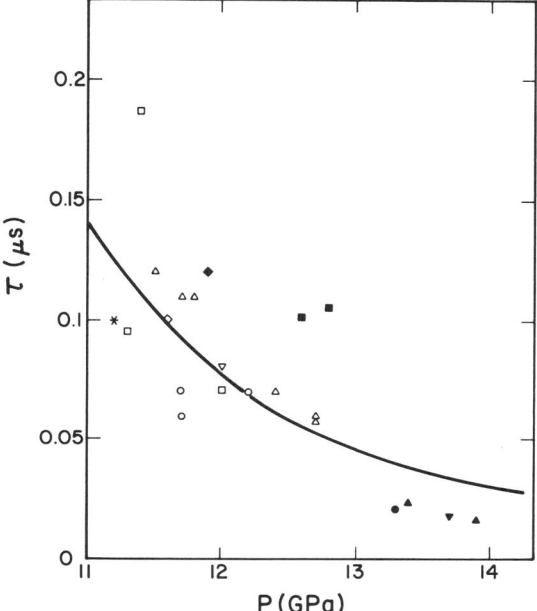

Additive/χ	0.05	0.1	0.2	0.3	0.4	0.5
$CHCl_3$	*	□	◆	△	▽	○
$CHBr_3$		■	●	▲	▼	
C_3H_6O			◇			

Fig. 9 Induction delay for nitromethane and nitromethane mixtures.

Trofimov. The detonation failure diameters which are calculated with the induction delays determined from streak camera records agree within 30% with experimental results.

Failure diameters calculated with induction delay determined from direct measurements are about one order of magnitude larger than the experimental results. Our experimental results show that the dependence of the detonation failure diameter of NM-chloroform and NM-bromoform mixtures on additive concentration is less pronounced than that of NM-acetone mixtures.

This confirms, in particular, the results of Kurbangalina (1947) who found that a mixture of glycol dinitrate with 15-16% of acetone has the same critical diameter as a mixture of glycol dinitrate with 50-60% chloroform or bromoform.

The calculation of the detonation failure diameter of an explosive with the model of Dremin and Trofimov is difficult because of the lack of precision in the induction delay. However, the calculation leads to a good agreement with experimental results when the question is restricted to the influence of additives.

Table 4 Induction delay for nitromethane and nitromethane with additives from streak camera records

Mixtures	$CH_3NO_2 + \chi CH Cl_3$					$CH_3NO_2 + \chi CH Br_3$			
χ	0	0.1	0.15	0.2	0.3	0.1	0.15	0.2	0.3
$\tau_3, \mu s$	0.28	0.39	0.47	0.55	0.55	0.76	0.57	0.8	0.7	0.9

As the detonation failure in a liquid explosive induced by a rarefaction wave is a very reproducible phenomenon and depends greatly on the state variables of the explosive (composition, temperature), the failure diameter could be chosen as a test to compare the different liquid explosives. Furthermore, the detonation failure diameter of gaseous explosives mixtures may be used to evaluate the critical energy for their detonation initiation, a measure of the sensitivity of explosives (Lee and Matsui, 1977).

References

Berke J.G., Shaw R., Tegg D., and Seely L.B. (1970) Shock initiation of nitromethane, methyl nitrite and some bis difluoramino alkanes. *5th Symposium (Intl.) on Detonation*, p. 163. ONR Rept., ACR 184.
Brochet C., Presles H. N., and Cheret R. (1974) Detonation characteristics of some liquid mixtures of nitromethane and chloroform or bromoform. *5th Symposium (Intl.) on Combustion*, p. 29. The Combustion Institute, Pittsburgh, Pa.
Campbell A.W., Davis W.C., and Travis J.R. (1961) Shock initiation of detonation in liquid explosives. *Phys. Fluids* 4, 498.
Chaiken R.F. (1958) Thesis, Brooklyn Polytechnic Institute, New York.
Chaiken R.F. (1978) Correlation of shock pressure, shock temperature, and detonation induction time in N.M. *Actes du Sym.*, p. 41. H.D.P., Paris.
Dremin A.N., Rozanov O.K., and Trofimov, V.S. (1963) On the detonation of nitromethane. *Combustion and Flame* 7, 153.
Dremin A.N. and Trofimov V.S. (1964) Analysis of critical diameter of detonation of liquid explosives. *Zhur, Prikl Mek. i Tekh. Fiz.* 1, 126.
Enig J.W. and Petrone F.J. (1965) Equation of state and derived shock initiation criticality conditions for liquid explosives. *Phys. Fluids* 8, 771.
Enig J.W. and Petrone F.J. (1970) The failure diameter theory of Dremin. *5th Symposium (Intl.) on Detonation*, p. 99. ONR Rept., ACR 184.
Evans M.W. (1962) Detonation sensitivity and failure diameter in homogeneous condensed materials. *J. Chem. Phys.* 36 (1) 193.
Eyring H., Powell R.F., Duffey G.H., and Parlin R.B. (1949) The stability of detonation. *Chem. Rev.* 45, 69.
Hardesty D.R. (1976) An investigation of the shock initiation of liquid nitromethane. *Combustion and Flame* 27, 229.
Hayes B. (1965) On electrical conductivity in detonation products. *J. Appl. Phys.* 38, 507.
de Icaza Herrera M., Presles H.N. and Brochet C. (1978) Polarisation du nitrométhane sous choc. *Rev. Phys. Appl.* 547.
Jones H. (1947) *Proc. Royal Soc.* A189, 415.
Khariton Yu.B. (1957) *Problems in the Theory of Explosives*. Academy of Sciences Press, U.S.S.R.

Kurbangalina R. Kh. (1947) Candidate Dissertation, Inst. of Chem. and Phys. Academy of Sciences, U.S.S.R.

Kusakabe M. and Fujiwara S. (1976) Effects of liquid diluents on detonation propagation in nitromethane. *5th Symposium (Intl.) on Detonation,* p. 133. ONR Rept. ACR 184.

Lee J.H. and Matsui H. (1977) A comparison of the critical energies for direct initiation of spherical detonations in acetylene-oxygen mixtures. *Combustion and Flame* **28**, 61-66.

de Longueville Y., Fauquignon C., and Moulard H. (1976) Initiation of several condensed explosives by a given duration shock wave. *6th Symposium (Intl.) on Detonation p. 105.* ONR Rept. ACR 221.

Presles H.N. and Brochet C. (1976) Instabilités de la détonation dans les mélanges nitrométhane-chloroforme et nitrométhane-bromoforme. *Acta Astronautica* **3**, 531.

Presles H.N. (1978) Polaire de choc et température de méanges nitrométhane-bromoforme et nitrométhane-chloroforme sous choc. *Actes du Sym.,* p. 55. H. D. P., Paris.

Presles H.N. (1979) Contribution a l'étude de la détonation de mélanges liquides binaires a base de nitrométhane. Thèse Doct. es. Sc. Phys., Poitiers.

Tarver C.M., Shaw R., and Cowperthwaite M. (1975) Detonation failure diameter studies of four liquid nitro alkanes. *J. Chem. Phys.* **64**, 2665.

Travis I.R. (1965) Electrical transducer studies of initiation of liquid explosives. *4th Symposium on Detonation,* p. 609. ONR Rept. ACR 126.

Voskoboinikov M., Bogomolov V., Margolin A.D., and Apin A.Y. (1966) Determination of the decomposition times of explosives materials in a shock wave. *Dokl. Akad. Nauk.* **167**, 610.

Walker F.E. and Wasley R.J. (1970) Initiation of nitromethane with relatively long duration low-amplitude shock waves. *Combustion and Flame* **15**, 223.

Walker F.E. (1979) Initiation and detonation studies in sensitized nitromethane. *Astronautica Acta* **6**, 807.

Walsh J.M. and Christian R.H. (1955) Equation of state of metals from shock wave measurements. *Phys. Rev.* **97**, 1544.

Wood W.W. and Kirwood J.G. (1954) Diameter effect in condensed explosives. The relation between velocity and radius of curvature of the detonation wave. *J. Chem. Phys.* **22**, 1915.

Critical Area Concept for the Initiation of a Solid High Explosive by the Impact of Small Projectiles

Henry Moulard
French-German Institute of Saint-Louis, Saint-Louis, France

Targets of a cast explosive (60% RDX/40% TNT by weight) are impacted by small steel projectiles. Experimental measurements concern the critical impact velocity for the shock-detonation transition. The studied parameter is the geometry of the projectile (shape and dimensions). Our experimental results show that the concept of critical area must be introduced to understand the sensitivity of solid high explosives.

Introduction

Sensitivity of high explosives to the impact of small projectiles is of practical interest. The current work at the Institute of Saint-Louis (ISL) is the natural continuation of earlier work on the sensitivity of explosives to plane shock waves. One of the primary efforts was an attempt to demonstrate a correlation for the critical conditions for a shock-detonation transition.

Results for a cast explosive (60% RDX/40% TNT by weight, density 1.73 g/cm^3, hereafter called composition B3) are presented here and support a new criterion for shock initiation in which not only the pressure and the duration are involved, but also a minimal shock-loaded surface area has to be taken into account.

Initiation by Plane Shock Waves

Previously, we have studied sensitivity of solid high explosives to plane shock waves of calibrated intensity and duration (de Longueville et al., 1976). In one-dimensional geometry, the shock-detonation transition is controlled by two parameters: shock pressure p and shock duration t (Gittings, 1965). At ISL, an aluminum flyer-plate impacts the explosive test cylinder. From the measurement of the flyer-plate velocity and thickness; the intensity and the duration of the overpressure created in the explosive target at impact are computed. The experimental sensitivity of composition B3 to plane shock waves is shown in Fig. 1. Sensitivity predictions based on the criterion for the shock-detonation transition proposed by Walker and Wasley (1968) and a constant critical energy value equal to 140 J/cm^2 fits the data well.

Presented at the 7th ICOGER, Göttingen, Federal Republic of Germany, Aug. 20-24, 1979.
Copyright © American Institute of Aeronautics and Astronautics, Inc., 1981. All rights reserved.

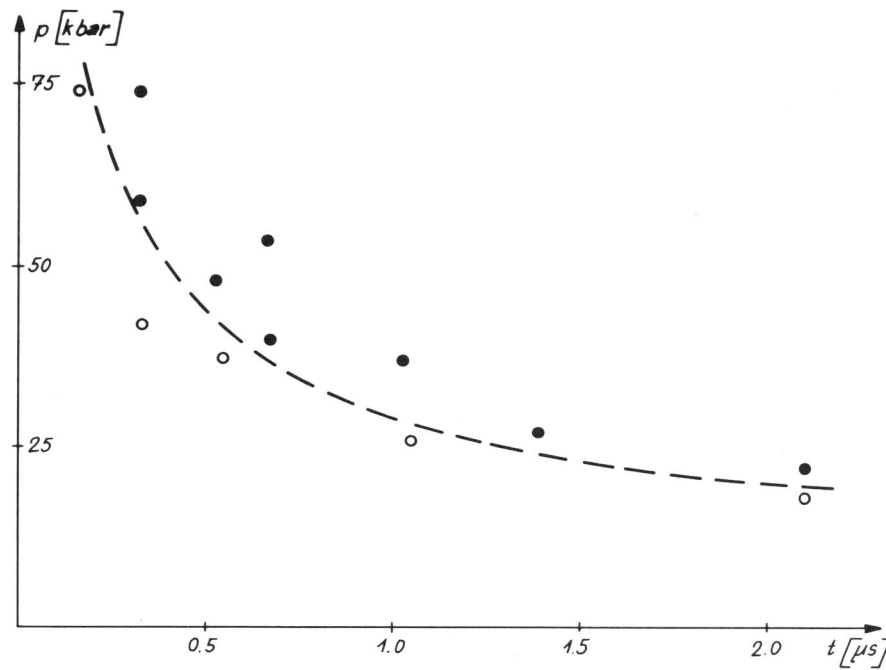

Fig. 1 Experimental sensitivity for the initiation of ISL composition B3 by plane shock waves: ○, no-detonation; ●, detonation; ---, computed sensitivity using constant critical energy value of 140 J/cm^2.

Initiation by Small Projectiles: First Approach

The impact by a small projectile or by a primary fragment is a three-dimensional phenomenon. As a consequence, data obtained with plane shockwave experiments cannot be used for solving the problem of vulnerability of explosive warheads. Some essential features of initiation by a small projectile are revealed by a two-dimensional model (Fig. 2). Elements of a quantitative approach to this problem have been suggested previously (Frey et al., 1976). Briefly, when an explosive cylinder, either unconfined or covered only by a thin metallic plate on the impact face, is impacted by a small projectile, either shock-detonation transition or a mechanical destruction occurs.

Moreover, an analogy with the plane shock-wave initiation may be invoked to estimate the critical conditions in the two-dimensional impact geometry, if the one-dimensional critical conditions are known. The analogy is based on the following: 1) the velocity and the nature of the projectile determines impact shock pressure and 2) the shock duration is now defined (Fig. 2) as the time for the lateral release waves to propagate from the periphery of the projectile to a point situated along the axis of the projectile and at the interface between the high explosive (H.E.) and the projectile.

The validity of this predictive model is supported by the fact that the variation of experimental critical velocity with projectile diameter can be

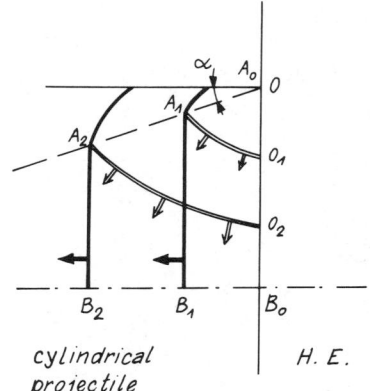

Fig. 2 Modeling of two-dimensional impact; lateral release waves control transmitted shock-wave profile: $O_n A_n$, first release wave at time n; $\overline{A_n B_n}$, undisturbed shock front at time n; $B_2 B_1$, projectile axis; OB_o, projectile radius.

correlated with a constant critical energy value. But, as the published critical energy values determined from small-projectile impact initiation experiments are much larger than one-dimensional critical energy values, the following experimental program has been initiated.

ISL Experiments and Results

Cylindrical Projectiles

Steel projectiles, of various diameters and length-to-diameter ratios greater than one, were launched by a powder gun on a composition B3 cylinder target located at the muzzle of the gun. Projectile velocity just before impact, the transit time through the explosive cylinder, and the local value of the wave speed at the end of the explosive target were measured. The measured critical velocity vs the diameter of cylindrical projectiles are shown in Fig. 3. Above the curve, transition to detonation was observed. Numbers in parenthesis are the local critical energy value computed by the aforementioned analogy. The local critical energy for two-dimensional experiments is about 6-8 times that determined for plane shock-wave initiation.

Same Overall Shock Duration Projectiles

This discrepancy may be ascribed to the requirement that, as a minimum number of particles must be involved in the initiation process, the critical conditions must consider off-the-axis particles. To elucidate further the two-dimensional initiation phenomenon, the following experiments were made with the projectiles shown in Fig. 4. These projectiles have the same overall shock duration, since at the same time, 500 ns after impact, lateral release waves have completely destroyed the overpressure created at impact. Consequently, the previously mentioned analogy would logically predict the same critical velocity for these projectiles.

A significant difference exists in the number of explosive particles which are subject to very high pressure at the end of the overall shock duration. For the circular projectile there is only one such explosive particle at that instant. For

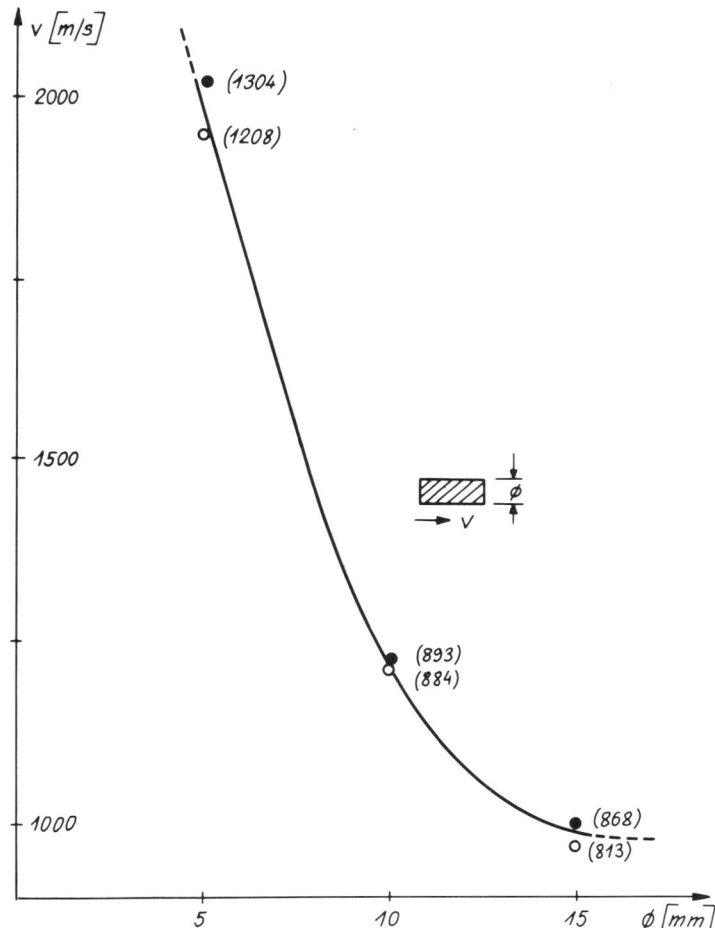

Fig. 3 Sensitivity curve of ISL composition B3 to impact of cylindrical steel projectiles: ○, no-detonation; ●, detonation; numbers in parenthesis along the curve correspond to critical energy value, J/cm^2. For one-dimensional experiments, it is 140 J/cm^2.

the annular and rectangular projectiles there is a great number of particles situated along a circle or a line still submitted to the high pressure created at that instant. The critical velocity measured with the different types of projectiles designated in Fig. 4. are shown in Table 1. The corresponding critical energy values as well as the overall shock durations are given to emphasize the very important discrepancy with the simple analogy.

Proposal for a New Criterion

With these new experimental results, it becomes necessary to introduce a new approach to the sensitivity of high explosives. The critical conditions for initiation of detonation by plane shock waves are controlled by the intensity

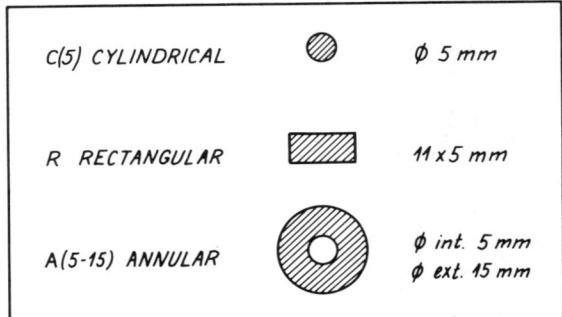

Fig. 4 Symbol, design, and dimensions of different projectile impact surface with same overall shock duration.

Table 1 Toward a new criterion

Projectile type	Critical velocity,[a] m/s	p.u.t. criterion,[b] J/cm^2	Overall shock duration, μs
C(5)	2020 ●	1300	0.530
	1950 ○	1200	0.535
R(×11)	1415 ●	607	0.568
	1320 ○	520	0.574
A(15-5)	1105 ●	357	0.588
	1060 ○	327	0.591
C(15)	1000 ●	866	1.785
	970 ○	813	1.791

[a] ●, detonation; ○, no detonation.   [b] p.u.t., local critical energy

and the duration of the plane shock wave. In the initiation caused by the impact by a small projectile, a third control parameter, the critical surface area, must be introduced. It is supposed that, in the critical conditions at least, the local mechanism for transition is identical for plane shock-wave initiation and for small projectile impact initiation.

Consequently, we propose that critical conditions for the initiation by a small projectile be defined as follows. At a given shock pressure, the shock-detonation transition occurs only if a sufficient area of the explosive target is loaded at high pressure for a time equal or greater than the critical shock duration $t(p)$ measured in plane shock-wave experiments at the same shock pressure.

Determination of the Critical Surface Area

The critical surface area is measured as follows. For a cylindrical projectile (Fig. 2), the explosive particle defined by position 0_n on the impact face is subjected to the high shock pressure created at impact, only during the time

$$t = (R-r)/c(p)$$

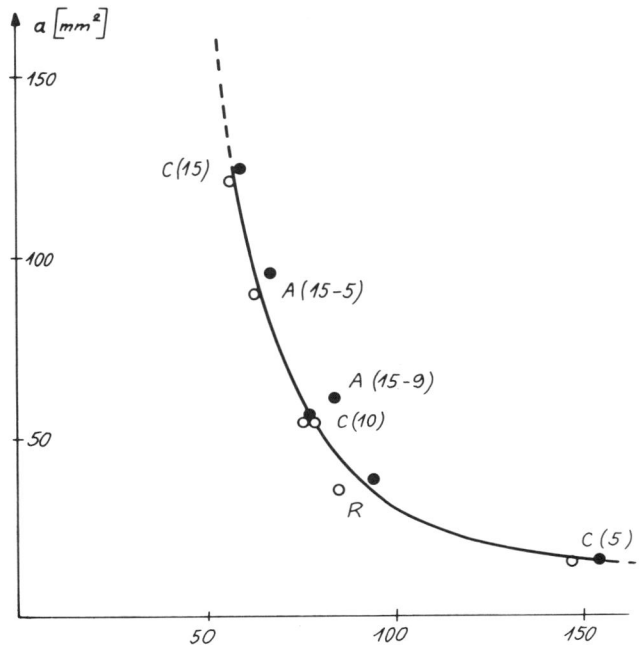

Fig. 5 Critical area vs shock pressure in ISL composition B3 impact initiation: ○, no-detonation; ●, detonation; for C(x), R, A(x,x), see Fig. 4.

where $c(p)$ is the velocity of sound in the steel projectile, a function of the shock pressure p; $\partial R = 0B_0$, projectile radius; and $r = 0_n B_0$, distance of the explosive particle to projectile axis. The overall shock duration is defined only for the particle situated B_0, and the particle situated in 0 is never submitted to the high pressure.

Then we measure the critical velocity for this given cylindrical projectile and compute the impact shock pressure p. The experiments on initiation by plane shock waves (Fig. 1) give the critical shock duration $t(p)$ corresponding to this shock pressure. The critical surface area at this given shock pressure is computed as the area subjected to high shock pressure during the time $t(p)$. Thus, we can write

$$t(p) = (R - r^*)/c(p)$$

Where r^* is the critical radius. Thus, the critical surface area depends on the impact shock pressure. This computation is easily transposable for annular or rectangular projectiles.

Consistency of Experimental Results

The critical surface area is a consequence of the minimum number of hydrodynamic interactions between hot spots necessary to build up the detonation through the solid high explosive (Fauquignon and Moulard, 1978). With this new criterion, the critical velocity measured for different projectiles (Table 1, Fig. 2) are correlated by a single curve $a(p)$ (Fig. 5). The two projectiles A(15-9) and C(10) have been designed to have a nominal area ratio of 1.44, but the same critical surface area. Experimental results show that initiation occurs at the same critical velocity.

Conclusion

For the prediction of sensitivity of high explosives to impact, our results show that it is now necessary to introduce a third parameter: a critical surface area. Our experiments with composition B3 also indicate that the critical surface area must increase as the impact shock pressure decreases (Fig. 5). This observation is very important for designing one-dimensional plane shock-wave initiation experiments. The critical energy value is useful for comparative study of the one-dimensional sensitivity curve of different high explosives. For a given explosive, this critical energy value is not strictly a constant, but varies with the shock conditions (de Longueville et al., 1976). If the critical energy value increases linearly with the shock duration, that means the experimental conditions are effectively two-dimensional, and the dimensions of the explosive target are below the critical surface area, which depends on the impact shock pressure.

Acknowledgment

The work reported here was performed under contract of the Société Nationale de Poudres et Explosifs, France.

References

de Longueville Y., Fauquignon C., and Moulard H. (1976) Initiation of several condensed explosives by a given duration shock wave. *6th Symposium on Detonation,* pp. 105-114. ONR Rept. ACR-221.

Fauquignon C. and Moulard H. (1978) Shock sensitivity of nitromethane with well defined hot spots distribution. *Acta Astronautica* **5,** 1035-1040.

Frey R., Melani G., Chawla M., and Trimble J. (1976) Initiation of violent reaction by projectile impact. *6th Symposium on Detonation,* pp. 325-335. ONR Rept. ACR-221.

Gittings E. F. (1965) Initiation of a solid high explosive by a short-duration shock. *4th Symposium on Detonation,* pp. 373-380. ONR Rept. ACR-126.

Walker F. E. and Wasley R. J. (1968) Critical energy of shock initiation of heterogeneous explosives. UCRL Rept. 70891.

Shock Initiation of Hydrazine Mononitrate

Masatake Yoshida* and Tadao Yoshida*
University of Tokyo, Tokyo, Japan
and
Katsumi Tanaka,† Shuzo Fujiwara,† and Masao Kusakabe†
National Chemical Laboratory for Industry, Ibaragi, Japan

An explosive wedge technique using streak camera photography was employed to investigate the shock-initiation characteristics of hydrazine mononitrate (HN), which is one of the group 2 explosives.

The Hugoniot obtained by the impedance-matching method had two remarkable features. One is that the Hugoniot did not pass the detonation pressure point, the other is that it had a break at about 8 GPa. The Hugoniot of HN was fitted by two linear relations: U (mm/μs) $= 1.2 + 4.9u$ ($u < 0.86$), and U (mm/μs) $= 5.0 + 0.53u$ ($u > 0.86$).

Introduction

PLANAR shock-initiation characteristics of solid heterogeneous explosives have been studied by many workers [e.g., see Campbell et al. (1961), Lindstrom (1966, 1970), Kennedy (1973), and Halleck and Wackerle (1976)]. Observations with a streak camera of shock fronts propagating through explosive wedges have made it clear that, in solid explosives, the incident shock-wave strength gradually increases as it propagates until an abrupt transition to detonation occurs. The growth of the shock front is ascribed to formation of local hot spots where chemical reaction due to high temperature occurs and the resulting energy release accelerates the shock front.

This model has been confirmed in many heterogeneous explosives, most of which belong to group 1 explosives, but planar shock-initiation studies on group 2 explosives are rather few. This grouping of explosives into two classes was made by Price (1967). Group 1 explosives (such as PETN, tetryl, and TNT) exhibit a decrease in critical diameter with an increase in loading density; while group 2 explosives (such as ammonium nitrate, hydrazine mononitrate, and ammonium perchlorate) exhibit an increase in critical diameters with an increase of loading density.

Presented at the 7th ICOGER, Göttingen, Federal Republic of Germany, Aug. 20-24, 1979. Copyright © American Institute of Aeronautics and Astronautics, Inc., 1981. All rights reserved
*Department of Reaction Chemistry, Faculty of Engineering.
†Tsukuba Research Center.

In this work, planar shock-initiation characteristics of hydrazine mononitrate (HN) were investigated by an explosive wedge technique and streak camera photography.

Experimental Techniques

A schematic of the experimental setup is shown in Fig. 1. The streak camera and explosive wedge technique was used to observe the shock initiation characteristics in HN. Observations were also made with explosive pellet samples.

The HN was pressed into a 20 mm diameter cylinder with a density range of 1.650-1.676 g/cm^3, which corresponds to about 99% of the crystal density, and it was shaped into a wedge or a pellet of required thickness. (The crystal density of HN was cited to be 1.685 g/cm^3 by Sarner (1966), but Robinson and McCrone (1958) reported the value of 1.661 g/cm^3.) The angle of a wedge was nominally 20 deg. Explosive samples were kept in a vacuum desiccator until just before firing. As soon as it was taken out of the desiccator, the sample was coated with grease to prevent moisture absorption.

A streak camera, either Hitachi SP-1 or Cordin Model 116, was used. The maximum writing speed is 4 and 20 mm/μs, respectively. Usually a speed of 3.5 mm/μs was used for SP-1 and of 10 mm/μs for Model 116. An argon flash was used as a light source for both wedge- and pellet-type experiments. Argon

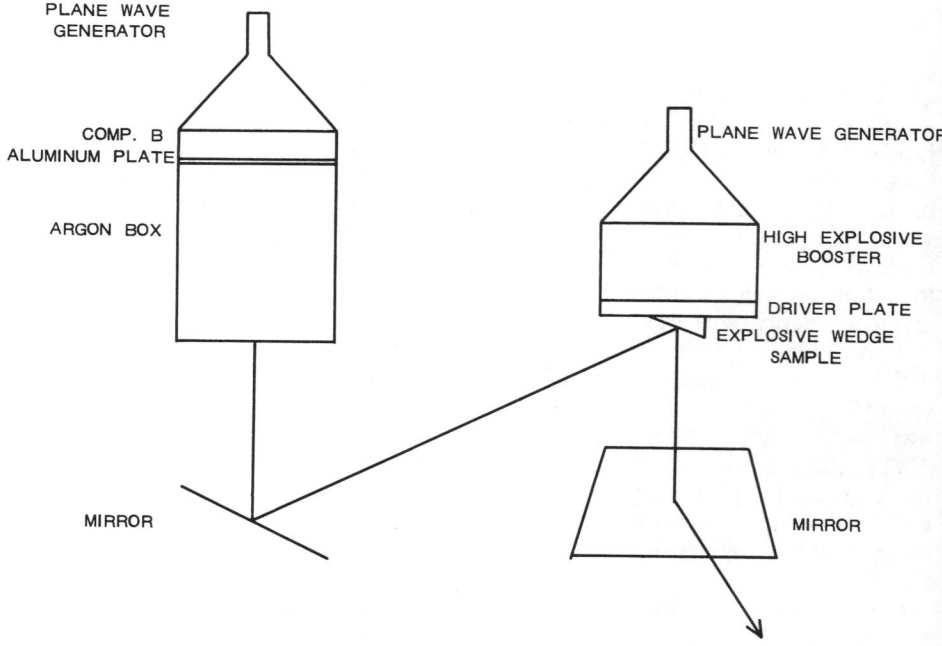

Fig. 1 Schematics of experimental setup.

was contained in a 70 mm diameter and 100 mm long tube of polyvinyl chloride and was shocked by a 0.5 mm thick aluminum flyer plate which was accelerated by the explosion of a 10 mm thick piece of composition B with planar initiation. The duration of light emission of this argon flash was about 16 μs.

The shock systems, each of which consisted of a plane wave explosive lens, high explosive booster, and one or three layers of inert materials, were employed to generate the plane shock waves of different amplitudes in explosive samples. A hydrazine hydrate solution of hydrazine nitrate was used as a higher velocity explosive for a plane wave explosive lens and nitromethane was used as a lower velocity one. This plane wave explosive lens has a 60 mm effective diameter of plane area with 20 ns accuracy.

Nitromethane, hydrazine hydrate solution of HN, tetryl, composition B (60% RDX/40% TNT), and Octol (80% HMX/20% TNT) were used as booster explosives. The thickness of each booster explosive was 30 mm.

To generate low-amplitude shock waves in the sample, three layers of inert materials (copper, liquid, and copper) were used. The thickness of each layer was 5 mm. A single layer, either 5 mm thick copper or 4 mm thick aluminum alloy (density = 2.68 g/cm^3) was used for higher pressures.

The flyer plate methods and overdriven detonation waves were also used to generate very high shock pressure in HN samples. These methods are described briefly in Table 1.

The effective diameter of plane area of the last driver plate was measured by argon flash gap. A typical effective diameter was 40 mm for single-driver plate systems. The free surface velocity of the driver plate was measured within the effective plane area either by the electrical pin contactors technique or by the inclined mirror technique. For the shock systems where flyer plate or overdriven detonation wave was employed, the effective diameter was less than 20 mm and the shock velocity measurement of the sample was made within the first 1 mm run in the sample.

Figure 2 is a streak camera record of a wedge-type experiment taken by SP-1. The booster explosive was nitromethane and the driver plate was copper.

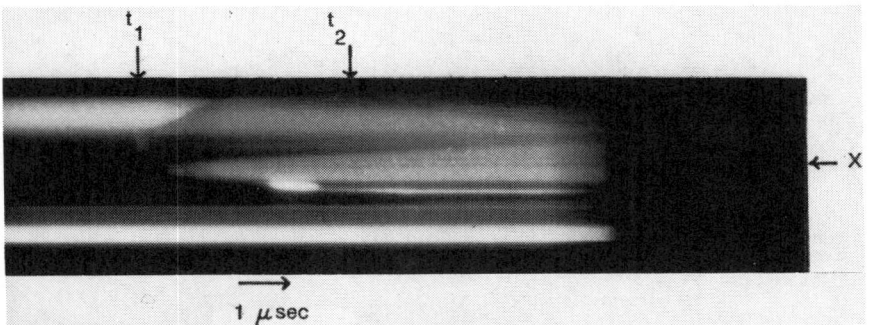

Fig. 2 Streak camera record of wedge-type experiment: incident shock pressure was about 7 GPa, shock wave entered the sample at time t_1, second wave which was accompanied by weak light emission X was observed from time t_2.

Table 1 Summary of wedge-type experiments

Booster explosive	Driver plate[a]	Free surface velocity of driver plate, mm/μs	HN shock parameters			Initiation type (see text)	Reaction delay time, μs[b]
			Shock velocity, mm/μs	Particle velocity, mm/μs	Pressure, GPa		
NM	Cu[c]	0.45	2.9	0.40	1.9	—[d]	—
NM	Cu[e]	0.54	3.75	0.457	2.9	—[d]	—
NM	Cu	1.00	3.4	0.46	2.6	—[d]	—
			5.25	0.809	7.1	A	4.3
			4.83	0.820	6.6	A	3.9
Comp. B[f]	Cu	1.92	4.91	1.21	9.9	B	3.9
NM	Al	1.47	5.59	1.24	11.6	B	2.7
			5.40	1.26	11.4	A	3.4
HN/HH	Al[g]	3.81	5.61	2.58	24	C	0.56
Comp. B	PMMA	[h]	9.0	4.4	65	C	0.25

[a] One inert layer was used except as noted.
[b] The time between the shock entrance into the sample explosive and the appearance of the second shock wave at the free surface of the sample.
[c] Three inert layers were used: copper, water, and copper.
[d] No reaction was observed.
[e] Three inert layers were used: copper, carbon tetrachloride, and copper.
[f] 60%RDX/40%TNT, with an initial density of 1.60 g/cm^3.
[g] Flyer-plate method was used. A 2.5 mm thick aluminum alloy was accelerated by hydrazine hydrate solution of hydrazine nitrate (75%). The experimental error in HN shock parameters obtained by this shock system was about 5%.
[h] Overdriven detonation wave in Comp. B shocked a PMMA plate of 5 mm thickness. The shock velocity in PMMA was 9.6 mm/μs. The error in HN shock parameters was about 10%.

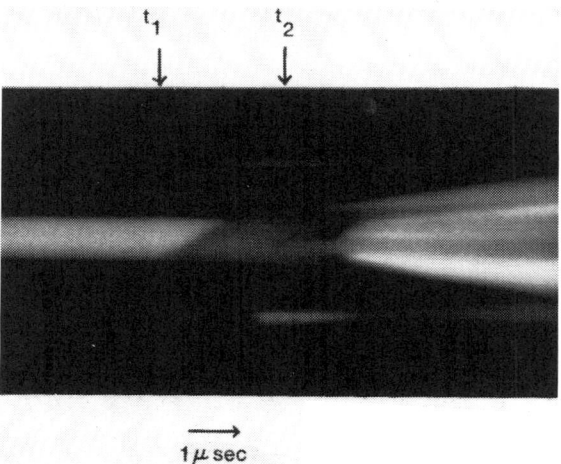

Fig. 3 Streak camera record of wedge-type experiment: incident shock pressure was about 11 GPa, shock entered the sample at time t_1, second shock wave with strong light emission appeared at time t_2.

The shock velocity and information on HN can be obtained from this record. From the shock velocity and the free surface velocity of the driver plate measured in this shock system the Hugoniot data of HN can be calculated by the impedance matching method. The shock pressure thus calculated was about 7 GPa, and a backward moving wave accompanied with very weak light emission which persisted for about 20 μs was observed. This delayed reaction was also observed in a pellet-type experiment of the same shock system. Whether the second wave moving backward was caused by real induction delay of reaction at this pressure region or by the dimensional effects (for example, reflected shock waves by surrounding materials) could not be determined. When the incident shock pressure was about 2-3 GPa delayed reactions were not observed.

Figure 3 was also taken by SP-1. The booster explosive was nitromethane and driver plate was aluminum alloy. The incident shock pressure was about 11 GPa. Strong light emission was observed at about 3 μs after the first shock entrance into the sample. Light emission persisted for several tens of microseconds.

Figure 4 is a streak camera record of a pellet-type experiment taken by Model 116. Tetryl was used as a booster explosive and the driver plate was aluminum alloy. The thicknesses of pellet samples were nominally 1 and 2 mm. The free surface velocity of HN was measured directly by an inclined mirror on 2 mm thick pellet sample. The samples were not coated with grease but were contained in a glass box with a drying agent when the inclined mirror method was employed. The free surface velocity of the driver plate was also measured by the inclined mirror method.

Experimental Results

The results of wedge-type experiments are listed in Table 1. The accuracy of shock velocities obtained from the streak camera records was about 5%.

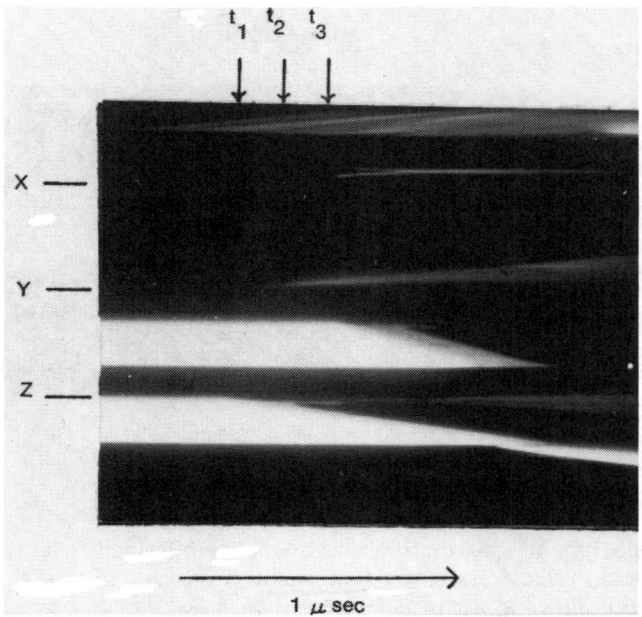

Fig. 4 Streak camera record of pellet-type experiment: incident shock pressure was about 15 GPa; XY and YZ correspond to 1 and 2 mm thick pellet samples, respectively; free surface velocities of driver plate and 2 mm thick sample were measured by inclined mirrors; shock entered two pellets at time t_1, arrived at their surfaces at time t_2 and t_3.

Particle velocities and shock pressures were calculated by the impedance matching method and the known Hugoniots of the driver plates. The accuracy of the free surface velocity measurement was about 1%. The Hugoniots of the driver plates were reported by Kinslow (1970) to be: $U = 3.940 + 1.489u$ for copper, $U = 5.328 + 1.338u$ for aluminum alloy, and $U = 2.260 + 1.816u$ for PMMA, where U and u are shock and particle velocity in mm/μs, respectively.

At shock pressures of less than 3 GPa, no significant light emission caused by the reaction was observed. While at shock pressures greater than 6 GPa, the following three reaction initiation types were observed as reported in Table 1:

Type A) A backward moving wave accompanied with very weak light emission (see Fig. 2). The duration of light emission was about 20 μs. This type was typical at the shock pressure of about 7 GPa.

Type B) A shock wave accompanied with strong light emission which occurred a few microseconds after the arrival of the first shock wave at the free surface of a wedge sample (see Fig. 3). This light emission persisted for several tens of microseconds. This type was typical at the shock pressure of about 11 GPa.

Type C) Shock-to-detonation transition at more than 24 GPa.

The second waves of the first two types are considered to be reactive because of the light emission. The time interval between the shock entrance into the sample and the appearance of the second wave at the free surface of the sample

Table 2 Summary of pellet-type experiments

Booster explosive	Driver plate	Free surface velocity of driver plate, mm/µs	HN shock parameters[a]				Nominal pellet thickness, mm
			Shock velocity, mm/µs	Particle velocity, mm/µs	Pressure, GPa	Free surface velocity, mm/µs	
NM	Cu	1.00	5.42	0.804	7.3	—[b]	4
NM	Al	1.47	5.45	1.26	11	—[b]	5
Tetryl	Al	2.25	6.49	1.38	15	4.20	1,2
HN/HH[c]	Al	2.32	6.06	1.46	15	3.92	1,2
Comp. B[d]	Al	3.59	6.34	2.25	24	5.60	1,2,4,8[e]
Octol[f]	Al	3.84	6.69	2.44	27	9.46	1,2[g]

[a] Shock velocity was an average value if pellets of different thickness were used, but the data of footnotes e and g were excluded. Particle velocity and pressure were calculated by the impedance matching method. Free surface velocity was measured by inclined mirror technique on a 2 mm thick pellet sample.
[b] Not measured.
[c] Hydrazine hydrate solution on hydrazine nitrate (63.4 wt%).
[d] 60%RDX/40% TNT, with initial density 1.66 g/cm^3.
[e] Shock-wave velocity obtained from a 8 mm thick pellet was much greater than shock velocities of other three pellets.
[f] 80%HMX/20% TNT.
[g] Shock-to-detonation transition was attained within 2 mm run.

was measured. The time intervals for types A and B, and shock-to-detonation induction times for type C are both listed in the last column of Table 1.

However, in types A and B, the reaction must have started in the region which had been affected by the rarefaction wave because our explosive wedge samples were very small. Therefore, the observed reaction delay time should be regarded to be peculiar to our shock systems.

The results of pellet-type experiments are listed in Table 2. The shock velocities obtained with the same shock system were averaged. The difference between the shock velocities in pellets was usually within experimental error, which was about 2% for this type of experiment, except for the two cases noted in the table. Reaction delay time was not observed because the pellets were contained in a glass box. The collision of the free surface with the glass plate made it impossible to observe the reaction initiation which would occur in the case of no collision. The free surface velocities of the 2 mm thick pellet samples were measured by inclined mirror method. One half of the free surface velocity, which would be approximately equal to the particle velocity, was always more than 20% greater than the particle velocity obtained from the impedance matching method.

Figure 5 is the Hugoniot of HN. The data from wedge- and pellet-type experiments are both plotted. The shock velocities from the pellet experiments are slightly higher than those from the wedge experiments. This discrepancy may be due to the fact that the shock velocities from the pellet experiments are average velocities over a run of a few millimeters.

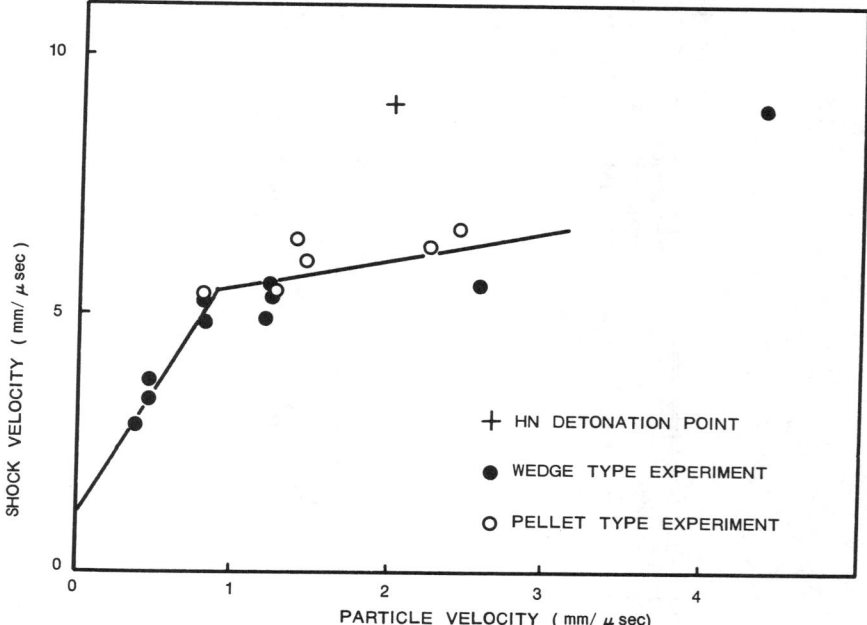

Fig. 5 Shock velocity vs particle velocity for HN.

The cross point in Fig. 5 is the Chapman-Jouguet (C-J) point of HN calculated from the Kihara-Hikita equation of state (Kihara and Hikita, 1953). The values of C-J parameters are 30.2 GPa for the detonation pressure, 9.11 mm/μs for the detonation velocity, 1.99 mm/μs for the particle velocity and 1506 K for the temperature of the gas products. The calculated detonation velocity was about 2% higher than that predicted from correlation due to Price et al. (1966).

$$D = 5.388\rho_0 - 0.100$$

where D is the detonation velocity in mm/μs and ρ_0 the initial loading density in g/cm^3.

The Hugoniot in Fig. 5 has two remarkable features. One is that it does not pass near the detonation point. The other is that it has a breaking point at about 1 mm/μs particle velocity and cannot be fitted by a single straight line.

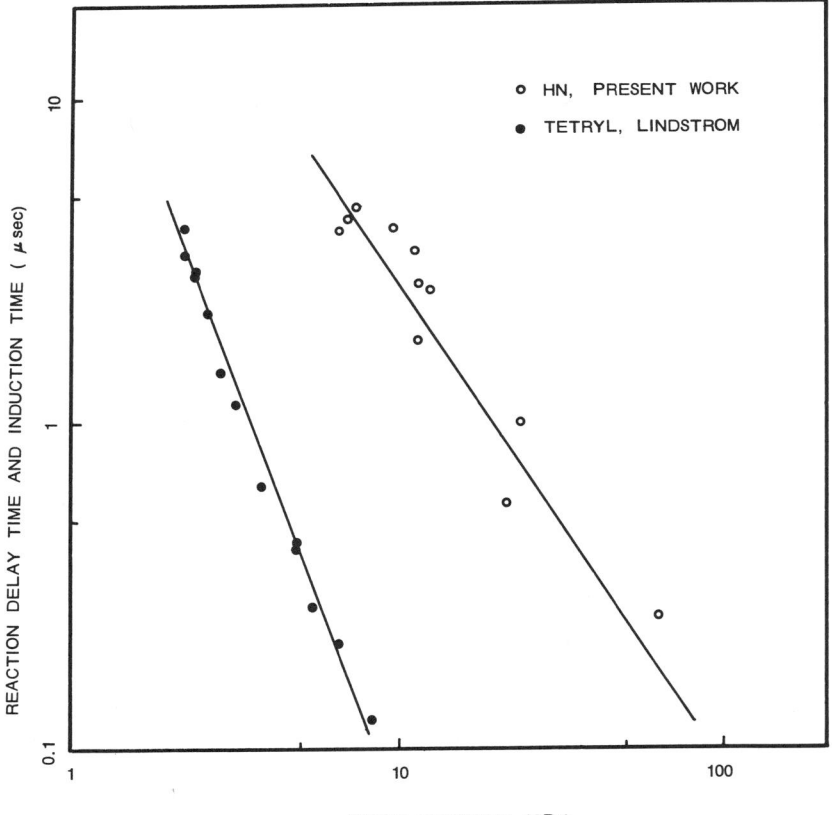

Fig. 6 Reaction delay time and shock-to-detonation induction time vs incident shock pressure for HN; Lindstrom's data for tetryl (Lindstrom, 1970) are shown for comparison purpose.

The data were fitted by two linear relations:

$$U = 1.2 + 4.9u \quad (u < 0.86) \tag{1}$$

$$U = 5.0 + 0.53u \quad (u > 0.86) \tag{2}$$

where U is the shock velocity and u is the particle velocity, both in mm/μs. The data obtained by the shock system that used an overdriven detonation wave were excluded because of its poor accuracy. Two straight lines intersect at the point where U is 5.4 mm/μs and u is 0.86 mm/μs. The shock pressure in HN at this intersection is about 8 GPa.

Reaction delay times and shock-to-detonation induction times as a function of incident shock pressure are both plotted in Fig. 6. The reaction delay time was defined as the interval between the first shock entrance into the sample and the appearance of the second shock wave at the free surface of the sample. Figure 6 gives us a rough estimate for the shock sensitivity of HN. The data reported by Lindstrom (1970) for tetryl at an initial density of 1.70 g/cm^3 are also shown for comparison.

Discussion

In this section, the features of the Hugoniot of HN are discussed, because the three initiation types and reaction delay times should have been affected by the scale of our shock systems.

The shock velocity U and the particle velocity u relationships of group 1 explosives, such as TNT (Ramsay and Popolato, 1965), PETN (Stirpe et al., 1970), and tetryl (Lindstrom, 1970) are well fitted by a single straight line when they are pressed to almost their crystal densities. Ramsay and Popolato (1965) showed that the linear fit of the U-u relationship extrapolates to the detonation pressure point rather than to the spike point, and the experimentally obtained Hugoniot of an explosive is not the Hugoniot of unreacted explosive but of a partially decomposed one.

However, the Hugoniot of HN had a break and did not pass the detonation pressure point.

Boyle et al. (1970) investigated the Hugoniot of several explosives and reported that there existed a break in the Hugoniot of TNT. They represented the Hugoniot of TNT by the following relations:

$$U \,(\text{mm}/\mu s) = 2.274 + 2.652u \quad (U < 3.7) \tag{3}$$

$$U \,(\text{mm}/\mu s) = 2.987 + 1.363u \quad (U > 3.7) \tag{4}$$

The pressure at the break was about 3.4 GPa. Boyle et al. (1970) and Craig (1970) noted the melting effects but they could not clarify whether the break is due to the melting effects or not.

The melting effects are also expected as a cause of the break in the Hugoniot of HN. [The melting point of HN at the atmospheric pressure was reported to be 70.71°C (Sommer, 1914).] If we presume the existence of melting for shock-compressed HN, the Hugoniot data will be considered to describe the mixture state of three components: decomposed gas, unreacted solid, and liquid explosive. The rate of transition, together with the behavior of the decomposed gas, should be taken into consideration, but it is complicated because we have neither the melting curve nor the rate of transition.

Our experiments were addressed mainly to the Hugoniot of HN. Shock-to-detonation induction time were generally not observed, except in a few cases, due to the scale of our shock systems. Larger scale shock systems are needed to investigate shock-to-detonation transition characteristics. Some other experimental methods such as pressure profile observations will give us a further detailed information about shock-initiation characteristics of HN.

Acknowledgments

The authors wish to thank Dr. K. Shiino for the help in synthesizing hydrazine nitrate and liquid explosives employed in this work.

References

Boyle V.M, Smother W. G., and Ervin L. H. (1970) The shock Hugoniot of unreacted explosives. *5th Symposium (Intl.) on Detonation,* pp. 251-257, ONR Rept. ACR-184.

Campbell A. W., Davis W. C., Ramsay J. B., and Travis J. R. (1961) Shock initiation of solid explosives. *Phys. Fluids* **4**, 511-521.

Craig B. G. (1970) *5th Symposium (Intl.) on Detonation,* pp. 257-258. ONR Rept. ACR-184.

Halleck P. M. and Wackerle J. (1976) Dynamic elastic-plastic properties of single-crystal pentaerythritol tetranitrate. *J. Appl. Phys.* **47**, 976-982.

Kennedy J. E. (1973) Pressure field in a shock-compressed high explosive. *14th Symposium (Intl.) on Combustion,* pp. 1251-1258. The Combustion Institute, Pittsburgh, Pa.

Kihara T. and Hikita T. (1953) Equation of state for hot dense gases and molecular theory of detonation. *4th Symposium (Intl.) on Combustion,* pp. 458-464. The Combustion Institute, Pittsburgh, Pa.

Kinslow R. (1970) *High Velocity Impact Phenomena.* Academic Press, New York.

Lindstrom I. E. (1966) Plane shock initiation of an RDX plastic bonded explosive. *J. Appl. Phys.* **37**, 4873-4880.

Lindstrom I. E. (1970) Planar shock initiation of porpus tetryl. *J. Appl.Phys.* **41**, 337-350.

Price D. (1967) Contrasting patterns in the behavior of high explosive. *11th Symposium (Intl.) on Combustion,* pp. 693-702. The Combustion Institute, Pittsburgh, Pa.

Price., Liddiard T. P., Jr., and Drosd R. D. (1966) The detonation behavior of hydrazine mononitrate. U. S. Naval Ord. Lab. Rept. NOLTR-66-31.

Ramsay J. B. and Popolato A. (1965) Analysis of shock wave and initiation data for solid explosives. *4th Symposium (Intl.) on Detonation,* pp. 233-238. ONR ACR-126.

Robinson R. J. and McCrone W. C. (1958) Hydrazine nitrate (1). *Anal. Chem.* **30**, 1014-1015.

Sarner S. G. (1966) *Propellant Chemistry,* p. 276. Reinhold Publishing Co., New York.

Sommer F. (1914) Studien uber das Hydrazin und seine anorganishchen Derivate. II. *z. anorg. allgem. Chem.* **86**, 71-87.

Stirpe D., Johnson J. O., and Wackerle J. (1970) Shock initiation of XTX-8003 and pressed PETN. *J. Appl. Phys.* **41**, 3884-3893.

Action of Charges with Axial Cavities on Rocks

V. V. Mitrofanov* and I.T. Bakirov†
Academy of Sciences, Novosibirsk, U.S.S.R.
and
G. A. Voroteljak‡ and V. A. Salganik§
Academy of Sciences, Krivoi Rog, U.S.S.R.

This paper presents experimental data which show that the axial cavity in the well charge permits the best exploitation of explosive energy. The detonation regimes of longitudinal charges with axial channels (LCAC) are analyzed. The significant effect of LCAC in rocks is rationalized on the basis of the change in the form of the pressure pulse that affects the walls of the well.

Introduction

IT is known that a longitudinal channel inside the charge or between its outer surface and a rigid wall exerts a significant influence on the detonation wave. This is known as the channel effect. Recent investigations of this phenomenon have been described in a series of publications (Johansson and Persson, 1970; Dubnov et al., 1973; and Zagumennov et al., 1969). For industrial mixed explosives, the channel effect was considered to be a negative factor causing the detonation to damp in the holes. For high explosive charges, the channel effect leads to an increase in detonation velocity up to levels characteristic of denser explosives. In the case of high-porosity explosives, damping has been observed under some conditions by Golbinder and Tyshevich (1967). For high-sensitivity explosives, such as tetryl and PETN, the detonation velocity of tubular charges exceeded by about 10% that of solid charges with a crystallic density (Ahrens, 1965).

New Experimental Results

Recently, new aspects of the channel effect have been discovered. A high detonation velocity (up to 14 km/s) was observed in sensitive primary ex-

---
Presented at the 7th ICOGER, Göttingen, Federal Republic of Germany, Aug. 20-24, 1979.
Copyright © American Institute of Aeronautics and Astronautics, Inc., 1981. All rights reserved.
*Head of Laboratory of Detonation Processes, Institute of Hydrodynamics.
†Senior Engineer, Institute of Hydrodynamics.
‡Head of Laboratory of Rock Fragmentation, Research Mining Institute.
§Senior Researcher, Research Mining Institute.

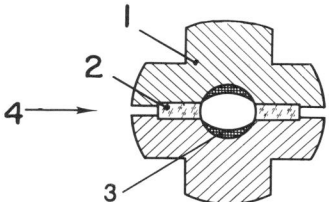

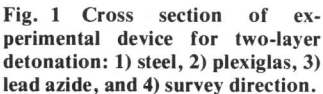
Fig. 1 Cross section of experimental device for two-layer detonation: 1) steel, 2) plexiglas, 3) lead azide, and 4) survey direction.

plosives with the channel filled with a light gas (Bakirov and Mitrofanov, 1976). The cross section of the channel inside which this process was observed is shown in Fig. 1 and a streak photograph in Fig. 2. The diameter of the channel is 10 mm, its length is 80 cm, the quantity of lead azide is about 80 mg/cm. The channel is filled with helium under the pressure of 500 Torr. In these experiments, the detonation velocity of the heterogeneous system was 2-2.5 times as high as that of a crystal-density explosive. The actual density of lead azide was about 1.2 g/cm^3; its detonation velocity out of the channel was 2.2 km/s. The process is called a two-layer detonation because the accelerated motion of a chemical reaction zone along the explosive layer is preceded by the initiating pulse transmittance through the adjacent gas layer (see also Mitrofanov, 1976).

The same type of two-layer process with a velocity exceeding 10 km/s in tubular charges of pressed RDX having an axial cavity filled with a liquid hydrogen has been observed by Silvestrov and Urushkin (1977).

At the same time, Salganik et al. (1977) discovered that the modern industrial granulated explosive charges, having longitudinal axial cavities, use more advantageously the explosive energy for rock fragmentation. In the meantime, a good deal of experience has been gained with the use of this type of charge, referred to hereafter as LCAC (longitudinal charges with axial channels).

The experiments were carried out with granular explosives of the type "zernogranulit 79/21" (79% ammonium nitrate and 21% TNT) and "granulit AC-8" (89% ammonium nitrate, 8% powdered aluminum, 3% mineral oil). The LCACs were formed in the holes pneumatically, using a hose with a special head. The explosive density was about 1.2 g/cm^3. The critical detonation diameter was 15-25 mm (steel tube confinement) and 60-100 mm (paper shell confinement) for the charges without cavities. Intermediate charges of more sensitive explosives without cavities were used for LCAC initiating by a standard detonator. The charge diameters d_1 were 40, 65, and 105 mm. The diameters d_2 of axial cavities were about 0.4 d_1. The charge length achieved was 30 m and even 50 m. Under these conditions the charge detonation always proceeded to its full extent, and the channel effect did not cause detonation damping.

As compared with solid charges, the above-mentioned LCAC produced a more uniform rock fragmentation of a larger volume. The effect of the axial cavity (expressed in terms of the diameter ratio) on the fragmentation volume is illustrated in Fig. 3. The results were obtained in the case of a singular

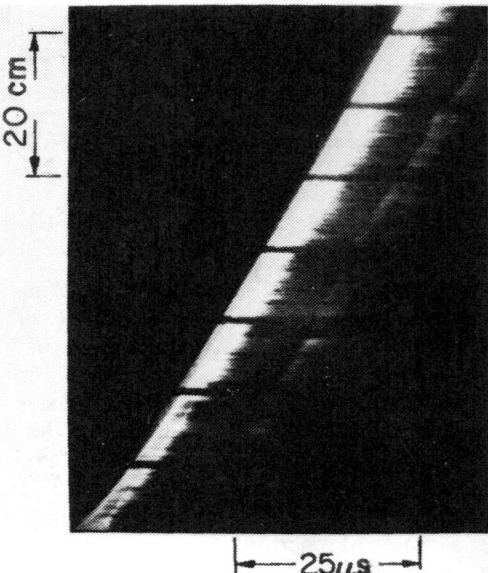

Fig. 2 Photoscan of two-layer detonation.

charge used in the hole of a rock with different rigidity. The axial cavity diameter was varied over a wide range. The fragmentation volume is largest when the ratio of the cavity diameter to the charge ranges 0.3-0.4. The optimum value of this ratio decreased for higher rock strengths.

The use of LCAC was effective only in the case when the charge was detonated from the end adjacent to the rock surface. If it is detonated in the opposite direction, the fragmentation volume becomes smaller than that for solid charges. Thus, LCAC has some new, unexpected advantages that enhance the efficiency of rock fragmentation by use of explosives.

Analysis and Discussion

Now let us try to unravel the various experimental observations of the channel effect and elucidate the specific features of the LCAC mechanism on the basis of current knowledge of detonation waves.

After the transition of detonation from the initiating solid charge to LCAC, a shock wave that is ahead of the detonation front in the explosive is established in the channel. We refer to it as the channel wave (CW). The CW is pushed by the "piston" formed by the explosion products (EP) closing the channel behind the detonation front, and sometimes overtaking the front in the form of a dense gas-cumulative jet (Fadeenko et al., 1974). For a charge without an inside facing, there is an additional flowfield behind the CW due to decomposition of the surface explosive layer within the zone of a hot shock-compressed gas (Johansson and Persson, 1970; Dubnov and Hotina, 1966; Dremin and Vanden Berghe, 1968). In the initial stage of the process, the CW velocity U is about 1.5 times as high as the detonation velocity D_0 of the solid

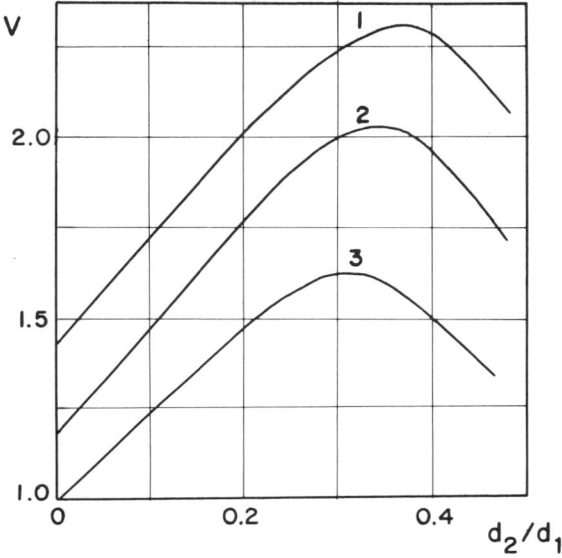

Fig. 3 Volume of rock outburst hopper V (in per unit value) vs ratio of internal and external diameters of LCAC ($d_1 = 40$ mm, hole depth = 1.5 mm): 1) for rock with strength $f = 4$-6 units in scale of Protodiakonov, 2) $f = 8$-10, 3) $f = 12$-14.

charge of the same density ρ_1. For typical industrial explosives a detonation velocity $D_0 = 3.5$-4.5 km/s, $U = 4.5$-7 km/s, the pressure immediately behind CW is $P_s \simeq 250$-500 bar, and the temperature $T_s \simeq 6 \times 10^3$-10^4 K.

The pattern of wave interactions associated with the channel effect is shown in Fig. 4. Under the action of the channel wave there appears an oblique compression wave occupying the conical region 3 between the channel wave and the detonation front AE in the uncompressed explosive. The dimensions of the region depend on the l distance between the channel wave and the detonation front. The compression wave separates into an elastic wave (elastic precursor) associated with a negligibly small density change, and a plastic packing wave, propagating with a velocity $W = \sqrt{(P_s - \sigma)/\rho_1(1 - \rho_1/\rho_3)}$ $\simeq 250$-400 m/s at $\rho_1 = 1$-1.2 g/cm^3, $\rho_1/\rho_3 \simeq 0.7$, and $\sigma = 0$. For the plastic wave to appear, the pressure behind the CW must be higher than the crushing strength σ of the explosive. As a rule, this condition is fulfilled. It should be noted, however, that for the charges obtained by "pneumofilling" with $\rho_1 \gtrsim 1.2$ g/cm^3, the crushing strength is $\sigma \sim 10^2$ bar. Under rapid dynamic loading, the strength can approach P_s; thus, the plastic wave can decrease in strength or even disappear altogether. In the last case, the CW almost does not affect the explosive. The explosive properties, and their change under the action of the CW, can produce different detonation structures, as depicted schematically in Fig. 4.

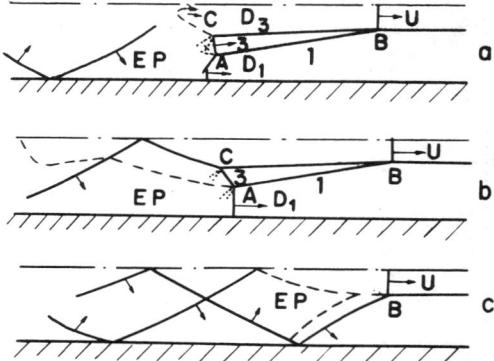

Fig. 4 Three types of structure of explosion wave in LCAC: 0B = shock wave in the channel, 1 = initial explosive, and 3 = compressed explosive; solid lines = shock and detonation jumps, explosive surface; dotted lines = EP boundaries; and dashed lines = Prandtl-Mayer waves.

Case (a)

The compressed layer 3 detonates together with the uncompressed layer 1. If the detonation velocity inside the layer is $D_3 > D_1$, the detonation pressure is higher. An intermediate oblique detonation front appears in the uncompressed layer (Woodhead and Titman, 1965). After its emergence on the charge surface and reflection from the wall, a narrow high-pressure region is created. At a distance $l = AB$ the compressed region 3 can cover the entire thickness of the explosive. Curve (a) in Fig. 5 represents an approximate time variation of pressure on the external charge wall. Curve (0) represents the pressure profile behind the detonation wave in the charge without a cavity; this wave creates a peak pressure $P_{W,0}$ on the wall. If $D_3 < D_1$ which is also possible under nonideal detonation regime, the detonation front in the compressed layer becomes oblique, and the head pressure peak decreases down to a level $P_{W,1} < P_{W,0}$. The initial peak is followed by the pressure decay due to expansion of explosion products into the channel. As a consequence of collision on the channel axis a reflected wave is created. Its radial motion produces an additional maximum pressure on the wall. The average pressure of such oscillations decreases due to the outflow of the explosive products (EP) from the detonation front and the expansion of the wall. When $D_3 < D_1$, the detonation velocity D_1 and peak pressure at the wall decreases as the thickness of the uncompressed layer thickness becomes smaller.

The detonation propagation regime corresponding to case (a) can be stationary or nonstationary, with periodic acceleration or retardation of CW at a great length of LCAC (Zagumenov et al., 1969). In the case under consideration, the generation mechanism of the channel wave pulsations is probably associated with a periodic disappearance of a cumulative jet and leakage of the shock-compressed gas along a near-axial region, when the channel pressure before the region of the EP collision increases due to the erosion of explosive and other boundary phenomena. In the presence of the

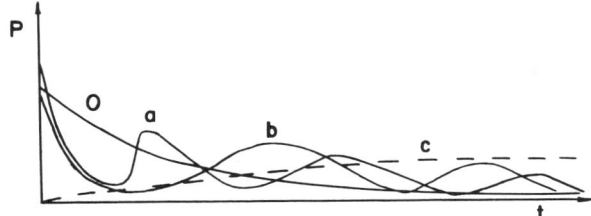

Fig. 5 Explosion wave pressure profiles on hole walls within regimes corresponding to Fig. 4 (0 indicates charge without cavity).

jet, the distance l cannot be constant for a long time, since gas is continuously being stored between the detonation front and the channel wave.

Case (a) is realized, as a rule, in the LCACs made of individual explosives, as well as in the industrial explosives, which are either susceptible to compression under the action of the CW or whose sensitivity to initiation is altered by the action of the shock waves.

Case (b)

In this case, only the uncompressed region 1 of the charge detonates. The compressed layer 3, passing through the oblique wave AC and the rarefaction wave, is deflected by the EP pressure to the near-axial region. After collision it reacts, releasing explosion heat with delay. This energy does not take part in sustaining the detonation wave within region 1, but can play a significant role in the effect exerted by the LCAC explosion on rock. This provides a more beneficial pressure redistribution in time (curve (b) in Fig. 5). In this case, the head pressure peak is lower in amplitude than that of the solid charge, and shorter in duration. After an initial decay, the pressure rises again for a longer period than in case (a), due to a longer heat release process and smaller initial wall expansion velocity. It should be noted that, at the channel axis, the conditions for ignition of the deflected compressed layer of explosive are more favorable than in the case of the "external channel" because, under implosion near the axis, the medium is subjected to an additional internal deformation and heating.

The state behind the oblique shock AC can be calculated from the conditions of equality of pressures and of turning angles of flows separated by a contact surface (in this case the flow immediately behind AC and that of the EP, which pass through the centered rarefaction wave beginning at point A). Assuming a plane approximation, a Hugoniot of porous TNT (Afanasenkov et al., 1970), and the adiabatic index of the EP $\eta = 3$, we find that P_{AC} falls within the range $20\text{-}40 \times 10^3$ bar when $\rho_1 = 1.2$ g/cm^3, $D_1 = 4\text{-}4.5$ km/s, and $\rho_3 = 1.3\text{-}1.6$ g/cm^3. If $\eta < 3$, the calculated value of P_{AC} is higher. Thus the parameters of shock wave AC are higher than their critical threshold for the initiation of a detonation wave if the duration of the process is sufficiently long. Although chemical reaction undoubtedly occurs in "hot points" behind AC, its rate is insufficient for transition to self-sustained detonation when the extent of AC is small. When shock wave AC reaches some critical thickness h_3 of layer 3, however, it gives rise to a detonation wave, and can be sustained due to its own energy release independently of the detonation shock AE within the uncompressed layer (similar results were suggested by Dremin and Vanden Berghe, 1968).

With increasing distance l the extent of AC increases and that of AE decreases. There exists a critical value h_1 of the thickness of AE below which the self-sustained detonation cannot be supported within the layer. If the extent of AE becomes smaller than h_1 before the thickness of layer 3 reaches h_3, failure of the detonation of LCAC occurs. If h_3 is attained before the extent of AE shrinks to h_1, the detonation is continued in the regime corresponding to that considered in case (a). The dimensions of LCAC for which h_1 and h_3 are reached simultaneously are critical for the propagation of the detonation wave.

The maximum displacement of the detonation front produced by the channel wave in the regime corresponding to case (b) can be estimated from the formula $l_m \simeq 0.5(d_1 - d_2)U/W$. When $d_2 = 0.4\ d_1$ and U/W is 15-20, $l_m \simeq (10\text{-}15)d_2$. When $l = l_m$, the detonation front is at a distance $X_m \simeq l_m \bar{D}_1 / (\bar{U} - \bar{D}_1)$ from the charge initiation plane. The velocities in this latter formula are taken as an average over the whole period of the motion, while in the formula for l_m the velocities are those for l_m long. Usually $\bar{D}_1 / (\bar{U} - \bar{D}_1) \simeq 5\text{-}7$, therefore $X_m \simeq (50\text{-}100)\ d_2$. The detonation decays approximately over the same distance, if transition to regime (a) does not occur. With this value of X_m, an estimate of the maximum arbitrary period of detonation wave pulsations in LCAC can be made. It should be noted that with increasing l and decreasing AE the velocity of the channel wave decreases continuously to D_1. After the detonation wave is established in the compressed layer, its propagation velocity along the charge and the stream turning angle behind the front AC increase sharply, while the distance l decreases. This is associated with the generation of a second compression wave in the channel following the first channel wave. As they merge forming a transmitted channel wave, region 3 is reduced and the detonation decays. Consequently, the initial conditions of case (b) are recovered. When the length of the charge is large, this process can be repeated many times. A stationary detonation mode is also possible in regime (b) if the velocity of the channel wave decreases to D_1 and stabilizes before the thicknesses of layers 3 and 1 reach h_3 and h_1. The formation of a cumulative jet of EP in regime (b) presents difficulties, since EP are covered by a thick layer of a slowly reacting explosive. The LCAC detonation in regime (b) is most typical for industrial, easily compressed explosives.

Case (c)

The channel wave creates in the explosive an oblique wave with an intensive chemical reaction behind its front. The channel wave is sustained by the expansion of gaseous products of this reaction, and it can run away from the primary detonation front AE. The latter decays when the quantity of unreacted explosive in front of it becomes small. This is the regime of a two-layer heterogeneous detonation that has been observed for highly sensitive explosives. A similar regime with a relatively low propagation velocity (1.5-2 km/s) is observed in long powder channels. In LCAC of industrial explosives regime (c) probably can exist only in the presence of a sturdy shell, such as thick steel walls and in rocks. Such possibility was emphasized by Johansson and Persson (1970). Because of the low sensitivity of explosives, regime (c) can be realized only with an oblique convective combustion wave associated with a delayed pressure increase inside the well (curve (c) in Fig. 5).

Thus, all the regimes of LCAC considered herein can be realized under certain conditions, with the use of both HE and mixed industrial explosives. For LCAC exploded in rocks, there are insufficient data on the conditions under which one or another regime can be realized. Most probably, a nonstationary process takes place. This would be consistent with the results of experiments in which LCAC was placed parallel to a free surface of a rock. The rock surface after explosion in the vicinity of the well sometimes retains the traces of a periodical pressure variation along the well. This phenomenon can result from the transitions between regimes (a) and (b) described above.

Conclusions

The detonation process, and the particular effects of the LCAC on rocks, are not clearly understood and require more exploratory research. However, on the basis of available data, it is possible to specify the following factors that increase the efficiency of the LCAC effect in comparison to that of a solid charge:

1) The presence of an axial cavity in the charge reduces the high-amplitude front of the pressure pulse that is responsible for the initial expansion velocity of the wall and the plastic deformation zone around the well. As a result of a reduction of this zone, the fraction of energy consumed unproductively is also reduced. The total time of the process at the average pressure level increases as a consequence of deeper cracks formed in the rock.

2) As the expansion velocity of the walls of a well decreases at the initial stage of fragmentation, the full pressure momentum increases and, thus, the final velocity of rock fragments is increased. This result has been recorded cinematographically.

3) The variation of the pressure pulse in the regime of nonstationary detonation enhances fragmentation.

4) By virtue of similar effects of those described in item 1, the delay in pressurization at the mouth of the hole is increased. Thus, the loss of explosion energy in the products is decreased. As a result of the decrease of the plastic deformation zone, the penetration of the explosive gas into the growing cracks is enhanced.

The LCAC effect is similar to the action of charges with cavities of other shapes as, in particular, charges with transverse air gaps (Melnikov et al., 1978). The main advantage of charges with axial cavities is the ease of fabrication by a practically standard method of pneumatic forming of the charge.

References

Afanasenkov A. N., Bogomolov V. M., and Voskoboinikov I. M. (1970) Kriticheskoe davlenie initsiirovaniia vzryvchatyh veshchestv. *Vzryvnoie delo* No. 68/25, 68-92.

Ahrens H. (1965) Uber den detonations vorgang bei zylindrishen sprengstoffladungen mit axialer Höhlung. *Explosivstoffe* 5, 6, 10, 11, S113-123, 155-166, 267-276, 295-309.

Bakirov I. T. and Mitrofanov V. V. (1976) Vysokoskorostnaya dvusloinaia detonatsiia v sisteme "VV-Gas." *Dokl. Akad. Nauk* 231 (6), 1315-1318.

Dremin A. N. and Vanden Berghe R. (1968) Etude sur l'arret de la detonation par l'effet de canal. *Explosifs* 1, 5-22.

Dubnov L. V., Baharevich I. E., and Romanov A. I. (1978) *Promyshlennye vzryvchatye veshchestva*, pp. 227-232. Nedra, Moscow.

Dubnov L. B. and Hotina L. D. (1966) O mekhanizme kanalnogo effekta pri detonatsii kondensirovannyh VV. *Fizika gorenia i vzryva* 4, 97-104.

Fadeenko Yu. I., Lobanov V. F., Silvestrov V. V. and Titov V. M. (1974) High speed gas flows in explosions of cavitated explosives. *Acta Astronautica* 1, 1171-1180.

Golbinder A. I. and Tyshevich V. F. (1967) Dalneishee issledovaniie kanalnogo effekta. *Teoriia vzryvchatyh veshchestv*, pp. 349-362. Vysshaya shkola, Moscow.

Johansson C. H. and Persson P. A. (1970) *Detonics of High Explosives*, Academic Press, London and New York.

Melnikov N. V., Marchenko L. N., Zharikov I. F., and Seinov N. P. (1978) Blasting methods to improve rock fragmentation. *Acta Astronautica* 5, 1113-1127.

Mitrofanov V. V. (1976) Detonation on two layer systems. *Acta Astronautica* 3, 995-1005.

Salganic V. A., Zheleznyak S. S. and Vorotelyak G. A. (1977) Metod povysheniia poleznoi raboty vzryva na gornorudnyh predpriiatiiah. *Gornyi zhurnal* 2, 54-57.

Silvestrov V. V. and Urushkin V. P. (1977) Detonatsiia trubchatyh zaryadov VV v zhidkom vodorode. *Fizika goreniia i vzryva* 1, 78-84.

Woodhead D. M. and Timan H. (1965) Detonation phenomena in a tubular charge of explosive. *Explosivstoffe* Nos. 5-6, S124-134, 141-155.

Zagumennov A. S., Titova N. S., Fadeenko Yu. I., and Christyakov V. P. (1969) Detonatsiia udlinnennyh zaryadov s polostyami. *Prik. mekh. tekh. fiz.* 2, 79-88.

Characterization of Mass Flow Rates for Various Percussion Primers

K. K. Kuo,* B. B. Moore,† and D. Y. Chen‡
The Pennsylvania State University, University Park, Pa.

Percussion primers have been used widely for ignition in various propulsion systems. However, the mass and energy fluxes generated through a primer blast are strongly time dependent, and are not well characterized. It has been observed that the ignition and flame spreading processes of solid propellants greatly depend upon the primer output. Therefore, the objective of this study is to determine the instantaneous gaseous mass flow rate, the energy fluxes, and the percentage of product in the condensed phases. A test rig has been constructed to conduct primer characterization studies. A theoretical model based upon two-phase conservation equations has been developed. The gaseous mass flow rates for a number of primers have been obtained from the theoretical model which uses the recorded P-t traces, and the adiabatic flame temperature based upon thermochemistry calculations. Experimentally, good reproducibility in the uprising portion of the P-t traces has been observed. The instantaneous gaseous mass flow rates for two different percussion primers have been obtained. Results also indicate significant percentages ($\sim 40\%$) of product in the condensed phases. The potential importance of the condensed phase product to the ignition mechanism suggests the need for further study.

Nomenclature

A_t = exit throat area, cm^2
c_p = constant-pressure specific heat, cal/g·K
c_v = constant-volume specific heat, cal/g·K
E = total stored energy, cal
g_c = gravitational constant, cm·dyne/g·s^2

Presented at the 7th ICOGER, Göttingen, Federal Republic of Germany, Aug. 20-24, 1979.
Copyright © American Institute of Aeronautics and Astronautics, Inc, 1980. All rights reserved.
*Associate Professor, Department of Mechanical Engineering.
†Graduate Assistant, Department of Mechanical Engineering; presently with Aerojet Solid Propulsion Company, Sacramento, Calif.
‡Graduate Assistant, Department of Mechanical Engineering; presently at University of California, Berkeley.

h = specific enthalpy, cal/g
J = heat-to-work conversion factor, 4.184×10^7 dyne·cm/cal
KE   = kinetic energy per unit mass, cal/g
$\dot{m}$ = mass flow rate, g/s
M = total instantaneous mass, integrated, g
P = pressure, dyne/cm^2 or atm
$\dot{Q}_{\text{loss}}$ = rate of heat loss to the surrounding of the control volume, cal/s
R = gas constant, cm·dyne/g·K
t = time, s
T = temperature, K
T_f = adiabatic flame temperature of the primer mix, K
u = gas velocity, cm/s
V = instantaneous volume, cm^3
γ = specific heat ratio
ρ = density, g/cm^3
$\bar{\rho}_s$ = average density of condensed phases in combustion product, g/cm^3
Ψ_g = mass fraction of the combustion product in the gas phase
Ψ_s = mass fraction of combustion product in the condensed phases, $\Psi_s \equiv 1 - \Psi_g$

Subscripts

c = properties evaluated in the test section
cv   = control volume
g = properties associated with the gas phase
in   = quantities influxing to the test section
0 = properties evaluated at initial condition
out   = quantities effluxing from the test section
p = properties associated with the primer mix
s = properties associated with the condensed phases

Introduction

FOR a given percussion primer of known initial ingredients and weight fractions, many unknown parameters are needed, either in experimental design considerations of the overall combustion system or in theoretical predictions of the combustion process. These parameters include the duration of the primer blast, the instantaneous mass flow rate, the percentage of the condensed phases in the primer product, the energy fluxes associated with the product gases, the average molecular weight of the primer product, the specific heat ratio of the product gases, etc. Experimental evidence shows that these parameters can have a significant influence upon the combustion performance of the overall system, such as a packed bed of granular propellants.

In the case of a granular propellant bed, the aforementioned parameters resulting from a primer blast have pronounced effects on the subsequent combustion event. The mass and energy fluxes from the primer are the necessary physical input data as boundary conditions for granular bed combustion predictions. The magnitude of the percentage of condensed species in the primer product also influences the heat-transfer mechanism involved in the ignition phase. If the percentage of the condensed species is negligible in comparison with that of the gaseous species in the primer product, the ignition of the granular propellants in a packed bed is achieved essentially by convective heating of the propellant grains. On the other hand, if there is a significant percentage of primer product in the condensed phase, conduction and radiation heat transfer must also be considered in the ignition mechanism.

A literature survey was conducted to determine the state-of-the-art concerning the characterization of primers. Very limited information is available. Davis (1977) performed a preliminary investigation of primer characterization for small arms. However, his theoretical model was based on a single-phase flow. The measured pressure-time traces exhibited multiple peaks, indicative of pressure wave phenomena within the primer characterization test rig. The necessary thermochemical data were lacking at the time of his investigation, and assumed values of adiabatic flame temperature and specific heat ratio were varied around 2700 K and 1.2, respectively. Summarizing his findings, Davis concluded that:

1) Predicted flow rates are primarily controlled by the slope of the pressure-time trace. The mass flow rate reaches a maximum when the pressurization rate attains its highest value. The predicted flow rates drop to zero shortly after the pressure slope becomes negative.

2) Increases in T_f, holding all other operating conditions constant in a parametric study, result in only slight decreases in the igniter mass flow rates. In general, mass flow rates are relatively unaffected by limited changes in T_f.

3) Similarly, primer mass flow rate decreases slightly with an increase in γ.

4) The time required for total consumption of the primer was found to be around 70-90 μs. The primer discharge time was significantly shorter than the characteristic time associated with the granular propellant combustion process.

Work concerning primer vent geometry and primer strength was investigated by Gerri et al. (1977). Their interest focused on the pressure-time trace of the granular propellant bed resulting from primer vent geometry and primer weight variations. They found that the propellant-primer interface condition is a complex function of primer strength, the manner of delivery. The venting geometry of the primer exhibit an immediate and fundamental influence on subsequent combustion events. In addition, they noted that the degree of bed compaction provided by the initial primer blast greatly influences the subsequent processes of the combustion wave and, in general, the overall combustion phenomenon within the granular bed. Their work, however, does not characterize the mass and energy fluxes from any given percussion primer.

The specific objectives of this research are: 1) to determine the mass flow rates of various primers so that the information can be used as input for granular propellant bed combustion studies, 2) to test the reproducibility of the pressure-time traces for a given primer type, 3) to study the effects of various operating conditions on the primer mass flow rate, 4) to determine the fraction of primer product in condensed phases, 5) to develop a scheme which can be used by other researchers in characterizing percussion primers, and 6) to suggest topics for future research in this area, based on the technological gaps disclosed by the present study.

Method of Approach

Experimental Setup for Primer Characterization

The experimental setup used in the primer characterization studies is shown in Fig. 1. Upon closing the solenoid switch, the solenoid armature passes through a series of coil windings. This displacement creates the sudden impact of the firing pin against the percussion primer installed in the primer housing insert at the upstream location of the test section. The test section consists of the divergent portion of the primer housing insert, a cylindrical void section, and a pressure wave absorber. In order to achieve chamber depressurization after a primer blast, a convergent-divergent exit nozzle (with throat diameter of 0.508 mm) was located at the downstream portion of the test section. A choked flow condition was maintained at the exit nozzle during the transient pressurization interval.

The nature of the rapid primer blast introduces the problem of complex wave interactions. In order to eliminate the multiple wave reflections within the chamber, a pressure wave absorber was designed, fabricated, and positioned anterior to the throat of the exit nozzle. The subsequent pressure-time traces had fewer multiple peaks and were a more accurate representation of the average overall pressurization-depressurization process. A detailed sketch of the pressure wave absorber is shown in Fig. 2.

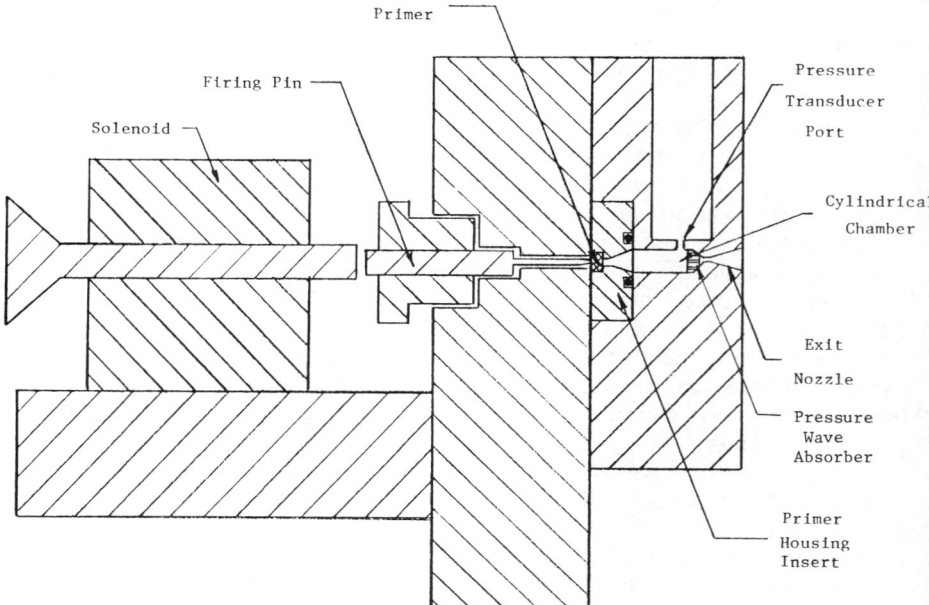

Fig. 1 Schematic diagram of experimental setup for primer characterization tests.

Table 1 Physical, compositional, and thermochemical data of FA-34 and FA-41 primers[a]

Ingredient	% by weight	Density, g/cm^3	Heat of formation, ΔH_f°, cal/mole
PETN ($C_5H_8N_4O_{12}$)	5.0 ± 1	1.77	−121,028
Lead styphnate (PbC$_6$H$_3$N$_3$O$_9$)	37.0 ± 5	3.02	−43,056
Tetracene, $C_2H_8N_{10}O$	4.0 ± 1	1.05	45,190
Aluminum powder (Al)	7.0 ± 1	2.70	0
Antimony sulfide (Sb$_2$S$_3$)	15.0 ± 2	4.12	−43,500
Barium nitrate, Ba(NO$_3$)$_2$	32.0 ± 5	3.24	−237,000

[a] Mix specification = FA-956, average density of the mixture = 3.092 g/cm^3, weight of FA-34 primer mix ≅ 0.0358 g, weight of FA-41 primer mix ≅ 0.0224 g, weight of CCI-200 primer mix ≅ 0.0369 g. (CCI-200 primer is manufactured by Omark Industries. Its composition is considered proprietary information. Normal lead styphnate and tetracene are used as initiators. Barium nitrate serves as the oxidizer, antimony sulfide and aluminum as the fuel.)

Three different primers were investigated. Table 1 presents the physical and compositional data for FA-34 and FA-41 primer mixes, as well as some comments on the commercially available CCI-200 primer; its exact composition could not be disclosed by the manufacturer.

A block diagram of the data acquisition system for the primer characterization study is shown in Fig. 3. The data acquisition system consists of a single Kistler pressure transducer (601B), a Kistler 504E charge amplifier, a Hewlett-Packard 3924B high-speed magnetic tape recorder, and a Hewlett-Packard 7044A X-Y plotter and Tektronix 535A oscilloscope as display devices.

Pressure was recorded using the 601B Kistler pressure transducer, located at a position approximately midpoint in the test section. Nominally, the transducer has a natural resonance frequency of 300 kHz, which corresponds to an effective rise time approximately equal to 1.5 μs. It is, therefore, satisfactory for use in the primer characterization studies that typically have rise times in the order of 60 times longer. This gage measures pressures accurately in a range of 1–1000 atm with a maximum allowable pressure measurable of 1225 atm. To ensure that the Kistler gage was not damaged by high-temperature combustion gases, the gage was slightly recess mounted and equipped with a water-cooled adapter (Kistler 628C).

The signal recorded on the tape recorder can be displayed by one of two methods: electronically by a cathode-ray oscilloscope or mechanically by an X-Y plotter for hard-copy purposes. The Biomation 1015 serves as an analog-digital converter which stores the information in digital form. It provides four-channel data storage, 1024 word memory per channel or 4096 word memory for a single channel, and is capable of a real-time resolution of 10 μs.

Theoretical Analysis

A theoretical model was constructed, based upon a multiphase flow consideration. The theoretical model was developed to use the measured data as input for the determination of the mass flow rates of a given percussion primer. Figure 4 gives a schematic diagram of the control volume for the gas

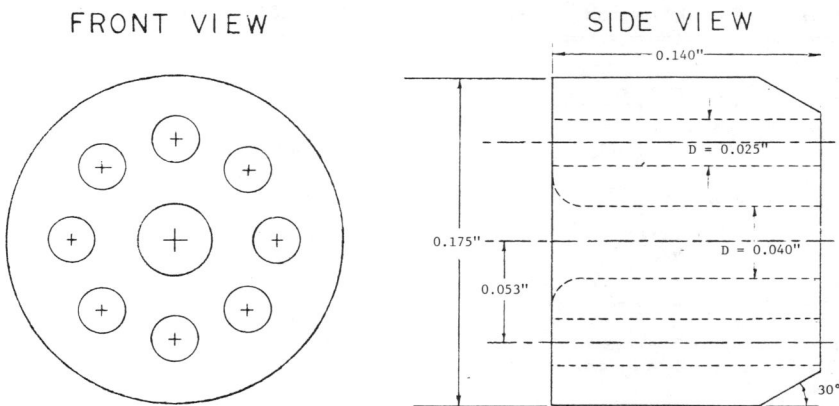

Fig. 2 Stainless steel pressure wave absorber insert for primer characterization test rig; scaling: 20 = 1.

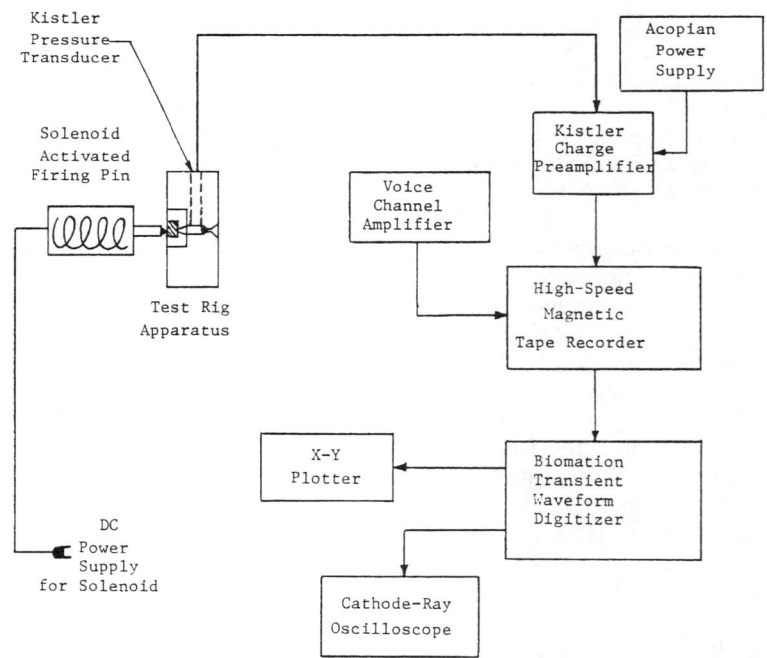

Fig. 3 Block diagram of the data acquisition system for the primer characterization study.

phase used in the analysis. The control volume is a multiply connected spatial region encompassed by the control surface, as indicated by dashed lines. Since the control volume is quite small (~ 0.26 cm^3), its gaseous properties can be treated essentially as lumped parameters. The use of the lumped-parameter approach is further warranted in view of the fact that a pressure wave absorber

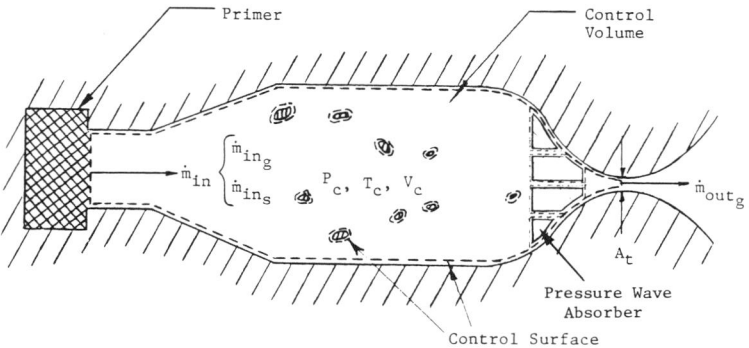

Fig. 4 Schematic diagram of the control volume used in the primer characterization analysis.

is used at the downstream location to minimize complex wave reflection phenomena.

The conservation of gaseous mass for the control volume shown in Fig. 4 is

$$\frac{dM_g}{dt}\bigg|_{cv} = \dot{m}_{in_g} - \dot{m}_{out_g} \tag{1}$$

where the instantaneous gaseous mass M_g in the control volume is equal to the product of $\rho_c V_c$.

Due to the mass consumption of the primer material and the accumulation of condensed phases in the control volume, the instantaneous chamber volume can be expressed as

$$V_c(t) = V_{c0} + \int_0^t \frac{\dot{m}_{in}}{\rho_p} dt - \int_0^t \frac{\dot{m}_{in_s}}{\bar{\rho}_s} dt \tag{2}$$

Differentiating Eq. (2) with respect to time, we have

$$\frac{dV_c}{dt} = \frac{\dot{m}_{in}}{\rho_p} - \frac{\dot{m}_{in_s}}{\bar{\rho}_s} \tag{3}$$

Substituting Eq. (3) into Eq. (1), and using the ideal gas law, the gaseous mass conservation equation becomes

$$\frac{V_c}{R}\frac{d[P_c/T_c]}{dt} = \dot{m}_{in_g} - \dot{m}_{out_g} - \frac{P_c \dot{m}_{in}}{RT_c \rho_p} + \frac{P_c}{RT_c}\frac{\dot{m}_{in_s}}{\bar{\rho}_s} \tag{4}$$

A different form of Eq. (4) for computing the rate of change of T_c is given as

$$\frac{dT_c}{dt} = \frac{RT_c^2}{V_c P_c}\left(\dot{m}_{out_g} - \dot{m}_{in_g}\right) + \frac{\dot{m}_{in} T_c}{V_c \rho_p} + \frac{T_c}{P_c}\frac{dP_c}{dt} - \frac{\dot{m}_{in_s} T_c}{V_c \bar{\rho}_s} \tag{5}$$

Since the area ratio of the test section to the exit nozzle throat is very large ($\simeq 82$), it is reasonable to assume that the throat is not blocked by solid particles during the short pressurization interval ($\simeq 100\ \mu s$). It is further assumed that the exit nozzle throat is choked during the transient interval of interest. The validity of this assumption is justified by the elevated chamber pressure that is considerably greater (about two orders of magnitude) than the ambient atmospheric pressure. The gaseous mass flow rate issuing from the exit nozzle can then be expressed as

$$\dot{m}_{\text{outg}} = \frac{\Gamma(\gamma) P_c A_t}{\sqrt{RT_c/g_c}} \qquad (6)$$

where

$$\Gamma(\gamma) \neq \sqrt{\gamma} \left[2/(\gamma+1) \right]^{(\gamma+1)/2(\gamma-1)} \qquad (7)$$

Upon substitution of Eq. (6), Eq. (5) now becomes

$$\frac{dT_c}{dt} = \frac{RT_c^2}{V_c P_c} \left[\frac{\Gamma(\gamma) P_c A_t}{\sqrt{RT_c/g_c}} - \dot{m}_{\text{ing}} \right] + \frac{\dot{m}_{\text{in}} T_c}{V_c \rho_p} + \frac{T_c}{P_c} \frac{dP_c}{dt} - \frac{\dot{m}_{\text{ins}} T_c}{V_c \bar{\rho}_s} \qquad (8)$$

Considering the lumped-parameter approach mentioned earlier the general form of the energy equation in the gaseous phase is

$$\left. \frac{dE}{dt} \right|_{cv} = \dot{m}_{\text{ing}} (h_{\text{in}} + \cancel{KE}_{\text{in}})_g - \dot{m}_{\text{outg}} (h_{\text{out}} + KE_{\text{out}})_g - \cancel{\phi}_{\text{loss}} - \frac{P_c}{J} \cancel{\frac{dV_c}{dt}} \qquad (9)$$

$$\phantom{\left. \frac{dE}{dt} \right|_{cv} =}\text{small}\text{small}\text{small}$$

The amount of heat lost during the extremely rapid combustion process can be considered negligible. This is justified through an order-of-magnitude analysis check on the heat lost term. In addition, the kinetic energy of the gas phase entering the control volume is several orders of magnitude smaller than the gaseous enthalpy and, hence, can be ignored. The work term, $(P_c/J)(dV_c/dt)$, is three orders of magnitude smaller than the leading term, $\dot{m}_{\text{ing}} h_{\text{in}}$, on the right-hand side of Eq. (9) and can therefore be neglected. Equation (9) then becomes

$$\frac{d(\rho_c V_c c_v T_c)}{dt} = \dot{m}_{\text{ing}} h_{\text{ing}} - \dot{m}_{\text{outg}} (h_{\text{out}} + KE_{\text{out}})_g \qquad (10)$$

Assuming that the ideal gas law is applicable, and noting that

$$h_{\text{ing}} = c_p T_{\text{in}} = c_p T_f, \qquad h_{\text{outg}} = c_p T_{\text{out}} \qquad (11)$$

$$KE_{\text{out}} = \frac{u_{\text{out}}^2}{2g_c J} \qquad (12)$$

With the substitutions given above, Eq. (10) becomes

$$\frac{1}{R} \frac{d(P_c c_v V_c)}{dt} = \dot{m}_{\text{ing}} c_p T_f - \dot{m}_{\text{outg}} \left(c_p T_{\text{out}} + \frac{u_{\text{out}}^2}{2g_c J} \right) \qquad (13)$$

For choked flow at the throat, both T_{out} and u_{out} can be expressed in terms of T_c; these relationships are readily derived from the adiabatic energy equation of gasdynamics.

$$T_{out} = 2T_c/(\gamma+1) \tag{14}$$

$$u_{out}^2 = \gamma g_c R T_{out} = 2\gamma g_c R T_c/(\gamma+1) \tag{15}$$

Substituting Eqs. (14) and (15) into Eq. (13), and further combining Eq. (13) with Eq. (3), we have

$$-\frac{P_c}{\bar{\rho}_s}\dot{m}_{ins} + \frac{P_c}{\rho_p}\dot{m}_{in} + V_c\frac{dP_c}{dt} = J(\gamma-1)c_P T_f \dot{m}_{in_g} - (\gamma-1)\dot{m}_{out}\left(\frac{2c_P T_c J}{\gamma+1} + \frac{\gamma R T_c}{\gamma+1}\right) \tag{16}$$

By defining the mass flow rate ratios in the following manner,

$$\dot{m}_{in_g}/\dot{m}_{in} \equiv \Psi_g \qquad \dot{m}_{ins}/\dot{m}_{in} \equiv \Psi_s \equiv 1 - \Psi_g \tag{17}$$

and solving $\dot{m}_{in}$ from Eq. (16), we have

$$\dot{m}_{in} = \frac{[(\gamma-1)\Gamma(\gamma)P_c A_t/\sqrt{RT_c/g_c}](Jc_P T_c) + V_c(dP_c/dt)}{\Psi_g(\gamma-1)c_P T_f J - \frac{P_c}{\rho_p} + \Psi_s \frac{P_c}{\bar{\rho}_s}} \tag{18}$$

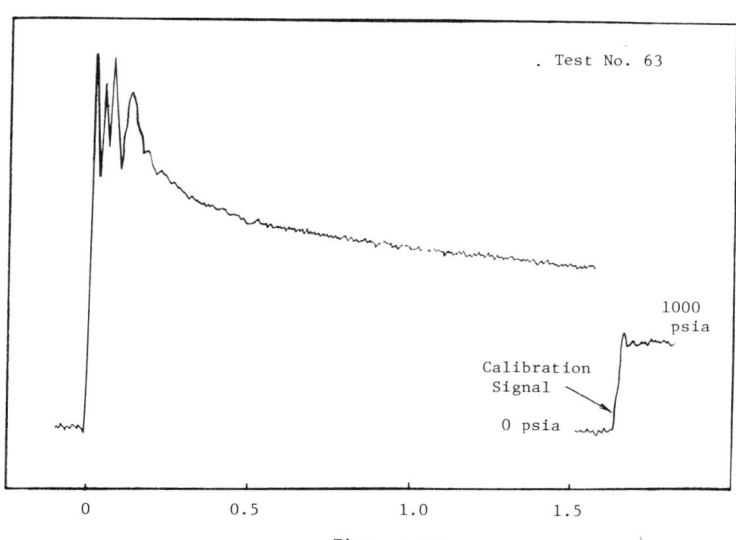

Fig. 5 Pressure-time trace for FA-34 showing characteristic wave phenomenon before insertion of pressure wave absorber.

In addition to the assumptions mentioned earlier in the development of the analysis, two other assumptions require further explanation:

1) Ideal gas behavior is assumed in the product gases. The peak chamber pressures experienced are around 300 atm. The average pressures during the pressurization process are low enough to warrant the ideal gas assumption.

2) Although the ratio of the product mass in the condensed form to the total product mass, as defined by Ψ_s, is in general a function of time, Ψ_s has been treated as a time-independent unknown parameter in the solution procedure. This assumption is supported by the thermochemistry calculations which revealed a very weak dependence of Ψ_s on pressure for the primer mix considered in this paper. This assumption drastically reduced the complexity of the computation.

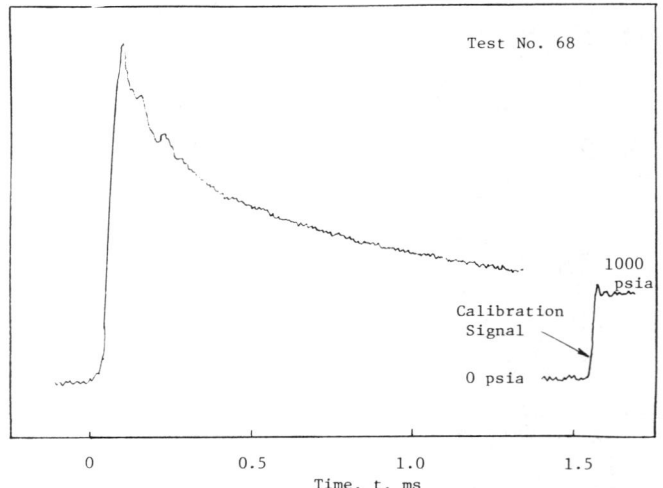

Fig. 6 Typical FA-34 primer pressure-time trace with calibration step.

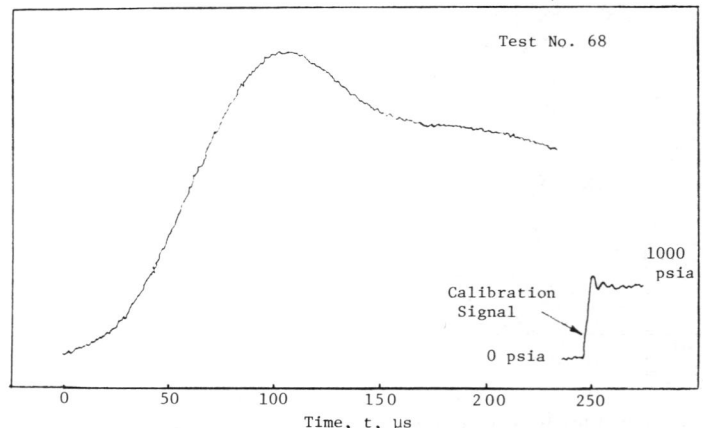

Fig. 7 Typical time-stretched P-t trace for FA-34 primer.

The theoretical model was coded into a computer program. The essential physical inputs for this program are: 1) the recorded pressure-time trace, 2) the adiabatic flame temperature of the primer mix, 3) the ratio of specific heats for the product gases, and 4) the mass of the primer mix.

The calculated mass fluxes from the primer depend strongly upon the input of the measured P-t traces. A typical P-t trace of FA-34 primer, before installation of the pressure wave absorber is shown in Fig. 5. The P-t trace clearly indicates a strong pressure wave reflection phenomenon with a frequency of 28 kHz that coincides closely with the frequency of sound wave reflection in the test section. This pressure oscillation presents serious problems in determining the instantaneous mass burning rate of the primer mix.

Figure 6 shows the effect on the P-t trace of the same primer by using a pressure wave absorber in the downstream portion of the test section. It is obvious that the pronounced pressure oscillation near the peak pressure region has been largely eliminated. A peak pressure in the neighborhood of 265 atm was recorded for this test run. The pressure rise time is approximately 108.5 μs. The average pressurization rate is in the order of 2.6×10^6 atm/s. The P-t trace shown in Fig. 6 represents the maximum time stretched data attainable by using the Biomation 1015 alone. Because accurate determination of the pressure-time history in the uprising portion was required, additional time stretching necessitated the use of a tape-dubbing procedure, in which an additional time-stretching factor of 16 was achieved. Figure 7 shows the resulting time-expanded P-t trace for the same test run as Fig. 6. It should be noted that the vertical axis has not been distorted by the tape-dubbing process.

Using a Tektronix 4662 plotter, the time-expanded P-t traces were digitized into the pressure-time raw data. Since the pressurization rate of P-t traces dominates the calculated gaseous mass flow rates, a polynomial curve fit for a given set of data is essential. POLY2, a computer program available in The Pennsylvania State University computer library, was used for the polynomial regression analysis. In statistical tests, the degree of the polynomial regression functions for primers FA-34, FA-41, and CCI-200 (a commercially available primer) was found to be 7, 8, and 7, respectively. Tests for normality, constant variance, and time independence were also made to check the suitability of the orders of the polynomial.

Adiabatic flame temperature T_f and the ratio of specific heats γ for the product gases of the primer are also important parameters in the primer characterization study. The primer gaseous product mass flow rates are somewhat affected by the changes in T_f and γ. Since the thermochemical properties of the primer constituents and its product are known, these two parameters can be calculated by using the CEC 72 thermochemistry program developed by Gordon and McBride (1971). Depending upon the pressure, the calculated values for adiabatic flame temperature and the ratio of the specific heats for the product gases of the primer are in the range of 2900 K and 1.13, respectively.

The unknown parameter Ψ_s, representing the percentage of the condensed species in the primer product, is determined by the following procedure: 1) record the P-t trace of the test firing; 2) the total mass burned is measured

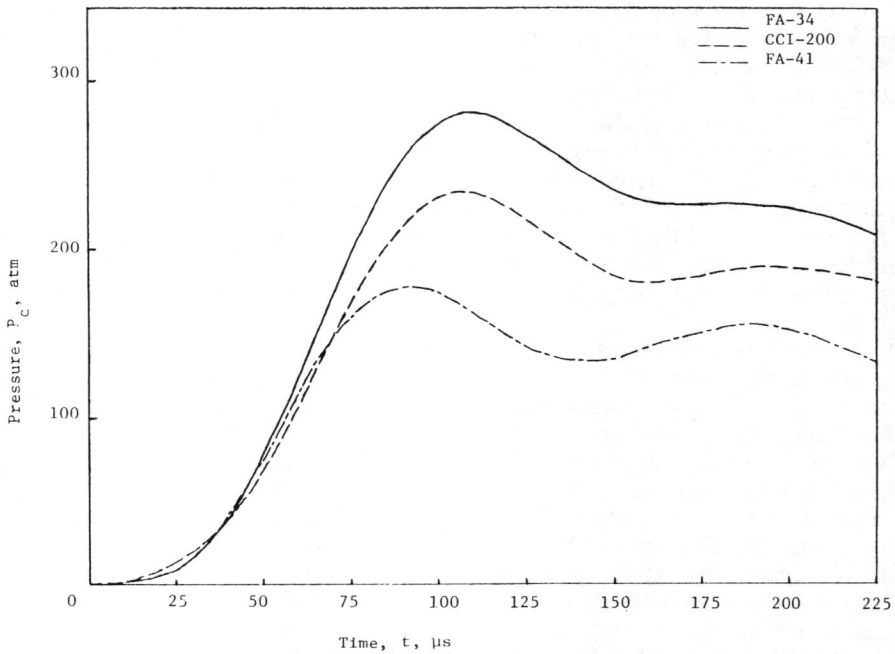

Fig. 8 Comparison of time-stretched P-t traces for FA-41, FA-34, and CCI-200 primers.

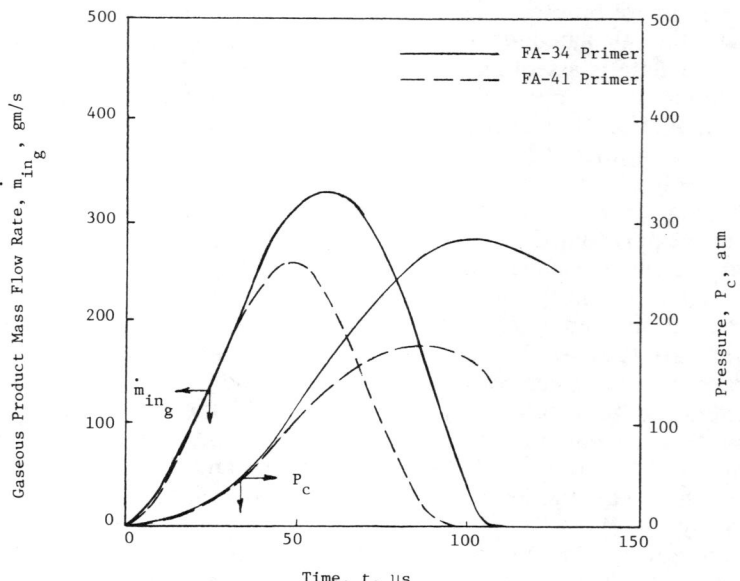

Fig. 9 Predicted gaseous product mass flow rates for FA-34 and FA-41 primers.

Table 2 Input data for primer characterization computations

V_{c0}	=	0.2622 cm^3
R	=	$1.2755 \times 10^6 \text{ cm} \cdot \text{dyne/g} \cdot \text{K}$
ρ_p	=	3.0916 g/cm^3
$\bar{\rho}_s$	=	6.007 g/cm^3
T_{c0}	=	294 K
P_{c0}	=	1.0 atm
Δt	=	$1 \times 10^{-6} \text{ s}$
$M_{\text{primer mix}}$	=	0.0343 g (for FA-34, test 68)
		0.0194 g (for FA-41, test 67)
T_f	=	2900 K
γ	=	1.13
A_t	=	$3.295 \times 10^{-3} \text{ cm}^2$

from the difference between the tare weight of the primer and the weight of the initially unburned primer; 3) assume a value Ψ_s and compute the instantaneous mass burning rate of the primer from Eq. (18); 4) the total mass burned is integrated for comparison with the measured value; and 5) reassume a value of Ψ_s and recompute $\dot{m}_{\text{in}}$ until the difference between the measured and the calculated mass is within a specified tolerance.

The accurately measured weight of the percussion primers is important in the determination of the percentage of the condensed phases in the primer product. The typically measured weights for FA-34, FA-41, and CCI-200 primers are 0.0343, 0.0194, and 0.0369 g, respectively. The comparison of time-stretched P-t traces for these three primers is shown in Fig. 8. The data used in primer characterization calculations are given in Table 2. In the primer characterization program, Eqs. (3), (8), and (18) are integrated by using a fourth-order Runge-Kutta method. The computation is iterated by gradually increasing Ψ_s by 0.01 at each step, and checking the calculated total mass burned with the measured value until convergence is reached.

Discussion of Results

The calculated gaseous mass flow rates, together with the curve fitted P-t data for FA-34 and FA-41 primers, are shown in Fig. 9. The peak of the calculated product mass flow rate curve for the FA-34 primer is higher than that of the FA-41 primer. In addition, the time duration required for the primer to be completely consumed is longer for the FA-34 than for the FA-41. Therefore, the FA-34 primer is a stronger convective heat generator than the FA-41 primer. It is interesting to note that the primer product mass flow rates are highly dominated by the pressurization rate of the pressure-time traces. The primer product mass flow rates peak when the pressurization rate is at its maximum value. Shortly after the pressure reaches its peak, the gaseous mass flow rate drops abruptly to zero.

The entire time duration required for the FA-34 and FA-41 primers to consume completely is in the range of 100 μs, or about one order of magnitude less than the characteristic time required for combustion of solid propellants in mobile granular beds in typical small arm cartridges. Although the primer function time is rather short, it has a significant influence on the combustion processes in granular beds; the primer-granular bed interface and coupling effect are not fully understood at the present time. It is recommended that further studies be made.

Through the computations, it is found that the percentage of the condensed phases in the primer product for FA-34 and FA-41 are in the range of 45 and 40%, respectively. This implies that the multiphase flow considerations for the primer characterization studies are essential. The presence of large percentages of condensed phases in the primer product implies that the contribution of the condensed phases to the ignition process in granular bed combustion cannot be ignored. Ignition mechanisms caused by condensed phases are not well established; these subjects deserve further investigation. It is interesting to point out that a significant amount of solids was found after each test firing. However, no comparison can be made between the residue mass and the calculated values of the mass of the condensed phases for the following reasons:

1) Although the chance of losing the mass of condensed phases through the exit nozzle is very small in the extremely short time interval (less than 0.1 ms) during pressurization, the particles of condensed phases can still leave the test chamber during the relatively long period of depressurization ($\simeq 5$ ms). Therefore, the mass of the condensed phases is not conserved in the test chamber.

2) The condensed phase products usually coat the interior surfaces of the test chamber, the pressure wave absorber, and the primer cap. Because geometry of the interior surfaces of the test rig is quite complex, it is very difficult to collect the entire coating. Even if it were possible to collect the weight of the residual mass, it would still be less than the actual mass of the condensed phase generated in the combustion product, since the test chamber is open to ambient atmosphere.

Summary and Conclusions

Based upon the present study on primer characterization, the following comments can be made:

1) An effective technique for characterizing various primers has been developed. The P-t traces for any given primer have been shown to be reproducible with respect to the uprising portion of the curve, the peak pressure, and the approximate rise time.

2) An effective means of treating the data has been developed, consisting of multiple time expansion combinations of the Biomation analog-digital converter and a tape-dubbing procedure. This permits more accurate digitizing of the uprising portion of the P-t trace.

3) From a theoretical analysis involving the conservation of mass and energy in the test section, the mass flux issuing from the primer can be expressed by Eq. (18).

4) Results from FA-34 firings indicate values of $\Psi_s \cong 0.45$. Those of FA-41 firings show $\Psi_s \cong 0.40$. CCI-200 could not be characterized in this manner due to lack of fractional composition data. These findings indicate that significant fractions of the product are in condensed-phase form, presumably as metals and metal oxides.

5) The resulting high percentages of condensed products are responsible for a less than initially expected flow rate entering the boundary of the granular propellant bed.

6) The mass flow rate is highly influenced by the pressurization rate, and attains its peak value when the pressure-time slope is a maximum. Shortly after the pressure peaks, the igniter mass flow rate drops abruptly to zero.

7) Based upon a parametric study, the effect of T_f on the gaseous mass flow rate can be examined as a trend. An increased value of T_f corresponds to a slightly lowered gaseous mass flow rate. Similarly, an increased value of γ corresponds to a slight decrease in the gaseous mass flow rate.

8) The rise time required for the pressure to peak, as seen on the P-t trace, ranges between 90 and 105 μs for three different primers considered. These times are an order of magnitude less than the time involved in granular bed combustion events.

9) The high percentages of condensed phases in the product indicate that, in addition to convective heat transfer, the conduction and radiation heat transfer may also play an important part in the ignition process of the granular propellant bed. The exact contributions of conduction and radiation are not known; they deserve further studies.

10) It is suggested that the interface coupling between the primer blast and the bed compaction at the breech end of the granular bed also be studied further.

Acknowledgments

This work represents a part of the results obtained in the research program sponsored under Grant DAAG29-77-0163 by the Engineering Sciences Division, Army Research Office, Research Triangle Park, N.C. The technical advice and information supplied by L. Stiefel of ARRADCOM-Dover, E. Freedman of the Ballistic Research Laboratories of the U.S. Army, and C.S. Leveritt of Teledyne McCormick Selph are greatly appreciated.

References

Davis T. R. (1977) Experimental study of transient combustion processes in granular propellant beds. M.S. Thesis, The Pennsylvania State University.

Gerri N., Stansbury L. and Henry C. (1977) A parametric study of the gas flow and flame spreading in packed beds of ball propellants, Pt. I. Ballistic Research Laboratories Rept. 1988.

Gordon S. and McBride B.J. (1971) Computer program for calculation of complex chemical equilibrium compositions, rocket performance, incident and reflected shocks, and Chapman-Jouguet detonations. NASA SP-273.

III. Cellular Structure of Detonations

Diffraction of a Planar Detonation in Various Fuel-Oxygen Mixtures at an Area Change

D. H. Edwards * and G. O. Thomas†
University of Wales, Aberystwyth, Wales
and
M. A. Nettleton‡
Central Electricity Research Council, Leatherhead, U. K.

When a steady detonation wave propagating in a uniform tube encounters a large and sudden area change, it becomes quenched if the tube diameter is below a critical value for that particular system. Experimentally this critical diameter was found by Zeldovich et al. (1956) to depend on the induction-zone length of the system or, what is equivalent, the transverse wave spacing S. Mitrofanov and Soloukhin (1964) discovered that, for the oxyacetylene system, the critical diamters are $13S$ and $10S$ for tubes of circular and rectangular section, respectively. This result has been confirmed in the present work, for both oxyacetylene and oxyhydrogen mixtures. Preliminary work with detonations in mixtures of ethane, propane, methane, and acetone with oxygen in a rectangular channel indicates that the number of transverse waves required for sustenance of the diffracted wave lies between roughly 14 and 18. In general, these systems exhibit a larger and more irregular cellular structure than both oxyacetylene and oxyhydrogen. Smoked-foil and Schlieren records show that, for supercritical waves, reignition occurs at sites along the wedge formed by the head of the expansion from the diffracting aperture and criticality is attained when the site is located at the apex of the wedge. These observations enable the conclusion of Mitrofanov and Soloukhin to be stated in a more general form which accounts for supercritical as well as critical conditions. In a previous paper (Edwards et al., 1979) the authors proposed a semi-empirical criterion for the reignition of a diffracted wavefront which is based on Whitham's theory (1957) of the diffraction of a nonreactive shock. Although a discrepancy arises between the form of the detonation wavefront predicted by the theory and observation, at large distances from the diffracting aperture, the criterion is shown to agree reasonably well with measurements of transverse wave spacings derived from the smoked-foil records.

Presented at the 7th ICOGER, Göttingen, Federal Rupublic of Germany, Aug. 20-24, 1979.
Copyright © American Institute of Aeronautics and Astronautics, Inc., 1980. All rights reserved.
*Reader in Physics, Department of Physics.
‡Research Assistant, Department of Physics.
‡Research Officer.

Introduction

WHEN a self-sustaining detonation wave propagating in a uniform tube encounters an abrupt area change, then for tube diameters below a critical value the wave is quenched; whereas for diameters greater than the critical, a cylindrical or spherical wave is established in the expanded region. Dremin and Trofimov (1965) showed that this behavior is not unique to the gas phase but is also found in solid and liquid-phase detonations. They arrived at the important conclusion that the critical diameter of a weakly confined solid or liquid explosive corresponds to the quenching diameter of a detonation wave passing into a larger volume of explosive. They further showed that the equivalence of critical and quenching diameters was not fortuitous and that previous theories of critical explosive diameters of unconfined charges, which assume that quenching results when rarefraction waves from the charge periphery reach the charge axis before the completion of the reaction, are essentially incorrect because they ignore the influence of frontal structures. More recently, Urtiew (1975) and Presles and Brochet (1972) have adduced further evidence of the cellular structure in liquid explosives, similar to that found in gaseous detonations. Moreover, Urtiew was able to explain the origin and growth of failure waves in unconfined explosives on the basis of the transverse motion and reflection of the triple-wave intersections that are present in the wavefront.

The first comprehensive examination of the behavior of a detonation wave as it emerges a narrow tube into a larger one was provided by Zeldovich et al. (1956). These authors employed a total of 14 gas mixtures at an initial pressure 800 Torr and a range of the tube diameters from 2 to 33 mm. They concluded that each mixture is characterized by a critical tube diameter d_c for which spherical detonation was only just initiated in the larger diameter tube. From the data they present for the system $C_2H_2 + 2.5O_2 + ZN_2$, we have derived the dependence of the one-dimensional induction zone thickness L_i on the values of d_c, as shown in Fig. 1. If τ_i is the one-dimensional steady-state induction reaction time, then $L_i = \tau_i M_{C-J} a_0 \rho_1 / \rho_2$ where M_{C-J} is the C-J Mach number, a_0 the speed of sound in the mixture, and ρ_2/ρ_1 the density ratio across the shock front. Values of τ_i where derived from the shock tube data given by White (1967) for very dilute systems:

$$\log_{10}\{[O_2]^{1/3}[C_2H_2]^{2/3}\tau_i\} = -10.81 + 17{,}300/4.58T \qquad (1)$$

The next significant attempts to understand the diffraction of a detonation wave at a change of cross section were those of Gvozdeva (1961) and Mitrofanov and Soloukhin (1964). The former described the gas dynamic processes involved in the diffraction and the latter discovered a fundamentally important result that relates the critical diameter to the equilibrium transverse wave spacing S. For the oxyacetylene system this is $d_c = 13S$ for a circular section tube and $d_c = 10S$ for a planar channel. A comparison of this observation with those of Zeldovich et al. is possible since $S \simeq 0.6 L_c$, where L_c is the cell length, which can be related to L_i by

$$L_c = n \times 10 L_i \qquad (2)$$

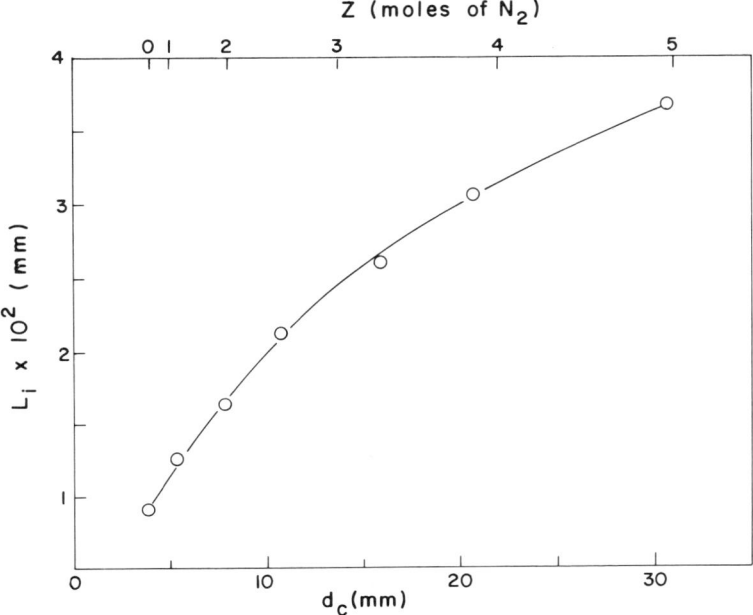

Fig. 1 Dependence of critical diameter d_c on one-dimensional induction zone thickness L_i derived from data of Zeldovich et al. (1956).

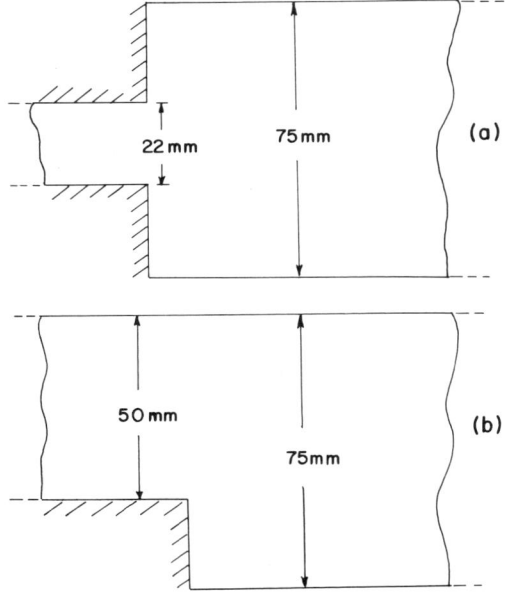

Fig. 2 Diagram of the two diffracting apertures used in the experimental work.

The value of n varies from one system to another and also between confined and unconfined waves, but the overall evidence indicates that $3 \le n \le 10$. Assuming a mean value of $n \simeq 6$ it can be seen, referring to Fig. 1, that for cylindrical tubes the data of Zeldovich et al. is in reasonably good agreement with the conclusion of Mitrofanov and Soloukhin. Further experimental evidence of the transition from planar to spherical detonation was given by Soloukhin and Ragland (1969) in their work on the diffraction from an open-ended circular tube. They showed that, at criticality, reignition occurs through localized explosions behind a smooth shock front, similar to the phenomenon that has been observed by Bach et al. (1965) in their laser ignition studies at critical power-density levels. The aim of the present work is to enlarge on the previous studies reported by the authors (Edwards et al., 1979) on the diffraction of a planar detonation and to further examine a criterion for criticality, based on Whitham's (1957) theory of the diffraction of a nonreactive shock, that was proposed there.

Experimental Details

Steady detonation waves were generated in a rectangular channel, of cross-sectional dimensions 22×6 mm^2, and length 4 m. This section was fixed into a right-angled expansion section, of internal dimensions 75×6 mm^2, as shown in the sketch of Fig. 2a. In order to allow systems with larger cell sizes to be studied a second channel was employed, in which there is only one diffracting corner, giving a half-aperture of dimension 50 mm, as indicated in the sketch Fig. 2b. As in Fig. 2a, the thickness of this channel is 6 mm. A narrow channel was chosen in preference to a circular section tube so as to provide as near a two-dimensional wavefront as possible, thus greatly facilitating the interpretation of optical records of the wavefront. Following Mitrofanov and Soloukhin (1964), the initial gas pressures were varied while keeping the tube dimensions constant. In this way the influence of varying the kinetics is studied, and consequently L_i and L_c, while the gasdynamic parameters remain fairly constant.

The motion of the shock front and reaction zones were recorded by both spark and streak photography while the trajectories of the triple-point interactions with the tube wall were obtained from smoked foil records. Pressure measurements were made with bar gages at various points in the expanded flowfield and also some preliminary studies were made of the density field using a Mach-Zehnder interferometer.

Experimental Results

Smoked-Foil Records

Four examples of smoked-foil records that were obtained with $C_2H_2 + 2.5O_2$ at initial pressures of 180, 150, 110 and 60 Torr are given in Fig. 3. These illustrate the type of cellular pattern that is observed under supercritical, critical, and failure conditions. In order to highlight the main features of the supercritical regime a sketch of one of these records is shown in Fig. 4. First, it can be seen that reignition of the detonation is achieved only after several attempts indicated by the shaded areas, and they correspond to strong scouring of the soot. Second, it will be noted that the reignition centers

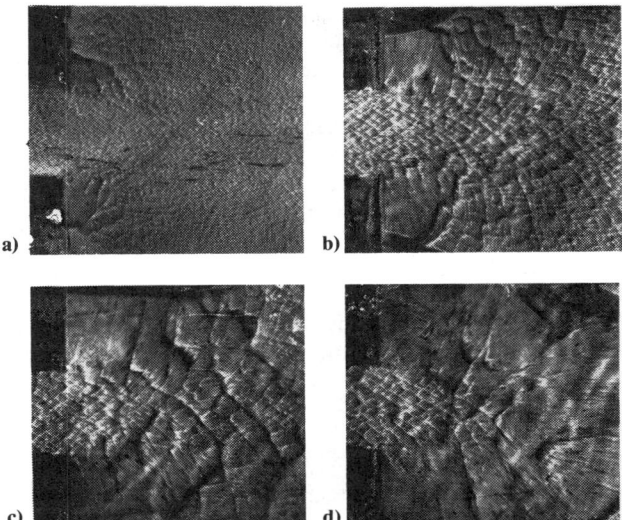

Fig. 3 Smoked-foil records obtained with stoichiometric oxyacetylene at different initial pressures in channel shown in Fig. 2a: a) 180 Torr and b) 150 Torr, supercritical wave; c) 110 Torr, critical wave, and d) 60 Torr, failing wave.

originate just outside the wedge formed by the lines AD and A'D, that represent the head of the expansion fan centered on the diffracting corners. These two features are present in all the smoked foil records. However, in order to render the process more tractable we have idealized the smoked foil pattern in Fig. 5 by envisaging the reignition to occur only at two opposite points E and E' along the wedge AD, A'D. Inside the area ADA' normal wave structure is obtained and along the two arcs EF and E'F' the fine triple-point writing is the result of an overdriven transverse detonation that has propagated through the unreacted gas behind the frontal shock. Beyond the arcs the structure rapidly approaches the normal cell spacing characteristic of the mixture. If the idealized picture given in Fig. 5 is acceptable as an adequate representation of the behavior of the diffracted wavefont, then we may generalize the result of Mitrofanov and Soloukhin. Thus, if the reignition process is similar at all points along the wedge, including the apex D, then an angle $2m$ (where m is the angle between the leading characteristic AD, A'D and the horizontal) drawn at any point on the wedge will subtend a length of $10S$ at the aperture. Expressing this in terms of the x coordinate of the shock front we obtain

$$x_s/S = 5\cot m \qquad (3)$$

Since $\cot m \simeq 2$ for detonation waves, $x_s/S \simeq 10$, which is the Mitrofanov and Soloukhin observation at criticality. Equation (3) thus expresses a generalized experimental result for all points along the wedge; this is confirmed by the experimental points plotted in Fig. 6 obtained for smoked foil records at different initial pressures in $C_2H_2 + 2.5O_2$.

346   D.H. EDWARDS, G.O. THOMAS, AND M.A. NETTLETON

Further experimental evidence was sought for the validity of Eq. (3) in systems which exhibit a less regular cellular structure than oxyacetylene. In this investigation the detonation tube shown in Fig. 2b was used. The systems studied were stoichiometric compositions, with oxygen, of ethane, methane, propane and acetone, at various subatmospheric initial pressures. To varying degrees, in all these systems the cell structure is complex, consisting of a large cell size which is superposed on a much finer and irregular pattern. The interpretation and measurement of the larger structure, that alone is relevant to the wavefront behavior, is usually difficult so that no great accuracy can be claimed for the results shown in Fig. 7. Here the measured cell size in the undiffracted wave is plotted against the initial pressure for each system, and the arrows indicate the uncertainty in the determined critical pressure at which failure occurs. Thus for ethane and propane the average transverse wave

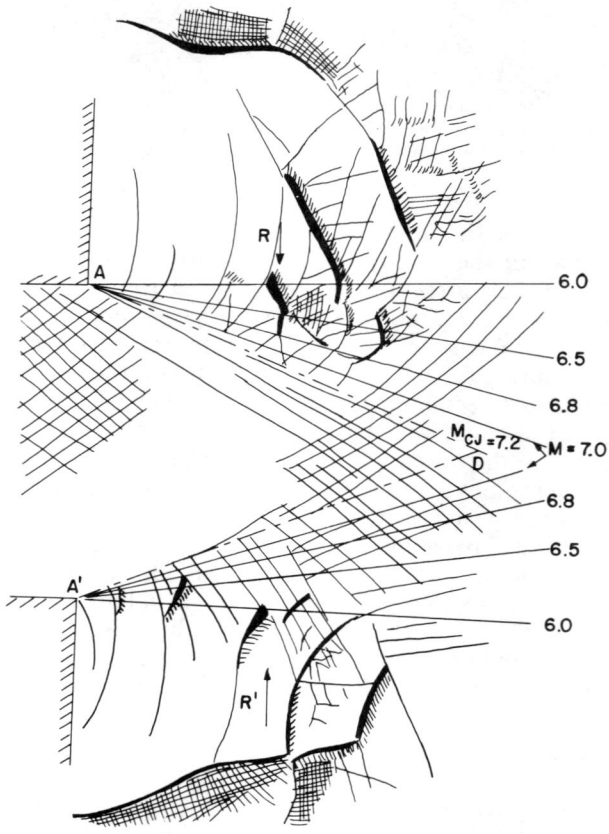

Fig. 4 Sketch of smoked-foil record given in Fig. 3b for supercritical wave in oxyacetylene at 150 Torr initial pressure. Transverse wave space $S = 1.3$ mm. Characteristic lines, labeled with constant Mach numbers, are derived from Whitham's theory and RR' is the position predicted by Eq. (10) for reignition. AD and A'D represent the head of the expansion fan originating at A and A'.

spacing is about 7 ± 1 mm, whereas for methane and acetone it is 5.5 ± 1 mm. This means, since d_c for this tube is 100 mm, that the Mitrofanov and Soloukhin result becomes $d_c = 14S$ and $18S$, respectively, for these systems, compared with $10S$ observed for the oxyacetylene system. Whether the boundary layer in the narrow channel exerts a controlling influence in systems with larger cell size is a moot point and further work in larger tubes is clearly required in order to establish more precise data. Two experimental points that add some confidence to the methane and acetone data have been derived from Zeldovich et al. (1956); their measured critical diameters at 800 Torr pressure for these systems are 20.6 and 32 mm, respectively. Assuming that for round tubes $d_c = 13S$, the two values of S corresponding to these critical diameters are plotted on the graph; these are seen to lie, within the experimental error, near the extrapolated curve obtained in the present work.

Schlieren Studies

Examples of spark Schlieren records of a diffracted detonation front under supercritical, critical, and subcritical conditions have been reported in a previous study by Edwards et al. (1979) Sketches of the wavefront at three

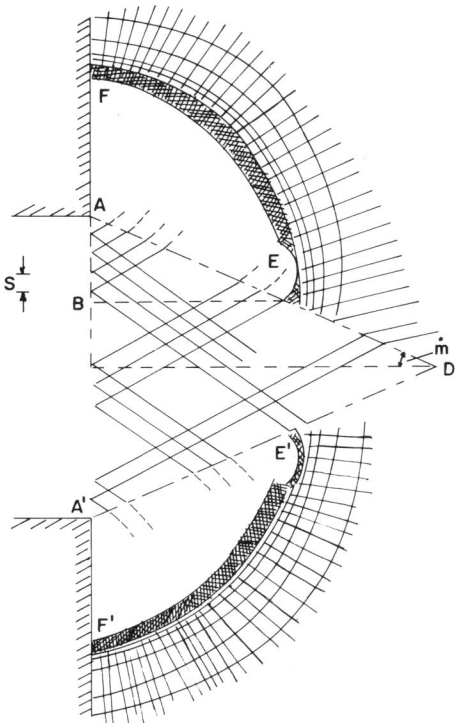

Fig. 5 Idealized sketch of smoked-foil record obtained under supercritical conditions of reinitiation.

distances from the aperture, for $C_2H_2 + 2.5O_2 + 3.5Ar$ at 70 Torr initial pressure, are given in Fig. 8. These confirm the deductions made from the smoked foil records, namely, that within the "wedge" formed by the head of the expansion wave originating at the diffracting aperture, the structure and velocity of the wavefront is unchanged. Outside this region, however, the transverse waves cease to support reaction and also the reaction zone detaches itself from the frontal shock. When reignition occurs it is highly localized, as we noted in the smoked foil records, at a distance of one or two cell lengths outside the head of the expansion behind a transverse shock such as TS in the diagram. The transverse wave then propagates around the wavefront through the shock heated gas between the wavefront and the main reaction zone. As a result, the wavefront is temporarily overdriven and detonation is re-established along the whole of the diffracted front. At criticality, of course, the reignition is delayed up to a point in the vicinity of the intersections of the limiting characteristics.

The shape of the diffracted detonation front was determined over a range of distances from the aperture for the 22×6 mm channel; values of the frontal position along the axis X and along the wall Y are plotted in Fig. 9. These values show that the front velocity along the wall is always one half that along the axis.

Streak Schlieren photographs were obtained of the wavefront, along the axis of symmetry, near critical conditions. An example of such a velocity profile is given in Fig. 10. These profiles are reminiscent of the behavior of the front in

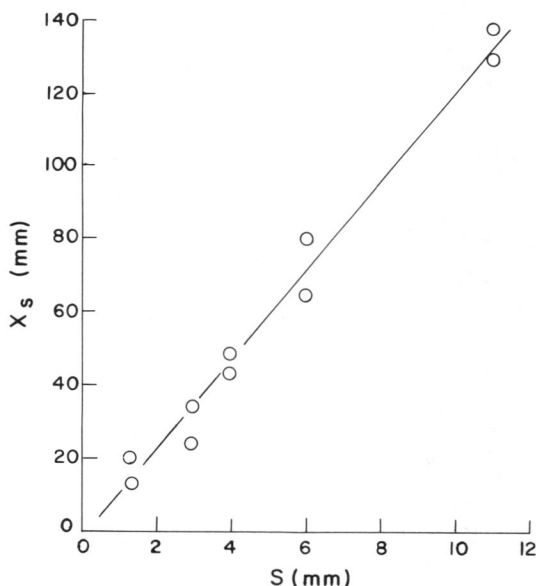

Fig. 6 Graph of shock front position x_s at reignition plotted against transverse wave spacing S for $C_2H_2 + 2.5O_2$ at different initial pressures.

the ignition of a spherical detonation by a blast wave (Edwards et al., 1978; Bull et al., 1976). At criticality it is seen that the velocity along the axis is sensibly constant along AB, at the C-J value of 2400 ms^{-1}. At B, which is a point just beyond the apex of the wedge formed by the expansion heads, there is a sudden drop in velocity to about 0.6 M_{C-J} at C. This is immediately followed by a rapid recovery to a velocity slightly greater than C-J at D. Decay of the front velocity occurs again before a further reinitiation establishes a steady wave.

Pressure and Density Measurements

Pressure profiles were obtained of both the undiffracted and diffracted wavefronts at several locations upstream of the aperture. An example of these pressure recordings is shown in Fig. 11a. Apart from the more rapid rate of decay of the average pressure behind the diffracted wavefront the two profiles are very similar, the amplitude of the transverse waves remaining virtually the same. On this evidence it would appear that a transverse wave will lose little or

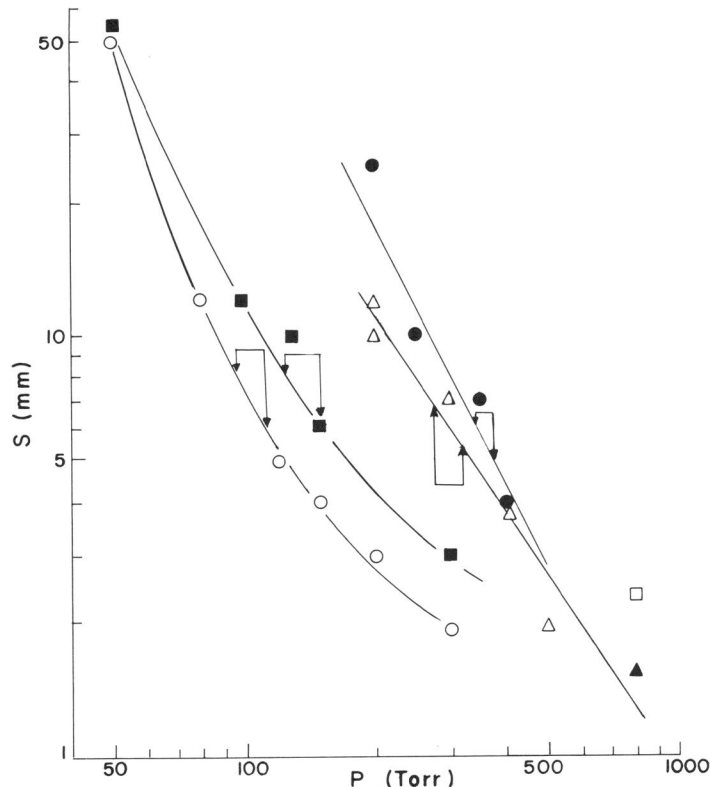

Fig. 7 Variation of transverse wave spacing S with initial pressure p_0 for stoichiometric mixtures with oxygen, of ethane ○, propane ■, acetone △, and methane ● . Results derived from Zeldovich et al.: methane □, acetone ▲. Arrows indicate range of p_0 over which criticality is observed.

none of its strength as it moves into the expansion wave though the reaction zone behind it rapidly recedes.

Because of the narrowness of the tube and, hence, the small optical path length, it proved very difficult to obtain density measurements over the whole flowfield. No information was obtained on the transverse waves, which is of crucial interest near reignition, but only on the density jump across the main shock front and the reaction zone. A typical density profile is shown in Fig. 11b.

Criterion for Critical Conditions

Failure of the diffracted portion of a detonation front occurs through the lack of transverse wave collisions, that are essential for the sustenance of a C-J detonation wave. Normally the requisite number of transverse waves per unit area of the wavefront is maintained through wall collisions, in confined waves, and also through the generation of new waves during triple-point intersections. In the presence of sufficiently strong rarefractions, however, this generative process of new transverse waves does not occur. As the wave moves away from the area change the rarefraction weakens, until at a certain distance along the

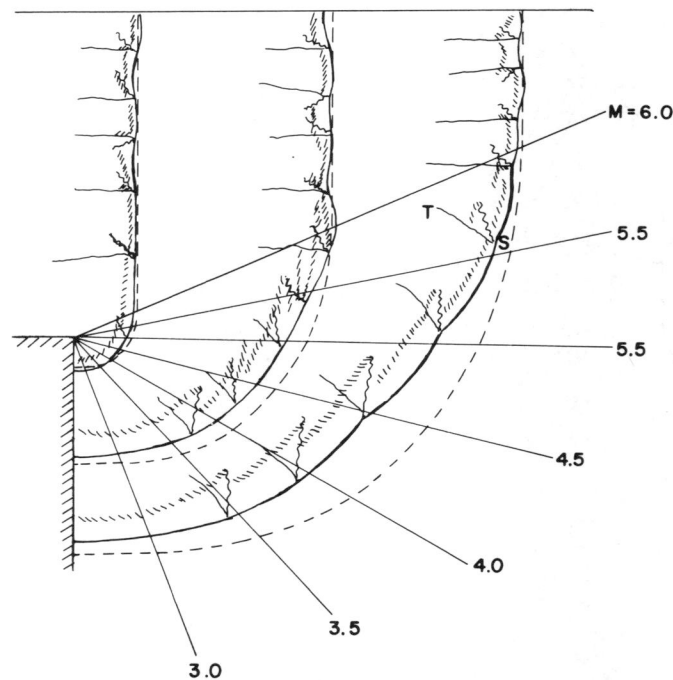

Fig. 8 Sketch of three position of diffracted waverong in $C_2H_2 + 2.5O_2 + 3.5Ar$ at $p_0 = 70$ Torr in detonation tube of Fig. 2b. Wavefront, indicated by broken lines, and expansion fan are calculated from Whitham theory. TS is a typical position of a tranverse wave at which localized reignition occurs. $M_{CJ} = 6.05$.

expansion head, sudden explosive reignition occurs. The point at which this happens must correspond to some value of the transverse velocity gradient $\partial M_s/\partial y_s$, where y_s is the y coordinate of the shock front, that is characteristic of the chemical system. From this observation we can stipulate that the condition for reignition can be expressed as

$$\frac{\partial M_s}{\partial y_s} \lesssim \frac{\Delta M_{C\text{-}J}}{\epsilon L_c} \tag{4}$$

where, from the evidence of smoked-foil records, ϵ is a small number $1 \leq \epsilon \leq 3$, and $\Delta M_{C\text{-}J}$ is the deceleration in Mach number, which must be determined by some criterion that governs wavefront stability. Thus one way of establishing the onset of the decoupling of the reaction zone in a decaying wavefront is through the Shchelkin (1959) instability criterion. This semiqualitative criterion simply states, for a one-dimensional wavefront, if an increase $\Delta \tau_i$ occurs in the steady-state induction-reaction time τ_i, such that

$$\frac{\Delta \tau_i}{\tau_i} \gtrsim 1$$

then the detonation wavefront decays. It can be shown (e.g., Soloukhin and Ragland (1969); Edwards et al., 1979) that under the conditions obtained in detonation waves Eq. (5) becomes

$$\frac{\Delta M_{C\text{-}J}}{M_{C\text{-}J}} \gtrsim -0.1 \tag{6}$$

In order to evaluate the transverse velocity gradient in the rarefractoin, $\partial M_s/\partial y_s$, we have used the analysis of Whittman (1957) for a nonreactive shock. This theory shows that the Mach number M is constant along each $C+$ characteristic of the simple expansion wave at the corners of the diffracting aperture and the inclination $\theta(M)$, of the particle path near, the shock front, with respect to the x axis, is constant along each C^+ and given by

$$\theta(M) = n^{\frac{1}{2}}(\cosh^{-1}M - \cosh^{-1}M_{C\text{-}J}) \tag{7}$$

where $n = 5.0743$ for large values of M. An expression for the angle m between the C^+ characteristic and a particle path is given by Whitman (1957) and Skews (1967) as

$$\tan m = 1/M(M^2 - 1/n)^{\frac{1}{2}} \tag{8}$$

If x_s, y_s are the shock front coordinates, then from Eq. (8) by direct integration along a C^+ characteristic we obtain

$$x_s = M_s \frac{\cos(\theta + m)}{\cos m} \alpha; \qquad y_s = \frac{M_s \sin(\theta + m)}{\cos m} \alpha \tag{9}$$

where $\alpha = a_0 t$, the shock position coordinate a_0 being the sound speed in the undisturbed gas. It is shown by Edwards et al. (1979) that Eqs. (9) and (4) lead to the following expression for the condition for reignition:

$$\frac{1}{[tan(\theta + m) + n^{1/2}]} \lesssim \frac{1}{\epsilon}\left(\frac{x_s}{S}\right)\frac{\Delta M_{C-J}}{M_{C-J}} \qquad (10)$$

The justification for using the Whitham theory in the present problem rests on one of its predictions, namely, that the ratio of the axial shock to the wall shock velocities is 0.5 which is almost exactly what is found in the experimental results quoted in Fig. 9. However, precise agreement between theory and observation on the shape of the wavefront is only obtained for small values of α, as can be seen from Fig. 8. As the value of α increases the discrepancy becomes larger, which means that the theoretical prediction of $\partial M_s / \partial y_s$, required in Eq. (4), would be in error as the dimensions of the diffracting aperture become large.

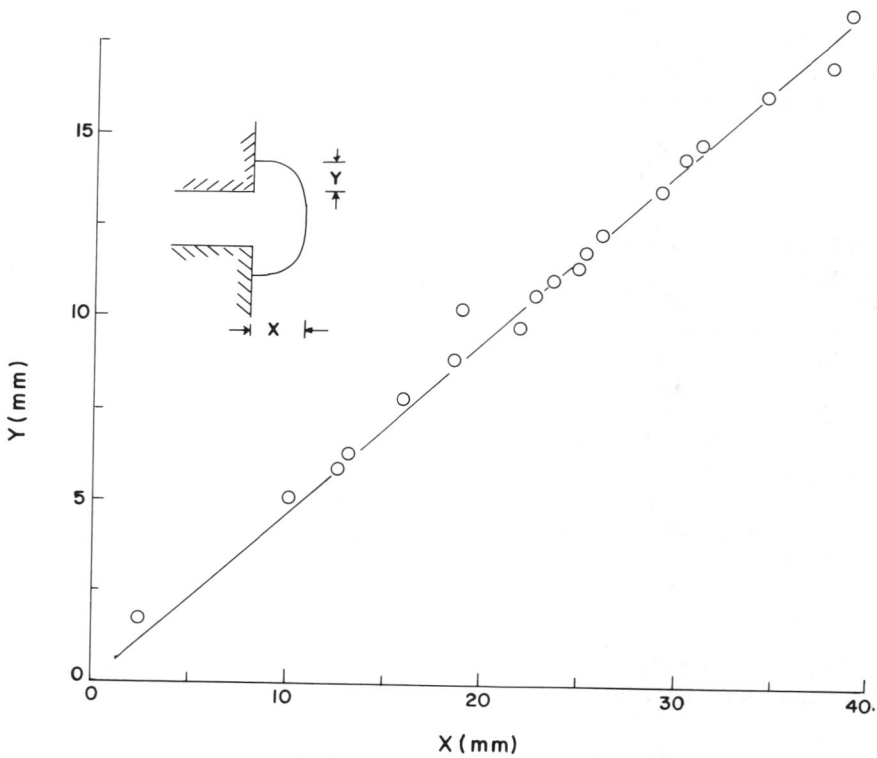

Fig. 9 Values of the frontal position along axis X and wall Y.

Application of the Criterion

For the oxyacetylene system the value of M_{C-J} remains fairly constant at approximately 7 as the initial pressure was varied. From Eq. (8) it follows that the angle m has the constant value of approximately 23 deg. Hence if the ratio $\Delta M_{C-J}/M_{C-J}$ is prescribed the value of -0.1, then the condition for reignition is obtained by substitution in Eq. (10)

$$\left(\frac{x_s}{S}\right)_{theor.} = 8\epsilon \qquad (11)$$

where, it will be recalled, ϵ is the number of transverse wave spacing separating the expansion head and the point of reignition. This point is difficult to define with precision because, as a rule, reignition will occur at more than one point before final re-establishment is achieved. Nevertheless, it is possible to give rough limits for ϵ, namely, $1 \leq \epsilon \leq 3$. With the value of $m = 23$ deg, Eq. (3) gives

$$(x_s/S)_{exp} = 5\cot m \simeq 11 \qquad (12)$$

Bearing in mind the uncertainties in evaluating both $\Delta M_{CJ}/M_{CJ}$ and ϵ, the agreement between Eqs. (11) and (12) is satisfactory and suggests that the basis of the criterion for reignition is correct.

A further example of a smoked-foil record, for comparison with that given in Fig. 4 for the oxyacetylene system, is shown in Fig. 12; this was obtained for

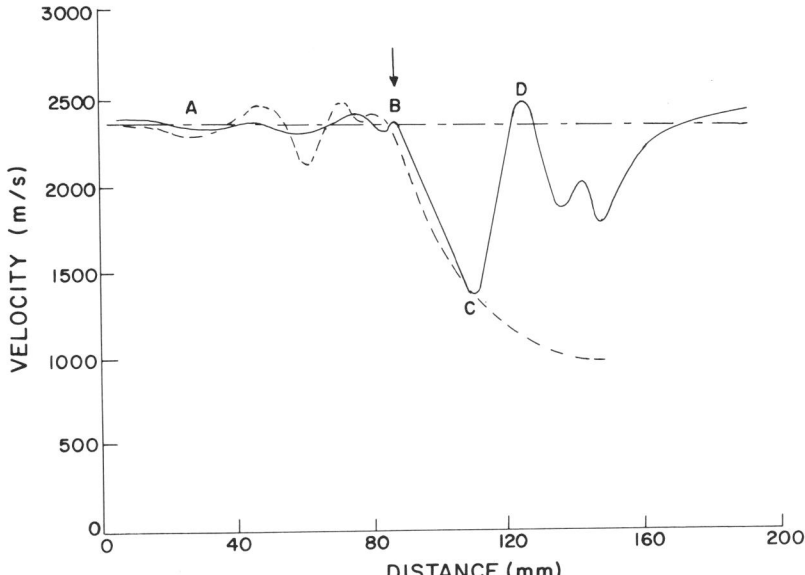

Fig. 10 Variation of frontal velocity along the axis, obtained from streak photographs. Solid line corresponds to critical reinitiation and broken curve to a failing wave. Arrow denotes position of intersection of expansion heads.

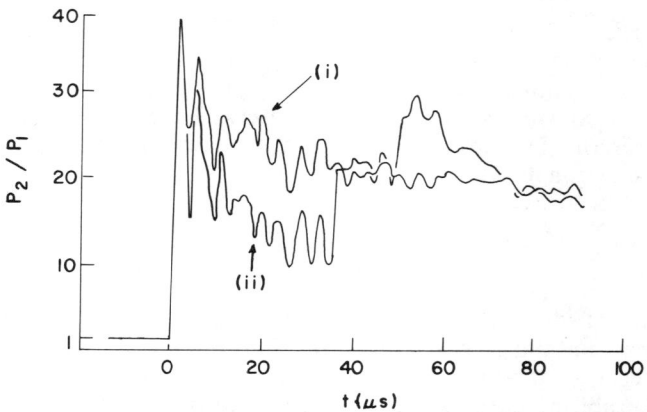

Fig. 11a) Pressure profiles behind 1) undiffracted and 2) diffracted wavefront in $C_2 + 2.5O_2$ at p_0 of Torr.

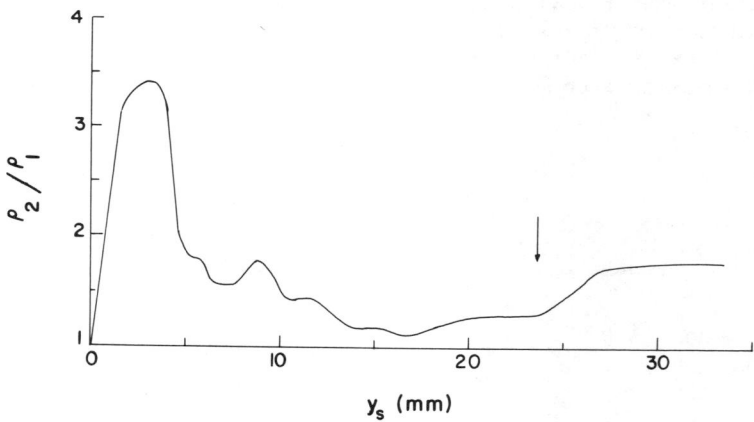

Fig. 11b) Transverse density profile across a diffracted shock front in $C_2H_2 + 2.5O_2 + 3.5Ar$. Arrow indicates the calculated position of head of expansion wave.

$2H_2 + O_2$ at an initial pressure of 390 Torr. This is a detonation close to critical conditions because the reignition, signaled by the appearance of new transverse waves, occurs near the end of the "wedge" at A and B. The straight lines representing the expansion fan are labeled with their associated constant value Mach number, that is derived from Eq. (7) of Whitham's theory. In this particular record reignition occurs for a value $\Delta M_{C-J}/M_{C-J}$ of 0.06 and for $\epsilon \sim 2$; these values again lead to an acceptable agreement between theory and experiment.

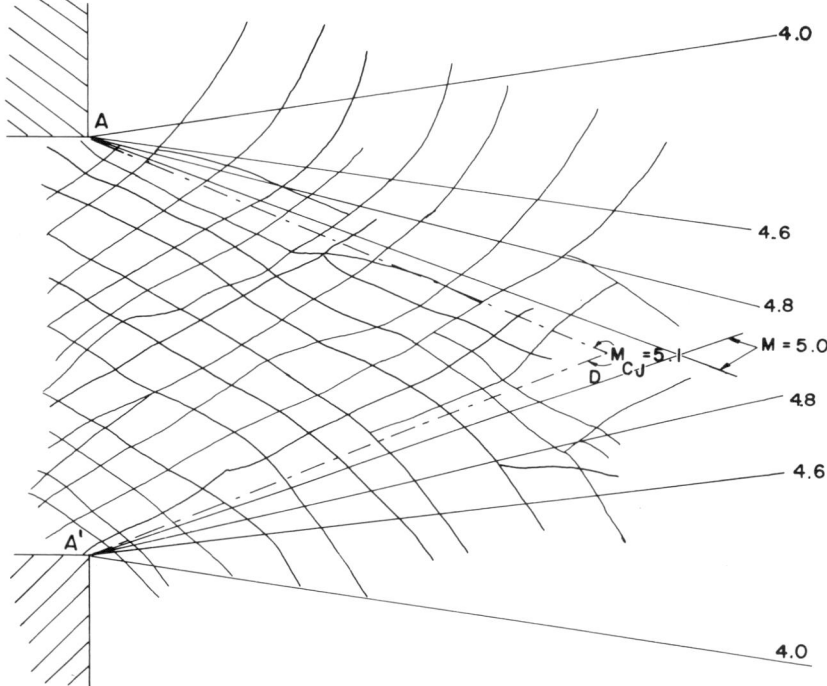

Fig. 12 Tracing of smoked-foil record obtained for $2H_2 + O_2$ at $p_0 = 390$ Torr in tube of Fig. 2a. $M_{C-J} = 5.1$. Conditions are near critical and reignition occurs near apex D of expansion heads.

Conclusions

Experiments carried out in a rectangular channel, of dimensions 22×6 mm^2, with detonation waves in stoichiometric oxyacetylene diluted with nitrogen, confirm the observations of Mitrofanov and Soloukhin (1964) that the principal dimension of a rectangular channel must be at least ten transverse wave spacings S in order to ensure the reignition of the detonation when it is diffracted at a large area change. The same observation was also found to hold true for stoichiometric oxyhydrogen diluted with varying amounts of argon. However, for detonations in systems with large and less regular cell structure, it would appear that more than ten transverse waves may be necessary to ensure sustenance of the diffracted wave. Stoichiometric mixtures, with oxygen, of ethane, propane, acetone and methane were studied in a 50×6 mm^2 rectangular channel, and although some difficulty was encountered in measuring the cell size in these systems, it was found that the average values of S at criticality were related to d_c by $d_c \sim 14S$ for ethane and propane, and $d_c \sim 18S$ for acetone and methane. The significance for these results is not understood at present but they clearly indicate that more data are required for the diffraction process in large diameter tubes so that the validity of the

Mitrofanov and Soloukhin result may be examined for systems with large cell structure.

For both critical and supercritical conditions reignition occurs at sites along the head of the expansion originating at the area change. Furthermore, if x_s is the shock position at reignition then (x_s/S) is invariant for oxyacetylene and oxyhydrogen systems and equal to $5\cot m$, where m is the angle between the head of the expansion wave and the tube axis.

A simple criterion for criticality is examined which demands that the transverse gradient of the shock front velocity $\partial M_s \leq \Delta M_{C-J}/\epsilon L_c$ where experimentally $1 < \epsilon \leq 3$ and ΔM_{C-J} is given by the Shchelkin criterion $\sim -0.1 M_{C-J}$. For the systems studied the criterion is in reasonably good agreement with the experimental data.

Further progress in the elucidation of re-establishment of the detonation wave will only be achieved through a better understanding of the nature of the reignition process itself. This, as we have already noted, bears a resemblance to the explosive reignition behind a smooth shock front which has been observed by Bach et al. (1968) during initiation by a blast wave produced by a laser source at critical power density. These authors attribute the ignition to inhomogenities in the reacting gas produced by turbulence when it is in a critical thermodynamic state corresponding to the autoignition limit. Similar ignition is observed behind reflected shock waves in a reactive system in which boundary-layer generated turbulence leads to reactive centers which, when they act coherently, produce a 'hot-spot' ignition (e.g., Voevodsky and Soloukhin, 1965; Meyer and Oppenheim, 1971). Another possible mechanism which may lead to explosive ignition is the time coherence of reacting particles produced by a nonsteady shock front. In the present case this could be brought about by the acceleration of a transverse wave as it moves into the centered rarefraction. Hot-spot formation, through the coherence of reacting particles has been demonstrated both experimentally and theoretically in the case of a slowly accelerating shock wave by Strehlow et al. (1967). There is some indication from the pressure records obtained behind the diffracted wave and also from the trajectories of the triple points, that there may be a slight enhancement of the transverse wave strength as it interacts with the expansion fan. At present, this explanation can only be offered in a very tentative manner but it may repay further investigation.

References

Bach G., Knystautas R., and Lee J. H. (1968) Direct initiation of spherical detonations in gaseous explosives. *Twelfth Sumposium (Intl.) on Combustion,* pp. 853-867. The Combustion Institute, Pittsburgh, Pa.

Bull D. C., Elsworth J. E., Hooper G., and Quinn C. P. (1976) Study of spherical detonation in mixtures of methane and oxygen diluted by nitrogen, *J. Phys. (D: Appl. Phys.)* **9**, 1991.

Dremin A. N. and Trofimov V. A. (1965) On the nature of the critical diameter. *Tenth Symposium (Intl.) on Combustion,* pp. 839-843. The Combustion Institute, Pittsburgh, Pa.

Edwards D. H., Hooper G., Morgan J. M., and Thomas G. O. (1978) The quasisteady regime in critically initiated detonation waves. *J. Phys. (D: Appl. Phys.)* **11**, 2103.

Edwards D. H., Nettleton M. A., and Thomas G. O. (1979) The diffraction of a planar detonation wave at an abrupt area change. *J. Fluid Mech.* **95,** 79-96.

Gvozdeva L. G. (1961) On the diffraction of a detonation wave. *Prykladnaya Mekhanika* **33,** 731-739.

Meyer J. W. and Oppenheim A. K. (1971) On the shock-induced ignition of explosive gases *Thirteenth Symposium (Intl.) on Combustion,* pp. 1153-1164. The Combustion Institute, Pittsburgh, Pa.

Mitrovanov V. V. and Soloukhin R. I. (1964) The diffraction of multifront detonation waves. *Sov. Phys.—Dokl.* **9,** 1055.

Presles H. and Brochet C. (1972) Detonation cylindrique divergente du nitromethane. *Astronautica Acta* **17,** 567.

Shchelkin K. I. (1959) Two cases of unstable combustion. *Sov. Phys. JETP.* **9,** 146.

Skews B. W. (1967) The shape of a diffracting shock wave. *J. Fluid Mech.* **29,** 297.

Soloukhin R. I. and Ragland K. W. (1969) Ignition processes in expanding detonations. *Combustion and Flame* **13,** 295-351.

Strehlow R. A., Cooker A. J., and Cusey R. E. (1967) Detonation initiation behind an accelerating shock wave. *Combustion and Flame* **11,** 339-351.

Urtiew P. A. (1975) From cellular structure to failure waves in liquid detonations, *Combustion and Flame* **25** 241.

Voevodsky V. V. and Soloukhin R. I. (1965) On the mechanism of explosion limits of hydrogen-oxygen chain self-ignition in shock waves, *Tenth Symposium (Intl.) on Combustion,* pp. 279-283. The Combustion Institute, Pittsburgh, Pa.

White D. R. (1976) Density induction times in very lean mixtures of D_2, H_2, C_2H_4 with O_2. *Eleventh Symposium (Intl.) on Combustion,* p. 147. The Combustion Institute, Pittsburgh, Pa.

Whitham G. B. (1957) A new approach to problems of shock dynamics, Part I: Two-dimensional problems. *J. Fluid Mech.* **2,** 145.

Zeldovich Y. B., Kogarko S. M, and Simonov N. N. (1956) An experimental investigation of spherical detonation in gases. *Sov. Phys.—Tech. Phys.* **1,** 1689.

Reinitiation Process at the End of the Detonation Cell

J-C. Libouton,* M. Dormal,† and P. J. Van Tiggelen†
Université Catholique de Louvain, Louvain-la-Neuve, Belgium

Experimental measurements of the leading shock velocity were used with a numerical solution of the detailed chemical kinetic mechanism for nonisothermal and nonisochoric conditions to predict induction times for shock-heated particles in the detonation cell. Because of shock-wave decay within the cell, the predicted induction times increased dramatically. For a gas overtaken by the leading shock at distances larger than three-quarters of the overall cell length, the induction time is so long that, in the absence of other gasdynamic effects, chemical reaction will not occur within the same cell. This paradox is explicable if the fresh gas mixture, already preheated and precompressed by the incident leading shock, is assumed to undergo a thermal perturbation that induces self-ignition. If the flow velocity of fresh gases behind the incident shock is recognized in the calculation, a computed Chapman-Jouguet detonation velocity for a preheated and precompressed medium compares quite favorably with the measured shock velocity at the beginning of the next cell. The results of the model are also in accord with the speed of the second pressure pulse observed experimentally at the end of the cell. Moreover, the occurrence of such a perturbation explains the enhancement of the emission intensity of the excited OH radicals.

Introduction

AS has been ascertained previously (Libouton et al., 1975), the cell structure of a detonation wave is linked to the chemical kinetics of the process. A simple correlation has been utilized between the characteristic time (t_{car}) of the cell structure and the induction time (t_{ind}) computed at the experimentally observed mid-cell conditions. The characteristic time (t_{car}) is defined as the ratio L/D where L and D are the cell length and the experimental detonation velocity, respectively (Libouton and Van Tiggelen, 1976).

Such macroscopic investigations based on average parameters of the detonation wave have been extended to the study of the detailed evolution of

---

Presented at the 7th ICOGER, Göttingen, Federal Republic of Germany, Aug. 20-24, 1979. Copyright © American Institute of Aeronautics and Astronautics, Inc., 1981. All rights reserved.
*Centre de Recherches de P.R.B. Nobel Explosifs.
†Laboratoire de Physico-Chimie de la Combustion.

induction times inside the detonation cell (Dormal et al., 1979). The data clearly indicate a continuous decoupling of the reaction zone from the leading shock wave during the whole progression of the wave throughout the cell. Only at the apex of the following cell is the tight coupling observable. None of the induction time measurements, however, could explain the origin of the sudden reinitiation process at the apex of the cell.

To investigate the sequence of events, it is convenient to employ the space-time diagram to specify the successive and relative locations of the leading shock and of the reaction zone inside the cell. It allows a display of the consequence of the progression of the unsteady leading shock.

The unsteady character of the leading shock throughout a detonation cell has been recognized previously by several investigators (see Steel et al., 1966; Crooker, 1969; and Strehlow and Crooker, 1974). However, a blast decay as fast as the one encountered by Lundstrom and Oppenheim (1969) has not been observed in our data (Dormal et al., 1979). The differences in behavior can be ascribed to the largely overdriven character of Lundstrom and Oppenheim's experiment. It explains the faster decay of the leading shock strength over the cell length, but it is surprising that such a marginal detonation does not exhibit a larger departure from the Chapman-Jouguet (C-J) velocity.

In this paper, we deal successively with the evolution of the induction time inside a detonation cell, and with the location of the reaction zone when the front shock arrives at the end of the cell. The discontinuity of the shock strength at the apex of the next cell is related to the appearance of a reacting shock behind the leading shock at a point about three-quarters of the length of the previous cell.

Induction Time

Induction times have been measured locally and systematically by using the emission intensity of the OH radical (Dormal et al., 1979, to be published). The induction times are computed at several planes in the cell from the solution of a kinetic model which accounts for reactant consumption as well as the nonisothermal and nonisochoric character of the overall reaction. Seven species are involved in the kinetic mechanism: H_2, O_2, H, O, OH, H_2O, and Ar. The subroutines of the computation program are schematized in Fig. 1 and the kinetic rate constants used for the computation of species profiles are shown in Table 1.

For a specified plane in the cell at time $t=0$, the temperature and the density are computed from the conditions at the von Neumann spike behind the leading shock, which propagates at the experimentally observed velocity V. During the time interval Δt, a Runge-Kutta numerical calculation is used to calculate the concentrations of the different species i (subroutine 1). From the starting point of the reaction, the heat release per unit mass is estimated at each step (subroutine 2). The instantaneous degree of conversion $\phi(t)$ is defined as the ratio of $Q(t)$, the heat released at time t, to Q_{C-J} the maximum heat which can be released locally in the process. The latter is a function of the local detonation velocity: $Q_{C-J} = V^2/2(\gamma_e^2 - 1)$, where the isentropic expansion coefficient γ_e is computed at the Chapman-Jouguet condition for a frozen

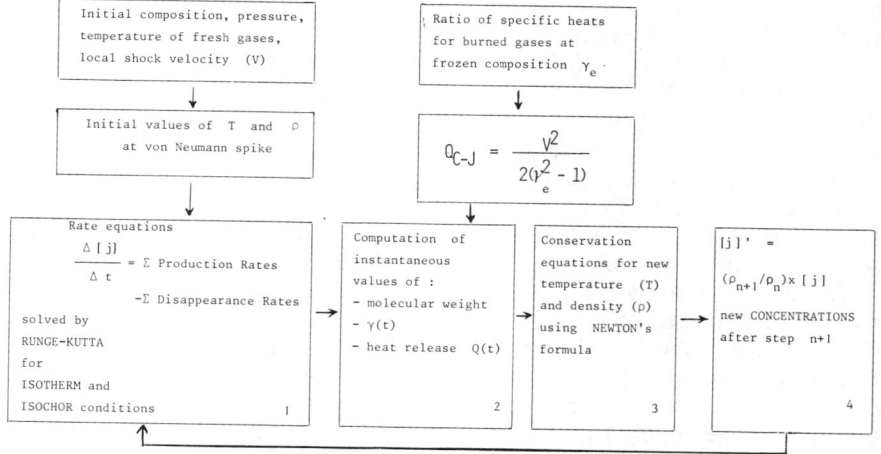

Fig. 1 Computation program of induction times.

Table 1 Kinetic mechanism

No.	Reaction	Rate constants, cm^{+3}·mole^{-1}·s^{-1}
0	$H_2 + O_2 \rightarrow 2OH$	$k_0 = 1 \times 10^{11} \exp(-40,000/RT)$
1	$H + O_2 \rightarrow HO + O$	$k_1 = 2.24 \times 10^{14} \exp(-16,800/RT)$
2	$O + OH \rightarrow H + O_2$	$k_2 = 1.13 \times 10^{13}$
3	$O + H_2 \rightarrow OH + H$	$k_3 = 1.74 \times 10^{13} \exp(-9450/RT)$
4	$H + OH \rightarrow O + H_2$	$k_4 = 7.33 \times 10^{12} \exp(-7300/RT)$
5	$OH + H_2 \rightarrow H_2O + H$	$k_5 = 2.19 \times 10^{13} \exp(-5150/RT)$
6	$H + H_2O \rightarrow OH + H_2$	$k_6 = 8.41 \times 10^{13} \exp(-20,600/RT)$
7	$H_2O + O \rightarrow OH + OH$	$k_7 = 5.75 \times 10^{13} \exp(-18,000/RT)$
8	$OH + OH \rightarrow H_2O + O$	$k_8 = 5.75 \times 10^{12} \exp(-780/RT)$
9	$H + OH + M \rightarrow H_2O + M$	$k_9 = 1.17 \times 10^{17a}$ (with $M = H_2O$)
10	$H_2O + M \rightarrow H + OH + M$	$k_{10} = 2.2 \times 10^{16} \exp(-105,000/RT)$

[a] Units: cm^6·mole^{-1}·s^{-1}.
(Baulch D. L., Drysdale D. D., and Lloyd A. C. (1968-1969) *Critical Evaluation of Rate Data for Homogeneous Gas Phase Reactions*, Vols. 2 and 3. Butterworths Leeds, England.)

composition. During the heat-release process, γ is assumed to be a linear function of $\phi(t)$, the degree of conversion:

$$\gamma(t) = \gamma_f - (\gamma_f - \gamma_e)\phi(t) \tag{1}$$

where: γ_f is γ of the fresh gaseous mixtures.

From $\gamma(t)$ and $Q(t)$, the increments of density ρ and temperature T are estimated from the solution of the conservation equations with Newton's formula (subroutine 3). New values of the concentrations are estimated from

REINITIATION PROCESS AT END OF DETONATION CELL 361

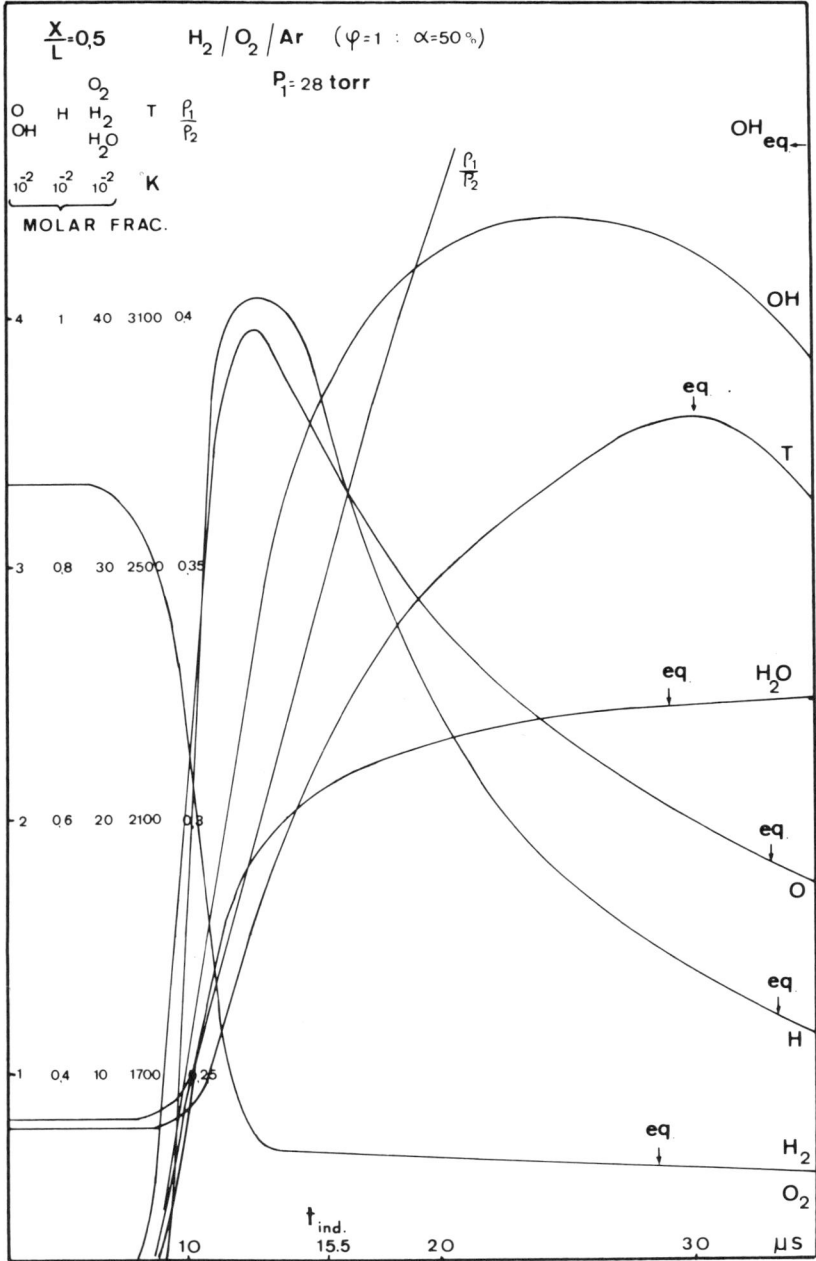

Fig. 2 Temperature and species mole fractions transients at mid-cell position (arrows indicate the results of equilibrium composition for C-J detonation propagating in investigated mixture, ϕ = equivalence ratio, α = percentage of argon).

the density change after a time interval Δt (subroutine 4). These new concentrations are fed back to subroutine 1 to start the next step. The computed induction time is defined as the time elapsed between the shock and the maximum OH concentration in the ground state (the maximum of OH concentration is reached before the maximum of OH mole fraction). Furthermore, the temperature is still increasing after the maximum of OH concentration and the temperature reaches the equilibrium value at about twice the induction time.

The value of this approach can be checked easily by comparing the profile of the mole fraction for each individual species with the Chapman-Jouguet equilibrium mole fraction (see Fig. 2). The equilibrium mole fractions for a detonation in stoichiometric hydrogen/oxygen mixture diluted with 50% argon at an initial pressure of 28 Torr are marked with arrows on the computed curves of the mole fractions vs time at a mid-cell position ($X/L = 0.5$). All the individual mole fractions at equilibrium are reached at about the same time plane, typically 30 μs. At that time plane for $X/L = 0.5$, the departure of mole fractions from their equilibrium values is around 10%.

Location of the Reaction Zone

The trajectory of the reaction zone is specified in the space-time diagram (reduced coordinates) by a dashed line (upper part of Fig. 3). For instance, at location $X/L = 0.7$ at time A, the gas particle is heated by the leading shock and set in motion along the path AD (dotted line). The gas particle reacts and

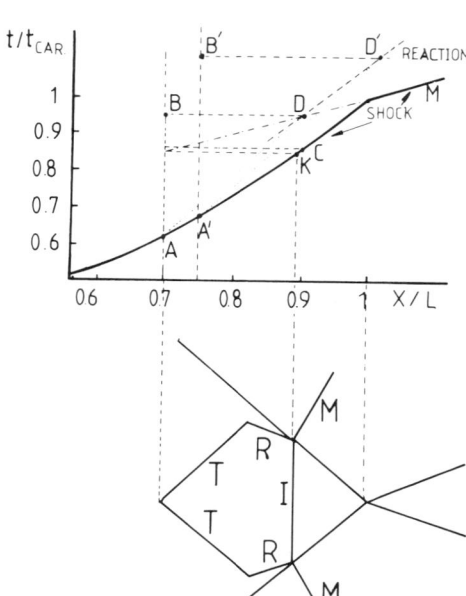

Fig. 3 Top: wave trajectories in reduced coordinates at the end of the cell; bottom: instantaneous shock waves pattern (M = Mach stem, R = reflected shock, T = transverse wave, I = incident shock).

reaches the maximum concentration of OH ground state on location D, i.e., after a reduced induction time (t/t_{car}) equals AB. The bulk of the reaction takes place, then, at a position $X/L = 0.9$, downstream with respect to the point where the gas particle has been heated up $(X/L = 0.7)$. Therefore, the usual correction $t_{ind} = t_{exp}(\rho_2/\rho_1)$, applied to change the laboratory coordinates in gas particle coordinates, for a steady shock is appropriate but not sufficient in this case, since the shock at location $X/L = 0.7$ is stronger than at $X/L = 0.9$. The nonsteadiness of the leading shock is, indeed, characterized by the decay of shock strength from the apex $(X/L = 0)$ to the end $(X/L = 1.0)$ of the cell.

If the proper time correction is made, it is apparent that a gas particle initially at $X/L > 0.75$ will have insufficient time, after shock heating, to react before it reaches the end of the cell $X/L = 1.0$. This observation is rather surprising, as it would appear to require a mixing of two successive reaction zones behind the leading shock at the beginning of the next cell. Such an event has never been noticed to the best of our knowledge, and suggests a different ignition behavior at the end of the cell. However, according to Edwards et al. (1969) and Nikolaev et al. (1978), the shock configuration at the end of the cell is schematized on the lower part of Fig. 3. A thermal perturbation, generated by interaction of transverse waves, could lead to a reinitiation of the detonation wave in a precompressed medium, whose state is characterized by the von Neumann state behind the leading shock.

Reinitiation Model

The plausibility of reignition by thermal perturbation is shown by the development of a reinitiation model. It is postulated that the velocity of detonation, which propagates in a precompressed and preheated medium at a plane in the cell close to its ending $(X/L = 0.95)$, should be essentially the speed of the Mach stem generated just after the apex of the following cell. As an example, the parameters were calculated of a detonation propagating though a stoichiometric mixture H_2/O_2 diluted with 50% argon at an initial pressure of 28 Torr. For this mixture the Chapman-Jouguet detonation velocity is 1806 m/s.

For the experimental apparatus, see Libouton et al. (1975). The detonation mode at these conditions is two (mode definition due to Strehlow, 1970). The length L of the cell is 180 mm. The experimental detonation velocity D is 1658 m/s which corresponds to a velocity averaged over the length of several cells and is reached, in fact, at mid-cell. At a plane $X/L = 0.95$, the local experimental shock velocity is 0.8 times the average detonation velocity D, i.e., 1326 m/s. For any stable detonation of mode two, the closing angle of the cell has been observed to remain constant and the half-angle is 40 ± 0.5 deg. If the triple-point velocity V_{TP} along its trajectory matches the detonation velocity D and if the incident shock is planar, the velocity of the incident shock V_{in} is the projection of V_{TP} along the propagation axis of the incident shock. Then:

$$V_{in} = V_{TP} \cos 40 \deg = D \cos 40 \deg = 0.77 D$$

is very close to the observed velocity at the end of the cell, $0.8\,D$. If the observed V_{in} is used with the equations of unreactive shocks:

$$\frac{P_2}{P_1} = 1 + \frac{2\gamma}{\gamma+1}(M_1^2 - 1) \tag{2}$$

$$\frac{T_2}{T_1} = 1 + \frac{2(\gamma-1)}{(\gamma+1)^2}\frac{\gamma M_1^2 + 1}{M_1^2}(M_1^2 - 1) \tag{3}$$

$$\frac{\rho_2}{\rho_1} = \frac{P_2}{P_1} \cdot \frac{T_1}{T_2} \tag{4}$$

the parameters of the von Neumann's spike at $X/L = 0.95$ are: $P_2 = 414$ Torr, $T_2 = 1146$ K, and $\rho_2/\rho_1 = 3.78$. From mass conservation, $\rho_1 U_1 = \rho_1 V_{in} = \rho_2 U_2$, the velocity of preheated (or shocked) gases with respect to the shock is 351 m/s, and in laboratory coordinates the preheated gases velocity V_2 is 975 m/s (from $V_2 = 0.8\,D(1 - \rho_1/\rho_2)$).

With the temperature of 1146 K and the pressure of 414 Torr, the induction time (computed by the procedure described in the preceding section) is 90 μs. Therefore, it is obvious that during such a long time delay, any shock perturbation due to the interaction of transverse waves, for instance, will ignite a new reactive shock in that precompressed medium. The problem reduces to the calculation of the velocity of a detonation that propagates into the fresh gas mixtures which have been previously shocked and heated to 1146 K and 414 Torr. For this purpose a program of Guirao et al. (1972) was used with slight modifications. The computed velocity U_{RS} of the reactive shock (detonation) is 1758 m/s; but since the reactive shock moves in a gas already in motion, the net velocity with respect to the laboratory coordinates becomes $V_{RS} = V_2 + U_{RS} = 2733$ m/s. Such a velocity is much larger than the local velocity of the leading shock (incident shock: $V_{in} = 1326$ m/s). The newly formed detonation will overtake the incident shock at the apex of the consecutive cell. The strength of the shock generated by the interaction of the incident and reactive shocks can be deduced from the product of pressure ratios across each of them. The pressure ratio across the incident shock is 14.8 and across the reactive shock 6.64 for the von Neumann spike and 4.45 for the Chapman-Jouguet condition. The final state of the reactive shock, more probably, will be close to the Chapman-Jouguet state. The pressure ratio of the emerging shock will be between 98.3 and 65.8; its Mach number, 9.06 and 7.42 from Eq. (2); and its velocity scale between 2.05 and 1.68 D. The Mach number measured at the apex of the cell is 7.9, and the shock velocity is 1.8 D, which agrees with the computed value.

It can be inferred, therefore, that a detonation is not only a self-sustaining ignition behind an unsteady shock, but, more precisely, a periodic reinitiation of reactive shocks at a frequency identical to the reciprocal of the characteristic time.

REINITIATION PROCESS AT END OF DETONATION CELL

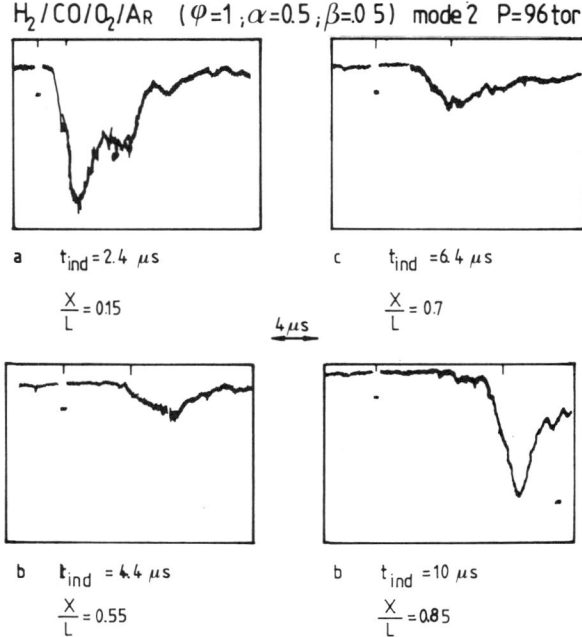

Fig. 4 OH emission at several positions inside the cell (ϕ and α as in Fig. 2, β = percentage of hydrogen in the fuel).

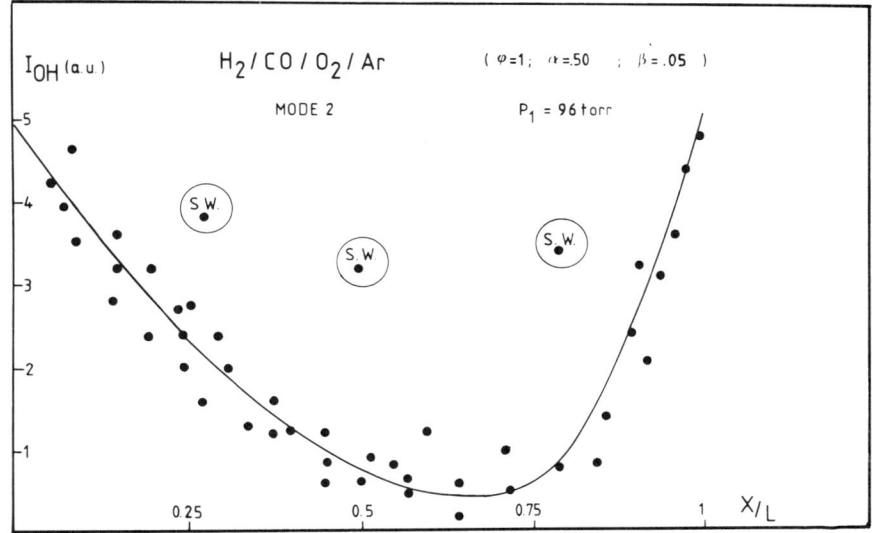

Fig. 5 Maximum OH emission intensity inside the cell. SW corresponds to runs with a slapping wave perturbating the record (same mixture as Fig. 4).

Interpretation of the Experimental Evidence

The experimental evidence of the occurrence of a second, reactive shock near the end of the cell are: 1) a progressive increase of the maximum excited OH emission for relative positions in the cell larger than 0.75, and 2) the presence of a second pressure jump behind the incident shock increasing in strength toward the end of the cell.

1) Profiles of induction times and variations of the maximum intensity of OH emission have been deduced from the characteristic emission transients of excited OH radicals along the cell axis; see Fig. 4. The induction time is defined as the time elapsed between the leading shock (pressure peak) and half the value of the maximum intensity of OH emission (Dormal et al., to be published). The induction time increases continuously from the apex to the ending of the cell as a consequence of the decreasing shock strength. Similarly, the maximum intensity of emission decreases from the apex, but goes through a minimum at around three-quarters of the cell length and, at the ending of the cell, it reaches a value close to the one noticed at the apex of the next cell (see Fig. 5). This observation is in accord with the open-shutter photograph of a detonation where the overall emission luminous intensity increases before the triple-point intersections (Oppenheim, 1970). Without the progressive development of the reinitiation process behind the incident shock, these observations appear anomalous because of the temperature and pressure conditions produced by the incident shock toward the end of the cell.

2) The local pressure jump induced by shocks has been observed with four flush-mounted piezoelectric gages at several points located along the cell axis (Dormal, to be published). The results indicate that a continuous decrease of

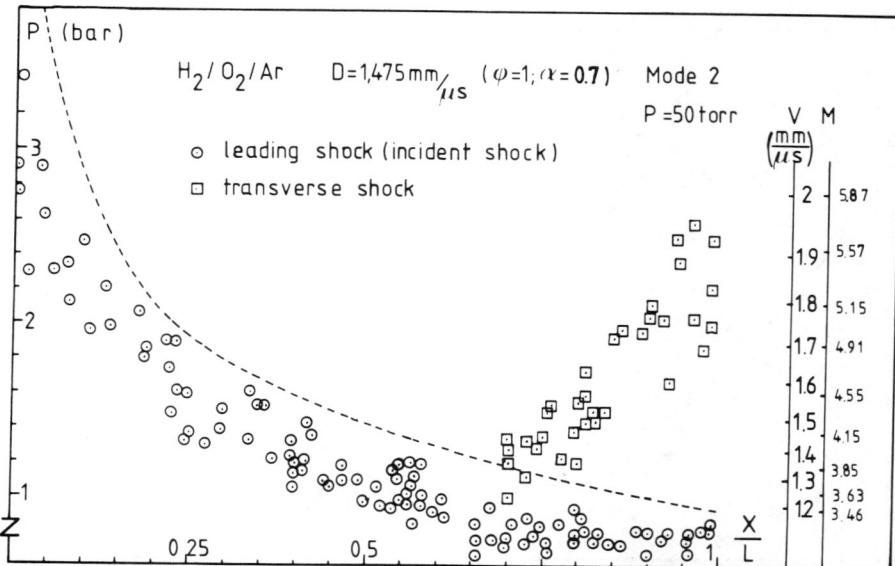

Fig. 6 Comparison of local pressure inside cell to that of von Neumann state (V and M = measured values).

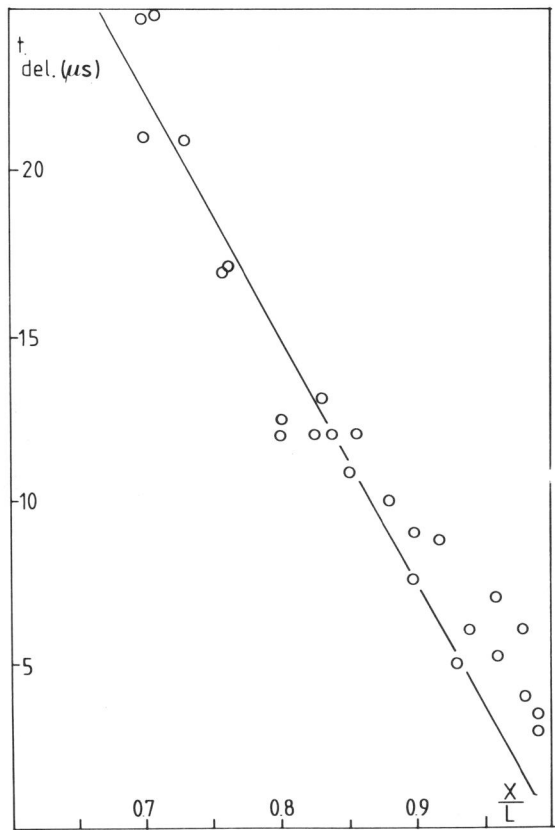

Fig. 7 Time delay t_{del} between both pressure jumps occurring at end of cell (same mixture as Fig. 6).

the incident shock pressure pulse occurs along the cell axis (see Fig. 6). However, from a position located at $X/L = 0.75$ and beyond, two successive pressure jumps are detected by a gage. The amplitude of the first pressure jump continues to decrease toward the end of the cell, and is in agreement with the decay of the incident shock (leading shock) strength. But the increasing intensity of the second pressure jump cannot be accounted for. The product of these pressure ratios is equivalent to the pressure ratio measured across the Mach stem at the apex of the consecutive cell. The delay time t_{del} between those two shocks decreases continuously along the cell axis and vanishes at the apex of the following cell (Fig. 7). From the local pressure records one can measure the local delays between both shocks. These delays are then plotted on a space-time diagram in reduced coordinates (t/t_{car} and X/L). Such a procedure is valid for any detonation propagating in mixtures with variable fuel compositions, dilutions, and initial pressures. The cells have similar shapes, in fact, particularly the half-angles of the cell ending that remain constant irrespective of pressure and dilution conditions. Measurements have been performed on a stoichiometric H_2/O_2 mixture diluted with 70% Ar, and at an

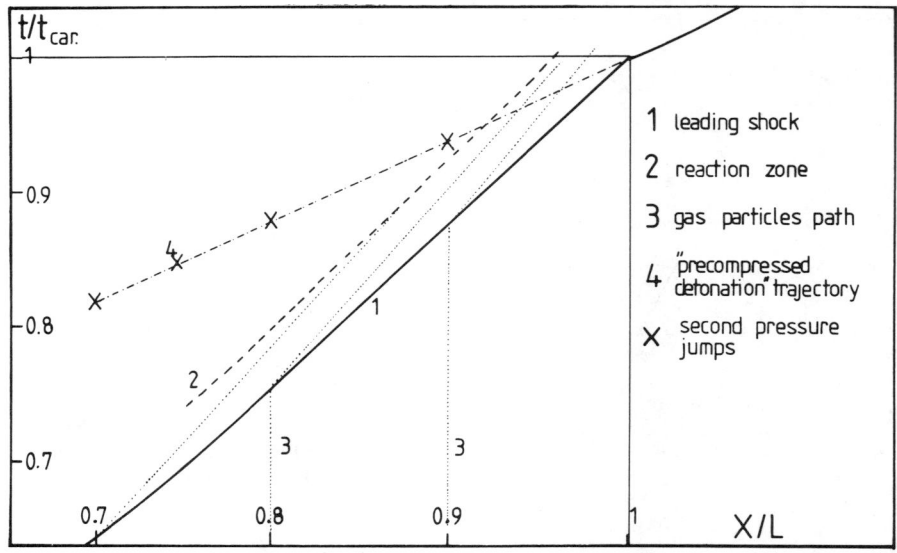

Fig. 8 Wave trajectories at end of cell; "precompressed detonation" coincides with experimental delay between pressure jumps (crosses).

initial pressure of 50 Torr. The characteristic time of the structure for this run is 122 μs. At a position $X/L = 0.7$, the time delay t_{del} between the two shocks is 22.5 μs, i.e., in reduced time scale 0.184. Since both shocks have to coincide at the ending of the cell, the reduced distance from position $X/L = 0.7$ to be covered by the second shock is 0.3. The available reduced duration τ for such a travel time is $(t_{car} - t_{0.7} - t_{del})/t_{car}$ where $t_{0.7}$ is the time elapsed when the incident shock reaches the location $X/L = 0.7$. The origin of the time scale is the apex of the cell. For the above numerical example at $X/L = 0.7$, $t_{0.7}$ equals 0.638 and yields $\tau = 0.178$. The reduced velocity of the second pressure jump, in laboratory coordinates, is then $0.3/\tau = 1.69$, and the effective value for the second shock velocity will be 1.69 times the experimental detonation velocity D, i.e., 1.69×1658 m/s = 2802 m/s. This result also means that the Mach number at the apex of the cell should be 1.69 times the average Mach number, i.e., the one measured at mid-cell. This value of 1.69 compares fairly well with the experimental data, 1.8, and with the computed value from the above developed model, 1.68.

The diagram in reduced coordinates (Fig. 8) summarizes all of the conclusions. The trajectories of the leading shock, of the reaction zone, and of the gas particles paths are indicated by curve 1 (continuous line), curve 2 (dashed line), and curve 3 (dotted line), respectively. The crosses correspond to the time at which the second shock reaches three different locations on the cell axis: 0.7, 0.8, and 0.9, respectively. The straight line (curve 4) (dashed-dotted line) is the trajectory of a detonation wave that propagates through a mixture precompressed and preheated by the incident shock as computed previously from the model that yields a detonation velocity of 1.68 D. The merging of shocks 1 and 3 induces the Mach stem, and occurs at the apex of the consecutive cell when the triple points intersect. That merging shock encounters the fresh gases at initial temperature. The divergence of flow after the apex explains the decay of the strength of the Mach stem.

In conclusion and in agreement with the experimental results, the reinforcement at the apex of the cell derives from a continuous amplification process starting at about three-quarters of the cell length. Although the ultimate origin of the amplification has not been determined, the available data suggest that the process results in the formation of a detonation wave that corresponds to an equilibrium detonation wave progressing in an already shocked medium.

Acknowledgments

We express our gratitude to Dr. C. Brochet for allowing us to use calibrated pressure gages built at the ENSMA, Poitiers, France. We also acknowledge financial support of FRFC, Contract 2.9009.76.

References

Crooker A. J. (1969) Phenomenological investigation of low mode marginal planar detonation. Ph.D. Thesis, Univ. of Illinois. Tech. Rept. AAE 69-2.

Dormal M., Libouton J-C., and Van Tiggelen, P.J. (1979) Evolution of induction time in detonation cell. *Acta Astronautica* **6**, 875-884.

Dormal M., Libouton J-C., and Van Tiggelen P. J. Etude expérimentale des paramètres à l'intérieur d'une maille. (Submitted for publication in *Acta Astronautica*.)

Edwards D.H. and Parry D.J. (1969) The structure of transverse waves in detonations. *Acta Astronautica* **14**, 533-537.

Guirao C.M., Knystautas R., and Lee S. H. (1972) Detonation study in H_2/Cl_2, CS_2/O_2 and $CO/H_2/O_2/N_2$ mixtures. MERL Rept. 72-6.

Libouton J-C., Dormal M. and Van Tiggelen P. J. (1975) The role of chemical kinetics on structure of detonation waves. *15th Symposium (Intl.) on Combustion*, pp. 79-86. The Combustion Institute, Pittsburgh, Pa.

Libouton J-C., and Van Tiggelen P. J. (1976) Influence de la composition du mélange gazeux sur la structure des ondes de détonation. *Acta Astronautica* **3**, 759-769.

Lundstrom E. A. and Oppenheim A. K. (1969) On the influence of non-steadiness of the thickness of the detonation waves. *Proc. Roy. Soc.* (London) **A310**, 463-478.

Nikolaev V., Topchian M. E., and Ioulanski B. V. (1978) Experimental study and computation of triple configurations of spin detonation. *Fizika Goreniia i Vzryva* **14** (6), 106 (in Russian).

Oppenheim A. K. (1970) Introduction to gasdynamics of explosions. *Intl. Centre for Mechanical Sciences* **48**, 34. Springer-Verlag, Berlin.

Steel G. B., Urtiew P. A., and Oppenheim A. K. (1966) Experimental study of the wave structure of marginal detonation in rectangular tube. Report A5-66-4. Univ. of California, Berkeley.

Strehlow R. A. (1970) Multidimensional detonation wave structure. *Acta Astronautica* **15**, 345.

Strehlow R. A. and Crooker A. J. (1974) The structure of marginal detonation waves. *Acta Astronautica* **1**, 303-315.

Effects of Cellular Structure on the Behavior of Gaseous Detonation Waves under Transient Conditions

Paul A. Urtiew* and Craig M. Tarver†
*Lawrence Livermore National Laboratory,
University of California, Livermore, Calif.*

> Increased concern about the detonability of gaseous mixtures in unconfined clouds renewed our interest in the behavior of the detonation process under various transient conditions, such as a nonuniform distribution of pressure, temperature, or composition. In readjusting itself during such a transient state, the process may either survive and continue as a detonation wave or may weaken so fast that it transforms into a simple deflagration wave. A similar transient condition may be imposed on the wave by letting it pass through an area change. Here again, the wave may either survive or weaken to a point of its extinction. Assuming that the cellular structure of the detonation wave is responsible for its existence and its uniqueness, we investigated the effect of that structure on the transition of the process into expanded geometries. Comparison was made with the results of other investigators who performed similar experiments for different reasons. The results show that it is not the tube diameter nor the initial pressure of the medium that controls the survival of the wave but that it is 1) the number of cells available to endure losses from flow effects generated by the transient conditions and 2) the number of cells remaining to regenerate new cells to revive the detonation process.

Introduction

RECENT advances in the field of gaseous detonation have renewed interest in one of the relatively untouched areas of the detonative process, that is, its behavior under transient conditions when one or more of the controlling variables is not constant. Such transient conditions are met when a detonation wave experiences a change in composition, pressure, temperature, or confinement. The process begins to adjust itself to a new set of conditions and

---

Presented at the 7th ICOGER, Göttingen, Federal Republic of Germany, Aug. 20-24, 1979. Copyright © American Institute of Aeronautics and Astronautics, Inc., 1980. All rights reserved.
*Engineer.
†Chemist.

undergoes a change in structure. It also begins to reveal its characteristic features. Knowledge of such details of the character of the wave may lead us to a better understanding of the process and, in particular, of the link between chemistry and hydrodynamics in a detonation wave.

The last few decades have been very fruitful in providing information about the nature of the detonative process and the essential features of its cellular structure. Full details about the subject may be found in the comprehensive reviews of Oppenheim et al. (1963), Strehlow (1968), Lee et al. (1969), and Lee. (1977). It is believed that the cellular structure of gaseous detonations has been proved beyond any doubt and that any new findings connected with the behavior of the detonation process must be associated, one way or another, with this cellular structure.

One important finding that emerged from all investigations during the last few decades is that the cellular structure not only exists but is essential to the detonative process. Its size and regularity are characteristics of a particular combination of initial conditions, such as composition, pressure, and temperature. Variation of these conditions causes the cell size to change and attain a new size that is characteristic of the new set of conditions. Differences in confinement (such as the tube diameter) or the lack of any confinement (as in the case of spherical detonations), if kept constant during the process, will have no effect on the characteristic cell size. However, should the detonation wave enter an area change, the cell size will immediately be affected by the change and will reveal some of the transient features of the phenomenon. Local variations in these conditions, including those of the cross-sectional area, will cause enhancement of the detonative process or weakening of it with a possibility of a complete extinction.

The first records of a detonation wave traveling into both gradual and abrupt changes of the cross-sectional area were shown by Voitsekhovsky et al. (1963). Their records were taken with an open-shutter technique, and changes in cell size and even complete loss of cellular structure are quite evident. Strehlow and his group at the Univ. of Illinois have gathered an enormous amount of valuable information on the cellular structure and have contributed significantly to our present understanding of the phenomenon. Their experiments (Strehlow et al., 1972) with an adjustable wedge inside the tube also reveal the effects caused by the change in the cross-sectional area and by the variation of the cell size as the detonation wave propagates through such an area change.

Our earlier unpublished experiments, in which an area change in the path of a detonation wave was produced by a wedge inserted into the 2.54 × 3.81 cm rectangular tube, also revealed some interesting results such as different threshold pressures for transition at different orientations of the wedge (Urtiew and Oppenheim, 1965). Similar experiments with a convergent-divergent tube were performed the same year by Lee et al. (1965). However, it is his recent study with Matsui (Matsui and Lee, 1978) on the transition of detonation from a planar geometry into a spherical one that stimulated the present work and led us to study the effects of cellular structure of the detonation wave on its transition from a tube of constant cross section into an expanded geometry.

Effects

If we examine the records of Strehlow et al. (1972) and Voitsekhovsky et al. (1963), as well as unpublished experiments of Urtiew and Oppenheim (1965), we find that as the detonation wave progresses into an area change the cell size either gets smaller or larger depending on whether the detonation wave propagates, respectively, into a contraction or expansion of the tube. This process is qualitatively illustrated in Fig. 1.

When the wave enters a convergent section (Fig. 1a), an oblique shock or a compression fan is generated at the corner or at its gradual convergence. The flow behind the front is affected and the wave becomes slightly overdriven. At the same time at the wave front, the triple-intersection point nearest the converging wall meets a reflecting surface sooner than it would under constant area flow, and it reflects at higher pressure. Therefore, the cells become smaller. The effect is propagated radially into the tube, continuously changing the cell size.

When the wave enters a divergent section (Fig. 1b), it produces a rarefaction wave centered at the corner or spread around a gradual expansion (not shown). At the same time, the triple-intersection point nearest the wall does not find a reflecting surface in time for a regular enforcement of the detonation process and thus becomes weaker and weaker. Both of these effects must work hand in hand because loss of reflecting surface alone would not prevent the wave from generating a new explosion point near the would-be reflection point and thereby continue the process, as it does in spherical

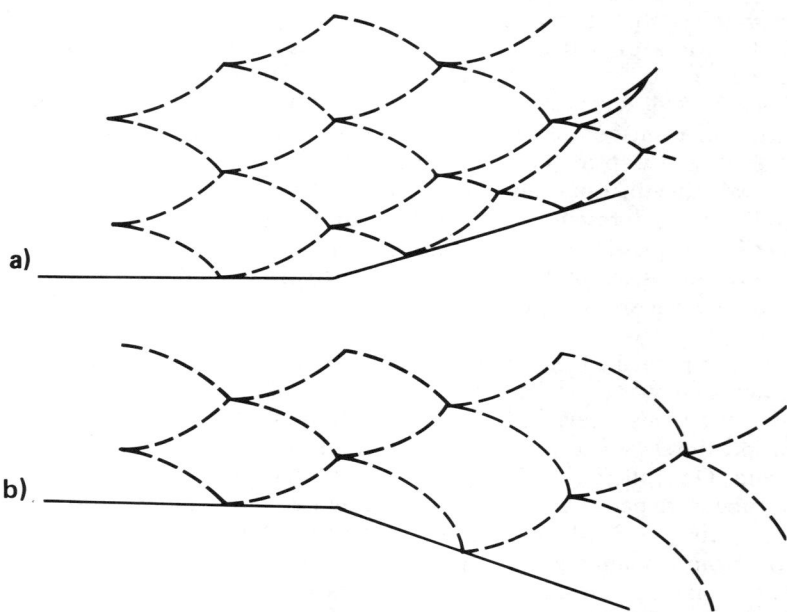

Fig. 1 Sketch of a cellular detonation wave entering a) convergent section and b) divergent section.

detonations. On the other hand, rarefaction alone (depending on its strength) may only weaken the process, which will still carry on until marginal conditions are attained.

In the case of tube expansion, the strength of the rarefaction, namely, the amount of pressure drop across the wave, depends on the amount of expansion, i.e.,

$$\Delta p = f(\delta) \qquad (1)$$

where δ is the angle of divergence. Rarefaction is an isentropic process propagating at a local velocity of sound. In a uniform flow, the sound velocity would be constant and the rarefaction wave would travel along a characteristic curve toward the center of the tube. In the case of a detonation wave with the cellular structure, local velocities of sound vary so much within the cell that interaction between the detonation front and the incoming rarefaction wave becomes very complex. The question of original strength becomes very important. The rarefaction wave may not be sufficiently strong to overcome the effect of burning as it propagates through individual cells and may cease to exist at some point along its path toward the center of the flow.

Transition of Detonation from Planar to Spherical Geometry

Consider the case of transition from a plane geometry into a hemispherical one, which represents the case where $\delta = 90$ deg. This case does not produce the strongest rarefaction wave at the corner, but it represents a simple case that

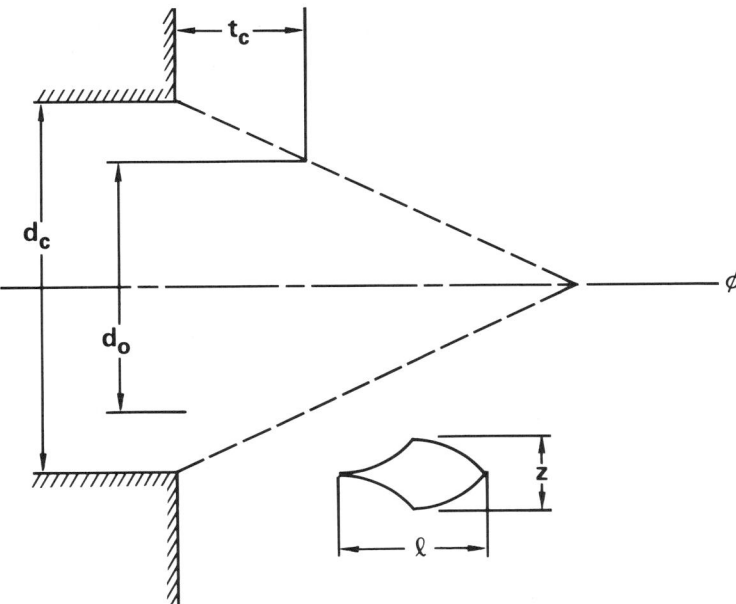

Fig. 2 Sketch of geometry for determining the work done, according to Eq. (3).

may be compared with experimental studies of Matsui and Lee (1978). Unlike the rarefaction caused by a plane shock wave, the effects of a rarefaction wave generated by a detonation wave will gradually diminish as it propagates toward the centerline of the hemisphere until, if the original tube diameter is large enough, the effects become so small that the wave begins to regenerate new cells as if it were a portion of a spherical detonation wave. The criterion for survival is, then, the diameter of the tube or, rather, how much of the inner core of the detonation wave will remain unaffected by the rarefaction wave and how well it will be able to re-establish the detonation process in a spherical geometry.

To determine the relationship of the inner-core diameter d_o with the critical diameter of the tube d_c let us relate the energy required for initiation of a spherical wave with the work done by the core of the wave on the undisturbed gas outside of the tube. If we consider the geometry as illustrated in Fig. 2, the work done (WD) by the core on the outside gas will have the following approximate expression:

$$\text{WD} = \int_0^{t_c} PuA\, dt \tag{2}$$

where P and u represent the Chapman-Jouguet (C-J) pressure and particle velocity, A the area of the inner core, and

$$t_c = \frac{l}{z}\left(\frac{r_c - r_o}{D}\right) \tag{3}$$

represents the time required by the head of the rarefaction wave to reach the outer edge of the core at r_o. There, l and z are the cell size dimensions in length and width, respectively, and D the detonation velocity of the wave.

Integrating Eq. (2), one obtains an expression for the work done:

$$\text{WD} = \pi P_{C-J} u_{C-J} r_o^2 (l/z)[(r_c - r_o)/D] \tag{4}$$

or

$$\text{WD} = \left(\frac{\pi P_{C-J} u_{C-J}}{8D}\right)\left(\frac{l}{z}\right)\left(\frac{d_c}{d_o} - 1\right) d_o^3 \tag{5}$$

In essence, this work done on the outside gas by the inner core of the wave is the energy required to initiate a detonation process in a hemispherical geometry. To compare its value with the experimental value of Matsui and Lee (1978) for the spherical case, we must take 2WD and express it in terms of d_c as follows:

$$E_c = 2\text{WD} = \frac{\pi P_{C-J} u_{C-J}}{4D} \frac{l}{z}\left[\left(\frac{d_c}{d_o} - 1\right)\left(\frac{d_o}{d_c}\right)^3\right] d_c^3 \tag{6}$$

As expected, E_c is proportional to d_c^3 provided the term in the square brackets remains constant. If we assume various values for the ratio of d_c/d_o, the expression in the square brackets attains a maximum value of 0.15 when the ratio of d_c/d_o is set equal to 1.5, which agrees quite well with experimental records of Lee (1979). Using this value for the ratio of d_c/d_o, setting $l/z = 2$, and calculating the C-J parameters for each of the mixtures investigated by Matsui and Lee (1978), we can compare our calculated energy values with theirs by plotting both values against their experimentally determined values of the critical diameter d_c. All calculated values are tabulated in Table 1, while the plot of E_c vs d_c is shown in Fig. 3.

The agreement between the two sets of values is quite evident, indicating that not only our expression for the critical energy is correct but also that the assumption of $d_c/d_o = 1.5$ is valid. Figure 3 also reveals that, while all the points of the fuel-air mixtures lie on the same straight line, they do fall below the extended fuel-oxygen line. This may be because of the chemical difference between the fuel-oxygen and fuel-air mixtures or simply because of erroneous, although consistent, extrapolation of the d_c vs N_2/O_2 curves to N_2/O_2 for air (Matsui and Lee, 1978). The critical energy for initiation of methane-air detonations resulted in the value of 1.74×10^8 J. which falls slightly below the

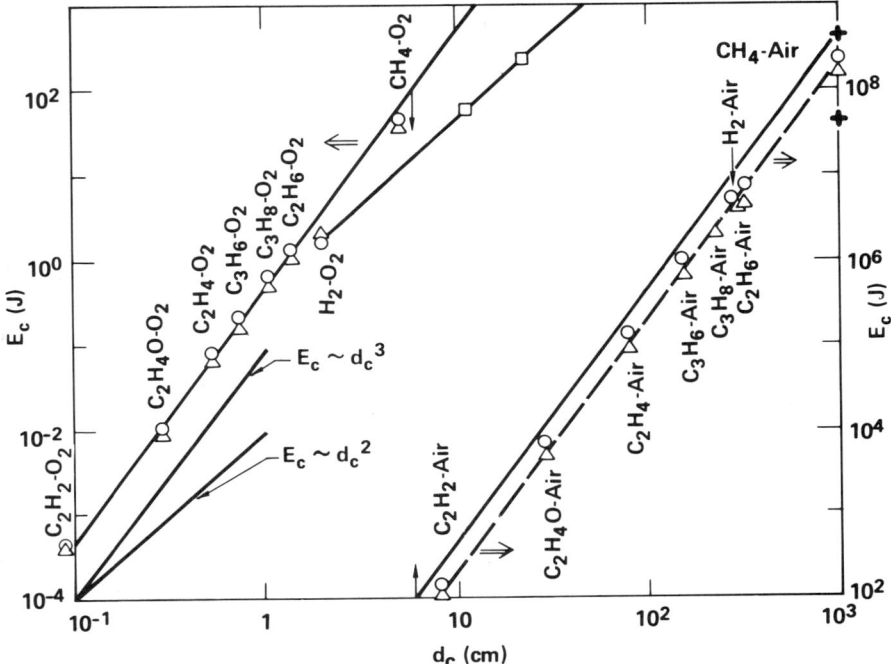

Fig. 3 Critical energy of initiation as a function of the critical diameter at S.T.P.: O, values reported by Matsui and Lee (1978); △, present values obtained by Eq. (3); □, effect of P_o on $E_c - d_c$ variation; +, range of prediction between Bull et al. (1976) and Westbrook and Haselman (1979).

Table 1 Detonation parameters and critical values of energy and diameter for various explosive mixtures.[a]

Fuel source	Oxid	Fuel, vol %	E_c^a, J	d_c^a, cm	P, atm	D, m/s	D, m/s	c, m/s	γ	E_c, J
C_2H_2	O_2	40	3.83×10^{-4}	$(0.25)^b$	41.42	2728.54	1235.08	1493.46	1.18	3.6×10^{-4}
			5.9×10^{-3}	0.09						
			10×10^{-3}							
C_2H_4O	O_2	40	1.2×10^{-2}	0.28	43.55	2499.78	1145.73	1354.05	1.16	1.07×10^{-2}
C_2H_4	O_2	33.3	7.2×10^{-2}	0.52	38.78	25.84	1182.18	1402.38	1.16	6.16×10^{-2}
C_3H_6	O_2		2.03×10^{-1}	0.7	43.22	2559.75	1172.33	1387.43	1.16	1.66×10^{-1}
C_3H_8	O_2	22.2	5.77×10^{-1}	1.0	42.72	2539.57	1165.69	1373.87	1.17	4.76×10^{-1}
C_2H_6	O_2	66.7	1.07	1.3	38.85	2544.80	1167.99	1376.81	1.15	9.52×10^{-1}
H_2	O_2	66.7	1.58	(1.9)	18.90	2848.83	1297.15	1551.68	1.13	1.66
CH_4	O_2	40	5.07×10	2.0	31.73	2544.78	1167.33	1377.45	1.14	4.42×10
				(3.2)						
				5.0						
C_2H_2	Air	12.5	1.29×10^2	8.0	20.01	1927.90	856.38	1071.52	1.19	1.08×10^2
C_2H_4O	Air	12.3	7.62×10^3	30.0	19.79	1852.26	818.05	1034.21	1.20	5.56×10^3
C_2H_4	Air	9.5	1.2×10^5	80.0	17.82	1816.10	797.98	1018.12	1.20	9.46×10^4
C_3H_6	Air	6.6	7.55×10^5	150.0	17.82	1786.99	782.28	1004.71	1.21	6.2×10^5
C_3H_8	Air	5.7	2.52×10^6	220.0	17.26	1759.78	768.54	991.23	1.22	1.89×10^6
C_2H_6	Air	5.7	5.09×10^6	280.0	16.96	1766.56	772.49	994.07	1.21	3.84×10^6
H_2	Air	29.6	4.16×10^6	280.0	14.80	1915.61	853.54	1062.07	1.16	3.42×10^6
CH_4	Air	12.3	2.28×10^8	1020.0	15.96	1763.88	770.11	993.77	1.21	1.74×10^8

[a]Values taken from Matsui and Lee (1978). [b]Values in parentheses taken from Voitsekhovsky, et al. (1963).

line but within the range of the predictions made by Bull et al. (1976) and recently by Westbrook and Haselman (1979). However, the resulting value falls far below that of Boni et al. (1978).

Recently Vasiliev (1978) suggested a formula for an estimate of the energy required to initiate a cylindrical detonation wave. Taking his value of 0.1 J/cm for a stoichiometric oxy-acetylene mixture at initial pressure of 0.1 atm and scaling it to spherical geometry at $P_o = 1$ atm with the cell size $z = 2$ mm, we get a number that agrees very closely with our number of $E_c = 3.6 \times 10^{-4}$ J. Furthermore, his relative values of energy required to initiate cylindrical detonation in other hydrocarbon mixtures, including that of methane-air, when plotted against Lee's experimental d_c, fall on a straight line with a quadratic slope, as one would expect for cylindrical geometry. This all suggests that our estimate of energy for initiating a spherical detonation is in good agreement with those found by others.

Quantity of Cells Needed for Transition

In their experimental investigation, Matsui and Lee (1978) found that, for a transition of a gaseous detonation from a planar geometry into a spherical one, the critical diameter of the tube is related to the initial pressure by

$$P_o = a d_c^\alpha \tag{7}$$

where a and α are constants for each particular mixture. Earlier studies on the cell size dependence on initial pressure (Strehlow and Engel, 1969) yielded

$$P_o = b z^\beta \tag{8}$$

where again b and β are constants for each particular mixture. An algebraic manipulation leads to a combined expression

$$n = d_c/z = K P^\nu \tag{9}$$

where n stands for the number of cells present across the critical diameter of the tube, and K and ν are new constants related to the previous ones by

$$K = a^{-1/\alpha} b^{1/\beta} \quad \text{and} \quad \nu = 1/\alpha - 1/\beta \tag{10}$$

For the four mixtures common to both studies, values of all constants are listed in Table 2. The numbers in parentheses for the oxy-hydrogen system represent a slight correction, which may be introduced by drawing another line through the experimental points of Strehlow and Engel (1969). The values of ν in Table 2 are much less than one, and therefore the effect of initial pressure is minimal. If this is true for the other explosive mixtures as well, then it apparently is not the initial pressure or the diameter of the tube that determines the criterion for the transition but the cell size z. Also important is the number of cells available to suppress the effect of the incoming rarefaction wave and to generate new cells which will revive the detonation process. The above results

Table 2 Parameters of Eqs. (4-7) for four of the explosive mixtures common to studies of Matsui and Lee (1978) and Strehlow and Engel (1969)

Fuel	Oxid	a	α	b	β	K^a	ν
C_2H_2	O_2	0.167	−0.882	0.168	−1.057	4.31	−0.188
C_2H_4	O_2	0.668	−0.918	0.0265	−1.44	8.0	−0.395
H_2	O_2	1.91	−0.928	0.135	−0.74 (−0.83)	30.0 (22.4)	+0.274 (+0.13)
CH_4	O_2	4.6	−0.95	0.29[b]	−1.0[b]	17.2	−0.05

[a] $K = n$ at $P = 1$ atm.   [b] Estimated on basis of other mixtures diluted with 50% argon.

seem to indicate that for each mixture a particular number of cells is required for that purpose.

Of interest here is the sign before the exponent ν. In contrast to acetylene-oxygen, ethylene-oxygen, and methane-oxygen mixtures (where a lower initial pressure increases the number of cells required across the critical diameter), the hydrogen-oxygen mixture has the opposite effect, i.e., the number of cells decreases as the initial pressure decreases.

Extrapolation of f and z to CH_4-Air Mixture

At one atmosphere initial pressure, the value of K in Eq. (9) represents the number of cells needed across the critical diameter for a transition of detonation to take place. As may be noted, this value of n varies inversely with the susceptibility to detonation. A plot of this value of n (see Table 2) against the critical diameter reveals the interesting fact that the values for all the hydrocarbon fuels indeed fall on a straight line. This line may then be extended to the critical diameter of the least detonable mixture, methane-air. This remarkable correlation of n and z for the available hydrocarbon-oxygen mixtures is illustrated in Fig. 4. The extrapolation of the straight lines to the critical diameter of CH_4-air results in $n = 105$ and $z = 9$ cm for that mixture.

Effect of Geometry and Initial Pressure

Thus, a transition of detonation from a cylinder to a hemisphere is assured if there is a core of the detonation wave which remains unaffected by the rarefaction wave and is large enough to supply the necessary amount of energy to the undisturbed gas for a hemispherical initiation (i.e., $E_c \sim d_c^3$, with $d_o = 0.67\,d_c$ and a different n for each particular gaseous system).

If, instead of an open hemisphere, the cylinder ends in a circular cone with a solid angle smaller than 2π steradians (sr) (i.e., where $\delta < 90$ deg), the rarefaction generated at the corner will be weaker and therefore both d_c and d_o will also become smaller. To the first approximation, this new critical core diameter may then be expressed as

$$d_{o\delta} = d_{oo} + (d_o - d_{oo})\frac{\delta}{90} \qquad (11)$$

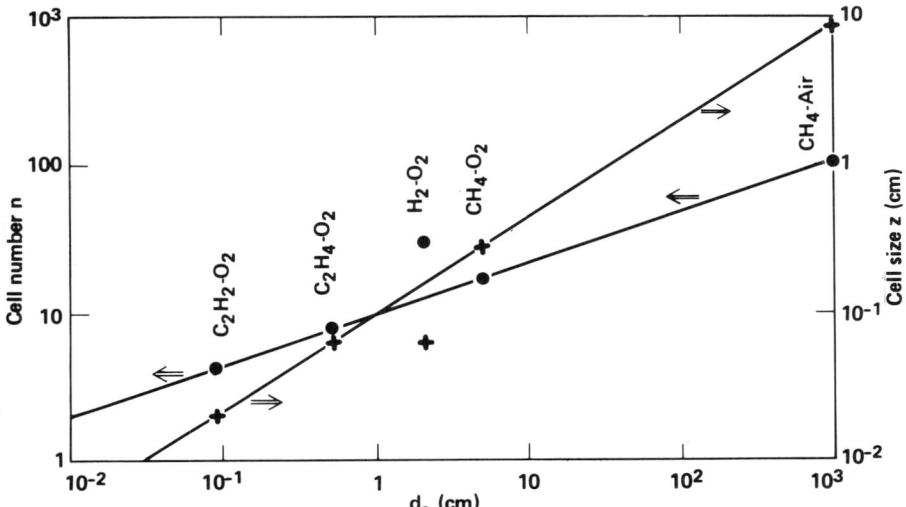

Fig. 4 Plot of the cell size z and cell number n as a function of the critical diameter for various mixtures.

where $d_{o\delta}$ is the core diameter for any expansion where $0 < \delta < 90$, and d_{oo} is the minimum-constant-area tube diameter ($\delta = 0$) capable of supporting a detonation wave, which in terms of cell size can be expected to be in the order of $z/2$.

If the geometry is kept the same while the initial pressure P_o is lowered, then, as noted earlier by Matsui and Lee (1978), one needs a larger d_c. Because during the derivation of Eq. (6) no particular value of initial pressure was stipulated, the ratio of $d_c/d_o = 1.5$ must still hold and, therefore, d_o will also be larger. Following Strehlow and Engel (1969), z will also be larger, but the number of cells, $n = d_c/z$ and $k = d_o/z$, will remain in the same proportion, although slightly different from the atmospheric condition because of a small effect of initial pressure in Eq. (9). On the E_c-d_c plot of Fig. 3, the effect of initial pressure shows a quadratic behavior. This is so because, while D and u remain almost invariant with P_o, $P_{C\text{-}J}$ and d_c vary as P_o and $1/P_o$, respectively. This causes $P_{C\text{-}J} \sim 1/d_c$ and $E_c \sim d_c^2$. The effect of P_o on the E_c-d_c relationship is illustrated in Fig. 3 for the hydrogen-oxygen system. The other systems are expected to follow suit.

Comparison of the Detonation Kernel to the Detonation Cell

It is of interest to compare the expression for the critical energy with that derived earlier by Lee and Ramamurthi (1976) for the spherical geometry and to correlate their findings of the critical detonation kernel R_s^* with the dimensions of the detonation cell.

Taking Lee's expression for the critical energy,

$$E_c = \tfrac{1}{3} [\pi/(\gamma^2 - 1)] \rho_o c_o^2 M_s^{*2} R_s^{*3} \tag{12}$$

and equating it to our critical energy Eq. (3), we get the expression for the critical diameter

$$d_c^3 = \frac{4D\rho_o c_o^2 M_s^{*2} R_s^{*3}}{3Pu(\gamma^2 - 1)l/z(d_o/d_c)^3[(d_c/d_o) - 1]} \qquad (13)$$

This expression can be simplified by setting

$$P_{C\text{-}J} = \rho_o D^2/(\gamma + 1) \qquad (14)$$

$$U_{C\text{-}J} = D/(\gamma + 1) \qquad (15)$$

$$M_{C\text{-}J} = D/c_o \qquad (16)$$

Then

$$d_c^3 = \frac{4}{3} \frac{(\gamma+1)}{(\gamma-1)} \frac{z}{l} \frac{1}{\{[d_c/d_c]^3[(d_c/d_o) - 1]\}} \frac{M_s^* R_s^*}{M_{C\text{-}J}^2} \qquad (17)$$

Introducing typical values for a hydrocarbon fuel, we may set

$$\gamma \cong 1.2; \quad \frac{l}{z} \cong 2; \quad \left[\left(\frac{d_o}{d_c}\right)^3 \left(\frac{d_c}{d_o} - 1\right)\right] = 0.15; \quad \frac{M_s^*}{M_{C\text{-}J}} \cong 0.7$$

and get the value of the detonation kernel in terms of the critical diameter d_c,

$$R_s^* = 0.3469 d_c \qquad (18)$$

If we now take Lee's argument that the detonation kernel size, R_s^*, is comparable to the cell length l we would get only one set of d_c and d_o, i.e.,

$$\text{if } R_s^* = l \text{ and } l = 2z, \quad \text{then } d_c = 5.77z \text{ and } d_o = 3.84z \qquad (19)$$

However, if $d_c > 5.8z$, as it appears from the records of Lee, and is approximately equal to $12z$, as suggested by Voitsekhovsky et al. (1963), then R_s^* is on the order of $4.16z$ or $2.08l$. This does not diminish the significance of the detonation kernel concept; however, it does change its relationship to the characteristic cell size.

Table 3 lists the values of R_s^* for the five mixtures described earlier. The values of the detonation kernel are given in terms of the cell length l as obtained from combining Eqs. (9) and (18), with the assumption that $l/z = 2$. Note that for the acetylene-oxygen mixture, which was the main working medium for Lee and his group, the critical size of the detonation kernel R_s^* is indeed very close to the cell length l. The other four mixtures (ethylene-oxygen, hydrogen-oxygen, methane-oxygen, and methane-air) result in r_s^* equal to approximately $1.39l$, $5.2l$, $2.98l$, and $18.21l$, respectively. Again note that following its direct proportionality with n, the critical size of the detonation

Table 3 Critical parameters evaluated in the present study for five explosive, including that of methane-air

Fuel source	Oxid	d_c^a cm,	n^b	$z=d_c/h$, cm	$l=2z$, cm	R_s^{*c}	R_s^{*d}, cm	N_c^e	θ^f
C_2H_2	O_2	0.09	4.3	0.021	0.042	0.74 l	0.031	9.27	34.11
C_2H_4	O_2	0.52	8.0	0.065	0.13	1.39 l	0.18	17.40	19.85
H_2	O_2	2.01	30.0	0.067	0.13	5.2 l	0.67	62.83	5.71
CH_4	O_2	5.0	17.2	0.29	0.58	2.98 l	1.72	37.27	9.57
CH_4	Air	1000.0	105.0	9.0	18.0	18.21 l	327.0	228.3	1.57

[a]M&L-78. [b]M&L-78, S&E-68. [c](0.347)n/2. [d]n(0.347)ln/2. [e]$2\pi R_s/z$. [f]$tg^{-1}(z/(R_s))$.

kernel R_s^* varies inversely with the susceptibility to detonation. This observation may be useful in determining proper values for other explosive mixtures in the series.

Spherical Initiation of Detonation

If, as previously noted in Eq. (18), the critical size of the detonation kernel R_s^* is uniquely related to the critical diameter d_c, one can use the values listed in Table 3 to evaluate the physical size of R_s^* and make a comparison with the blast initiation radius computed with the numerical techniques as used by Westbrook and Haselman (1978).

For this purpose, the numerical model described by Wilkins (1969) was used to calculate the distances at which a blast wave initiated by a charge of high explosives would decay to pressures covering the range of C-J pressures for most fuel-oxygen and fuel-air mixtures. The results are shown in a reduced form in Fig. 5, where pressure is plotted as a function of the radius normalized by the cube root of the critical energy. The solid line represents the decay of pressure following a release of energy from a charge into an inert atmosphere. When the explosive charge is set off in an explosive medium, then the energy of the medium itself also contributes to the total energy release and the decay of the blast wave is somewhat slower, increasing the radius at which a particular pressure is attained. This effect is illustrated in Fig. 5 by a broken line, that is somewhat arbitrary because it is based on estimates made for several pressures.

Also included in Fig. 5 are the points representing the values of the detonation kernel R_s^* reduced by $E_c^{1/3}$ for the five mixtures described above. The agreement is rather remarkable. This confirms the original notion of Lee and Ramamurthi (1976) about the R_s^*. They said that for each particular mixture, there is a certain minimum distance R_s^* wherein the chemical energy released by the medium is comparable to the energy released by the source; so that the subsequent shock motion will be strongly coupled to the chemical processes and a detonation wave is sustained. However, their postulate that the size of the detonation kernel is comparable to the cell length is true only for a mixture of acetylene-oxygen. Other explosive hydrocarbon mixtures have their relative R_s^*/l ratios in the order of their susceptibility to denotation.

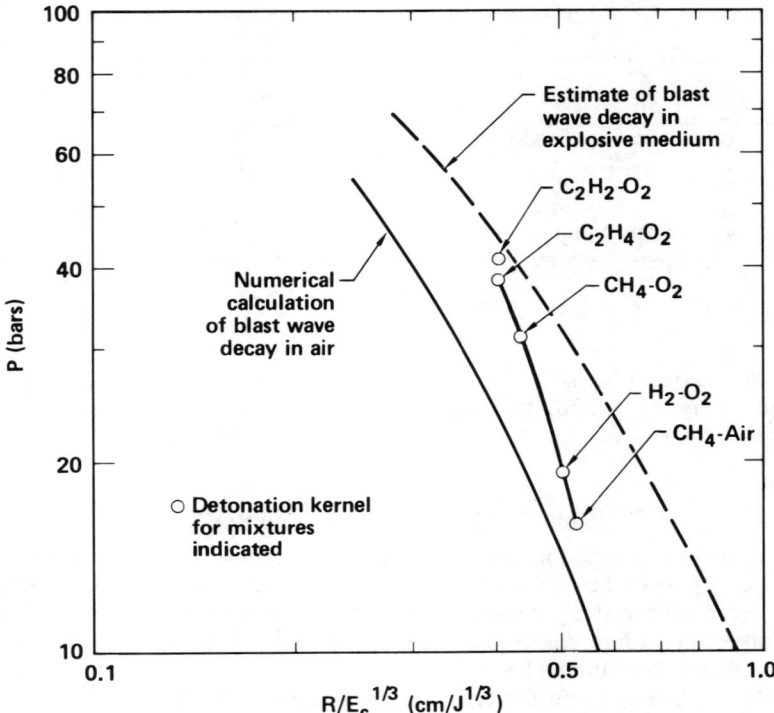

Fig. 5 Plot of blast pressure P vs the reduced radius R/E_c^{∂}.

In view of this finding, one can invoke another criterion for the self-sustenance of the detonation wave: to continue the existence of the detonative process, one must have, in each particular mixture, a certain number of cells around the periphery of the sphere and that number can be found from

$$N_c = 2\pi R_s^*/z \quad (20)$$

This number is also listed in Table 3.

Thus, initiation of spherical detonation will occur if, at the proper radius from the center of the blast wave, i.e., the radius of the detonation kernel R_s^*, the decay of the wave matches that of the decay through the individual cell of the detonation wave. If the initial charge is too small, then the decay will be too steep and the detonative process will fail.

This criterion for spherical initiation of detonation can also be attributed to the amount of angular expansion of the cell as the wave propagates radially outward. For each particular mixture, there is a certain maximum of such angular expansion at which a new cell must be initiated or the process will decay to extinction. Such angular expansion may be expressed in terms of

$$\theta = tg^{-1}(z/R) \quad (21)$$

whose value for each of the five mixtures is also listed in Table 3. Also note that plotting θ vs d_c on the log-log scale will result in a straight-line relationship.

Of course, these are purely physical observations of the phenomenon and cannot be considered as fundamental criteria for the process. Here the chemistry of the medium plays the most important role. The decay of the wave is inherently connected with the induction time of a certain molecule or group of molecules that, if retained in a thermodynamic state for a sufficient length of time, will autoignite and thereby start a new local explosion. This results in a new cell. Being an unsteady process, it does not lend itself to an easy analytical treatment. Therefore, observations such as those made here are considered very helpful and informative.

Conclusion

The object of this work was to study the effect of the cellular structure of the detonation wave on the transition of the process into expanded geometries. Previous observations revealed that for each initial pressure there is a characteristic cell size and a critical diameter from which a transition of detonation is possible. We have demonstrated the effect of the cellular structure by showing that there must be an inner core of the wave, consisting of a critical number of cells unaffected by side rarefactions, to ensure the transition of the detonation process.

In this study, the work done on the outer gas by that inner core of the wave was compared with the critical energy for spherical initiation. The agreement was remarkably good. Extrapolation of data on the log-log plot of critical energy vs critical diameter led to an estimate of the critical energy needed to initiate detonation in a pure methane-air mixture. A plot of the characteristic cell size and critical number of cells needed for transition also led to quantitative estimates of these values for the methane-air mixtures. In fact, if we consider values for the whole range of mixtures from the most to the least susceptible to detonation (acetylene-oxygen and methane-air, respectively) and find them to fall on a straight line over several decades on a log-log plot, we can provide a good estimate of the behavior of any other explosive mixture, provided one of the parameters becomes known.

Comparison of our results on the critical energy of initiation with those previously reported by Lee and Ramamurthi (1976) led to a slightly different view on the physical significance of the detonation kernel and, in particular, its correspondence to the length of the detonation cell. While the size of the kernel is essentially unique in that the chemical energy contained within is comparable to the source energy released by the initiator change, the size of the kernel is not the same as the characteristic length of the detonation cell. As reported here, the kernel size varies according to the susceptibility of the explosive mixture to detonation. The detonation kernel, however, agreed well with the size of the radius at which numerical calculations predicted the blast wave would decay to the detonation pressures of the gaseous mixture.

Acknowledgments

One of the authors (PAU) wishes to thank J.H. Lee for several fruitful discussions of the subject and, in particular, for his original suggestion to look into the "work-done" concept. We are also grateful to our colleagues L. Haselman and F. McMurphy for providing numerical calculations for Fig. 5 and Table 1, respectively.

This work was performed under the auspices of the U.S. Department of Energy by Lawrence Livermore Laboratory under Contract No. W-7405-Eng-48.

References

Boni A.A., Wilson C.W., Chapman M., and Cook J.L. (1978) A study of detonation in methane/air clouds. *Acta Astronautica* **5**, 1153-1169.

Bull D.C., Elsworth J.E., Hooper G., and Quinn C.P. (1976) A study of spherical detonation in mixtures of methane and oxygen diluted with nitrogen. *J. Phys. D.* **9**, 1991-2000.

Lee J.H.S. (1977) Initiation of gaseous detonation. *Ann. Rev. Phys. Chem.* **28**, 75-104.

Lee J.H.S. (1979) McGill University, private communication.

Lee J.H., Knystautas R. and Lee B.H.K. (1965) Structure of gaseous detonations in a convergent-divergent channel. *AIAA J.* **3**, 1785-1787.

Lee J.H., Soloukhin R.I., and Oppenheim A.K. (1969) Current views on gaseous detonation. *Acta Astronautica*, **14**, 565-586.

Lee J.H. and Ramamurthi K. (1976) On the concept of the critical size of a detonation kernel. *Combustion and Flame,* **27**, 331-340.

Matsui H. and Lee J.H. (1978) On the measure of the relative detonation hazards of gaseous fuel-oxygen and air mixtures. *Seventeenth Symposium (Intl.) on Combustion,* pp. 1269-1279. The Combustion Institute, Pittsburgh, Pa.

Oppenheim A.K., Manson N., and Wagner H.Gg. (1963) Recent progress in detonation research. *AIAA J.* **1**, 2243-2252.

Strehlow R.A. (1968) Gas phase detonations: recent developments. *Combustion and Flame* **12**, 81-101.

Strehlow R.A., Adamczyk A.A., and Stiles R.J. (1972) Transient studies of detonation waves, *Acta Astronautica* **17**, 509-527.

Strehlow R.A. and Engel C.D. (1969) Transverse waves in detonations: II. Structure and spacing in H_2-O_2, $C_2H_2-O_2$, $C_2H_4O_2$ and CH_4-O_2 systems. *AIAA J.* **7**, 492-496.

Urtiew P.A. and Oppenheim A.K. (1965) University of California at Berkeley, unpublished work.

Vasiliev A.A. (1978) Estimate of the energy to initiate a cylindrical detonation. *Fizika Gorenia. Vzryva* **14**, 154-155.

Voitsekhovsky B.V., Mitrofanov V.V. and, Topchian M.E. (1963) Struktura fronta detonatsii v gazakh. *Izd. Sib. Otd. AN SSSR,* pp. 97-103; translation: The structure of a detonation front in gases. Foreign Technol. Div. Rept. FTD-MT-64-527 (1966).

Westbrook C.K. and Haselman L.C. (1979) *Chemical Kinetics in LNG Detonations.* Lawrence Livermore National Laboratory, UCRL-82263.

Wilkins, M. (1969) *Calculations of Elastic-Plastic Flow.* Lawrence Livermore National Laboratory, UCRL-7322.

IV. Detonations at Moderate Pressures

Influence of the Heat-Release Function on the Detonation States

Henri Guénoche,* Patrice Le Diuzet,† and Chantal Sèdes‡
Université de Provence, Marseille, France

In the classical theory of detonations, the Chapman-Jouguet (C-J) velocity depends only on the heat release per unit mass Q_e of the mixture. The heat release Q_e is evaluated from a thermodynamic calculation of the equilibrium state of the burned gases. If we consider the heat release to be effected over a finite period, it is not evident that the Chapman-Jouguet state based on the equilibrium heat release Q_e can always be attained irrespective of the details of the heat-release function itself. For detonable mixtures, the heat-release function is a natural consequence of the detailed chemistry of the system and it is not certain that every mixture gives the same type of heat-release function of a monotonic increase from zero behind the shock to Q_e at the C-J plane. Certain systems may give to an overshoot in the energy release over the equilibrium value Q_e during the progress of the reactions. The aim of the present paper is to investigate the influence of these two detailed heat-release functions on the detonation states. In the first part, a constant γ ZDN model is used because, in this case, the governing equations can be straightforwardly integrated, giving simple relationships between the flow parameters and the heat release. When the heat-release function is not monotonic, the thermal shocking condition $|m_f| = 1$ is obtained in a chemical nonequilibrium region; and a detonation velocity greater than the C-J velocity results. In the second part, the model is illustrated by computer calculations in H_2-O_2 and H_2-Cl_2 mixtures.

Nomenclature

a = speed of sound
$\hat{C}_{p_j}$ = molar heat capacity at constant pressure for species j
D = detonation velocity

Presented at the 7th ICOGER, Göttingen, Federal Republic of Germany, Aug. 20-24, 1979. Copyright © American Institute of Aeronautics and Astronautics, Inc., 1981. All rights reserved.
*Professor, Director of the Laboratory, Laboratoire de Dynamique et Thermophysique des Fluides.
†Research Scientist, Laboratoire de Dynamique et Thermophysique des Fluides.
‡Research Student, Laboratoire de Dynamique et Thermophysique des Fluides.

$(\hat{D}_j)_0$	= molar dissociation energy at $T=0$ K for the dissociation of species j, atoms
$[\hat{H}_j]_0^T$	= variation of molar enthalpy of species j between 0 and T
k_{+r}	= forward rate constant for reaction number r
k_{-r}	= backward rate constant for reaction number r
m	= relative Mach number of the flow, $m = u/a$
M	= total mass of the mixture
M_s	= absolute Mach number of the shock front
n_j	= mole number of species j
p	= pressure
Q	= heat release per unit mass
$\hat{R}$	= molar gas constant, $\hat{R} = 1.98725$ cal (K)$^{-1}$
t	= time
T	= temperature, K
u	= flow velocity in shock-fixed coordinates
U	= absolute flow velocity
γ	= ratio of the specific heats at constant pressure and constant volume
ξ	= Q/Q_e
ρ	= density of the mixture

Subscripts

C-J	= Chapman-Jouguet
1	= unburned mixture
d	= state immediately behind the shock front
e	= burned gas in chemical equilibrium
f	= frozen composition
M	= correspond to the point where the heat release is maximum without chemical equilibrium

Introduction

DETONATION velocities greater than, as well as less than, the Chapman-Jouguet velocity have been observed experimentally. These discrepancies have been explained by: 1) the influence of wall boundary layers (Fay, 1959; Tsuge, 1971; Dove et al., 1974); 2) transient multiheaded structure of real detonations (the numerous contributions on this subject are not cited here); or 3) turbulence of the flow (White, 1961; Strehlow, 1971). These factors are not taken into account in the classical theory of detonations. In the classical theory, the Chapman-Jouguet (C-J) velocity depends only on the heat release per unit mass Q_e of the mixture. Q_e is evaluated from a thermodynamic calculation of the equilibrium state of the burned gases. Since the actual heat release occurs over a finite period, it is not evident that the C-J state based on the equilibrium heat release Q_e can always be attained irrespective of the details of the heat-release function. For real detonable mixtures, the heat-release function is a natural consequence of the detailed chemistry of the system, and does not necessarily increase monotonically from zero behind the shock to Q_e at the C-J plane (Fig. 1a). Certain systems may give rise to an overshoot in the energy release over the equilibrium value during the progress of the reactions (Fig. 1b). Our aim is to show that a detonation velocity greater than the C-J velocity can be obtained in this case. For this purpose, we shall investigate the influence of the two detailed heat-release functions on Fig. 1 on the detonation states. First, a constant γ and fixed composition model (denoted subsequently as CGFC model) will be used because, in this case, simple relations between the flow parameters and the heat addition can be obtained. In spite of its simplicity, this model gives a good qualitative description of the phenomena. Then, we shall study numerically the detonation states in hydrogen-oxygen and hydrogen-chlorine mixtures. These two systems were chosen because, first, the reaction mechanisms are well known and extensive kinetic data on the elementary reactions are available for each of them and, second, they illustrate quite well the time evolutions of the heat-release function shown in Fig. 1. The numerical model with variable γ and variable composition will be denoted subsequently as the VGVC model.

Constant γ and Fixed Composition Model

In the one-dimensional Zeldovich-Döring-Von Neumann (ZDN) model (Zeldovich, 1940; Döring 1943; Von Neumann 1942), a steady detonation wave can be represented by a nonreactive shock discontinuity followed by a region where overall exothermic chemical reactions take place at a finite rate. For simplicity, we shall assume that behind the shock front there are no composition variations of the mixture but the effect of the chemical reactions shall be accounted for by introducing in the equations a heat-addition term ξQ_e, the time evolution of which is given either by Fig. 1a or by Fig. 1b. Furthermore, we shall neglect the temperature variations of the heat capacities.

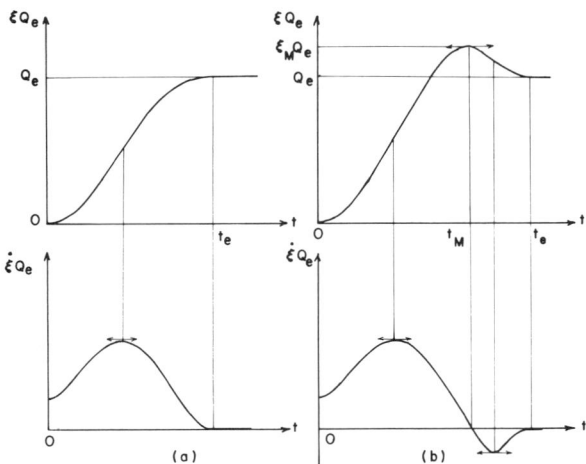

Fig. 1 Two possible time evolutions of heat addition in the reaction zone of a detonation wave: a) monotonically increasing heat addition; and b) heat addition with a maximum in the reaction zone.

Governing Equations of Flow with Heat Addition

The governing equations, shock-fixed coordinates, are

$$\rho u = \rho_1 u_1 \tag{1}$$

$$p + \rho u^2 = p_1 + \rho_1 u_1^2 \tag{2}$$

$$\tfrac{1}{2}(u_1^2 - u^2) = c_p(T - T_1) - Q \tag{3}$$

where $Q > 0$ represents the heat addition per unit mass. These governing equations can be transformed easily to represent the flow in the plane $([\Delta u]_1/a_1, p/p_1)$, where

$$[\Delta u]_1 = u - u_1 \tag{4}$$

The mass and momentum equations can be solved to yield the expression

$$p/p_1 = 1 - \gamma m_1([\Delta u]_1/a_1) \tag{5}$$

which is the Rayleigh line. The energy equation can be written as

$$\left(\frac{[\Delta u]_1}{a_1}\right)^2 - \frac{2}{\gamma+1}\frac{1-m_1^2}{m_1}\frac{[\Delta u]_1}{a_1} + \frac{2(\gamma-1)}{\gamma+1}\frac{Q}{a_1^2} = 0 \tag{6a}$$

or

$$\frac{[\Delta u]_1}{a_1} = \frac{1-m_1^2}{(\gamma+1)m_1}\left\{1 \pm \sqrt{1 - 2(\gamma^2-1)\frac{m_1^2}{(1-m_1^2)^2}\frac{Q}{a_1^2}}\right\} \tag{6b}$$

The negative sign on the radical in relation Eq. (6b) must be eliminated since, for $Q=0$, Eq. (6b) must reduce to the well-known shock relation

$$\frac{[\Delta u]_1}{a_1} = \frac{2}{\gamma+1}\frac{1-m_1^2}{m_1}$$

Substitution of Eq. (6b) into Eqs. (4) and (5) yields

$$\frac{p}{p_1} = 1 + \frac{\gamma(m_1^2 - 1)}{\gamma+1}\left\{1 + \sqrt{1 - 2(\gamma^2-1)\frac{m_1^2}{(1-m_1^2)^2}\frac{Q}{a_1^2}}\right\} \tag{7}$$

$$\frac{u}{a_1} = \frac{1+\gamma m_1^2}{(\gamma+1)m_1} + \frac{1-m_1^2}{(\gamma+1)m_1}\sqrt{1 - 2(\gamma^2-1)\frac{m_1^2}{(1-m_1^2)^2}\frac{Q}{a_1^2}} \tag{8}$$

From Eq. (8), together with the energy Eq. (3), we obtain

$$\frac{T}{T_1} = \left(\frac{a}{a_1}\right)^2 = \frac{\gamma-1}{(\gamma+1)^2 m_1^2}\left\{\gamma m_1^4 + \frac{2\gamma}{\gamma-1}\left[2 + (\gamma^2-1)\frac{Q}{a_1^2}\right]m_1^2 - 1 + \dots\right.$$

[Eq. (9) continued on next page.]

$$\ldots + (m_1^2 - 1)(1 + \gamma m_1^2)\sqrt{1 - 2(\gamma^2 - 1)\frac{m_1^2}{(1-m_1^2)^2}\frac{Q}{a_1^2}} \Bigg\} \quad (9)$$

The problem has a solution only if the term under the radical in Eqs. (6b-9) is greater than or equal to zero

$$1 - 2(\gamma^2 - 1)\frac{m_1^2}{(1-m_1^2)^2}\frac{Q}{a_1^2} \geq 0 \quad (10)$$

so that

$$0 \leq \frac{Q}{a_1^2} \leq \frac{(1-m_1^2)^2}{2(\gamma^2 - 1)m_1^2} = \frac{Q_{\max}}{a_1^2} \quad (11a)$$

or

$$m_1^2 \geq \left\{1 + (\gamma^2 - 1)\frac{Q}{a_1^2}\right\}\left\{1 + \sqrt{1 - \frac{1}{\left[1 + (\gamma^2 - 1)\frac{Q}{a_1^2}\right]^2}}\right\} \quad (11b)$$

Graphical Interpretation

For $Q = 0$, Eqs. (6b) and (7) reduce to

$$\frac{[\Delta u]_I}{a_1} = \frac{[\Delta u]_I^s}{a_1} = \frac{2}{\gamma + 1}\frac{1 - m_1^2}{m_1} \quad (12)$$

$$\frac{p}{p_1} = \frac{p^s}{p_1} = 1 + \frac{2\gamma}{\gamma + 1}\frac{1 - m_1^2}{m_1} \quad (13)$$

These equations define parametrically the shock polar in the plane

$$([\Delta u]_I/a_1, p/p_1)$$

For $Q = Q_{\max}$, Eqs. (6b) and (7) reduce to

$$\frac{[\Delta u]_I}{a_1} = \frac{[\Delta u]_I^*}{a_1} = \frac{1 - m_1^2}{(\gamma + 1)m_1} \quad (14)$$

$$\frac{p}{p_1} = \frac{p^*}{p_1} = 1 + \frac{\gamma}{\gamma + 1}(m_1^2 - 1) \quad (15)$$

It is easy to see from Eqs. (8) and (9) that for $Q = Q_{\max}$,

$$u^* = \pm a^*$$

the plus sign corresponding to left-facing waves and the minus sign to right-facing waves. Thereafter, we shall consider only right-facing waves, that is to say, waves for which $u_1 < 0$, $m_1 < 0$, and $[\Delta u]_I > 0$. In the plane $([\Delta u]_I/a_1, p/p_1)$, Eqs. (14) and (15) define parametrically a curve that we shall call the sonic line. The sonic line can be obtained from the shock polar by homothetic transformation of center (0,1) and ratio ½. The sonic line is also the maximum heat-addition curve.

For constant Q Eqs. (6b) and (7) define parametrically curves of constant heat addition that we shall call isoenergetic lines. These isoenergetic curves are evidently limited by the sonic line and move upward when Q increases. It can be shown that the tangent on an isoenergetic line at the point of intersection with the sonic curve is a Rayleigh line.

To study the evolution of the flow parameters in the reaction zone of the detonation wave, we must introduce a time-dependent heat addition

$$Q = \xi Q_e = \xi(t) Q_e$$

The value of Q is zero immediately behind the shock front (no chemical reaction) and Q_e when chemical equilibrium is reached. Between these two values, we have considered two possible modes of time evolution of the heat addition in the reaction zone. In the first one, the heat addition increases monotonically with time until it reaches its equilibrium value Q_e (Fig. 1a); and in the second one, the heat addition goes through a maximum before it attains the equilibrium value Q_e (Fig. 1b). In the two cases, for each ξ value in the reaction zone, the heat addition corresponds to an isoenergetic curve lying between the shock polar ($\xi = 0$) and the isoenergetic line $Q = \xi_M Q_e$ which corresponds to the maximum heat addition. The condition Eq. (10) must apply to every value of the heat addition. in the reaction zone and, in particular, for its maximum value $\xi_M Q_e$ so that

$$1 - 2(\gamma^2 - 1) \frac{m_1^2}{(1 - m_1^2)^2} \frac{\xi_M Q_e}{a_1^2} \geq 0$$

and

$$m_1^2 \geq m_1'^2 = \left\{ 1 + (\gamma^2 - 1) \frac{\xi_M Q_e}{a_1^2} \right\} \left\{ 1 + \sqrt{1 - \frac{1}{[1 + (\gamma^2 - 1)(\xi_M Q_e/a_1^2)]^2}} \right\} \quad (16)$$

Thus the different states of the flow in the reaction zone will be represented in the plane $([\Delta u]_1/a_1, p/p_1)$ by a Rayleigh line the slope of which is either equal to or greater than $-\gamma m_1^*$. The point of intersection of this Rayleigh line with the shock polar $\xi Q_e = 0$ determines the state immediately behind the shock front. The detonation state will be determined by the intersection of the Rayleigh line with the isoenergetic line $\xi_M Q_e$.

Let us consider first the case of a monotonically increasing heat addition (Fig. 1a). For this case, the maximum heat addition $\xi_M Q_e$ is reached only at chemical equilibrium, and $\xi_M = 1$. On Fig. 2a we have plotted in the plane $([\Delta u]_1/a_1, p/p_1)$ the shock polar, the sonic line, and some isoenergetic curves between $Q = 0$ and $Q = Q_e$. The solution behind the shock front is represented either by the straight line A'A, which corresponds to the C-J detonation because at point A where chemical equilibrium is achieved $|u| = a$, or by straight lines such as B'B, which correspond to strong (or overdriven) detonations. We can see that for the C-J detonation, pressure decreases from

p_{dC-J} immediately behind the shock front (point A′) to p_{eC-J} when equilibrium is reached (point A) and the absolute value of the relative flow velocity $|u|$ increases from $|u_d|_{C-J}$ behind the shock front to $|u_e|_{C-J} = a_{eC-J}$. The same evolution for these variables can be observed in the reaction zone of an overdriven wave except that $|u_e| < a_e$ (point B).

In the case of a time evolution of the heat addition similar to that represented on Fig. 1b, the maximum heat addition $\xi_M Q_e$ is reached before completion of chemical equilibrium and $\xi_M > 1$. The solution behind the shock front is represented (Fig. 2b) by the straight line A′AA″ which corresponds to a pathological detonation because at point A, $|u|_A = a_A$ without chemical equilibrium, or straight lines such as B′BB″ which correspond to overdriven detonations. For an overdriven detonation, pressure decreases from p_d immediately behind the shock front (point B′) to p_M for $\xi = \xi_M$ (point B) then increases up to p_e (point B″) when chemical equilibrium is achieved and the absolute value of the relative flow velocity increases from $|u_d|$ behind the shock to $|u_M|$ for $\xi = \xi_M$ and decreases to $|u_e|$ at point B″. For the pathological detonation, we have the same behavior of p and $|u|$ as observed in the reaction zone of an overdriven detonation except that $|u_M| = a_M$ for $\xi = \xi_M$ (point A). By similarity with flows through a Laval nozzle, Gruschka and Wecken (1971) suggested that the flow beyond the sonic point is supersonic, allowing a weak detonation with $|m_e| > 1$ to be stable. In fact this is not possible, as can be seen from Fig. 2b. When chemical equilibrium is reached

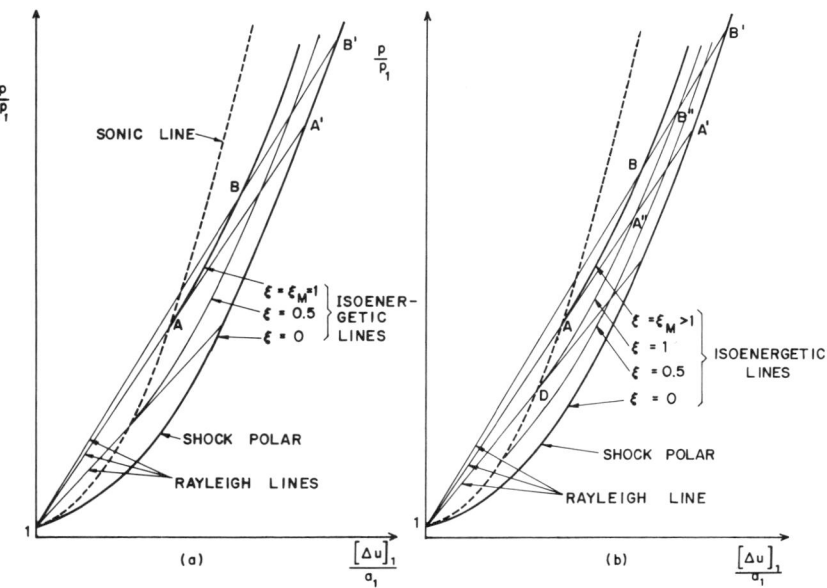

Fig. 2 Shock polar, sonic line, and isoenergetic lines in the plane ($[\Delta u]_1/a_1$, $p(p_1)$): a) monotonically increasing heat addition ($Q_e/a_1^2 = 14.73$); and b) heat addition similar to that represented on Fig. 1a ($\xi_M Q_e/a_1^2 = 14.73$, $Q_e/a_1^2 = 9.53$).

($\xi = 1$)

$$\left(\frac{[\Delta u]_I}{a_I}\right)_{A''} > \left(\frac{[\Delta u]_I}{a_I}\right)_A$$

and

$$|u|_{A''} < |u|_A = a_A$$

so that

$$|m|_{A''} = \frac{|u|_{A''}}{a_{A''}} < \frac{a_A}{a_{A''}} = \frac{p_A}{p_{A''}} \frac{1}{|m|_{A''}}$$

$$|m|^2_{A''} < \frac{p_A}{p_{A''}} < 1 \tag{17}$$

Thus, for ξ decreasing from ξ_M to 1 (part AA″ of the Rayleigh line), the flow in shock-fixed coordinates is subsonic. As the detonation wave is generally followed by a nonsteady rarefaction wave the head of which is moving with the velocity $U + a = D - |u| + a > D$ relative to the wall of the tube, this head can propagate through the reaction zone up to the sonic point, where $\xi = \xi_M$. So, in the part of the reaction zone where ξ decreases from ξ_M to unity, the flow is purely time dependent and it is no longer possible to represent with the part AA″ of the Rayleigh line. We can furthermore see on the curves that

$$|m_I|_A > |m_I|_D \tag{18}$$

The pathological detonation velocity is greater than the C-J velocity which corresponds to $|u| = a$ for $\xi = 1$.

Evolution of the Flow Parameters in the Reaction Zone

To study the evolution of all the flow parameters in the reaction zone, it is necessary to use the equations in differential form. It is possible to express the time derivatives of all the flow parameters in terms of $\dot{\xi} Q_e = (d\xi/dt) Q_e$, the instantaneous heat addition per unit mass

$$\frac{1}{|u|} \frac{d|u|}{dt} = \frac{\gamma - 1}{a^2} \frac{1}{1 - m^2} \dot{\xi} Q_e \tag{19}$$

$$\frac{1}{\rho} \frac{d\rho}{dt} = -\frac{\gamma - 1}{a^2} \frac{1}{1 - m^2} \dot{\xi} Q_e \tag{20}$$

$$\frac{1}{p} \frac{dp}{dt} = -\frac{\gamma(\gamma - 1)}{a^2} \frac{m^2}{1 - m^2} \dot{\xi} Q_e \tag{21}$$

$$\frac{1}{T} \frac{dT}{dt} = \frac{2}{a} \frac{da}{dt} = \frac{\gamma - 1}{a^2} \frac{1 - \gamma m^2}{1 - m^2} \dot{\xi} Q_e \tag{22}$$

$$\frac{1}{|m|} \frac{d|m|}{dt} = \frac{\gamma - 1}{2a^2} \frac{1 + \gamma m^2}{1 - m^2} \dot{\xi} Q_e \tag{23}$$

$$\frac{d(a-|u|)}{dt} = -\frac{\gamma(\gamma-1)}{2a} \frac{\left\{|m| - \frac{1}{\sqrt{\gamma+1}+1}\right\}\left\{|m| + \frac{1}{\sqrt{\gamma+1}-1}\right\}}{1-m^2} \dot{\xi}Q_e \quad (24)$$

The time evolution of $\dot{\xi}Q_e$ in the reaction zone is given on Fig. 1 for the two heat-addition models studied.

The relations of Eqs. (19-24) display a singularity at $|m|=1$. In order to achieve a continuous transition through the sonic point, the instantaneous heat addition per unit mass $\dot{\xi}Q_e$ should vanish at the same time as the flow reaches the sonic condition. As it can be seen on Fig. 1, $\dot{\xi}Q_e$ becomes zero at the end of the reaction zone, for $\xi=1$ in the case of a monotonically increasing heat addition and for $\xi=\xi_M>1$ in the case of a heat addition which exhibits a maximum before reaching its equilibrium value. The evolution of the different flow parameters is given on Table 1 for the first heat-addition model and on Table 2 for the second one. In this last case, as we have shown before, for $t_M<t\leq t_e$ the flow is time dependent and it cannot be represented by this purely steady model. For an overdriven detonation wave, the relative Mach number of the flow is less than unity throughout the reaction zone and for the different flow parameters one must have the evolution presented in Table 3 for a monotonically increasing heat addition and in Table 4 for a heat addition that goes through a maximum in the reaction zone.

Model with Variable γ and Variable Composition

Assumptions

Here also we use the one-dimensional ZDN model but take into account the changes of composition of the mixture in the reaction zone. So, the heat addition does not appear as external to the flow but as a natural consequence of the detailed chemistry of the considered system. Although it is well known that, during highly exothermic reactions, molecules are formed on excited vibrational levels (it is the case, for example, of HCl molecules formed by chemical reactions in H_2-Cl_2 mixtures), we presume that the mixture is in vibrational equilibrium. This assumption is not a fundamental one: it has been made only to enable us to write the equations in their simplest form and we have verified that it does not alter the results presented here. Furthermore, we assume that the mixture behaves as a perfect gas.

Differential Equations of the Flow

Because of the variations of the mixture composition, it is no longer possible to use in a simple manner the plane $([\Delta u]_1/a_1, p/p_1)$ to represent the flow. However, it is possible to show up the time evolution of the flow parameters from their time derivatives.

It can be shown that the differential flow equations lead to the following relations:

$$\frac{1}{|u|}\frac{d|u|}{dt} = \frac{\gamma_f-1}{a_f^2}\frac{1}{1-m_f^2}\dot{\psi} \quad (25)$$

Table 1 Evolution of the flow parameters in the reaction zone of an overdriven detonation (monotonically increasing heat addition)

		t	0		t_e
		ξ	0	→	1
		$\|u\|$	$\|u_d\|$	↗	$\|u_e\|$
		$\|m\|$	$\|m_d\|$	↗	$\|m_e\|$
		ρ	ρ_d	↘	ρ_e
		p	p_d	↘	p_e
	$\|m_e\| \leq \sqrt{1/\gamma}$	T	T_d	↗	T_e
T	$\|m_e\| > \sqrt{1/\gamma}$ [a]		T_d ↗	T_{max} [b] ↘	T_e
			$a_d - \|u_d\|$	↘	$a_e - \|u_e\|$
	$\|m_d\| \geq 1/(1+\sqrt{\gamma+1})$				
$a - \|u\|$	$\|m_d\| < 1/(1+\sqrt{\gamma+1}) < \|m_e\|$		$a_d - \|u_d\|$ ↗	$(a-\|u\|)_{max}$ [c] ↘	$a_e - \|u_e\|$
	$\|m_e\| \leq 1/(1+\sqrt{\gamma+1})$		$a_d - \|u_d\|$	↗	$a_e - \|u_e\|$

[a] The case $\|m_d\| > \sqrt{1/\gamma}$ has been disregarded because T_d is too low to permit a self-ignition of the mixture.
[b] For $\|m\| = \sqrt{1/\gamma}$.   [c] For $\|m\| = 1/(1+\sqrt{\gamma+1})$.

$$\frac{1}{\rho}\frac{d\rho}{dt} = -\frac{\gamma_f - 1}{a_f^2}\frac{1}{1-m_f^2}\dot\psi \qquad (26)$$

$$\frac{1}{p}\frac{dp}{dt} = -\frac{\gamma_f(\gamma_f - 1)}{a_f^2}\frac{m_f^2}{1-m_f^2}\dot\psi \qquad (27)$$

$$\frac{1}{T}\frac{dT}{dt} = \frac{\gamma_f - 1}{a_f^2}\frac{1-\gamma_f m_f^2}{1-m_f^2}\dot\psi' \qquad (28)$$

$$\frac{1}{\|m_f\|}\frac{d\|m_f\|}{dt} = -\frac{1}{2\gamma_f}\frac{d\gamma_f}{dt} + \frac{\gamma_f-1}{2a_f^2}\frac{1+\gamma_f m_f^2}{1-m_f^2}\dot\psi \qquad (29)$$

$$\frac{d(a_f - \|u\|)}{dt} = \frac{a_f}{2\gamma_f}\frac{d\gamma_f}{dt} - \frac{\gamma_f(\gamma_f - 1)}{2a_f}\frac{\left\{\|m_f\| - \frac{1}{\sqrt{\gamma_f+1}+1}\right\}\left\{\|m_f\| + \frac{1}{\sqrt{\gamma_f+1}-1}\right\}}{1-m_f^2}\dot\psi \qquad (30)$$

with

1)
$$\gamma_f = \left[\sum_j n_j \hat{C}_{p_j} \Big/ \sum_j n_j (\hat{C}_{p_j} - \hat{R})\right] \qquad (31)$$

$$a_f^2 = \gamma_f \sum_j n_j (\hat{R}/M) T \qquad (32)$$

Table 2 Evolution of the flow parameters in the reaction zone of a C-J detonation

	t	0		t_M		t_e
	ξ	0	↗	ξ_M	↗	1
	$\|u\|$	$\|u_d\|$	↗	$\|u_M\|$	↗	$\|u_e\|$
	$\|m\|$	$\|m_d\|$	↗	$\|m_M\|$	↗	$\|m_e\|$
	ρ	ρ_d	↘	ρ_M	↘	ρ_e
	p	p_d	↘	p_M	↘	p_e
T	$\|m_M\| \leq \sqrt{1/\gamma}$	T_d	↗	T_M	↗	T_e
	$\|m_d\| < \sqrt{1/\gamma} \leq \|m_e\| < \|m_M\|$	T_d	↗ $T_{max}^{\prime a}$ ↘	T_M	↗	T_e
	$\|m_d\| < \|m_e\| < \sqrt{1/\gamma} < \|m_M\|$	T_d	↗ $T_{max}^{\prime a}$ ↘	T_M ↗ $T_{max}^{\prime\prime a}$	↘	T_e
$a - \|u\|$	$\|m_d\| \geq 1/(1+\sqrt{\gamma+1})$	$a_d - \|u_d\|$	↘	$a_M - \|u_M\|$	↘	$a_e - \|u_e\|$
	$\|m_d\| < 1/(1+\sqrt{\gamma+1})$ $\|m_e\| < \|m_M\| \leq 1/(1+\sqrt{\gamma+1})$	$a_d - \|u_d\|$	↗ $(a-\|u\|)_{max}^{\prime b}$ ↘	$a_M - \|u_M\|$	↘	$a_e - \|u_e\|$
	$\|m_d\| < 1/(1+\sqrt{\gamma+1})$ $\|m_e\| < 1/(1+\sqrt{\gamma+1}) < \|m_M\|$	$a_d - \|u_d\|$	↗ $(a-\|u\|)_{max}^b$ ↘	$a_M - \|u_M\|$	↘	$a_e - \|u_e\|$
	$\|m_d\| < 1/(1+\sqrt{\gamma+1})$ $1/(1+\sqrt{\gamma+1}) \leq \|m_e\| < \|m_M\|$	$a_d - \|u_d\|$	↗	$a_M - \|u_M\|$ ↗ $(a-\|u\|)_{max}^{\prime\prime b}$	↘	$a_e - \|u_e\|$

[a] For $\|m\| = \sqrt{1/\gamma}$. [b] For $\|m\| = 1/(1+\sqrt{\gamma+1})$.

Table 3 Evolution of the flow parameters in the reaction zone of an overdriven detonation (heat addition similar to that presented on Fig. 1b)

	t	0		t_e
	ξ	0	↗	1
	$|u|$	$|u_d|$	↗	$|u_e| = a_e$
	$|m|$	$|m_d|$	↗	1
	ρ	ρ_d	↘	ρ_e
	p	p_d	↘	p_e
	T	T_d	↗ $T_{max}{}^a$ ↘	T_e
$a - |u|$, $|m_d| \geq 1/(1+\sqrt{\gamma+1})$		$a_d - |u_d|$	↘	0
$a - |u|$, $|m_d| < 1/(1+\sqrt{\gamma+1})$		$a_d - |u_d|$	↗ $(a-|u|)_{max}{}^b$ ↘	0

[a] For $|m| = \sqrt{1/\gamma}$.   [b] For $|m| = 1/(1+\sqrt{\gamma+1})$.

Table 4 Evolution of the flow parameters in the reaction zone of a pathological detonation

	t	0		t_M
	ξ	0	↗	$\xi_M > 1$
	$|u|$	$|u_d|$	↗	$|u_M| = a_M$
	$|m|$	$|m_d|$	↗	1
	ρ	ρ_d	↘	ρ_M
	p	p_d	↘	p_M
	T	T_d	↗ $T_{max}{}^a$ ↘	T_M
$a - |u|$, $|m_d| \geq 1/(1+\sqrt{\gamma+1})$		$a_d - |u_d|$	↘	0
$a - |u|$, $|m_d| < 1/(1+\sqrt{\gamma+1})$		$a_d - |u_d|$	↗ $(a-|u|)_{max}{}^b$ ↘	0

[a] For $|m| = \sqrt{1/\gamma}$.   [b] For $|m| = 1/(1+\sqrt{\gamma+1})$.

$$m_f = u/a_f \tag{33}$$

2) $$\dot{\psi} = \frac{1}{M}\sum_j\left\{\left[\Delta\hat{H}_j\right]_0^T - \epsilon_j(\hat{D}_j)_0 - \frac{\gamma_f}{\gamma_f-1}\hat{R}T\right\}\frac{dn_j}{dt} \tag{34}$$

where $(D_j)_0$ is the energy at $T = 0$ K for the dissociation of species j in atoms, and $\epsilon_j = 0$ if species j is an atom and $\epsilon_j = 1$ if species j is a molecule.

$$\dot{\psi}' = \dot{\psi} - \frac{\gamma_f}{\gamma_f-1}\frac{1-m_f^2}{\gamma_f m_f^2 - 1}\frac{\hat{R}T}{M}\sum_j\frac{dn_j}{dt} \tag{35}$$

3) $$\frac{dn_j}{dt} = \sum_{r=1}^n \left(\nu''_j{}^{(r)} - \nu'_j{}^{(r)}\right)\left(\frac{\rho}{M}\right)^{\sum_{i=1}^p \nu'_i{}^{(r)}-1} C_{(r)} \tag{36}$$

where

$$C_{(r)} = k_{+r}\left\{\prod_i (n_i)^{\nu_i'{}^{(r)}} - \left(\frac{\rho\hat{R}T}{M}\right)^{\sum_{i=1}^p (\nu_i''{}^{(r)} - \nu_i'{}^{(r)})} \frac{1}{K_{p_r}}\prod_i (n_i)^{\nu_i''{}^{(r)}}\right\} \tag{37}$$

if the kinetic scheme consists of N simultaneous reversible chemical reactions

$$\sum_{i=1}^p \nu_i'{}^{(r)} A_i \underset{k_{-r}}{\overset{k_{+r}}{\rightleftarrows}} \sum_{i=1}^p \nu_i''{}^{(r)} A_i \quad 1 \leq r \leq N$$

between p distinct chemical species

4) $$\frac{d\gamma_f}{dt} = -\frac{(\gamma_f-1)^2}{\sum_j n_j}\left\{\sum_j n_j \frac{d(\hat{C}_{p_j}/\hat{R})}{dT}\frac{dT}{dt} + \sum_j\left(\frac{\hat{C}_{p_j}}{\hat{R}} - \frac{\gamma_f}{\gamma_f-1}\right)\frac{dn_j}{dt}\right\} \tag{38}$$

Discussion

If we compare the relations Eqs. (25-30) with the equivalent ones obtained with the CGFC model, we can see that the time derivatives of $|u|$, ρ, and p are identical in the two models provided that

$$\dot{\psi} \equiv Q_e \frac{d\xi}{dt}$$

while additional terms due to changes in mixture composition or γ_f variations appear in the time derivatives of T, m_f, and $a_f - |u|$.

As for the CGFC model, the relations Eqs. (25-30) display a singularity at $|m_f| = 1$. To insure a continuous transition through the sonic plane, it is necessary that $\dot{\psi}$ vanishes at the same time as the flow reaches the condition $|m_f| = 1$. It can be seen from Eq. (35) that for $|m_f| = 1$, $\dot{\psi}' = \dot{\psi}$, so that if $\dot{\psi}$ vanishes for $|m_f| = 1$, $\dot{\psi}'$ should vanish also. Thus the behavior of the different flow parameters in the reaction zone depends on the time evolution of $\dot{\psi}$ which is related to the detailed chemistry of the system.

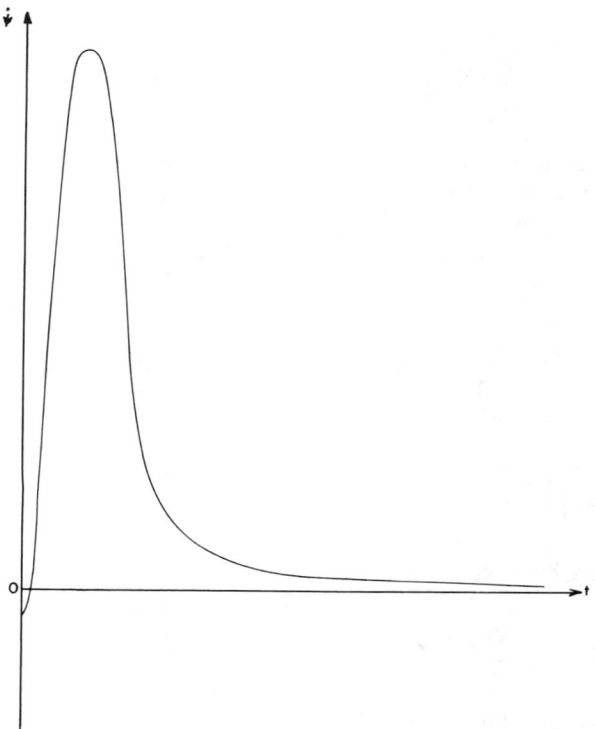

Fig. 3 Time evolution of $\dot{\psi}$ in the reaction zone of a detonation wave in the equimolar H_2-O_2 mixture.

To illustrate this discussion, we have chosen to present here the results of computations in H_2-O_2 and H_2-Cl_2 mixtures. The reaction schemes and the kinetic data used in the computations are given in Table 5 for the H_2-O_2 system and in Table 6 for the H_2-Cl_2 system. The method of computation of reactive shocks has been detailed previously by Sedes (1977) and may be outlined as follows. First, if the Mach number of the shock M_s and the initial pressure, temperature, and composition of the mixture are given, the state immediately behind the shock (state d), from which the chemical reactions take place and where it is assumed that equipartition of energy among the translation, rotation, and vibration of molecules is realized, can be computed from a solution of the nonreactive normal shock equations. The flow conditions in state d are then used as the starting conditions for a numerical integration of the kinetic equations. The integration is carried out by a fourth-order Runge-Kutta method in such a manner that continuity, momentum, energy, and state equations are satisfied. The sonic condition is searched for by an iterative process on the shock Mach number.

Table 5 H_2-O_2 system: kinetic scheme and reaction rate data used in the computations (from Dove et al., 1974) (only the forward rate constant for each reversible reaction is given; the backward rate constants are derived from forward rate constants and equilibrium constants)

	Kinetic model		$k = AT^n \exp(-\hat{E}/\hat{R}T)$, $cm^3 \cdot mole^{-1} \cdot s^{-1}$	A	n	$\hat{E}$, cal/mole	M
1	$H_2 + O_2$	$\rightleftarrows$ 2 OH	k_{+1}	2.1×10^{14}	0	57,500	—
2	$OH + H_2$	$\rightleftarrows H_2O + H$	k_{+2}	2.2×10^{13}	0	5,150	—
3	$H + O_2$	$\rightleftarrows OH + O$	k_{+3}	2.2×10^{14}	0	16,800	—
4	$O + H_2$	$\rightleftarrows OH + H$	k_{+4}	1.8×10^{10}	1	8,900	—
5	$O_2 + M$	$\rightleftarrows 2O + M$	$k_{+5, M}$	7.1×10^{14}	0	104,800	H, OH, H_2O, HO_2, H_2, O_2
				1.97×10^{15}	0	104,800	O
6	$H_2 + M$	$\rightleftarrows 2H + M$	$k_{+6, M}$	8.8×10^{14}	0	96,000	H, O, OH, H_2, O_2, H_2O, HO_2
7	$H_2O + M$	$\rightleftarrows H + OH + M$	$k_{+7, M}$	3.5×10^{15}	0	5,000	H, O, OH, H_2, O_2, HO_2
				2.1×10^{16}	0	105,000	H_2O
8	$OH + M$	$\rightleftarrows O + H + M$	$k_{+8, M}$	1.4×10^{14}	0.21	101,300	H, O, OH, H_2, O_2, H_2O, HO_2
9	2OH	$\rightleftarrows H_2O + O$	k_{+9}	6.3×10^{12}	0	1,100	—
10	$H + O_2 + M$	$\rightleftarrows HO_2 + M$	$k_{+10, M}{}^a$	5×10^{15}	0	−1,000	H, O, OH, HO_2, H_2
				3.2×10^{16}	0	−1,000	H_2O
				2×10^{15}	0	−1,000	O_2
11	$OH + HO_2$	$\rightleftarrows H_2O + O_2$	k_{+11}	3×10^{14}	0	0	—
12	$O + HO_2$	$\rightleftarrows OH + O_2$	k_{+12}	3×10^{14}	0	0	—
13	$H + HO_2$	$\rightleftarrows 2OH$	k_{+13}	2.5×10^{14}	0	1,900	—

$^a cm^6 \cdot mole^{-2} \cdot s^{-1}$.

Table 6 H_2-Cl_2 system: kinetic scheme and reaction rate data used in the computations (only one rate constant for each reversible reaction is given and the missing rate constant is derived from it and the corresponding equilibrium constant)

	Kinetic model	$k = AT^n \exp(-\hat{E}/\hat{R}T)$, $cm^3 \cdot mole^{-1} \cdot s^{-1}$	A	n	$\hat{E}$, cal/mole	M	References
1	$H_2 + M \rightleftharpoons 2H + M$	$k_{-1,M}$[a]	10^{18}	-1	0	HCl, Cl_2, Cl	Jacobs et al. (1976a)
			2×10^{19}	-1	0	H	
			9×10^{16}	-0.6	0	H_2	
2	$Cl_2 + M \rightleftharpoons 2Cl + M$	$k_{+2,M}$	6.15×10^{21}	-2.07	57,040	H_2, Cl_2, HCl, H	Jacobs and Giedt (1963)
			6.15×10^{22}	-2.07	57,050	Cl	
3	$HCl + M \rightleftharpoons H + Cl + M$	$k_{+3,M}$	6.76×10^{21}	-2	102,170	H_2, Cl_2, HCl, H, Cl	Jacobs et al. (1967b)
4	$Cl + H_2 \rightleftharpoons HCl + H$	k_{+4}	4.80×10^{13}	0	5260	—	Benson et al. (1969)
5	$H + Cl_2 \rightleftharpoons HCl + Cl$	k_{+5}	6.61×10^{11}	0.68	1090	—	Wilkins (1965)

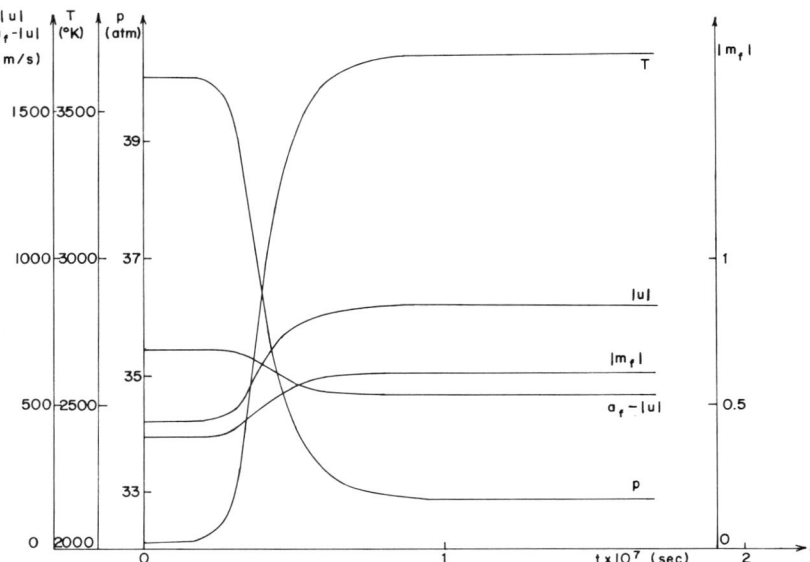

Fig. 4 Time evolution of the flow parameters in the reaction zone of the C-J detonation in the equimolar H_2-O_2 mixture ($p_1 = 1$ atm, $T_1 = 300$ K, $M_s = 5.8$).

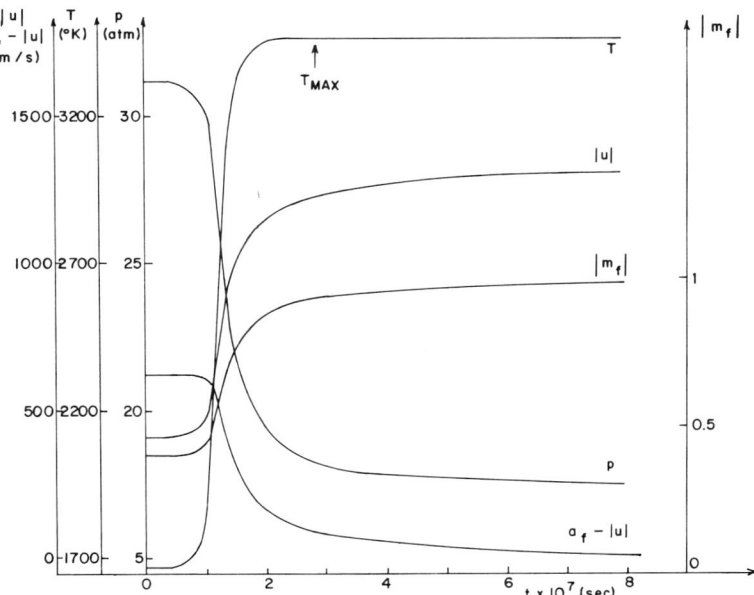

Fig. 5 Time evolution of the flow parameters in the reaction zone of the C-J detonation in the equimolar H_2-O_2 mixture ($p_1 = 1$ atm, $T_1 = 300$, $M_s = 5.1321$).

Table 7 Equimolar H_2-O_2 mixture, $p_1 = 1$ atm, $T_1 = 300$ K

	Method A	Method B	%
M_{oc}	5.1367	5.1321	0.09
D, m/s	2325.1	2323.0	0.09
p, atm	17.052	17.097	0.26
T, K	3457.6	3452.4	0.15
$|u|$, m/s	1312.2	1306.4	0.44
a_f, m/s	1313.2	1311.7	0.04
n_{HO_2}, moles	$0.4608 \; 10^{-3}$	$0.4588 \; 10^{-3}$	0.43
n_{H_2O}, moles	0.7926	0.7924	0.10
n_{OH}, moles	0.2381	0.2386	0.02
n_{H_2}, moles	$0.6609 \; 10^{-1}$	$0.6647 \; 10^{-1}$	0.57
n_{O_2}, moles	0.4384	0.4382	0.05
n_H, moles	$0.4278 \; 10^{-1}$	$0.4320 \; 10^{-1}$	0.98
n_O, moles	$0.9101 \; 10^{-1}$	$0.9161 \; 10^{-1}$	0.66

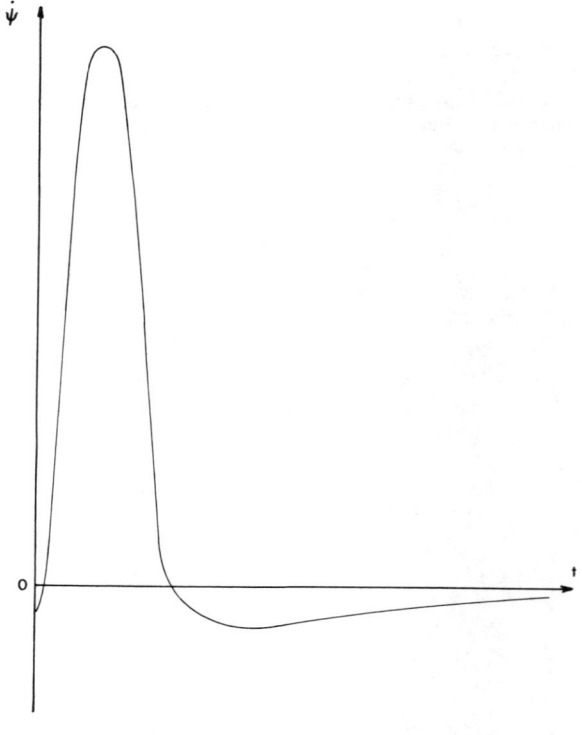

Fig. 6 Time evolution of $\dot{\psi}$ in the reaction zone of a detonation wave in the equimolar H_2-Cl_2 mixture.

HEAT-RELEASE FUNCTION ON DETONATION STATES 405

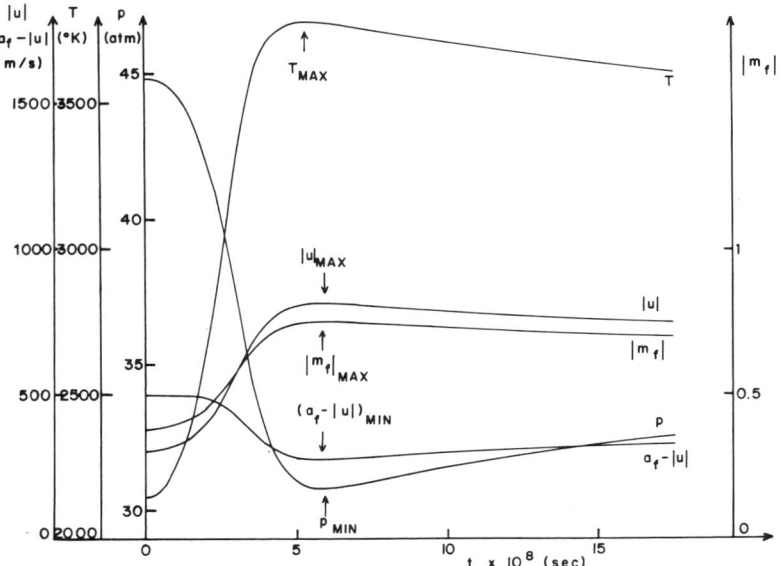

Fig. 7 Time evolution of the flow parameters in the reaction zone of an overdriven detonation in the equimolar H_2-Cl_2 mixture ($p_1 = 1$ atm, $T_1 = 300$ K, $M_S = 6.2$).

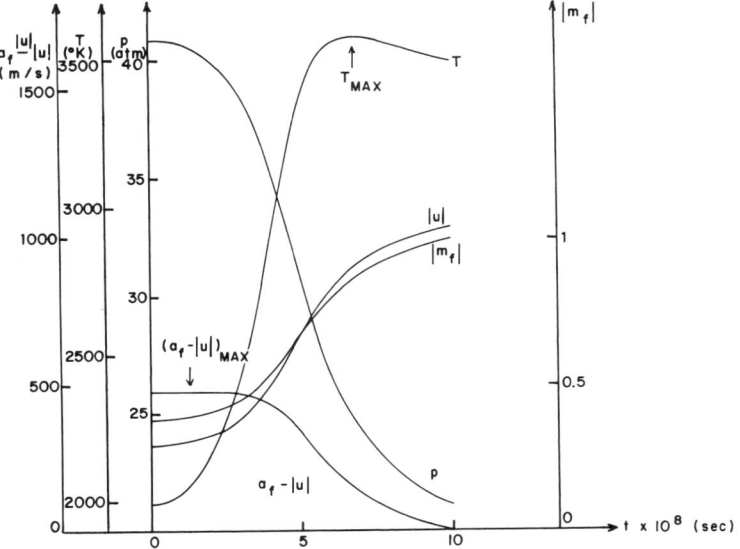

Fig. 8 Time evolution of the flow parameters in the reaction zone of the pathological detonation in the equimolar H_2-Cl_2 mixture ($p_1 = 1$ atm, $T_1 = 300$ K, $M_S = 5.9219$).

Table 8 Equimolar H_2-Cl_2 mixture, $p_I = 1$ atm, $T_I = 300$ K

	Method A	Method B	%
M_{oc}	5.5665	5.9219	6.0
D, m/s	1695.6	1804.9	6.1
p, atm	18.61	21.15	12.0
T, K	3015.4	3502.5	13.9
$|u|$, m/s	985.3	1041.0	5.4
a_f, m/s	985.3	1041.2	5.4
n_{HCl}, moles	1.701	1.801	5.6
n_{Cl_2}, moles	0.726×10^{-2}	0.1612×10^{-1}	55.0
n_{H_2}, moles	0.1391	0.9309×10^{-1}	49.4
n_{Cl}, moles	0.2846	0.1664	71.0
n_H, moles	0.2799×10^{-1}	0.1243×10^{-1}	125.2

H_2-O_2 System

For the H_2-O_2 systems, the time-evolution of $\dot{\psi}$ in the reaction zone is given on Fig. 3. It can be seen that the time evolution of $\dot{\psi}$ is identical with that of the instantaneous heat addition per unit mass $\dot{\xi}Q_e$ presented on Fig. 1a: if we disregard the very short (only a few nanoseconds) negative part of the curve, which corresponds to the endothermic initiation phase of the chemical reactions, $\dot{\psi}$ becomes zero only when chemical equilibrium is reached. Therefore, the sonic condition $|m_f| = 1$ will be reached at chemical equilibrium, and the self-sustained detonation wave in the H_2-O_2 system will be a C-J detonation wave. The computed profiles of the flow parameters in the reaction zone are plotted against time in Fig. 4 for an overdriven wave and in Fig. 5 for the C-J detonation wave. It can be seen that the behavior of the flow parameters is the same as that predicted by the CGFC model (Tables 1 and 3).

Since the sonic condition is reached at chemical equilibrium, one must find the same values for the C-J condition as these computed from the classical thermodynamic method (method A). The agreement between the C-J values calculated by the two methods is shown in Table 7.

H_2-Cl_2 System

The time-evolution of $\dot{\psi}$ in the reaction zone is given on Fig. 6 for the H_2-Cl_2 system. If we neglect the endothermic initiation phase of the reactions (the negative part of the curve at the beginning of the reaction zone), we can see that the evolution of $\dot{\psi}$ is identical with that of the instantaneous heat addition per unit mass $\dot{\xi}Q_e$ presented on Fig. 1b. Thus, the sonic condition $|m_f| = 1$ will be reached before completion of chemical equilibrium, and the self-sustained detonation wave in the H_2-Cl_2 system will be a pathological wave. The computed profiles of the flow parameters in the reaction zone are plotted against time in Fig. 7 for an overdriven wave and in Fig. 8 for the pathological wave. The behavior of the flow parameters is similar to that predicted by the CGFC model (Tables 2 and 4).

The sonic condition $|m_f| = 1$ being reached before completion of chemical equilibrium, one cannot expect a close agreement between the values of the flow parameters calculated with our model (method B) and those given by the classical thermodynamic method (method A) in which the sonic condition corresponds to chemical equilibrium. The discrepancies between the values computed by the two methods are shown in Table 8. In particular, as predicted by the CGFC model, the velocity of the pathological detonation is greater than the C-J velocity.

Conclusion

In the present investigation, we have shown the influence of the heat-release function on the detonation states. In particular if the heat addition in the reaction zone goes through a maximum before reaching its equilibrium value, the resulting free detonation propagates with a velocity greater than the equilibrium Chapman-Jouguet velocity.

Although the results may not be verifiable experimentally on a quantitative basis since other effects such as wall boundary layers, transient multiheaded, or turbulent structure of the wave may overshadow the heat-release function effect studied, it may be possible at least to indicate and explain those systems where the experimentally observed velocities are greater than the equilibrium C-J values in spite of losses.

References

Benson S.W., Cruickshank F.R., and Shaw R. (1969) Iodine monochloride as a thermal source of cholrine atoms: the reaction of chlorine atoms with hydrogen. *Intl. J. Chem. Kinet.* **1**, 29-43.

Döring W. (1943) Über den detonationsvorgang in gasen. *Annln. Phys.* **43**, 421-436.

Dove J.E., Scroggie B.J., and Semerdjian H. (1974) Velocity deficits and detonability limits of hydrogen-oxygen gas detonation at low initial pressures. *Acta Astronautica* **1**, 345-359.

Fay J.A. (1959) Two-dimensional gaseous detonation: velocity deficit. *Phys. Fluids* **2**, 283-289.

Gruschka H.D. and Wecken F. (1971) *Gasdynamic Theory of Detonations*, Chap. V, Pt.14. Gordon and Breach Sciences Publisher.

Jacobs T.A., Giedt R.R., and Cohen N. (1967a) Kinetics of hydrogen halides in shock waves. II A new measurement of the hydrogen dissociation rate. *J. Chem. Phys.* **47**, 54-57.

Jacobs T.A., Cohen N., and Giedt R.R. (1967b) Kinetics of hydrogen halides in shock waves: HCl and DCl. *J. Chem. Phys.* **46**, 1958-1968.

Jacobs T.A. and Giedt R.R. (1963) Dissociation of Cl_2 in shock waves. *J. Chem. Phys.* **39**, 749-756.

Sèdes C. (1977) Contribution à l'étude des mécanismes de création d'inversions de population entre niveaux de vibration de molécules formées par réactions chimiques au sein d'un écoulement. Thèse de Doctorat es-Science, Marseille.

Strehlow R.A. (1971) Detonation structure and gross properties. *Combustion Science Tech.* **4**, 65-71.

Tsuge S. (1971) The effect of boundaries on velocity deficit and the limit of gaseous detonations. *Combustion Science Tech.* **3**, 195-205.

Von Neumann J. (1942) Progress report on "Theory of detonation waves." OSRD Rept. 549.

White D.R. (1961) Turbulent structure of gaseous detonations. *Phys. Fluids* **4**, 465-480.

Wilkins R.L. (1965) Competitive reaction rates of hydrogen atoms with HCl and Cl_2. *J. Chem. Phys.* **42**, 806-807.

Zeldovich Y.B. (1940) On the theory of the propagation of detonations in gaseous systems. *Zh. Eksp. Teor. Fiz.* **10**, 542-568. Also NACA TM 1261 (1950).

Detonation Characteristics of Gaseous Ethylene, Oxygen, and Nitrogen at High Initial Pressures

P. Bauer,* S. Krishnan,† and C. Brochet ‡
Université de Poitiers, Poitiers, France

To better understand detonation at very high pressures (several hundreds of kilobars), a simplified fundamental thermodynamic study of the detonation products at high pressures (100-1000 bars) was undertaken. Explosive gaseous mixtures of ethylene, oxygen, and nitrogen (equivalence ratio, $r = 1.05$-1.60) were detonated at high initial pressures ($2 < p_i < 45$ bars). The detonation velocities were measured by ionization pickups. To study the diameter effect, the mixtures were detonated in two different tubes (15 and 52 mm diam) of 6 m length. The experimental results show that the diameter effect is reduced as the initial pressure increases. The measured detonation velocities are compared with the theoretical values calculated from solutions of the jump conditions with different equations of state. The validity of these equations of state for a detonation pressure range of 100-1000 bars is studied. The theoretical pressure and temperature of a detonation appear to be less sensitive to the equation of state than the detonation velocity.

Introduction

THE knowledge of the detonation behavior of gaseous mixtures at pressures up to 2000 bars is important for practical applications, e.g., rock blasting or experiments on strength of materials as well as for fundamental studies. This range of medium pressures can be considered as the first step for the understanding of more complex phenomena, such as the one occurring in the detonation of high explosives.

Following the early work of Schmidt (1935), Gealer and Churchill (1960) measured the detonation velocity in H_2-O_2 mixtures with initial pressures up to 70 bars. Hoelzer and Stobaugh (1954) measured the detonation velocity in H_2-O_2 and C_2H_6-O_2 mixtures for initial pressures in the range of 1-10 bars. Wolfson (1963) measured the detonation velocity and pressure in H_2-O_2

Presented at the 7th ICOGER, Göttingen, Federal Republic of Germany, Aug. 20-24, 1979. Copyright © American Institute of Aeronautics and Astronautics, Inc., 1981. All rights reserved.
*Laboratoire d'Energétique et de Détonique.
†Delegate from Indian Institute of Technology, Department of Aeronautical Engineering, Madras, India.
‡Laboratoire d'Engergétique et de Détonique.

mixtures with inert dilution and initial pressures up to 50 bars and compared the experimental results with the theoretical values based on the ideal-gas equation of state. Pujol (1968) studied the influence of initial pressure (1-3 bars) and temperature on the detonation velocity of C_3H_8-O_2-N_2 mixtures. Brossard (1971) measured the influence of initial pressure (up to 4 bars), for various H_2-O_2 mixtures, on the detonation velocities measured in a spherical vessel and compared these values to those obtained by calculation based on the ideal-gas equation of state.

No work has been done to compare experimental results with theoretical values based on various equations of state that describe the behavior of gaseous mixtures in a range of pressures where gas behavior is nonideal.

The results of the measurements of detonation velocity in C_2H_4-O_2-N_2 mixtures for initial pressures up to 50 bars are reported. These mixtures are fuel-rich and have a global chemical formula similar to that of typical CHNO condensed explosives. The experimental values are compared with those of theoretical computations based on two real-gas equations of state; one of the equations of state is semiempirical and is used primarily for computing the detonation characteristics of condensed explosives.

Experimental Apparatus

Two different tubes whose inside wall surfaces were perfectly smooth were used. Tube 1, made of stainless steel and specially designed to withstand very high pressures (several kbars), was 15 mm i.d. and 6.5 m long (Fig. 1). Tube 2, made of stainless steel and designed for medium pressures (< 250 bars), was 52 mm i.d. and 5.4 m long. To check the stability of the detonation wave (Manson et al., 1963), two sets of ionization probes were placed at two different positions in tube 1. Each set was composed of two gages placed at a distance of 50 ± 0.02 cm from each other (Fig. 1). Both chronometers were

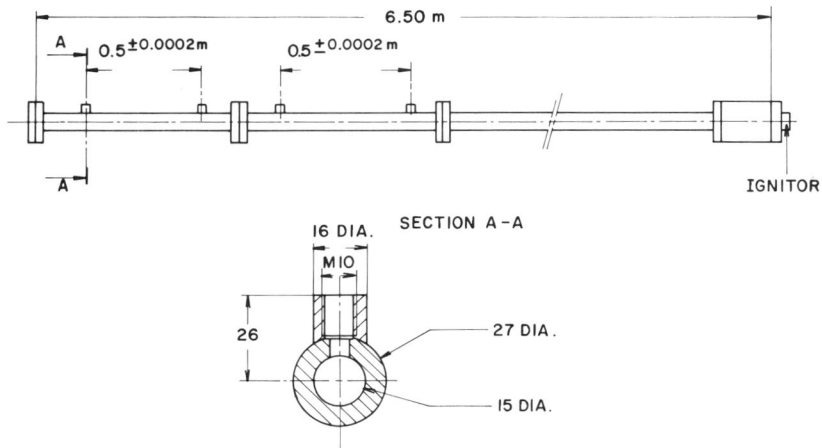

Fig. 1 Detonation tube.

Table 1 Gas purity

Gas	Purity
Ethylene N 25	$C_2H_4 > 99.5\%$
Oxygen	$O_2 > 99.5\%$
	$N_2 + A < 0.5\%$
	$CH_4 \simeq 40$ ppm
Nitrogen	$O_2 + H_2O < 50$ ppm
	$A \simeq 1000$ ppm
	$H_2O < 5$ ppm

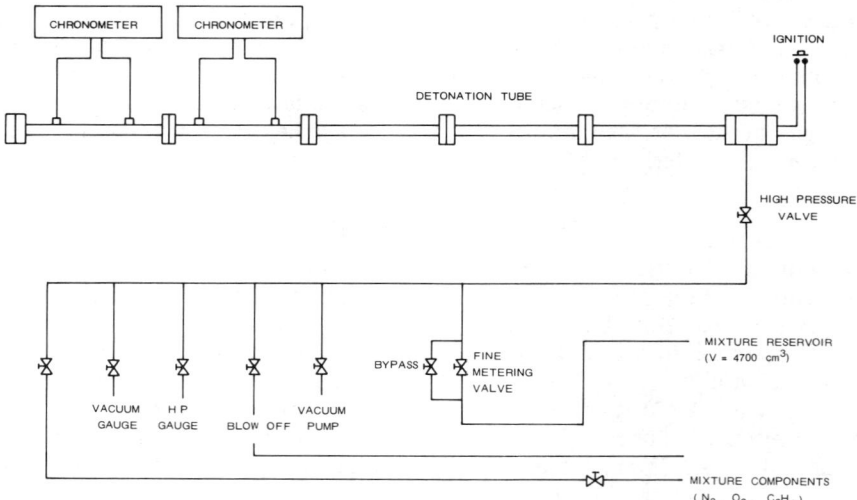

Fig. 2 Schematic diagram of apparatus.

calibrated on a quartz clock, which gave an uncertainty of 0.1 μs. The uncertainty of the measurement was 0.2-0.3%. On tube 2, the measurements of detonation velocity were made with only one set of gages. The mixtures were initiated by means of an electrical igniter.

The mixtures were prepared in a stainless-steel tank ($V = 4.7$ ℓ) at high pressure (60 bars). The hydrocarbon used should satisfy two conditions: 1) stability of the mixture at these initial pressures; and 2) possibility of detonation even when diluted with an inert gas. For these reasons, ethylene was used. The purities of the components are given in Table 1.

The setup used for mixture preparation is shown in Fig. 2. Proportions of each component were obtained by measurement of the partial pressure. A calibrated gage gave these values with an uncertainty of ±0.01 bar in the range 1-70 bars. For 1 mole of oxygen, the approximate number of moles of inert gas was: one for mixture D; and two for mixtures A, B, C, E, and F.

In the calculation of mole ratios from measurements of partial pressure, real-gas effects have been included by the introduction of the compressibility factor (Pujol, 1968):

$$PV = nzRT \tag{1}$$

Values of the compressibility factor z were taken from the Hougen and Watson (1947) diagram. This correction was about 4%. Because of the uncertainty in the determination of z, two samples taken from each prepared mixture were analyzed as follows: 1) before starting the set of experiments, the mixture was at an initial pressure of about 60 bars; and 2) after the experiments, the mixture was at an initial pressure of about 2 bars.

Analyses were performed by gas chromatography. Oxygen and nitrogen, separated in a 2.6 m long and 1/4 in. diam tube filled with 5 Å zeolithe (42 × 60 mesh), were measured by means of a catharometer. Ethylene, separated in a 2 m long and 1/8 in. diam tube filled with activated alumina (80 × 100 mesh), was measured by means of a flame ionization detector (Bauer, 1977). From these samplings the exact composition of the mixtures and the stability of the composition for the sampled mixtures were determined.

The difference between the values before and after the set of experiments was less than 1.5%, this value being smaller than the analysis uncertainty (2%). We report in Table 2 the compositions of each mixture (the equivalence ratio is defined as the ratio of the number of O_2 moles in the stoichiometric mixture to the number of moles corresponding to the present mixture).

Theoretical Calculations

The detonation characteristics (velocity, pressure, and temperature) of each mixture were computed using the three different programs explained below.

Calculation Based on the Ideal-Gas Equation of State

This program, originally written by Gordon and McBride (1971), has been slightly modified to accommodate it to a CII Iris 45 computer. Results from this program are in good agreement with those obtained from this laboratory's program "Kapten" (Johnson, 1968).

Table 2 Gas mixture compositions

| | | | Moles of species | | | |
| | Equivalence | | O_2 | | N_2 | |
Mixture	ratio	C_2H_4	a	b	a	b
A	1.30	1	2.31	—	4.60	—
B	1.29	1	2.29	2.32	4.55	4.46
C	1.38	1	2.03	2.17	4.05	3.85
D	1.05	1	2.59	2.85	2.58	2.48
E	1.55	1	1.93	—	3.90	—
F	1.60	1	1.87	—	3.79	—

[a] Composition calculated from measurement of partial pressure with real-gas effects accounted for by compressibility factor.
[b] Composition obtained from analysis. Average value of a high and low initial pressure sample.

Table 3   Constants for BKW equation of state

α	β	K	θ
0.5	0.16	10.91	400

Computation Based on the Becker, Kistiakowsky, and Wilson Equation of State

The Becker, Kistiakowsky, and Wilson (BKW) equation of state (Mader, 1963) has the following form:

$$\frac{PV_g}{RT} = 1 + X \exp(\beta X)$$

where

$$X = \frac{K \sum_i x_i k_i}{V_g (T+\theta)^\alpha} \qquad (2)$$

where V_g is the molar volume of the gas and x_i the molar fraction of the component i. The summation covers all the gaseous species present in the burned products. The parameters α, β, θ, K are semiempirical constants adjusted to fit the experimental detonation characteristics of condensed explosives. The values of these constants due to Mader (1963) are given in Table 3. The FORTRAN program for the BKW equation of state written by Mader (1963, 1967) was used on an IBM 370 computer to calculate the detonation characteristics of CH-NO and other mixtures.

Calculation Based on Boltzmann Equation of State

For simple and fast calculations, we have written a program with a HP 9820 desk computer. This program employs a method, proposed by Taylor (1952), built upon the Boltzmann equation of state to calculate the detonation characteristics of CH-NO mixtures:

$$\frac{PV}{nRT} = 1 + x + 0.625x^2 + 0.287x^3 + 0.193x^4 \qquad (3)$$

where

$$x = \frac{b}{V} \text{ and } b = \sum_i n_i b_i$$

The quantity x represents the second virial coefficient calculated (Pujol, 1968) from values of b_i given by Hirschfelder, Curtiss, and Bird (1954). The simplifying assumptions for the program are: 1) the detonation products do not include any condensed species; 2) the initial pressure can be neglected in comparison with the detonation pressure; and 3) the only species present in burned gases are CO_2, H_2O, H_2, CO, and N_2 in equilibrium. The thermodynamic data required for the computations are obtained from Gordon and McBride (1971).

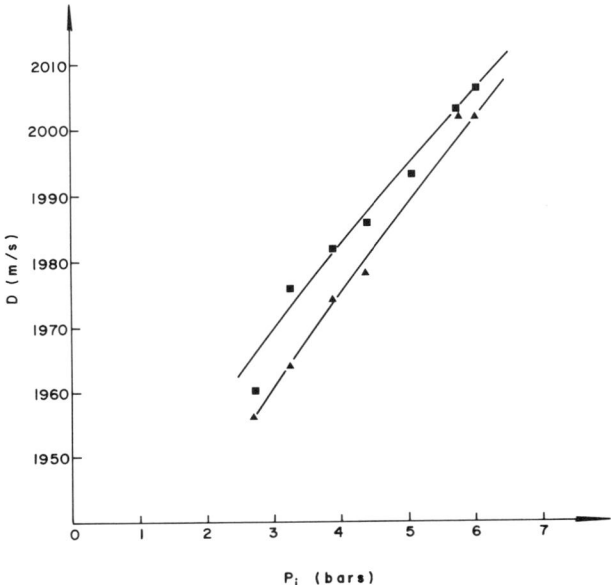

Fig. 3 Effect of diameter on the detonation velocity. Mixture: $C_2H_4 + 2.78O_2 + 7.26N_2$; $r = 1.08$; ▲ tube 1, $\phi = 15$ mm; ■ tube 2, $\phi = 52$ mm.

Experimental Results

Influence of the Tube Diameter

Up to the maximum initial pressure of 3 bars, experimental as well as theoretical investigations have shown that the diameter of detonation tube has a sensitive effect on the measured detonation velocity (e.g., Pujol, 1968; Brochet, 1966; and Guénoche and Manson, 1952). Therefore, we have tried to determine whether this effect persisted at high initial pressures. We have measured the detonation velocities for the mixture: $C_2H_4 + 2.78O_2 + 7.26N_2$ (equivalence ratio $r = 1.08$) in the two tubes described earlier for $2 < p_i < 7$ bars. The results are reported in Fig. 3. It appears that, for these conditions, the detonation velocity has almost the same value. The divergence is 0.5% for $p_i = 3$ bars and 0.3% for $p_i = 6$ bars.

We can find, graphically, for $p_i = 3.9$ bars and $p_i = 6.1$ bars (Fig. 4) the value D_∞ corresponding to the value that would be obtained in an infinite-diameter tube. The value $(D_\infty - D)/D_\infty$ (D being the detonation velocity measured in tube 1) is 0.5% at $p_i = 3.9$ bars and 0.25% at $p_i = 6.1$ bars.

Detonation Velocity as a Function of Initial Pressure

Measurements, in tube 1, of the detonation velocity of mixtures A-F (Table 2) for initial pressures $2 < p_i < 45$ bars are reported in Fig. 5.

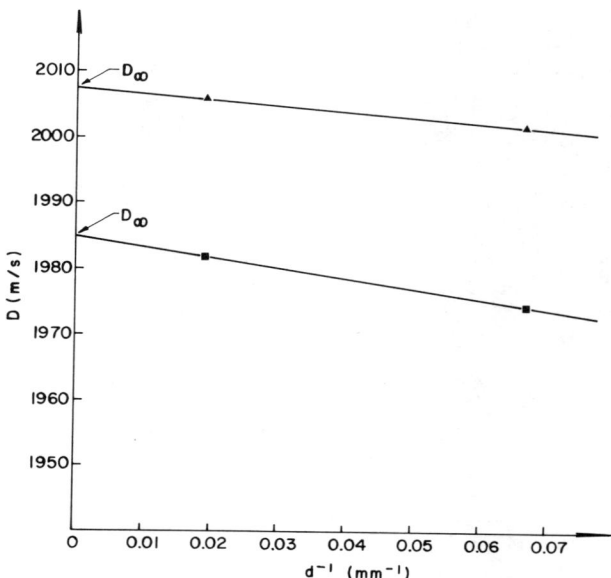

Fig. 4 Determination of D_∞, the detonation velocity for a tube of infinite diameter. Mixture: $C_2H_4 + 2.78O_2 + 7.26N_2$, $r = 1.08$; ▲ $P_i = 6.1$ bars, ■ $P_i = 3.9$ bars.

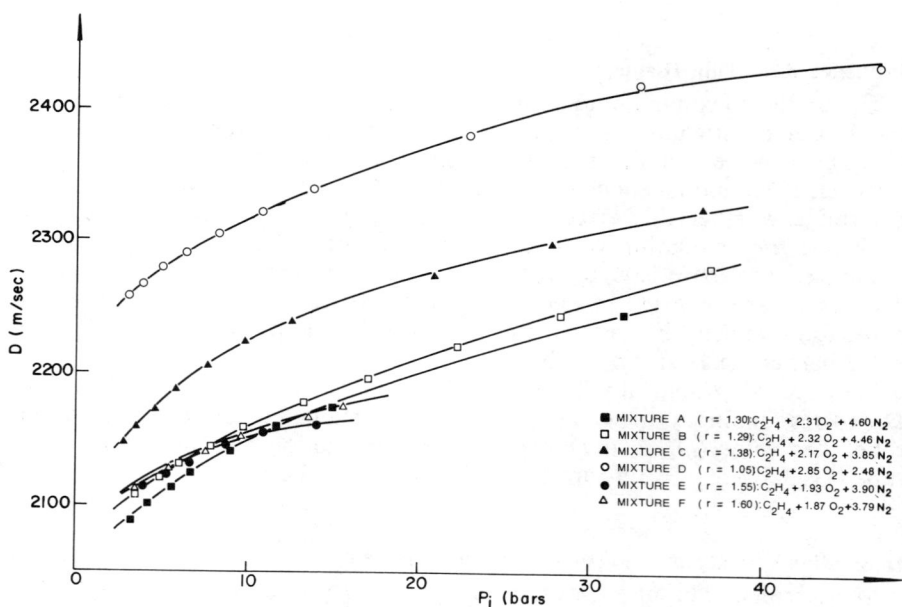

Fig. 5 Observed detonation velocities for mixtures ■ A, □ B, ▲ C, ○ D, ● E, and △ F.

Discussion

Validity of the Equations of State

The values of the detonation velocity, computed by the programs described above, are shown in Fig. 6. At a glance, three ranges of pressures can be noticed:

1) $p_i < 7$ bars. The calculation using the ideal-gas equation of state gives values of velocities very close to the experimental ones (better than 1%). For the richest mixtures (E and F) this range can be extended to the pressures $p_i = 12\text{-}14$ bars. The values computed with the BKW equation of state are also quite satisfactory (divergence less than 1-2%).

2) $7 < p_i < 18$ bars. The theoretical calculation based on the BKW equation of state gives detonation velocities in good agreement with the experimental values ($\Delta D/D < 1\%$). The difference is greater (3% at $p_i = 16$ bars) for the richest mixtures (E and F) or for equivalence ratio close to 1 (mixture D).

3) $p_i > 18$ bars. For this pressure range, none of the equations of state used leads to values in good agreement with the experimental results (the discrepancy is 6-7% for $p_i = 40\text{-}45$ bars). The curve resulting from the computation using the ideal gas equation of state has a profile that tends to an asymptotic value. The experimental values are, in this case, always greater than the calculated ones. Wolfson (1963) obtained this result for H_2-O_2 mixtures at initial pressures greater than 10 bars.

Investigation for an Adapted Equation of State in the Range $2 < p_i < 40$ bars

After obtaining the effects of varying the values of α, β, K, and θ of the BKW equation of state, we noticed that the only sensitive parameter was K. Taking $K = 8$ (BKWC), we had good agreement between computed and experimental values of the detonation velocity for $20 < p_i < 40$ bars with mixtures B, C, D, and F (the agreement is good over the whole range of pressure $3 < p_i < 40$ bars for mixture B).

The computation based on the Boltzmann equation of state gives highest values of detonation velocities. The values computed seem to suffer from the simplifying assumptions that we have used: 1) for $p_i < 5$ bars, the effect of neglecting the initial pressure gives a difference of 1.5-2%; 2) by accounting only for the species CO_2, CO, H_2O, H_2, and N_2, the dissociation effects have been ignored. This leads to an increasing discrepancy as the initial pressure decreases.

To determine the importance of the dissociation effects, the values of detonation velocities have been computed for mixture B with a reduced number of product species. The species involved in the programs (i.e., species present only at the reference temperature of 298 K)—ideal gas without dissociation (IG_{WD}) and BKW without dissociation (BKW_{WD})— are presented in Table 4 and the results are shown in Fig. 6b. The values obtained with IG_{WD}

Table 4 Species included in computations

Case	Equation of state	Species
I	IG_{WD}	$C_2H_4, O_2, H_2O, CO_2, CO, H_2, N_2$
II	BKW_{WD}	H_2O, CO_2, CO, H_2, N_2

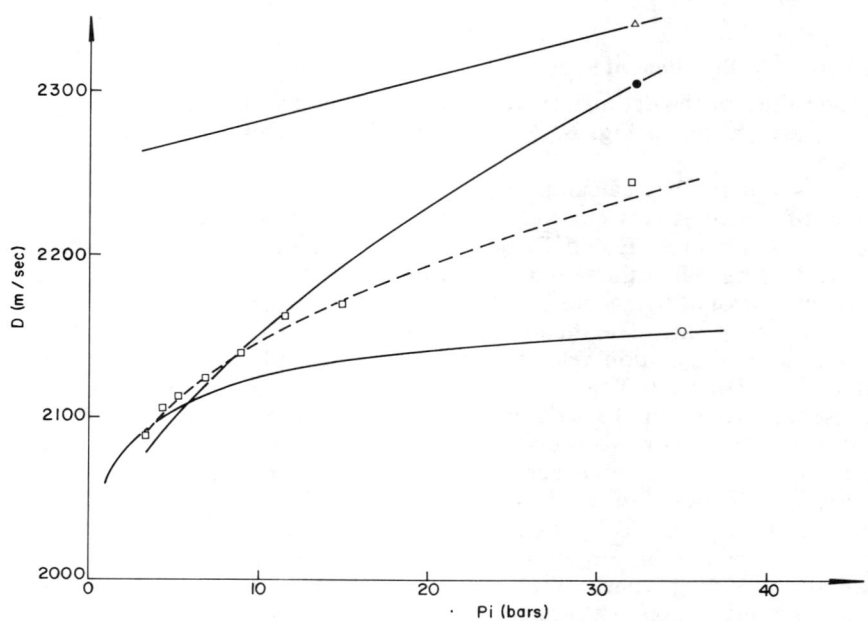

Fig. 6a Experimental and computed detonation velocities for mixture A: △ Boltzmann, ● BKW, □ experimental, ○ ideal gas.

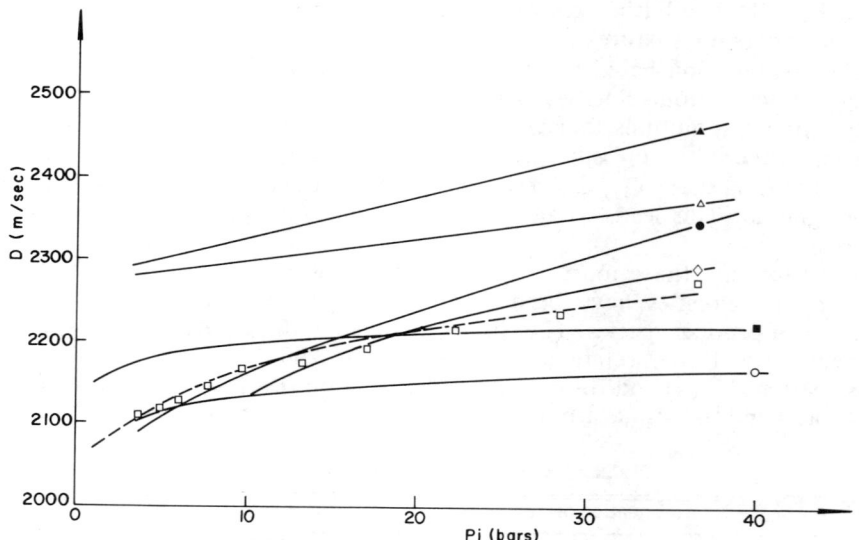

Fig. 6b Experimental and computed detonation velocities for mixture B: △ Boltzmann, ● BKW, □ experimental, ○ ideal gas, ■ IG_{WD}, ▲ BKW_{WD}, ◇ BKWC.

GASEOUS ETHYLENE, OXYGEN, AND NITROGEN MIXTURES 417

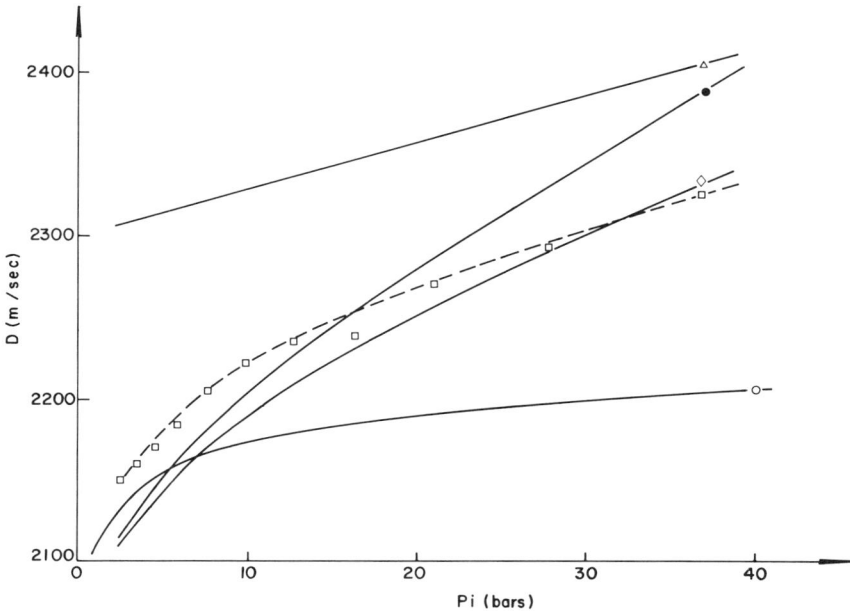

Fig. 6c Experimental and computed detonation velocities for mixture C: △ Boltzmann, ● BKW, □ experimental, ○ ideal gas, ◇ BKWC.

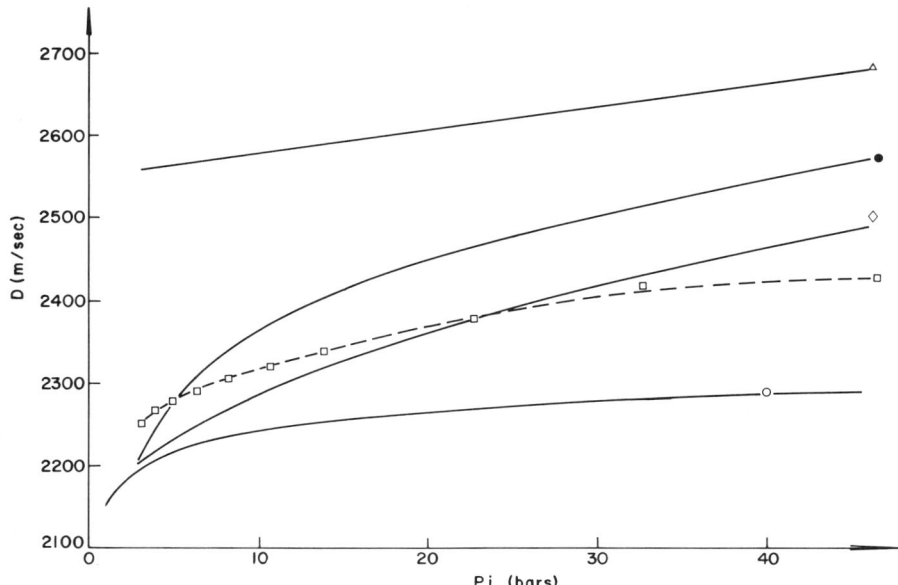

Fig. 6d Experimental and computed detonation velocities for mixture D: △ Boltzmann, ● BKW, □ experimental, ○ ideal gas, ◇ BKWC.

418 P. BAUER, S. KRISHNAN, AND C. BROCHET

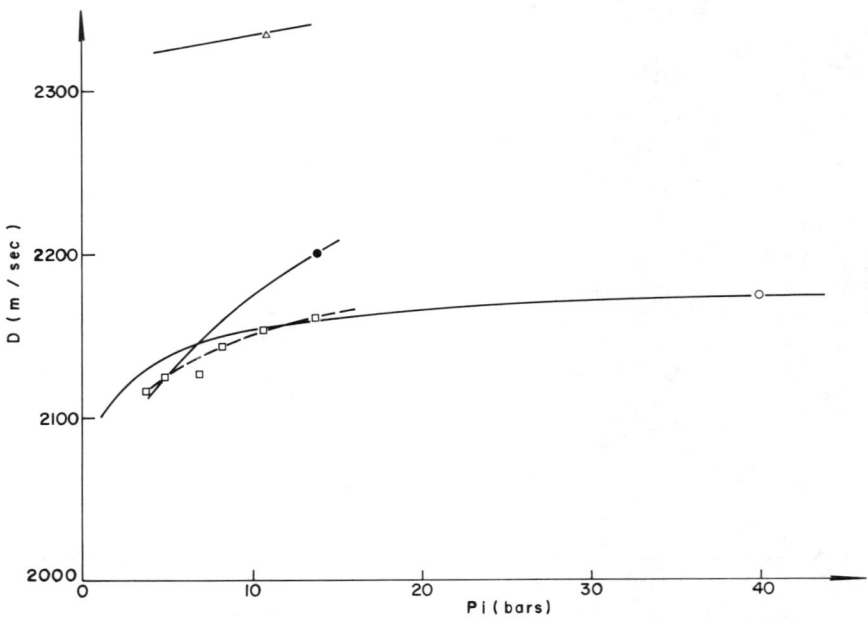

Fig. 6e Experimental and computed detonation velocities for mixture E: △ Boltzmann, ● BKW, □ experimental, ○ ideal gas.

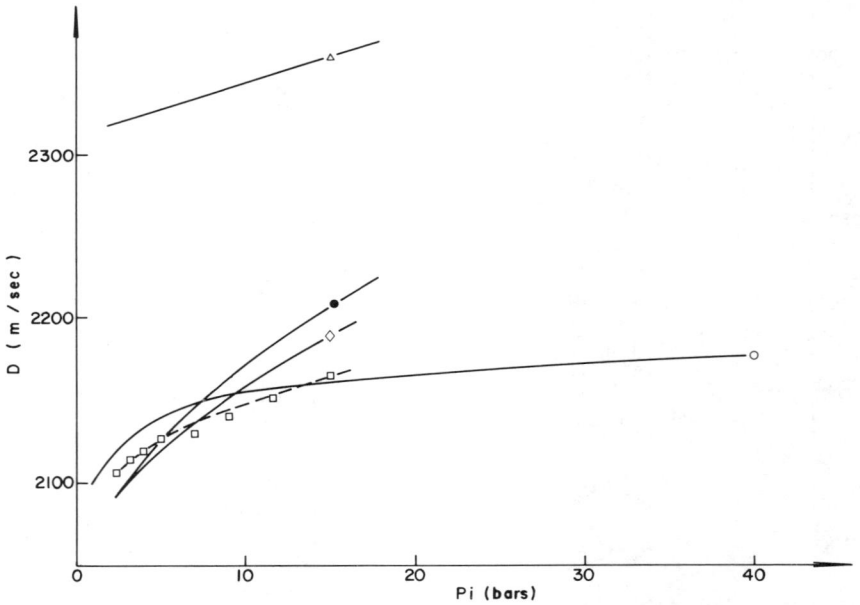

Fig. 6f Experimental and computed detonation velocities for mixture F: △ Boltzmann, ● BKW, □ experimental, ○ ideal gas, ◇ BKWC.

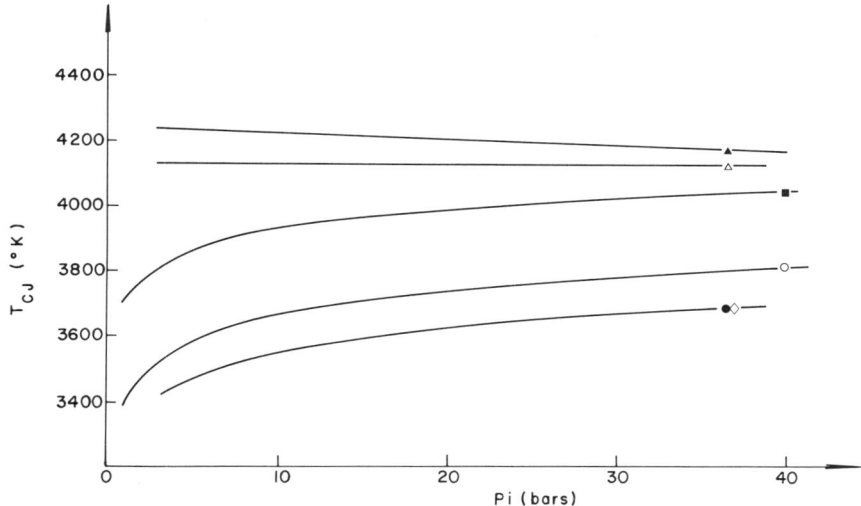

Fig. 7 Computed C-J temperatures for mixture B: △ Boltzmann, ● BKW, ○ ideal gas, ■ IG_{WD}, ▲ BKW_{WD}, ◇ BKWC.

are 4% greater than the values computed with all the species. This divergence decreases at higher initial pressures since the increased pressure of the Chapman-Jouguet (C-J) states depresses dissociation.

The divergence for the BKW_{WD} computation when compared with the calculation involving all species is 10%. The values are on a straight line close to the one corresponding to the Boltzmann equation of state calculation. If the profiles of BKW and BKW_{WD} curves and experimental and Boltzmann curves are compared, some similarity among the profiles can be noticed. It is reasonable to expect, therefore, that if the dissociation effects are adequately treated, the results of the Boltzmann calculation should better approximate the experimental results.

Parameters Sensitive to the Choice of the Equation of State

In Fig. 7, the computed C-J temperatures for mixture B are presented. The lowest temperatures were obtained using BKW equation of state. It appears that the parameter K has no influence on the value of the detonation temperature.

The effect of dissociation (i.e, the difference between BKW and BKW_{WD}) leads to a value of 700-800 K. The value obtained with BKW_{WD} is independent of initial pressure. The constant value obtained (4200 K) is close to the value predicted (4300 K) when the Boltzmann equation of state is used. The calculation with ideal-gas equation of state, with and without dissociation, leads to curves that have the same profile. The divergence in this case is 800 K and does not depend on initial pressure.

It seems that the computed values of temperatures are very much dependent on the assumptions concerning dissociation. When these phenomena are taken into account (curves IG and BKW), the temperature varies 3500-3750 ±50 K

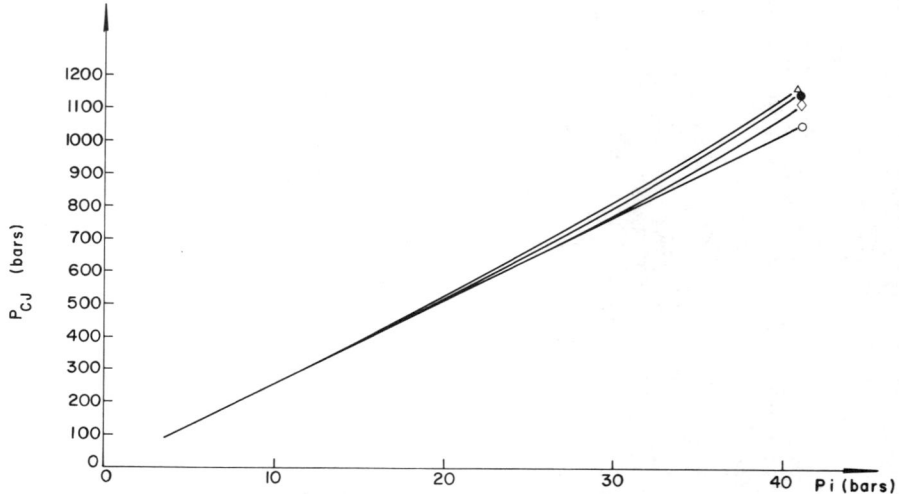

Fig. 8 Computed C-J pressures for mixture B: △ Boltzmann, ● BKW, ◇ BKWC, ○ ideal gas.

for $2 < p_i < 45$ bars. The temperature difference between these two curves is essentially constant around 100 K. It means that the calculated detonation temperature is less sensitive to the choice of the equations of state than the velocity, at least for BKW and IG. The effect of the equation of state on the calculated pressure P_{C-J} for mixture B is shown in Fig. 8. There is good agreement between the results obtained by these different means of computation. The difference becomes significant between BKW, BKWG, and Boltzmann for pressures higher than $p_i \simeq 25$ bars (4% for $p_i = 28$ bars and 8% for $p_i = 40$ bars).

For initial pressures lower than 25 bars, detonation pressure does not appear to be sensitive to the choice of the equation of state.

Conclusions

By measuring the detonation velocity for various C_2H_4-O_2-N_2 mixtures at high initial pressures ($2 < p_i < 45$ bars) with equivalence ratio $1.05 < r < 1.60$, we have pointed out that the detonation velocity was not affected by the tube diameter. We compared these experimental values to the theoretical values computed using three different equations of state. It appears that in the range $3 < p_i < 40$ bars, the ideal-gas equation of state gives values of detonation velocity that are much too low. The BKW equation in the form used by Mader (1963) is in better agreement with the experimental results, but only for initial pressures up to 20 bars. For higher initial pressure the computed value is too high. By changing the value of the adjustable parameter K in this equation of state, we changed the range of agreement, which has moved (for mixtures B, C, D, F) toward $20 < p_i < 30$ bars.

The velocities obtained using Boltzmann equation of state disagree with experimental results. We believe that this is due to the simplifying assumptions we used for the calculations. The program could be improved by involving more species in the calculation of detonation products, and by taking into consideration the initial pressure, which cannot be neglected when compared to the detonation pressure. This should lead to better agreement between experiment and theory.

Moreover, the theoretical values of detonation temperature and pressures do not seem to be sensitive to the choice of the equation of state, at least for $p_i < 25$ bars. For higher initial pressures, the difference among the computed values of detonation pressures becomes more significant.

We conclude that in our conditions of investigation the best means for testing the validity of these equations of state is to measure detonation velocity, more especially as this quantity may be measured with good accuracy. The measurement of detonation pressure or temperature, apart from the fact that it is less accurate, leads to results that may often be considered as physically ambiguous and, in our conditions of experiments, their value would have been less reliable than the measured values of detonation velocity.

References

Bauer P. (1977) Contribution á l'étude de la cinetique de la combustion du propane au foyer tubulaire. Thèse de Doctorat, Université de Poitiers, Poitiers, France.

Brochet C. (1966) Contribution á l'étude des detonations instables dans les mélanges gojeux. Thèse de Doctorat d'Etat, Université de Poitiers, Poitiers, France.

Brossard J. (1971) Etude expérimentale des detonations et des deflagrations rapides spherique divergentes dans quelques mélanges goyeux. *Revue de l'Institut Francis du Pétrole* **XXVI** (11), 1085-1102.

Gealer R.L. and Churchill S.N. (1960) Detonation characteristics of H_2-O_2 mixtures at high initial pressures. *AIChE J.* **6**, 501.

Gordon S. and McBride B. J. (1971) Computer program for calculation of complex chemical equilibrium compositions, rocket performance, incident and reflected shocks and Chapman-Jouguet detonations. NASA SP 273.

Guénoche H. and Manson N. (1952) Sur la variation de la célerité des oudes explosiones avec le diametre des tubes. *Comptes Rendus Acad. Sci.* (Paris) **235**, 1617.

Hirschfelder J. O., Curtiss C. F., and Bird R. B. (1954) *Molecular Theory of Gases and Liquids.* John Wiley and Sons, New York.

Hoelzer C. A. and Stobaugh W. K. (1954) Thesis, USAF Institute of Technology, Wright-Patterson Air Force Base, Ohio.

Hougen O. A. and Watson K. M. (1947) *Chemical Process Principles.* John Wiley and Sons, New York.

Johnson C. (1968) Contribution á l'étude des detonations dans les mélanges: hydrogène-oxygéne-azote. Thèse de Doctorat d'Etat, Unviersité de Poitiers, Poitiers, France.

Mader C. L. (1963) Detonation properties of condensed explosives computed using the BKW equation of state. Rept. LA 2900, Los Alamos Laboratory, N. Mex.

Mader, C. L. (1967) FORTRAN BKW: code for computing the detonation properties of explosives. Rept. LA 3704, Los Alamos Laboratory, N. Mex.

Manson N., Brochet C., Brossard J., and Pujol Y. (1963) Vibratory phenomena and instability of self-sustained detonation in gases. *9th (Intl.) Symposium on Combustion*. Academic Press, New York.

Pujol Y. (1968) Contribution a l'etude des detonations par la "methode inverse." Thèse de Doctorat d'Etat, Université de Poitiers, Poitiers, France.

Pujol Y. (1968) Determination de la composition et de l'enthalphic de melanges gogeux á haute temperature, compte tenu de leur second coefficient de viriel. *Entropie* **20**, 39.

Schmidt A. (l935) Über du detonation von spreug stoffen und die beziehung zwischen dicte und detonations geschwindigkeit. *Zs. Ges. Schiess. U. Sprengst* **30**, 364.

Taylor J. (l952) *Detonation in Condensed Explosives*, Oxford Press, Oxford, England.

Wolfson B.T. (l963) The effect of additives on the mechanism of detonation in gaseous systems. Ph. D. dissertaion, Rept. ARL 63-82, Aeronautical Research Laboratory, Ohio.

Detonation Characteristics of Two Ethylene-Oxygen-Nitrogen Mixtures Containing Aluminum Particles in Suspension

B. Veyssière,* R. Bouriannes,† and N. Manson‡
Université de Poitiers, Poitiers, France

The experimental apparatus described in this paper was conceived to determine the characteristics of the detonations propagating in gaseous mixtures which contain particles in suspension. The experiments were carried out with mixtures of ethylene and normal or suroxygenated air, the equivalence ratio of which was $r = 1.15$, with the concentration of aluminum particles (average diam 10 μm) being 30 g/Nm^3. Measured and calculated values of detonation velocity and maximum pressure behind the front were compared. Chronophotographs and luminosity records of the products of mixtures without and with aluminum were analyzed. As a result the ignition delay of aluminum has been evaluated, and a schematic description of the detonation, including three successive zones, is being proposed. The first zone is characterized by the complex of shock and combustion waves of gaseous components of the mixture. The second zone is where the aluminum particles warm up; and the last zone is where combustion of aluminum, in the reaction products of gaseous components, occurs.

Introduction

IN order to take the best advantage of the energy capable of being released by the combustion of metals, several works on combustion and detonation of condensed reactive systems (propellant powders and explosives) in which small particles of metal (especially aluminum) were incorporated have been performed for many years (Calzia, 1968). More recently, the details of the ignition and combustion of Al particles in gaseous oxidizing mixtures and in

---

Presented at the 7th ICOGER, Göttingen, Federal Republic of Germany, Aug. 20-24, 1979. Copyright © American Institute of Aeronautics and Astronautics, Inc., 1980. All rights reserved.
*Engineer, Laboratoire d'Énergétique et de Détonique.
† Chief of Research, Laboratoire d' Energétique et de Détonique.
‡Professor.

combustion products of gaseous combustible mixtures have also been studied (Bouriannes, 1971).

In spite of the increasing necessity of resolving safety problems related to prevention of explosions of dust suspensions or small metallic particle clouds (particularly formed from aluminum) and of oxidizing or, even, combustible atmospheres, there exist very few works that give much information on the possibility of a detonation in such media. So far as we know, only Strauss (1969) observed the propagation of a detonation in an oxygen-aluminum mixture and measured its characteristics (velocity and pressure).

The relative ease and excellent reproducibility with which a detonation in gaseous mixtures can be generated and observed caused us to examine the influence on its propagation by an aluminum suspension and to try to get new information on aluminum combustion. First, we describe the apparatus and techniques that were used for this purpose, and then present the first results obtained.

As will be noted, our experiments demonstrate that the presence of small aluminum particles in gaseous reactive mixtures modifies in a significant manner the structure and, consequently, the effects of the detonation.

Apparatus and Measurement Devices

Generation of the Suspension

In order to perform our observations in as reproducible a manner as possible, suspensions containing a constant particle diameter were used and a uniform mixture was achieved at least 10 s prior to ignition. We generated the suspension in a vertical detonation tube by means of a device that allowed us to realize such suspensions in a more satisfying manner than with techniques previously employed (Nettleton and Stirling, 1973 or Palmer and Tonkin, 1971).

In the technique used here, small particles of aluminum are put into suspension with the aid of a fluidized bed (Angelino et al., 1962; Melling and Whitelaw, 1975). This elutriation is performed by means of an apparatus (Fig. 1) (Guichard, 1967) including a bed compound of 95% glass balls of 200 μm average diameter, and 5% aluminum particles of 10 μm average diameter. With a gaseous (air) fluidizing flow of 72.9×10^{-2} g/s, the aluminum particles concentration obtained in these conditions was $\sigma = 35 \pm 15$ g/Nm^3 of gas.

Detonation Tube

The vertical detonation tube (see Fig. 2), 6 m long, includes two parts. The first part (X-Y in Fig. 2) is about 3.50 m in length and consists of: 1) the suspension generator G; 2) a valve R_1 for closing the tube, operated by an electropneumatic device; 3) the ignition chamber F (ignition by means of a blasting cap "Briska"); and 4) two elements of a circular (i.d. $d = 69$ mm) 1 m long tube.

The second part (Y-Z, Fig. 2) is about 2.50 m in length and the end is provided with a valve R_2 identical to R_1 for closing the tube. It includes either two elements of tube having the same internal diameter as the previous ones, with the second element having a longitudinal slit (1 mm wide, 100 mm long)

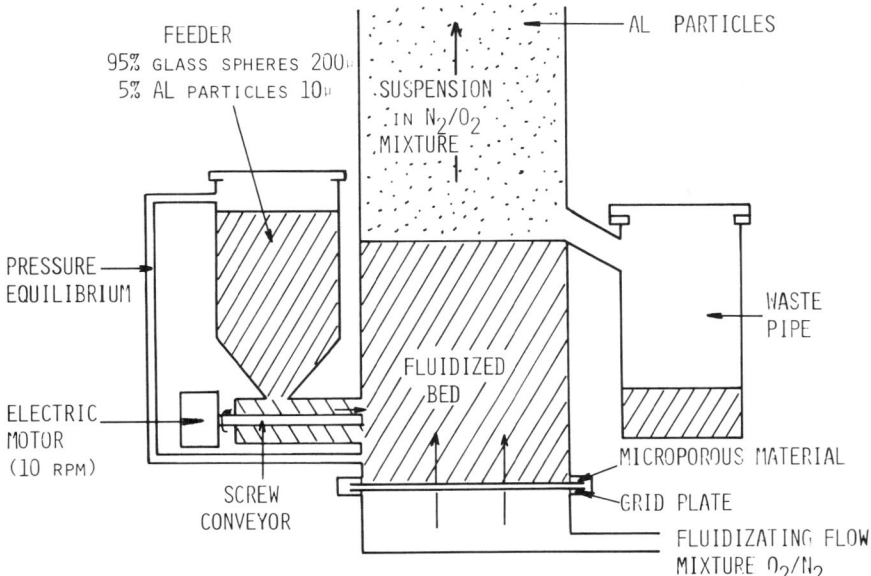

Fig. 1 Schematic diagram of the dust feed mechanism.

for chronophotography; or an element of transition from the circular to the square section, followed by two square elements (each 53 mm × 53 mm × 0.75 m) and ending in a visualization chamber of the same cross section equipped with two windows (400 mm long, 4 mm wide).

Ionization probes or pressure gages may be set out along the tube in $S_1, \ldots, S_7$ at distances of 30-50 cm.

Gases and Al Particles

In all our experiments, we used N 35 ethylene (purity >99.95%) from Air Liquide Industries and made O_2/N_2 mixtures by compressing atmospheric air (eventually enriched with Air Liquide N 48 oxygen, purity >99.998%).

The percentage of aluminum contained in the particles is more than 99% according to the supplier (Koch-Light). By proportioning samples (Bouriannes, 1971), each weighing 30 mg, we found that the particles contained more than 98% aluminum. Also, according to the electronic microscope examinations we conducted (Fig. 3), the shape of the particles was ellipsoidal and more than 90% of them had a particle size distribution between 8-15 μm. We thus assumed that their characteristic diameter was 10 μm.

Realization of Solid Particle-Gas Mixtures

The flow of the O_2/N_2 mixture enters, after expansion from 50-150 bars to about 1 bar, through a sonic flowmeter in the suspension generator G (see Fig. 2) where it loads up with aluminum.

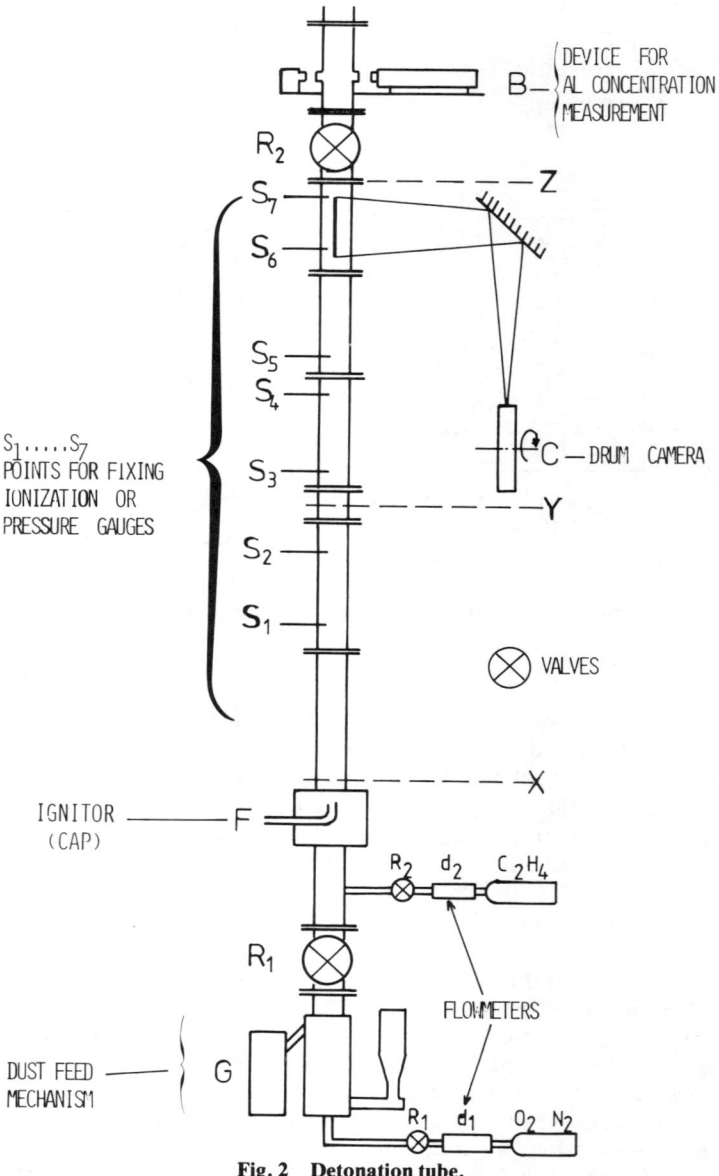

Fig. 2 Detonation tube.

Then, the ethylene flow, controlled with a rotameter, is admitted into the tube above valve R_1 (Fig. 2), and the whole mixture flows through the detonation tube to the top and outside. When the solid particles-gaseous mixture fills the tube completely, the valves R'_1, R'_2, R_1, and R_2 are shut simultaneously, so that the two-phase medium is immobilized just before ignition. Ignition occurs about 5 s after the closing of the valves.

$C_2H_4/O_2/N_2$/A1 PARTICLES MIXTURES

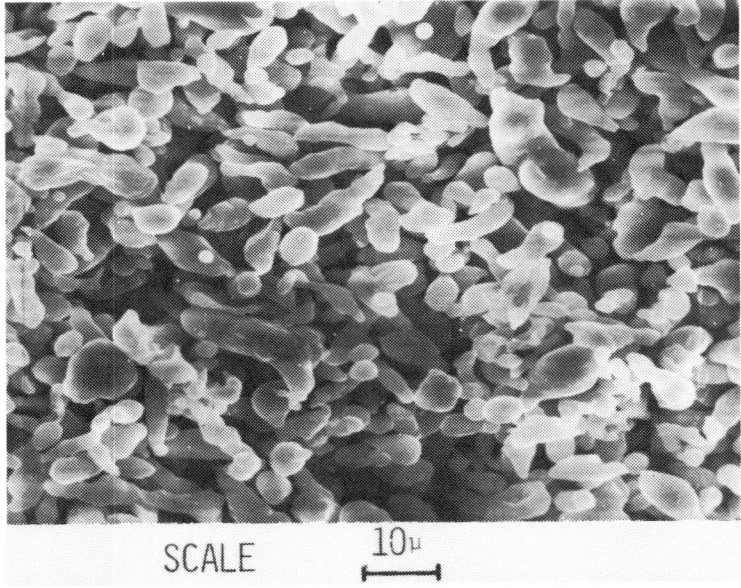

Fig. 3 Microphotograph with electronic microscope of aluminum particles.

Aluminum Particle Concentration Measurement

In order to know the aluminum concentration in the mixture, we built a device, the principle of which is based upon the determination of transmission coefficient, τ of light through a solid particle suspension (Dobbins et al., 1959; Leenaerts, 1966). The transmission coefficient τ (ratio of emergent luminous flux ϕ_I to incident luminous flux ϕ_0) is related to the mass concentration σ of particles by:

$$\tau = \frac{\phi_0}{\phi_I} = \exp(-A\sigma) \tag{1}$$

where A is a constant depending upon the nature of the particles.

In our device (Fig. 4), one compares two fluxes, ϕ_0 and ϕ_I, obtained by separating the light beam coming from a He-Ne laser source into two parts by means of a semitransparent mirror. The first flux ϕ_I traverses the entire two-phase medium, while the other flux ϕ_0 is outside of it. Using Eq. (1) the recorded resulting tension V

$$V = B\log\tau \tag{2}$$

is proportional to mass concentration σ

$$V = -AB\sigma = a\sigma \tag{3}$$

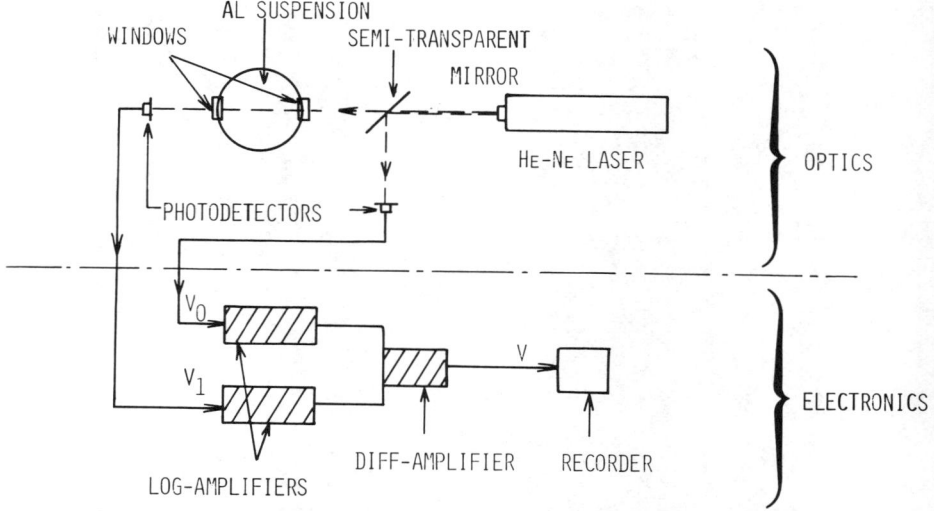

Fig. 4 **Diagram of opto-electronical device for measuring dust concentration.**

where the value of the constant a

$$a = 1.85 \pm 0.20 \, \text{mV}/(\text{g/m}^3)$$

was determined during preliminary experiments by weighing Al powder samples collected in the suspension flow produced by the generator.

With this device we established that during the filling of the detonation tube: 1) the stationary regime (average value of concentration constant when time is varying) is reached less than 5 min after the beginning of the sweeping of the tube, 2) concentration fluctuations around this average value were lower than 7%, and 3) in the conditions of our experiments, aluminum concentration was $\sigma = 30 \pm 10 \, \text{g/m}^3$. The uncertainty of the measurements was about 20-25%.

Velocity and Pressure Measurements

During each experiment the detonation velocity D was measured with Hewlett Packard (HP 5304 A—resolution 0.1 μs) chronographs, each one driven by starting and stopping probes. These probes were either ionization probes (Brossard, 1970) or pressure gages (Guerraud et al., 1967). The uncertainty of the velocity measurements was about 0.5%.

The pressure in the burned gases behind the detonation front was recorded by means of piezoelectrical pressure gages (**ANVAR 11568**) built in our laboratory (Guerraud et al., 1967).

Chronophotographs and Luminosity Records

The luminosity variation of the burned products behind the detonation front was observed with a monochromatic (wavelength $\lambda = 0.657 \, \mu$m), short rise-time (<40 ns) pyrometer. The pyrometer is a simplified version of the fast, several colors pyrometer (Bouriannes et al., 1977) built in our laboratory for studying detonations.

Records were made in the visualization chamber, at the same height as the chronophotographs, obtained by means of a drum-camera "Strobodrum."

Experimental Results

All experiments were performed at initial pressure $p_f = 1$ bar and temperature $T_f = 293 \pm 5$ K. As preliminary tests in mixtures without aluminium demonstrated that the detonation is stable only for rich mixtures $1 < r < 1.4$, we chose to study the influence of aluminum in the case of two mixtures:

$$C_2H_4 + XO_2 + ZN_2$$

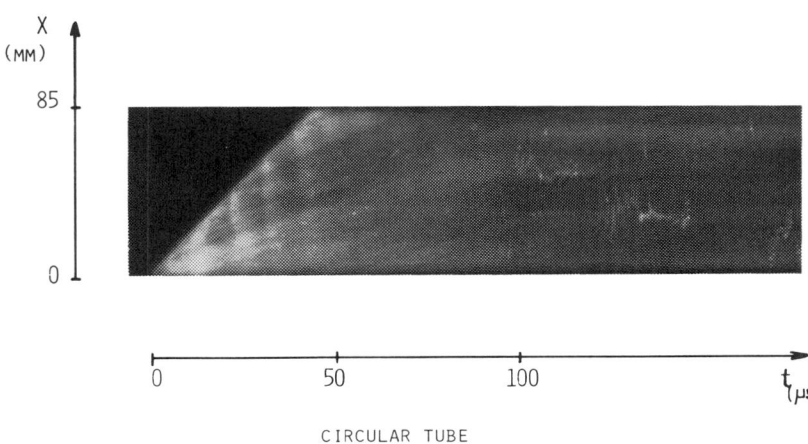

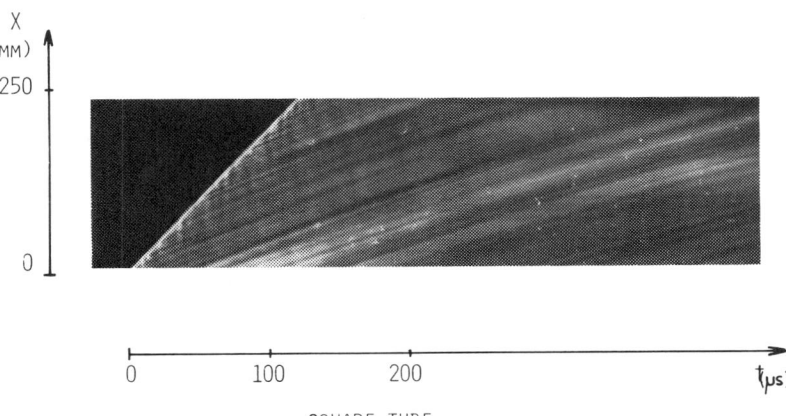

Fig. 5 Chronophotographs of detonation in mixture A without aluminum particles.

i.e., A mixture $X=2.6$, $Z=9.8$ ($Z/X=3.77$ normal air), B mixture $X=2.6$, $Z=8.15$ ($Z/X=3.14$ suroxygenated air), the equivalence ratio always being $r=1.15$.

These preliminary experiments showed also that for mixtures A and B: 1) chronophotographs of the detonation (compare Fig. 5) had the usual appearance of slightly unstable detonations in gaseous mixtures (Manson et al., 1963); and 2) in the terminal part of the tube (on about 1.50-2 m), the fluctuations of the velocity D were ranging about 1% which led us assume that, for sufficient Al concentrations (5-10 g/m^3), one could detect the influence of aluminum particles.

The systematic experiments with A and B mixtures with Al particles of average diam 10 µm and concentration $\sigma = 30 \pm 10$ g/Nm^3 demonstrated that:

1) The proportion of unburned aluminum in solid products collected on the detonation tube walls after experiments and analyzed by a chemical method (Bouriannes, 1971) is lower than 10%.

2) The chronophotographs of the detonation, recorded in mixtures with aluminum particles, differ from those of mixtures without particles (see Fig. 6) in: a) a sharp attenuation of luminosity of the detonation front A; and b) the appearance of a dark zone behind this front rather distinctly delineated, followed by a zone the luminosity of which is, on the contrary, very intense. The time interval τ_1 corresponding to the dark strip AB is: $10~\mu s \leq \tau_1 \leq 35~\mu s$ for mixture A and $40~\mu s \leq \tau_1 \leq 70~\mu s$ for mixture B.

3) In the presence of aluminum particles, the pyrometer delivers a signal (nonobservable when there are no Al particles) that exhibits the following characteristics (see Fig. 7): a) after a time interval AB corresponding to the delay τ_1 noticed on chronophotographs, there is a more rapid rise of luminosity; and b) after a time delay τ_2 of about 250 µs, there is a maximum of luminosity, and then a plateau which is maintained for a time duration τ_3 of about 300-400 µs.

These observations seem to indicate that the reaction of aluminum with gaseous detonation products occurs intensively only behind the detonation front after some "ignition delay" τ_i.

With regard to apparent delay τ_1 measured on records, the value of τ_i (taking account of the fact that particles more or less flow with the gases) is such that

$$\tau_1 \leq \tau_i \leq \mu \tau_1$$

where

$$\mu = \rho_b / \rho_f$$

is the ratio (about 1.8, B. Veyssière, 1978) of the density of burned gases ρ_b to that of unburned gases ρ_f.

Moreover the times τ_2 and τ_3 that one can consider as characteristic of the time interval necessary to reach maximum temperature behind the detonation front and of that necessary for the complete combustion of aluminum indicate that the duration of this combustion is far longer than the duration of the corresponding purely gaseous fast rate reaction (M. Veyssière, 1971).

$C_2H_4/O_2/N_2/Al$ PARTICLES MIXTURES

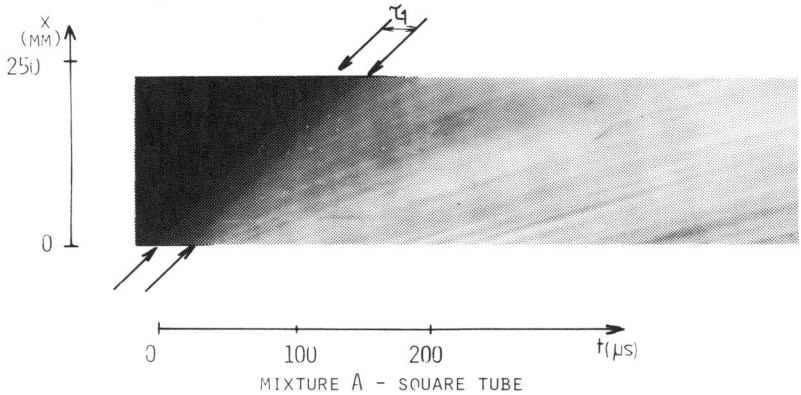

MIXTURE A - SQUARE TUBE

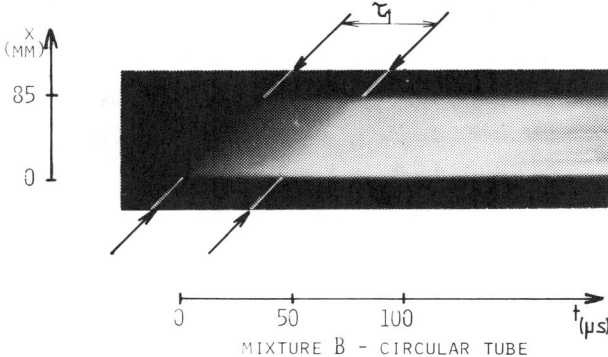

MIXTURE B - CIRCULAR TUBE

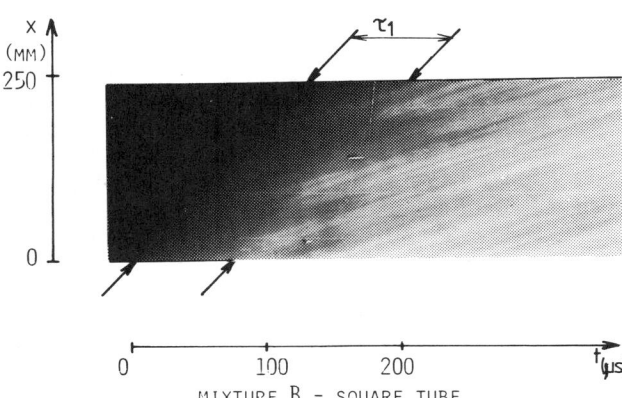

MIXTURE B - SQUARE TUBE

Fig. 6 Chronophotographs of detonation in mixtures A and B with aluminum particles.

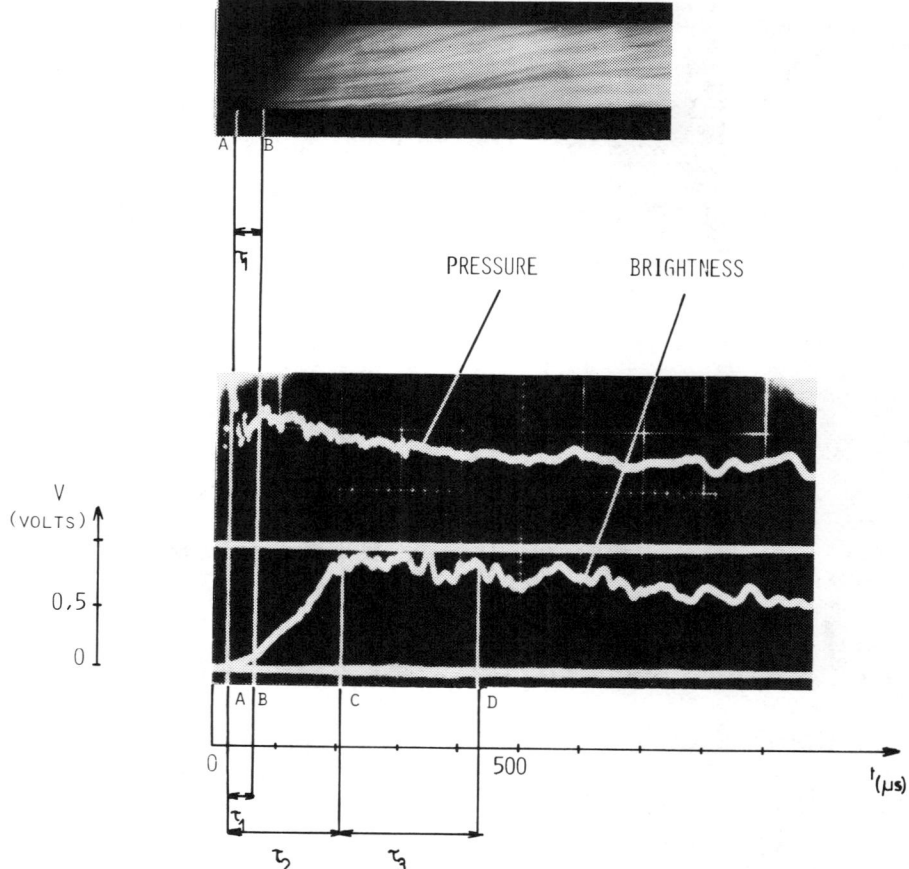

Fig. 7 Comparison of chronophotographs and pyrometric signal.

4) The presence of aluminum particles leads to a decrease of about 3% in detonation velocity (see Table 1), and this decrease is identical to that calculated (see Computational Results) when the assumption is made that the particles are chemically inert.

5) The average value of pressure during the first 50 μs behind the detonation front is practically the same as in mixtures without aluminum (see Table 1 and Fig. 8); but for 50 $\mu s \leq t \leq 200$ μs the pressure reaches a maximum value p_{max} (see Table 2 and Fig. 8) and then decreases later than in gaseous mixtures without particles.

Computational Results

We calculated detonation characteristics of mixtures A and B, assuming that aluminum particles, on the one hand, reacted with oxygen at the same time as the gaseous fuel, according to the reaction $4\,Al + 3\,O_2 \rightarrow 2\,Al_2O_3$, and, on the

Table 1 Detonation velocity measured D and calculated $D_{C\text{-}J}$ detonation pressure measured p for $0 \le t \le 50$ μs and calculated $p_{C\text{-}J}$ ($p_f = 1$ atm, $T_f = 293 \pm 5$ K.)

	Mixture			
	A[a]		B[b]	
Tube	Without Al	With Al	Without Al	With Al
Circular				
D, m/s	1879 ± 20	1811 ± 20	1936 ± 20	1863 ± 10
p, bar	$15.877 + 1.5$	15.7 ± 1.5	17.6 ± 1.15	17.9 ± 1.15
Square				
D, m/s	1865 ± 10	1816 ± 10	1924 ± 20	1874 ± 30
p, bar	16.10 ± 1.15	16.7 ± 1.5	18.15 ± 1.5	17.9 ± 1.5

	Without Al	Inert Al	Reactive Al	Without Al	Inert Al	Reactive Al
$D_{C\text{-}J}$, m/s	1866	1805	1920	1918	1866	1974
$p_{C\text{-}J}$, bar	19.4	18.1	20.7	20.1	19.5	21.8

[a] A mixture: $C_2H_4 + 2{,}6\,O_2 + 9{,}8\,N_2$.
[b] B mixture: $C_2H_4 + 2{,}6\,O_2 + 8{,}15\,N_2$.

Table 2 Maximum detonation pressure p_{max} measured for $50\,\mu s \le t \le 200\,\mu s$ and calculated $p_{C\text{-}J}$.

		Mixture	
Tube		A with Al	B with Al
Circular	p_{max}, bar	17 ± 3	21 ± 3
Square	p_{max}, bar	19 ± 3	20 ± 3
	$p_{C\text{-}J}$, bar (reactive Al)	20.7	21.8

other hand, was chemically inert. These calculations were made for aluminum concentrations σ varying 0-100 g/m³ by performing the program usually used at the Laboratoire d'Energétique et de Détonique (Johnson, 1968) and by taking into account the numerical values given by the JANAF tables.

As can be seen in Fig. 9 where the results of our calculations are summarized, the addition of $\sigma = 30$ g/m³ of aluminum to the gaseous mixtures studied determines with regard to velocity $D_{C\text{-}J}$, pressures $p_{C\text{-}J}$, and temperature $T_{C\text{-}J}$ an increase of, respectively, 4, 8, and 9% when assuming that Al reacts and a decrease of 3, 6, and 9% when assuming Al is chemically inert.

Discussion of Results

All of our experimental observations and computational results allow for a qualitative description of the detonation wave in gaseous mixtures containing small aluminum particles in suspension. This description can be summarized as follows.

1) During time interval τ_i which precedes the ignition behind front A of the aluminum particles, these particles behave as chemically inert. However, because they absorb energy released by reactions between gaseous products,

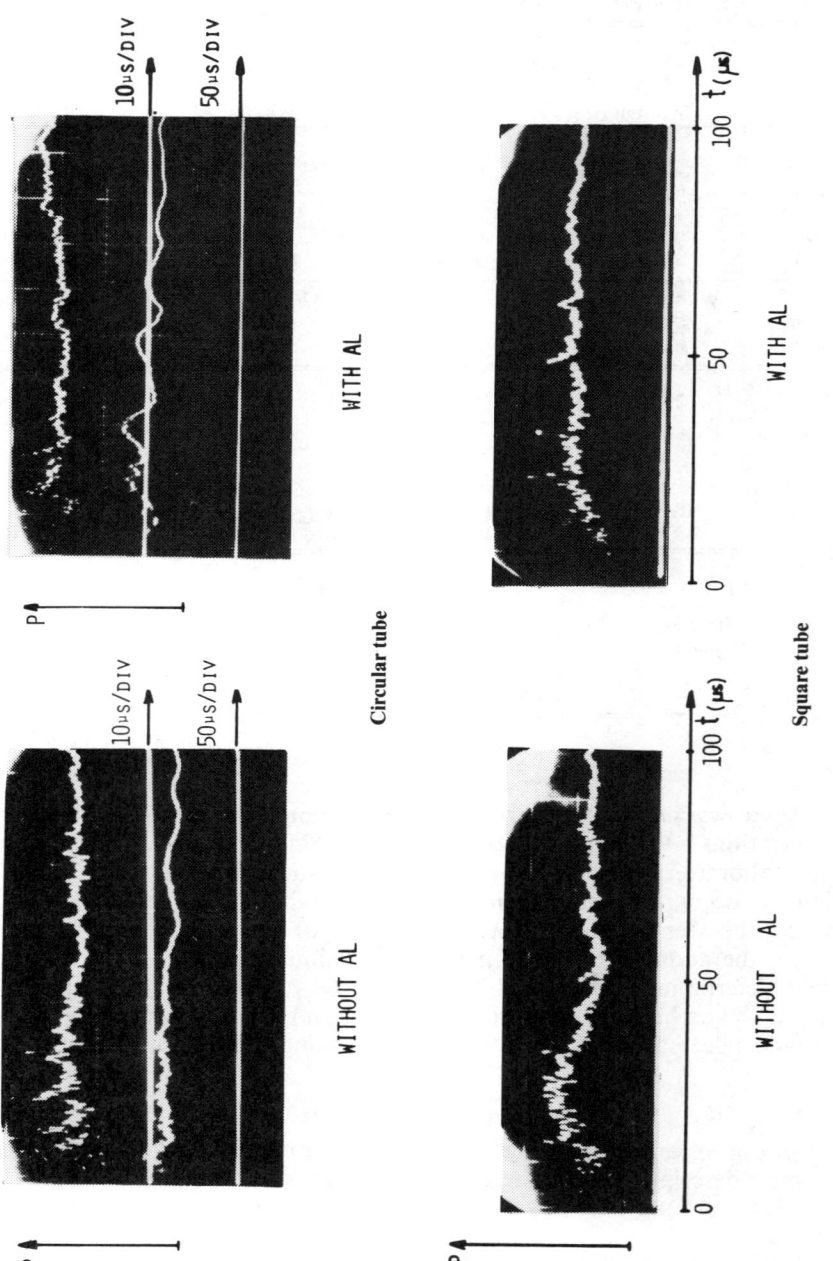

Fig. 8 Pressure records.

$C_2H_4/O_2/N_2/Al$ PARTICLES MIXTURES

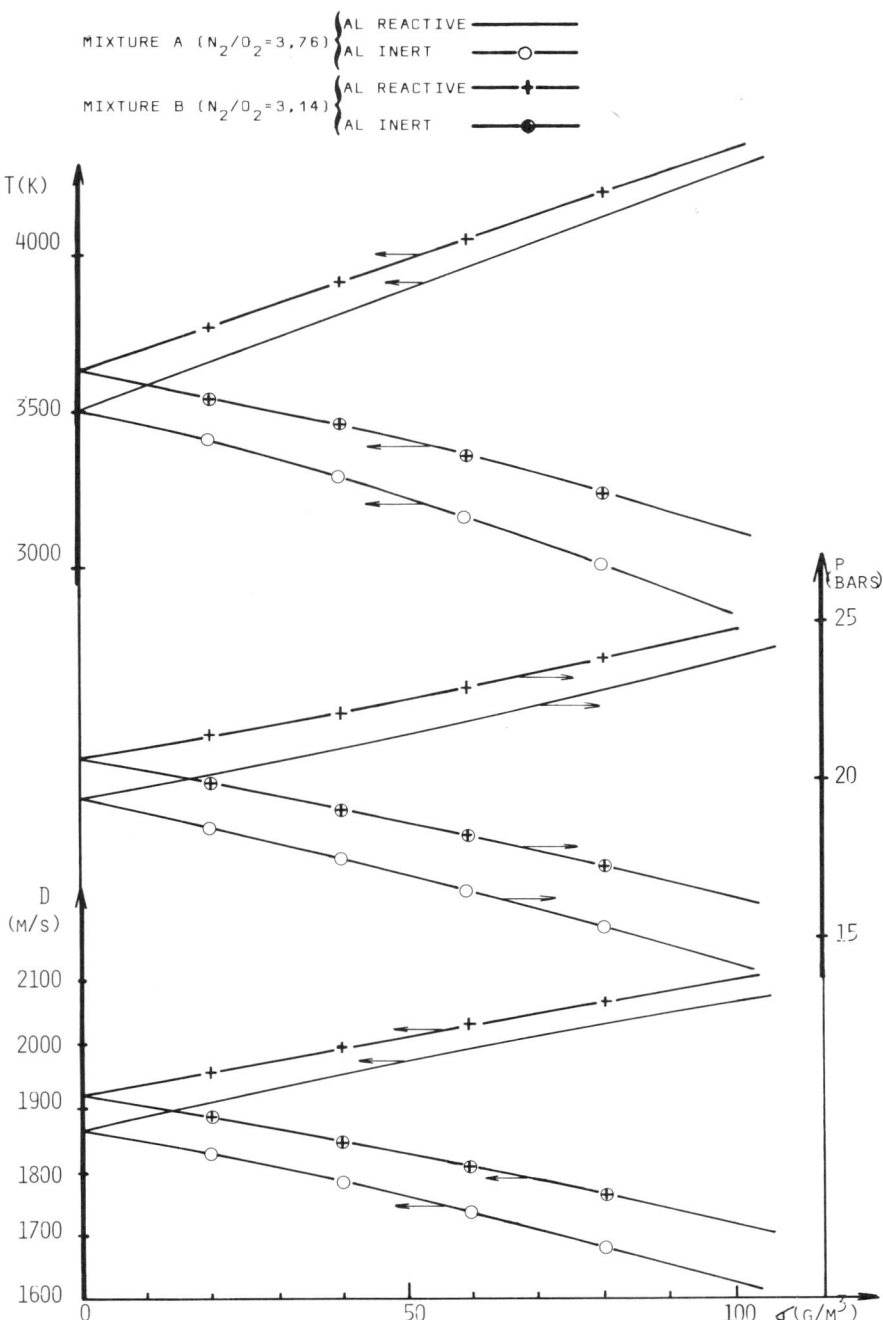

Fig. 9 Detonation velocity D_{C-J}, pressure p_{C-J}, and temperature T_{C-J} vs Al concentration mixtures A and B.

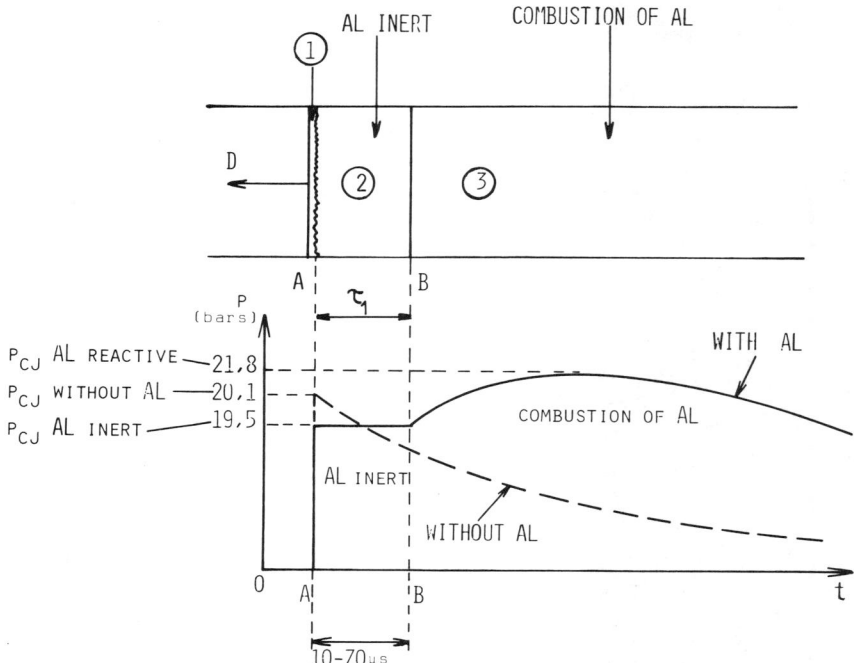

Fig. 10 Schematic structure of the detonation in a $C_2H_4/O_2/N_2$ mixture with Al particles.

the value of the detonation velocity is lower and the luminosity of the front weaker than those in the same mixtures without Al particles.

2) The rate of combustion of aluminum particles begins further behind the detonation front, but that combustion lasts far longer than in purely gaseous components and does not involve noticeable modifications of the average pressure during the first 50 μs following the crossing of the detonation front. However, for 50 $\mu s \leq t \leq 200$ μs this burning determines the existence of a pressure plateau and, even, a new increase of pressure. Finally, when aluminum particles are present in the mixture, the pressure decrease occurs distinctly further behind the detonation front.

All of these results can be considered as a confirmation of the hypothesis expressed in the past (Kast, 1901; Pokhil et al., 1972) concerning classical condensed explosives to which aluminum powder has been incorporated. According to this hypothesis, aluminum reacts with detonation products only behind the detonation front.

It is, thus, possible to imagine a schematic structure of the detonation wave in gaseous mixtures containing an aluminum particle suspension. This detonation is the juxtaposition of three zones (see Fig. 10):

1) The first zone ① is very narrow and one can consider it to be the front A, where the complex of shock waves initiates gaseous reactions.

2) The second zone ② is between the front A and the plane B where particles begin to ignite. In zone ② only the gases react at a fast rate, the aluminum particles behave as chemically inert, and the time interval between A and B is the apparent aluminum particles ignition delay τ_l.

3) The third zone ③, beyond B, is where aluminum reacts intensively with the gaseous detonation products of zone ②. The reactions in this zone apparently have no influence on detonation velocity, but modify pressure profile $p_b(t)$.

Conclusion

The presented results demonstrate that the aluminum particles react with gaseous products after the crossing of the detonation front with some delay (about tens of microseconds in the case of the C_2H_4/air mixtures we studied) and that the result of the fast rate combustion of aluminum is to maintain, for a longer time, a higher pressure in the detonation products.

References

Angelino H., Enjalbert, L., and Gardy, H. (1962) Emission des poussiéres par les lits fluidisés, Ed. Association Francaise de fluidisation, Paris.

Bouriannes R. (1971) Contribution á l'étude de la combustion de l'aluminium dans les mélanges oxygéne-argon, dans l'azote et dans l'air, These de Doctorat és Sciences Physiques, University of Poitiers.

Bouriannes R., Moreau M., and Martinet, J. (1977) Un pyrométre rapide a plusieurs couleurs, *Revue de Physique Appliquée*, **12**, 893-899.

Brossard J. (1970) Contribution á l'étude des ondes de choc et de combustion dans les gaz, These de Doctorat es Sciences Physiques, University of Poitiers.

Calzia J. (1969) *Les substances explosives et leurs nuisances,* edited by Dunod, Paris.

Dobbins R.A., Crocco, L., and Glassman, I. (1959) Further Studies on the light scattering technique for determination of size distributions in burning sprays—I, Rept. H63, Aeronautical Engineering Laboratory, Princeton University, 463.

Guerraud C., Leyer, J.C., and Brochet C. (1967) Mesure de la variation de la pression de détonation dans les mélanges gazeux. *C.R. Acad. Sci.* (Paris), **264B**, 5-8.

Guichard J.C. (1967) Rept. IRCHA, Verte-le-Petit, France..

Johnson C. (1968) Contribution á l'étude des détonations dans les mélanges hydrogéne-azote, Thése de Doctorat és Sciences Appliquées, University of-Poitiers.

Kast G. (1901) Detonation der Salpetersaure-Trinitrotoluol-Aluminum Explosifstoffe. *Zs. f. Schiess u. Sprengstw.*, **5**, 251.

Leenaerts R. (1966) Les techniques d'analyse granulométrique—Principes et appareillages, *Revue universelle des Sciences et des Arts appliqués á l'industrie*, **8**, 197-207.

Manson N., Brochet C., Brossard J., and Puljol Y. 1963 Vibratory phenomena and instability of self-sustained detonations in gases, *9th Symposium (Intl.) on Combustion*, pp. 461-469. Academic Press, Inc., New York.

Melling A. and Whitelaw, J.H. (1975) Optical and flow aspects of particles. *Proceedings of the L.D.A. Symposium*, Copenhagen.

Nettleton M.A. and Stirling, R. (1973) Detonations in Suspensions of Coal Dust in Oxygen, *Combustion and Flame*, **21**, 307-314.

Palmer K.N. and Tonkin, P.S. (1971) Coal Dust explosions in a large vertical tube apparatus, *Combustion and Flame,* **17,** 159-170.

Pokhil P.F., Beliaev A. F., Frolov Yu. V., Logatchev V. C., and Korotkov A. I. (1972) *Combustion des m*étaux pulvérulents dans des atmosphéres actives, Ed. Nauka, Moscow.

Strauss W. A. (1968) Detonation of aluminum powder oxygen mixures. *AIAA J.* **6,** 753-1756.

Veyssiere B. (1978) Caractéristiqes de la détonation dans des mélanges gazeux éthyléne-oxygéne-azote contenant des particules fines d'aluminum en suspension, Thése de Docteur-Ingénieur, University of Poitiers.

Veyssiere, M. (1971) Contribution a l'étude des caractéristiques physiques des produits de détonation dans les mélanges gazeux, These de Doctorat es Sciences Physiques, University of Poitiers.

Generation of Detonations by Two-Stage Burning

M. Zalesiński* and S. Wójcicki†
Warsaw Technical University, Warsaw, Poland

> Combustion of coal dust and gaseous oxidizer in a tube opened at one end can develop into an explosion at the closed end, creating a shock wave which follows the flame. The velocity of the flame can be accelerated and a detonation wave formed if the shock wave merges with the flame front. A physical model of this phenomenon, based on the concept of two-stage burning was developed and a number of experiments performed using a constant-volume combustion chamber, a detonation tube, and an apparatus capable of simulating blasts in unconfined space. The concept of two-stage burning can explain some of the hazardous effects that occur during explosions of coal dust or other industrial dusts.

Genesis of the Problem

THE starting point of this investigation was an analysis of the combustion process of coal dust and a gaseous oxygen mixture in a vertical tube with a square cross section (Fig. 1a). The tube was closed at the top and open at the bottom. The front wall of the tube was made of Plexiglass to permit direct streak photography of the combustion process (Fig. 1b). The motion of combustion products and coal particles burning right behind the flame front can be seen clearly. Initially, the combustion products follow the flame motion and subsequently their motion is opposite to that of the flame. Compression and fast (often explosive) burning of coal particles occur as a result of this flow in the closed-end region. The energy release results in a second flow reversal of the combustion products.

The explosion at the closed end of the tube can be so strong, under certain circumstances, that it generates a shock wave (seen in Fig. 1c) which follows the flame and which can transform into a detonation wave upon overtaking the flame.

Two-Stage Burning Model

A series of experiments was conceived to attain a better understanding of the phenomenon. The experiments were based on the following model of flame

---

Presented at the 7th ICOGER, Göttingen, Federal Republic of Germany, Aug. 20-24, 1979. Copyright © American Institute of Aeronautics and Astronautics, Inc., 1981. All rights reserved.
*Ph. D.
†Professor of Mechanical Engineering.

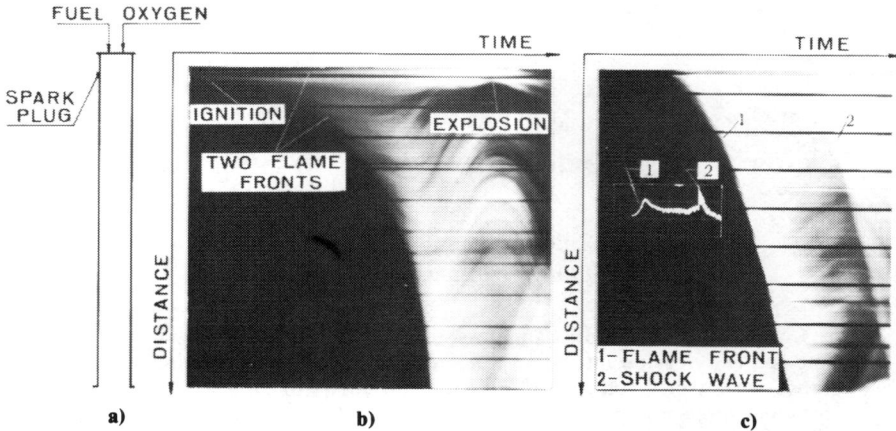

Fig. 1 Flame spreading in coal dust-oxygen mixture: a) experimental apparatus (detonation tube); b) streak photograph of the process of flame acceleration with the explosion at the closed end of the tube; and c) generation of the shock wave as a result of explosion from b. Pressure distribution curve (inset) confirms existence of shock wave.

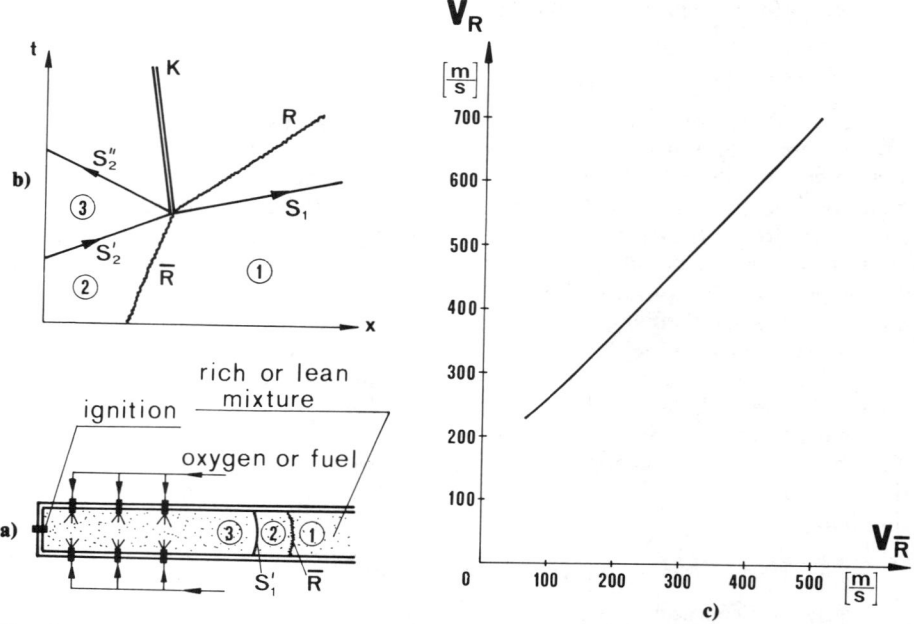

Fig. 2 Model of two-stage burning: a) detonation tube; b) course of phenomena in time t, distance x coordinates; and c) effect of initial flame speed $V_{\bar{R}}$ on its velocity V_R after the shock wave S_2' passed through; ① combustible mixture, ② products of primary combustion, and ③ primary combustion products after the shock wave S_2' passed through (K = contact surface, S_1 = shock wave after the flame $\bar{R}$ passed through, R = flame behind the shock wave, and S_2'' = reflected wave).

acceleration (Fig. 2). A tube closed at one end (Fig. 2a) is filled with a lean or rich gaseous combustible mixture. The mixture is ignited at the closed end by means of an electric spark after which the flame front develops and moves toward the open end of the tube. After an appropriate interval, a quantity of oxygen, in a rich-mixture case, or of gaseous fuel, in a lean-mixture case, is injected. The quantity injected is sufficient to create a stoichiometric mixture in the region of the injection. An explosion occurs upon mixing of the hot combustion products and injected media, which generates a pressure wave. The wave merges with the flame and promotes its acceleration. Under certain circumstances, a detonation wave is thereby formed. Figure 2b illustrates this phenomenon in the time-space domain. It can be seen that the pressure wave S_2' interacts with the combustion front $\bar{R}$ and causes the increase of the speed of the front from $V_{\bar{R}}$ to V_R.

The efficacy of this type of flame acceleration is illustrated for a primary flame which propagates in a propane-air mixture of excess air coefficient of 0.7. The influence of the initial flame-front speed on its velocity behind the pressure wave was evaluated (see Bartlmä, 1977). The results are presented in Fig. 2c. It is shown that the flame acceleration ($V_R - V_{\bar{R}}$), caused by the passing of a shock wave, increases with its initial velocity $V_{\bar{R}}$. For example, for $V_{\bar{R}} = 100$ m/s, $V_R - V_{\bar{R}} = 130$ m/s and for $V_{\bar{R}} = 500$ m/s, $V_R - V_{\bar{R}} = 200$ m/s.

Bomb Experiments

The experimental setup, shown in Fig. 3, was used to observe the acceleration of the flame associated with two-stage burning in a bomb, a constant-volume combustion chamber equipped with quartz windows. The fuel

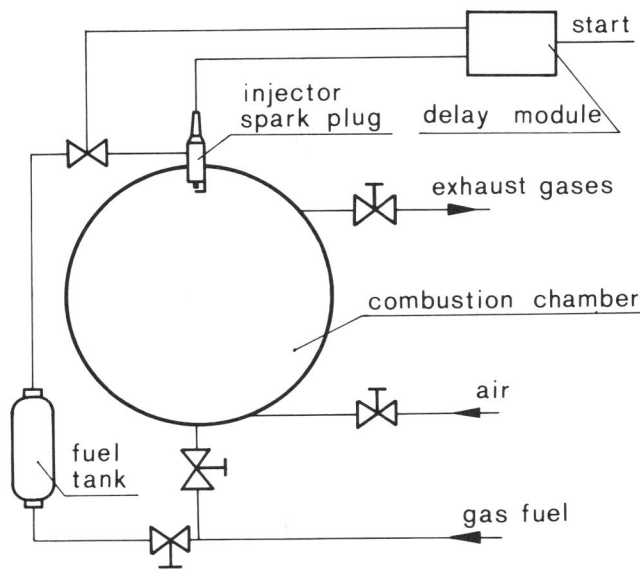

Fig. 3 Constant-volume combustion chamber.

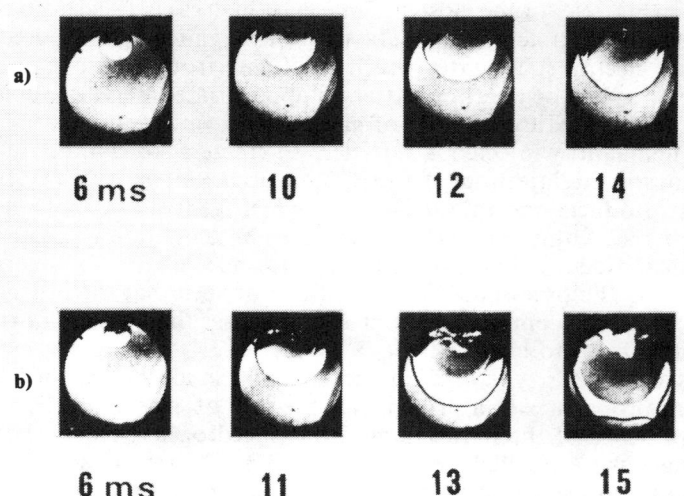

Fig. 4 Flame spreading in constant-volume combustion chamber: a) one-stage burning, and b) two-stage burning.

injector and spark plug were placed in the upper part of the chamber. The chamber was filled with a lean propane-air mixture corresponding to an excess air coefficient of 1.5. After ignition and when the flame had partially traversed the chamber, a predetermined quantity of propane was injected into the region filled with combustion products.

The flame spreading can be seen in Fig. 4, without (Fig. 4a) and with (Fig. 4b) propane injection.

In the latter case, the effects of interaction are shown behind the primary flame front, caused by the secondary combustion. These effects result in an increase in the velocity of the primary flame, as is evident in Fig. 5. In this case the flame speed was increased from 5 to 8.6 m/s.

Detonation Tube Experiments

The experimental setup shown schematically in Fig. 6, was used to observe the transition to detonation by the process of two-stage burning. The main element of the apparatus is the detonation tube. A propane-rich air mixture was introduced (excess air coefficient of 0.7) and was ignited by an electric spark at the top, closed end of the tube. After a suitable time interval, oxygen was injected into the region of combustion products and it reacted with the unburned fuel which was not consumed by the primary flame. The history of the process that leads to detonation is shown in Fig. 7b. A significant increase in flame velocity from 115 to 435 m/s was observed. Evidently it was caused by the blast wave formed behind the primary flame front. The flame acceleration, in turn, resulted in the generation of the detonation wave and of the accompanying retonation wave. For comparison, Fig. 7a shows the combustion process without oxygen injection.

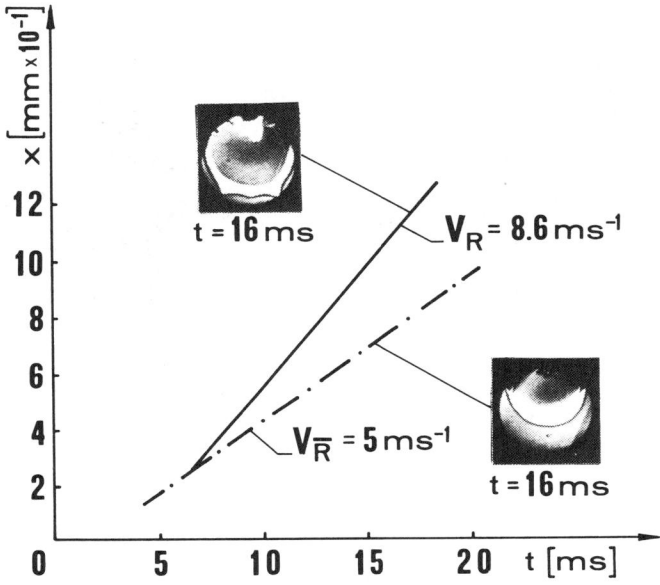

Fig. 5 Comparison of combustion velocities for one- and two-stage combustion: same description as for Fig. 2.

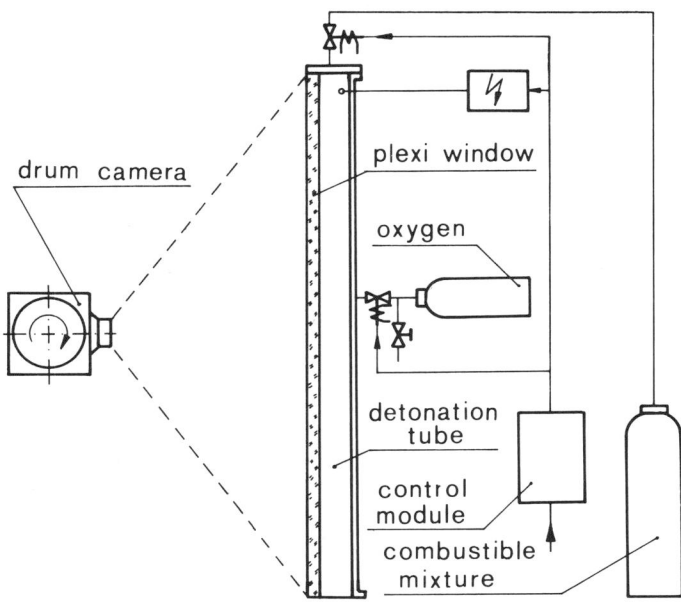

Fig. 6 Detonation tube.

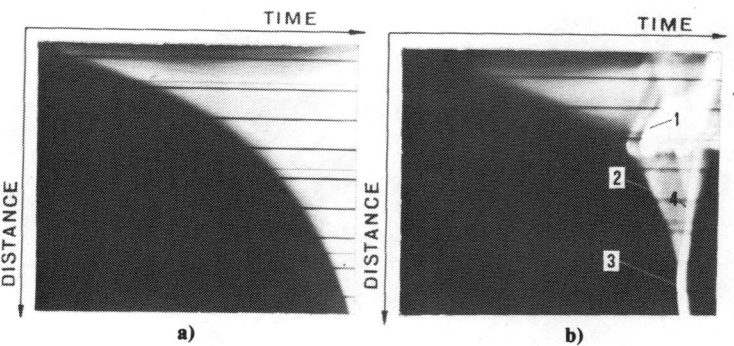

Fig. 7 Streak photographs of flame spreading in the detonation tube: a) one-stage burning, and b) two-stage burning (transition to detonation), with ① indicating explosive secondary combustion, ② accelerated primary flame, ③ transition into detonation, and ④ retonation wave.

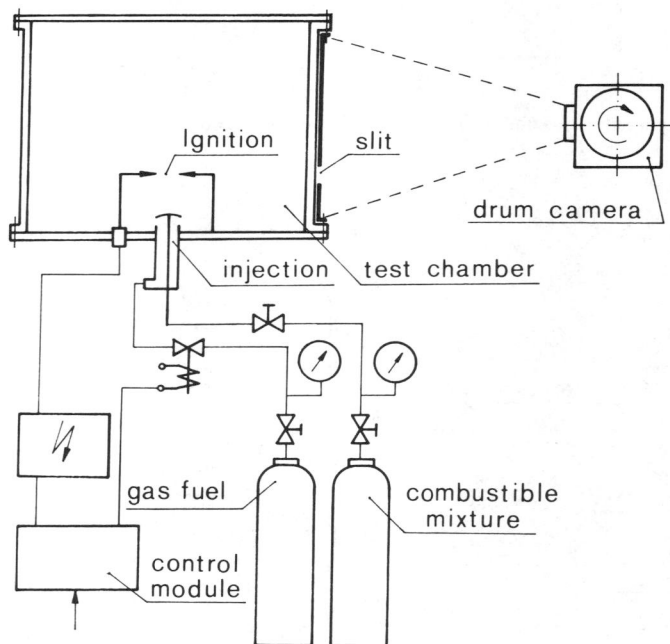

Fig. 8 Experimental setup for investigation of spherical flames.

Two-Stage Burning in Unconfined Space

The apparatus for investigation of spherical flames is shown in Fig. 8. The main part of the apparatus is the frame made of L-shaped steel beams. The base and one of the side walls were made of steel sheets. A slit in this wall

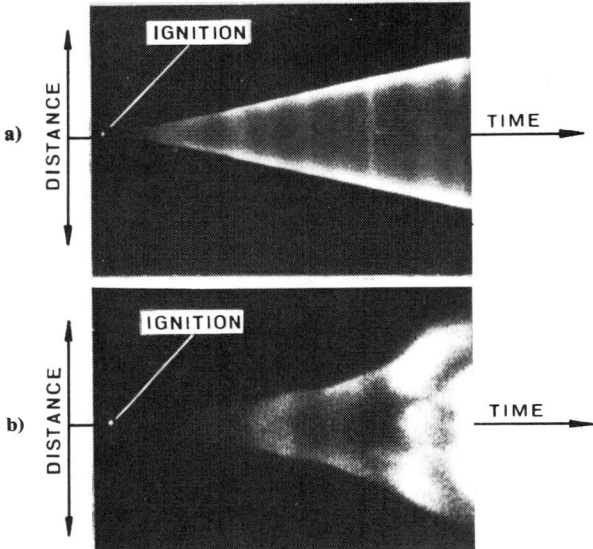

Fig. 9 Streak photographs of spherical flames: a) one-stage burning, and b) two-stage burning.

permitted streak photography of the combustion process. The feeding and ignition systems were located in the base sheet. All other walls were covered by cellophane film. The enclosed volume was filled with a lean propane-air mixture corresponding to an excess air coefficient of 1.5. The streak photography of one- and two-stage burning are shown in Fig. 9. Injection of propane during the two-stage burning process leads to the generation of a blast wave behind the primary combustion front and causes an increase in the velocity of the flame from 1.7 to 8.5 m/s.

Conclusions

The two-stage burning, in which the second stage has an explosive character, causes an acceleration of the flame, and, under certain conditions, leads to detonation. This process can be considered as a particular case of combustion with a secondary energy supply. An example of a similar process was investigated by Wojcicki (1972); in that study the electric discharge was used to supply energy to the flame.

Two-stage burning of combustible gaseous mixtures may be used to simulate selected cases of flame propagation in a coal dust. During the combustion of coal dust in a channel closed at one end, a strong shock wave is generated at the closed end of the channel. The pressure behind this wave is much higher than that in the combustion front (Fig. 1c). As a consequence, the occurrence of such a wave can be particularly hazardous under actual conditions encountered in coal mines. It is believed that similar effects can occur in a channel with a sudden change of direction or of cross-sectional area. Such conditions are often encountered in the galleries of coal mines as well as in the corridors and halls of industrial buildings.

Acknowledgment

This work was supported by the National Science Foundation under Grant INT 75-22131.

References

Bartlmä F. (1977) On the acceleration of flames in premixed gases. *Archives of Thermodynamics and Combustion* **8**(4).

Wójcicki S. (1972) Influence of electric discharge on flame propagation. *Acta Astronautica* **17**, 827-832.